ROCKFORD PUBLIC LIBRARY
3 1112 01090 0567

D0706490

14th EDITION

SOUTH-WESTERN

Applied BUSINESS Mathematics

Robert A. Schultheis
Professor, Department of
Management Information Systems
Southern Illinois University
Edwardsville, Illinois

Raymond M. Kaczmarski
Supervisor of Business Education
Detroit Public Schools
Detroit, Michigan

I(T)P® An International Thomson Publishing Company

Cincinnati • Albany • Boston • Detroit • London • Madrid • Melbourne • Mexico City • New York
Pacific Grove • San Francisco • Singapore • Tokyo • Washington

Editor-in-Chief: Peter McBride
Managing Editor: Eve Lewis
Marketing Manager: Colleen Thomas

Editorial and Production Coordinator: Patricia Matthews Boies
Editors: Tamara Jones, Marianne Miller, Edna Stroble
Judith A. Witt, Consulting Editor

Design and Photography Coordinator: Devore M. Nixon
Chapter Openers and Cover Design: Lou Ann Thesing
Photo Research: Meyers: Photo Art

ISBN: 0-538-65251-9

I(T)P
International Thomson Publishing

South-Western Educational Publishing is a division of International Thomson Publishing Inc.

The ITP trademark is used under license.

5 6 7 VH 03 02 01 00 99 98

Printed in the United States of America

Dear Student,

The material in *Applied Business Mathematics* will teach you mathematical and critical thinking skills that will help you be a smart shopper, an informed citizen, and a valued employee.

The text is divided into four main units.

Unit 1: **Managing Your Money.** In this unit, you learn personal cash management techniques and how to compare job offers, including salary and job benefits.

Unit 2: **Spending Wisely.** In this unit, you learn the basics of budgeting and spending. You'll learn how to make smart buying decisions for everyday purchases, such as food, and for major purchases, such as a car or home.

Unit 3: **Making Money Grow.** The focus of this unit is on interest—the time value of money. This unit will show you how to compare interest on your investment dollar as well as how to calculate the cost of interest when you borrow money.

Unit 4: **Business Mathematics.** This unit teaches you the basics of business management in a global economy. You will gain an understanding of how the world economy impacts business. You will learn how to read and analyze financial reports. You will also see how your mathematics skills can be applied to everyday business activities, such as pricing and record keeping.

Career profiles, lesson opening discussions, critical thinking questions, Put It All Together exercises, and other individual and group assignments have been designed to help you identify and strengthen critical work skills. These skills include:

- ***Interpersonal Skills*** These include the ability to lead, work on a team, serve customers knowledgeably and courteously, and work well with people from many different backgrounds. This book gives you the opportunity to observe and practice interpersonal skills, such as leading and negotiating.
- ***Information Skills*** This book includes many activities that call on you to gather, analyze, and present information in verbal, written, and visual ways. In addition, you will learn about database applications, which allow users to store and retrieve information electronically.
- ***Resource Management Skills*** In addition to helping you learn to manage financial resources effectively, you can learn important fundamentals of human resource management by keeping a Work Skills Portfolio to monitor your own performance.
- ***Technology Skills*** The Explore Technology lessons help you develop the ability to use spreadsheet and database software.

Some of the key features of this new edition are highlighted on the following pages. We hope you will enjoy working with this book and that the skills you learn here will help you build your own successful future.

ABOUT THE AUTHORS

Robert A. Schultheis
Professor, Department of Management Information Systems
Southern Illinois University
Edwardsville, Illinois

Raymond M. Kaczmarski
Supervisor of Business Education
Detroit Public Schools
Detroit, Michigan

Reviewers:

Millie Conrad
Mathematics Teacher
Logan Elm High School
Circleville, Ohio

Tom Kennedy
Retired Business Teacher
Perry Meridian High School
Indianapolis, Indiana

Carol Kontchegulian-Maggiore
Business Teacher
Williamsville North High School
Williamsville, New York

Timothy J. McNamara
K-12 Mathematics Supervisor
West Irondequoit Central School District
Rochester, New York

Dr. Jean Thiry
Math Department Chair
Glenbard West High School
Glen Ellyn, Illinois

Kimberly S. Wilson
Business Education Instructor
Greene County Career Center
Xenia, Ohio

Features Highlights

Career Profiles

Each chapter begins with a profile of an individual, a job, or an industry that may be of interest to you. Taken as a whole, the profiles identify many math and workplace skills necessary to be successful in different kinds of work. Even if you are not interested in the specific job, read the profiles. You'll learn what a work day may be like, how to match skills to work, and what kinds of questions you might ask if you were interested in a particular job.

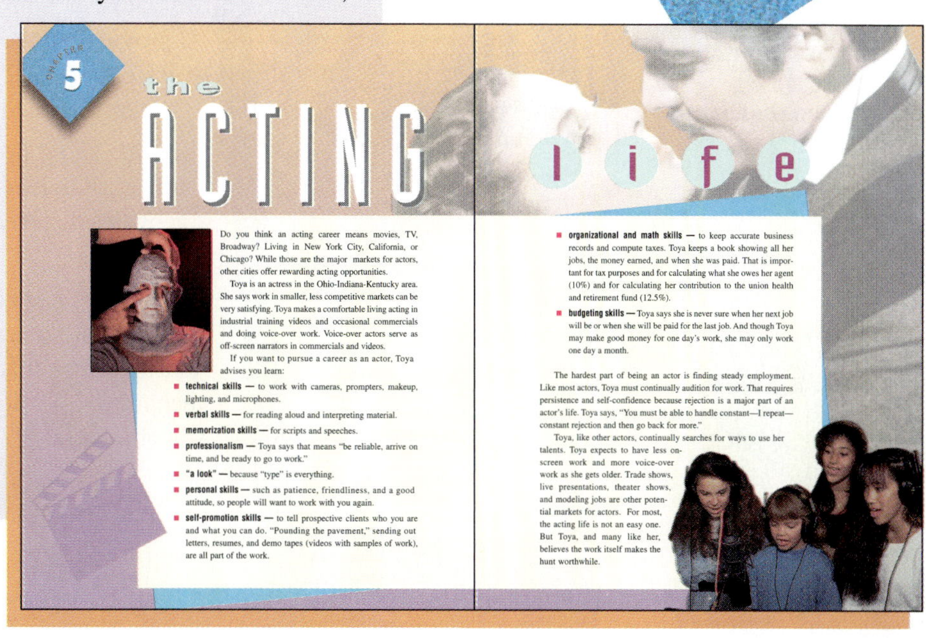
Chapter 5

the ACTING life

Do you think an acting career means movies, TV, Broadway? Living in New York City, California, or Chicago? While those are the major markets for actors, other cities offer rewarding acting opportunities.

Toya is an actress in the Ohio-Indiana-Kentucky area. She says work in smaller, less competitive markets can be very satisfying. Toya makes a comfortable living acting in industrial training videos and occasional commercials and doing voice-over work. Voice-over actors serve as off-screen narrators in commercials and videos.

If you want to pursue a career as an actor, Toya advises you learn:

- **technical skills** — to work with cameras, prompters, makeup, lighting, and microphones.
- **verbal skills** — for reading aloud and interpreting material.
- **memorization skills** — for scripts and speeches.
- **professionalism** — Toya says that means "be reliable, arrive on time, and be ready to go to work."
- **"a look"** — because "type" is everything.
- **personal skills** — such as patience, friendliness, and a good attitude, so people will want to work with you again.
- **self-promotion skills** — to tell prospective clients who you are and what you can do. "Pounding the pavement," sending out letters, resumes, and demo tapes (videos with samples of work), are all part of the work.
- **organizational and math skills** — to keep accurate business records and compute taxes. Toya keeps a book showing all her jobs, the money earned, and when she was paid. That is important for tax purposes and for calculating what she owes her agent (10%) and for calculating her contribution to the union health and retirement fund (12.5%).
- **budgeting skills** — Toya says she is never sure when her next job will be or when she will be paid for the last job. And though Toya may make good money for one day's work, she may only work one day a month.

The hardest part of being an actor is finding steady employment. Like most actors, Toya must continually audition for work. That requires persistence and self-confidence because rejection is a major part of an actor's life. Toya says, "You must be able to handle constant—I repeat—constant rejection and then go back for more."

Toya, like other actors, continually searches for ways to use her talents. Toya expects to have less on-screen work and more voice-over work as she gets older. Trade shows, live presentations, theater shows, and modeling jobs are other potential markets for actors. For most, the acting life is not an easy one. But Toya, and many like her, believes the work itself makes the hunt worthwhile.

PROBLEM SOLVING

DRAW A DIAGRAM

When solving a problem, drawing a diagram can be a useful strategy. Kyle is designing the offices for his boss and coworkers. There is a large square room to be used. His boss will require $\frac{1}{4}$ of the space for her office. He will need $\frac{1}{2}$ as much space as his boss. The rest of the room is to be used for 6 workstations. Kyle, the secretary, has drawn this first draft on the new floor plan. What is missing? How would you improve on Kyle's draft to make it more efficient?

Boss

Secretary

Problem-Solving Skill

Each chapter includes an explanation and an example of a classic problem-solving technique. When you do not know how to solve a problem, you can take definite steps and actions to find an answer. Knowing how to solve a problem is often more important than knowing the answer to a problem.

Put It All Together

Each chapter of the book includes one or more "Put It All Together" assignments. These assignments give you an opportunity to see how the variety of skills and information presented in a chapter fit together and can be used to solve real-world problems.

PUT IT ALL TOGETHER

Ikuko Goro's job pays $23,650 in yearly wages and 19% of her wages in yearly fringe benefits. She estimates that yearly job expenses are $1,354. Another job which she is looking at pays $22,490 in yearly wages and has estimated yearly fringe benefits of 26%, with job expenses of $1,080.

20. Find the net job benefits of Ikuko's current job.
21. Find the net job benefits of the other job.
22. Which job offers the greater net job benefits, and how much greater?
23. Georgio Bakalis can work for TRI Co. for $360 per week or Ellis, Inc. for $1,505 per month. Fringe benefits average 19% of yearly wages at TRI Co. and 24% at Ellis, Inc. Job expenses are estimated to be $956 per year at TRI Co. and $237 per year at Ellis, Inc. Which job would give Georgio more net job benefits? How much more?
24. Make a list of five job benefits you would like to receive. Next to each item, estimate the cost of each benefit and give one reason why an employer might want to offer such a benefit to their employees.

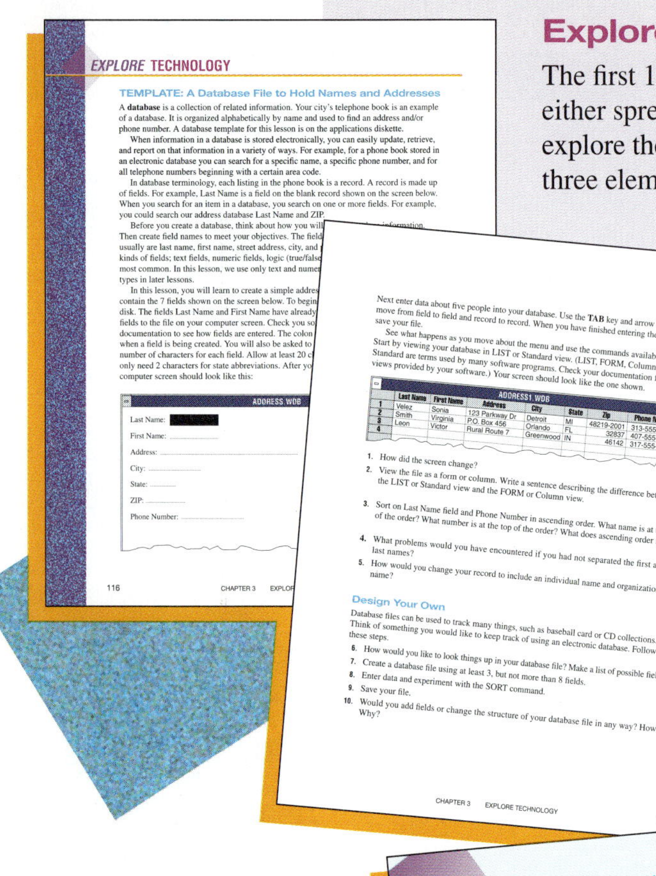

EXPLORE TECHNOLOGY

TEMPLATE: A Database File to Hold Names and Addresses

A **database** is a collection of related information. Your city's telephone book is an example of a database. It is organized alphabetically by name and used to find an address and/or phone number. A database template for this lesson is on the applications diskette.

When information in a database is stored electronically, you can easily update, retrieve, and report on that information in a variety of ways. For example, for a phone book stored in an electronic database you can search for a specific name, a specific phone number, and for all telephone numbers beginning with a certain area code.

In database terminology, each listing in the phone book is a record. A record is made up of fields. For example, Last Name is a field on the blank record shown on the screen below. When you search for an item in a database, you search on one or more fields. For example, you could search our address database Last Name and ZIP.

Before you create a database, think about how you will

Then create field names to meet your objectives. The field

usually are last name, first name, street address, city, and

kinds of fields; text fields, numeric fields, logic (true/false

most common. In this lesson, we use only text and numer

types in later lessons.

In this lesson, you will learn to create a simple addres

contain the 7 fields shown on the screen below. To begin

disk. The fields Last Name and First Name have already

fields to the file on your computer screen. Check you so

documentation to see how fields are entered. The colon

when a field is being created. You will also be asked to

number of characters for each field. Allow at least 20 c

only need 2 characters for state abbreviations. After yo

computer screen should look like this:

ADDRESS.WDB

Last Name:
First Name:
Address:
City:
State:
ZIP:
Phone Number:

116 CHAPTER 3 EXPLOR

Next enter data about five people into your database. Use the **TAB** key and arrow keys to move from field to field and record to record. When you have finished entering the records, save your file.

See what happens as you move about the menu and use the commands available to you. Start by viewing your database in LIST or Standard view. (LIST, FORM, Column and Standard are terms used by many software programs. Check your documentation for the views provided by your software.) Your screen should look like the one shown.

ADDRESS1.WDB

	Last Name	First Name	Address	City	State	Zip	Phone Number
1	Velez	Sonia	123 Parkway Dr	Detroit	MI	48219-2001	313-555-0100
2	Smith	Virginia	P.O. Box 456	Orlando	FL	32837	407-555-0132
3	Leon	Victor	Rural Route 7	Greenwood	IN	46142	317-555-0178
4							

1. How did the screen change?
2. View the file as a form or column. Write a sentence describing the difference between the LIST or Standard view and the FORM or Column view.
3. Sort on Last Name field and Phone Number in ascending order. What name is at the top of the order? What number is at the top of the order? What does ascending order mean?
4. What problems would you have encountered if you had not separated the first and last names?
5. How would you change your record to include an individual name and organization name?

Design Your Own

Database files can be used to track many things, such as baseball card or CD collections. Think of something you would like to keep track of using an electronic database. Follow these steps.

6. How would you like to look things up in your database file? Make a list of possible fields.
7. Create a database file using at least 3, but not more than 8 fields.
8. Enter data and experiment with the SORT command.
9. Save your file.
10. Would you add fields or change the structure of your database file in any way? How? Why?

CHAPTER 3 EXPLORE TECHNOLOGY 117

Explore Technology

The first 13 chapters conclude with a computer lesson that uses either spreadsheet or database software. In Chapter 14, you explore the Internet. The first 13 technology lessons have three elements:

Template Exercise: A template is a prepared spreadsheet file or database file that is set up for you to enter information provided in the activity. Many companies have custom applications that employees use to enter and access information.

Think About It: For each template, you are asked to identify the mathematical activities that were used to develop the application.

Design Your Own: In this last portion of the technology lesson, you design a solution to a specific situation. The ability to create technological solutions to problems is a valuable job skill.

Chapter Review

Each chapter concludes with a chapter review that highlights key points and formulas covered. Use the review to supplement your own notes when preparing for tests.

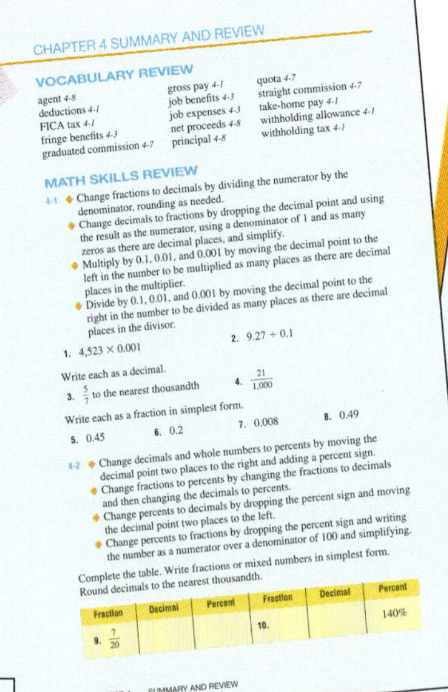

CHAPTER 4 SUMMARY AND REVIEW

VOCABULARY REVIEW

agent *4-8*
deductions *4-1*
FICA tax *4-1*
fringe benefits *4-3*
graduated commission *4-7*
gross pay *4-1*
job benefits *4-3*
job expenses *4-3*
net proceeds *4-8*
principal *4-8*
quota *4-7*
straight commission *4-7*
take-home pay *4-1*
withholding allowance *4-1*
withholding tax *4-1*

MATH SKILLS REVIEW

4-1
- Change fractions to decimals by dividing the numerator by the denominator, rounding as needed.
- Change decimals to fractions by dropping the decimal point and using the result as the numerator, using a denominator of 1 and as many zeros as there are decimal places, and simplify.
- Multiply by 0.1, 0.01, and 0.001 by moving the decimal point to the left in the number to be multiplied as many places as there are decimal places in the multiplier.
- Divide by 0.1, 0.01, and 0.001 by moving the decimal point to the right in the number to be divided as many places as there are decimal places in the divisor.

1. $4{,}523 \times 0.001$
2. $9.27 \div 0.1$

Write each as a decimal.

3. $\frac{5}{7}$ to the nearest thousandth
4. $\frac{21}{1{,}000}$

Write each as a fraction in simplest form.

5. 0.45
6. 0.2
7. 0.008
8. 0.49

4-2
- Change decimals and whole numbers to percents by moving the decimal point two places to the right and adding a percent sign.
- Change fractions to percents by changing the fractions to decimals and then changing the decimals to percents.
- Change percents to decimals by dropping the percent sign and moving the decimal point two places to the left.
- Change percents to fractions by dropping the percent sign and writing the number as a numerator over a denominator of 100 and simplifying.

Complete the table. Write fractions or mixed numbers in simplest form. Round decimals to the nearest thousandth.

Fraction	Decimal	Percent	Fraction	Decimal	Percent
9. $\frac{7}{20}$			10.		140%

CHAPTER 4 SUMMARY AND REVIEW

4-4
- Find a percent of a number by changing the percent to a decimal and multiplying.
- Find what percent one number is of another by dividing the number (the part) by the other number (the whole) and showing the result as a percent.
- Multiply by fractional equivalents of parts of 100% by mentally changing the percent to a fraction and multiplying.
- Find 1%, 10%, 100%, and 1,000% of a number by changing the percent to a decimal and multiplying.

11. $37\frac{1}{2}\%$ of $560
12. 1,000% of $2.30
13. $0.47 is what percent of $235

4-6
- Find a number that is a percent greater or smaller than another by multiplying by the percent and then either adding or subtracting the result to the number.
- Find what percent one number is greater or smaller than another by subtracting the smaller number from the larger to find the part and then dividing the part by the original number.

14. $8 is what percent greater than $6?
15. 1% less than $475

APPLICATIONS

4-3
- Multiply gross pay by the FICA rate to find FICA taxes.
- Add deductions to find total deductions, then subtract deductions from gross pay to find net pay.

16. Luisa Medina is paid $356 a week. Her employer deducts $28 for federal withholding tax, $14.58 for insurance, and 0.0765 for FICA tax. For the week, what is her FICA tax, total deductions, and net pay for the week?

4-5
- Add fringe benefits to find total fringe benefits, then add gross pay and fringe benefits to total job benefits.
- Add job expenses to find total job expenses. Subtract total job expenses from total job benefits to find net job benefits.

17. Karen Polk's job pays $45,650 per year plus 0.22 of wages in benefits. She estimates that yearly job expenses are $1,956. A job she has been offered pays $49,250 in yearly wages with estimated fringe benefits of 0.18 with job expenses of $2,122. What are her current job's net job benefits? What are the other job's net job benefits? Which job offers the greater job benefits? How much greater?

CHAPTER 4 SUMMARY AND REVIEW 163

WORKPLACE KNOWHOW: PARTICIPATES AS A TEAM MEMBER

Being able to work with other people to accomplish common objectives and goals is a very important skill. Many companies are organizing as teams to get more work done quickly. Team work skills include the ability to contribute ideas, build on members strengths, and be responsible for achieving the team goal. Reread the career profile that began this chapter. That profile discusses how teams work and are evaluated in a warehouse distribution center. What do you like about the team approach discussed? Is there anything that you do not like about it?

Assignment Working as a group, identify a goal that your class would like to achieve. You do not have to plan on achieving the goal. Simply identify one and define a plan for how the goal could be achieved. What team rules would you put into place?

When you are done, write a report describing how you determined what goal to strive for and how the goal could be accomplished. Discuss whether or not the class demonstrated an ability to work as a team on this assignment. Use specific examples to support your opinion.

WorkPlace KnowHow

Each chapter concludes with a WorkPlace KnowHow activity. These activities develop skills identified by the U.S. Department of Labor as critical for success in the workplace.

Work Skills Portfolio

We encourage every student to maintain a portfolio of his or her work skills. At the end of every chapter, the skills learned in the chapter are highlighted. You are encouraged to include samples of your work and an assessment of your skills at the end of each chapter. (We also encourage you to include material from other classes in the portfolio.) The work skills portfolio can help you identify skills that you have or need to acquire or enhance. It can also help you prepare a resume or for a job interview. The maintenance of a work skills portfolio can be a valuable, lifelong practice.

Work Skills Portfolio You have applied percentages and ratios to business situations and learned about the costs of doing business and business analysis. Include an assessment of your skills in this area. If you have ever run or considered owning your business, you may want to include a statement about how you would handle the finances of your business. Also include an assessment of your reasoning skills.

If you completed the integrated project in your student notebook, you may want to include a copy of the worksheets in your portfolio.

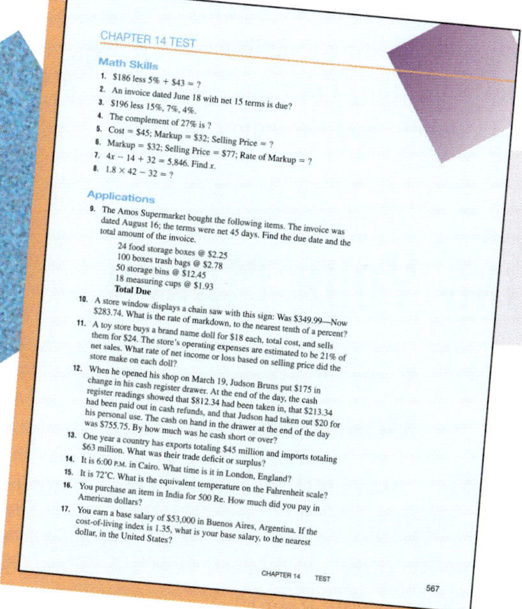

CHAPTER 14 TEST

Math Skills

1. \$186 less 5% + \$43 = ?
2. An invoice dated June 18 with net 15 terms is due?
3. \$196 less 15%, 7%, 4%.
4. The complement of 27% is ?
5. Cost = \$45; Markup = \$32; Selling Price = ?
6. Markup = \$32; Selling Price = \$77; Rate of Markup = ?
7. $4x - 14 + 32 = 5{,}846$. Find x.
8. $1.8 \times 42 - 32 = ?$

Applications

9. The Amos Supermarket bought the following items. The invoice was dated August 16; the terms were net 45 days. Find the due date and the total amount of the invoice.
   24 food storage boxes @ \$2.25
   100 boxes trash bags @ \$2.78
   50 storage bins @ \$12.45
   18 measuring cups @ \$1.93
   Total Due
10. A store window displays a chain saw with this sign: Was \$349.99—Now \$283.74. What is the rate of markdown, to the nearest tenth of a percent?
11. A toy store buys a brand name doll for \$18 each, total cost, and sells them for \$24. The store's operating expenses are estimated to be 21% of net sales. What rate of net income or loss based on selling price did the store make on each doll?
12. When he opened his shop on March 19, Judson Bruns put \$175 in change in his cash register drawer. At the end of the day, the cash register readings showed that \$812.34 had been taken in, that \$213.34 had been paid out in cash refunds, and that Judson had taken out \$20 for his personal use. The cash on hand in the drawer at the end of the day was \$755.75. By how much was he cash short or over?
13. One year a country has exports totaling \$45 million and imports totaling \$63 million. What was their trade deficit or surplus?
14. It is 6:00 P.M. in Cairo. What time is it in London, England?
15. It is 72°C. What is the equivalent temperature on the Fahrenheit scale?
16. You purchase an item in India for 500 Re. How much did you pay in American dollars?
17. You earn a base salary of \$53,000 in Buenos Aires, Argentina. If the cost-of-living index is 1.35, what is your base salary, to the nearest dollar, in the United States?

CHAPTER 14 TEST 567

Chapter Test

Each chapter concludes with a chapter test that checks your understanding of the material presented.

Additional Resources

In addition to the text, you may also have available a student workbook, which includes additional practice for each lesson. Every chapter has a vocabulary puzzle and an integrated project that lets you demonstrate your understanding of the skills and concepts taught.

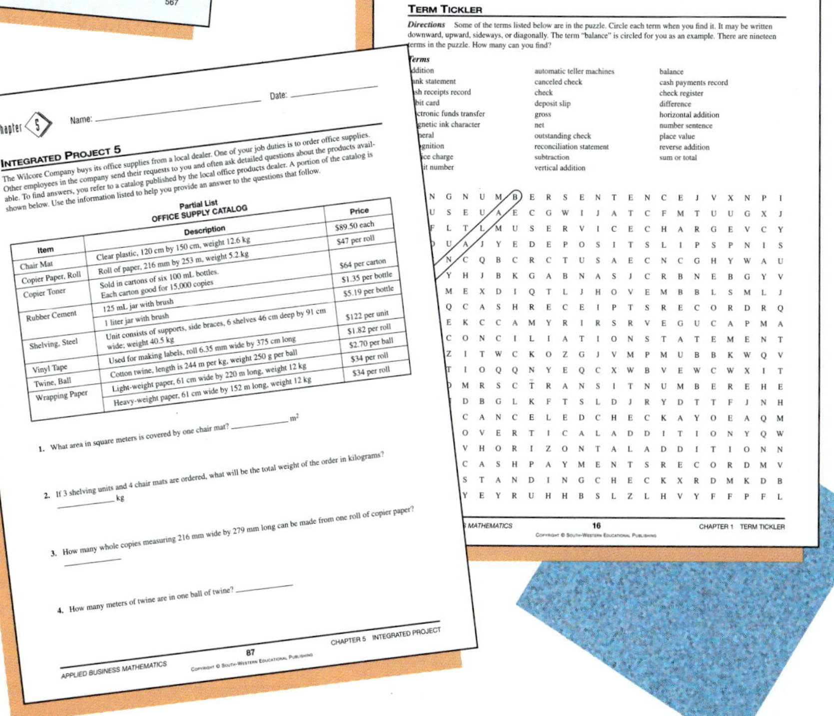

Chapter 5 Name: ____________ Date: ____________

INTEGRATED PROJECT 5

The Wilcore Company buys its office supplies from a local dealer. One of your job duties is to order office supplies. Other employees in the company send their requests to you and often ask detailed questions about the products available. To find answers, you refer to a catalog published by the local office products dealer. A portion of the catalog is shown below. Use the information listed to help you provide an answer to the questions that follow.

Partial List
OFFICE SUPPLY CATALOG

Item	Description	Price
Chair Mat	Clear plastic, 120 cm by 150 cm, weight 12.6 kg	\$89.50 each
Copier Paper, Roll	Roll of paper, 216 mm by 253 m, weight 5.2 kg	\$47 per roll
Copier Toner	Sold in cartons of six 100 mL bottles. Each carton good for 15,000 copies	\$64 per carton
Rubber Cement	125 mL jar with brush	\$1.35 per bottle
	1 liter jar with brush	\$5.19 per bottle
Shelving, Steel	Unit consists of supports, side braces, 6 shelves 46 cm deep by 91 cm wide; weight 40.5 kg	\$122 per unit
Vinyl Tape	Used for making labels, roll 6.35 mm wide by 375 cm long	\$1.82 per roll
Twine, Ball	Cotton twine, length is 244 m per kg, weight 250 g per ball	\$2.70 per ball
Wrapping Paper	Light-weight paper, 61 cm wide by 220 m long, weight 12 kg	\$34 per roll
	Heavy-weight paper, 61 cm wide by 152 m long, weight 12 kg	\$34 per roll

1. What area in square meters is covered by one chair mat? ____________ m²
2. If 3 shelving units and 4 chair mats are ordered, what will be the total weight of the order in kilograms? ____________ kg
3. How many whole copies measuring 216 mm wide by 279 mm long can be made from one roll of copier paper? ____________
4. How many meters of twine are in one ball of twine? ____________

APPLIED BUSINESS MATHEMATICS 87 CHAPTER 5 INTEGRATED PROJECT
Copyright © South-Western Educational Publishing

Chapter 1 Name: ____________ Date: ____________

TERM TICKLER

Directions Some of the terms listed below are in the puzzle. Circle each term when you find it. It may be written downward, upward, sideways, or diagonally. The term "balance" is circled for you as an example. There are nineteen terms in the puzzle. How many can you find?

Terms

ddition	automatic teller machines	balance
nk statement	canceled check	cash payments record
sh receipts record	check	check register
bit card	deposit slip	difference
ctronic funds transfer	gross	horizontal addition
gnetic ink character	net	number sentence
neral	outstanding check	place value
gnition	reconciliation statement	reverse addition
ice charge	subtraction	sum or total
it number	vertical addition	

N G N U M B E R S E N T E N C E J V X N P I
U S E U A E C G W I J A T C F M T U U G X J
F L T L M U S E R V I C E C H A R G E V C Y
U A J Y E D E P O S I T S L I P S P N I S
N C Q B C R C T U S A E C N C G H Y W A U
Y H J B K G A B N A S J C R B N E B G Y V
M E X D I Q T L J H O V E M B B L S M L J
Q C A S H R E C E I P T S R E C O R D R Q
E K C C A M Y R I R S R V E G U C A P M A
C O N C I L I A T I O N S T A T E M E N T
Z I T W C K O Z G J V M P M U B B K W Q V
T I O Q Q N Y E Q C X W B V E W C W X I T
M R S C T R A N S I T N U M B E R E H E
D B G L K F T S L D J R Y D T T F J N H
C A N C E L E D C H E C K A Y O E A Q M
O V E R T I C A L A D D I T I O N Y Q W
V H O R I Z O N T A L A D D I T I O N N
C A S H P A Y M E N T S R E C O R D M V
S T A N D I N G C H E C K X R D M K D B
Y E Y R U H H B S L Z L H V Y F F P F L

MATHEMATICS 16 CHAPTER 1 TERM TICKLER
Copyright © South-Western Educational Publishing

TABLE OF CONTENTS

Unit 1 MANAGING YOUR MONEY

CHAPTER 1 ♦ MONEY RECORDS

PROBLEM SOLVING

CHAPTER 2 ♦ GROSS AND AVERAGE PAY

CHAPTER 3 ♦ REGULAR AND OVERTIME PAY

CHAPTER 4 ♦ NET PAY, FRINGE BENEFITS, AND COMMISSION

Unit 2 SPENDING WISELY

CHAPTER 5 ♦ METRIC MEASUREMENT

PROBLEM SOLVING

CHAPTER 8 ♦ PAYING YOUR TAXES

CHAPTER 9 ♦ MANAGING YOUR INSURANCE NEEDS

MAKING MONEY GROW

Unit 3

CHAPTER 10 ♦ SAVING AND BORROWING MONEY

CHAPTER 11 ♦ INVESTMENTS

BUSINESS MATHEMATICS

CHAPTER 12 ♦ BUSINESS ANALYSIS AND STATISTICS

PROBLEM SOLVING

CHAPTER 13 ♦ Business Profit and Loss

CHAPTER 14 ♦ Doing Business in a Global Economy

PLACE VALUE AND ORDER

EXAMPLE Write 2,345,678.9123 in words.

millions	hundred thousands	ten thousands	thousands	hundreds	tens	ones	.	tenths	hundredths	thousandths	ten thousandths
2	3	4	5	6	7	8		9	1	2	3

The place-value chart shows the value of each digit. The value of each place is ten times the place to the right.

SOLUTION The number shown is *two million, three hundred forty-five thousand, six hundred seventy-eight* ***and*** *nine thousand one hundred twenty-three ten-thousandths.*

EXAMPLE Use < or > to make this sentence true.
6 • 2

SOLUTION Numbers can be graphed on a number line. The number farther to the right is the larger number. Remember, < means "less than" and > means "greater than".

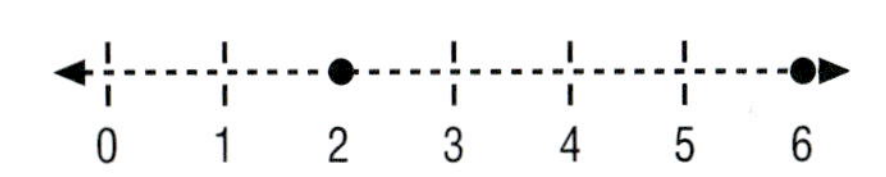

Six is greater than two.
$6 > 2$

Write each number in words.

1. 3,647
2. 6,004,300.002
3. 0.9001

Write each of the following as a number.

4. two million, one hundred fifty thousand, four hundred seventeen
5. five thousand, one hundred twenty and five hundred two thousandths
6. nine million, ninety thousand, nine hundred and ninety-nine ten-thousandths

Use < or > to make each sentence true.

7. 9 • 8
8. 164 • 246
9. 63,475 • 6,435
10. 52 • 50
11. 5.39 • 9.02
12. 43.94 • 53.69

SKILLS WORKSHOP 2

ADDITION AND SUBTRACTION OF WHOLE NUMBERS AND DECIMALS

To add or subtract whole numbers and decimals, write the digits so the place values line up. Add from right to left, renaming when necessary. When adding or subtracting decimals, be sure to place the decimal point directly below the aligned decimals in the problem.

EXAMPLE

Find the sum of 0.058, 25.39, 6,346, and 1.57.

The answer is called the **total** or **sum.**

SOLUTION

```
  1 2
    0.058   ← The red zero is used to
   25.39      show there no ones.
6,346.      ← The decimal point is at
+   1.57      the end of whole numbers.
6,373.018   ← Add from right to left.
```

EXAMPLE

Find the difference between 10,049 and 5,364.

The answer is called the **difference.**

SOLUTION

```
    9
  9 10 14
 1 0,0 4 9   ← Rename as needed to
−  5,3 6 4     subtract.
   4,6 8 5
```

EXAMPLE

Subtract 6.37 from 27.

The smaller number is subtracted from the larger.

SOLUTION

```
       9
    6 10 10
  2 7. 0 0   ← Add zeroes if that helps
−   6. 3 7     you complete the
  2 0. 6 3     subtraction.
```

Add or subtract.

1. 23.146 + 17.215
2. 46.48 − 6.57
3. 52 − 1.95
4. 0.86 + 0.75
5. 83 − 82.743
6. 9.45 + 13.2
7. 14.5 − 9.684
8. 6.4 + 54.2 + 938.05 + 3.7 + 47.3
9. 913.03 − 79
10. 1,765.36 + 1,587.50 + 1,400
11. 0.8523 − 0.794
12. 51,876.36 + 48,156.95 + 1,417.86
13. 17,347.85 − 12,516.90
14. 76.2 + 80 + 56 + 9.321
15. 107,285 − 61,500.25
16. 567.1 + 6 + 13.452 + 100

MULTIPLYING WHOLE NUMBERS AND DECIMALS

To multiply whole numbers, find each partial product and then add.

The red zeros are annexed to help align the answers.

Add. Then rewrite the answer with commas.

```
   5,754  ← factor
 ×   236  ← factor
   34524  ← 5,754 × 6
  172620  ← 5,754 × 30
 1150800  ← 5,754 × 200
 1357944  ← Add.
1,357,944 ← product
```

When multiplying decimals, locate the decimal point in the product so that there are as many decimal places in the product as the total number of decimal places in the factors.

EXAMPLE Multiply 2.6394 by 3,000.

SOLUTION

```
     2.6394
 ×    3,000
 7,918.2000   or 7,918.2
```

Zeros at end (far right) *after* the decimal point can be dropped because they are not *significant digits.*

EXAMPLE Multiply 3.92 by 0.023.

SOLUTION

```
    3.92  ←   2 decimal places
 × 0.023  ← +3 decimal places
    1176
    7840
 0.09016  ←   5 decimal places
```

The red zero is added *before* the nine, so the product will have five decimal places.

Multiply.

1. 36 × 45
2. 500 × 30
3. 17,000 × 230
4. 6.2 × 8
5. 950 × 1.6
6. 3.652 × 20
7. 179 × 83
8. 257 × 320
9. 8,560 × 275
10. 467 × 0.3
11. 2.63 × 183
12. 0.758 × 321.8
13. 49.3 × 1.6
14. 6.859 × 7.9
15. 794.4 × 321.8
16. 0.08 × 4
17. 0.062 × 0.5
18. 0.0135 × 0.003
19. 21.6 × 3.1
20. 8.76 × 0.005
21. 5.521 × 3.642

DIVIDING WHOLE NUMBERS AND DECIMALS

Dividing whole numbers and decimals involves a repetitive process of estimating a quotient, multiplying, and subtracting.

$$\begin{array}{rrl} & 34 & \leftarrow \textbf{quotient} \\ \text{divisor} \rightarrow & 7\overline{)239} & \leftarrow \textbf{dividend} \\ & \underline{21\downarrow} & \leftarrow 3 \times 7 \\ & 29 & \leftarrow \text{Subtract. Bring down the 9.} \\ & \underline{28} & \leftarrow 4 \times 7 \\ & 1 & \leftarrow \textbf{remainder} \end{array}$$

EXAMPLE Find 283.86 ÷ 5.7.

SOLUTION When dividing decimals, move the decimal point in the divisor to the right until it is a whole number. Move the decimal point the same number of places in the dividend. Then place the decimal point in the answer directly above the new location of the decimal point in the dividend.

$$5.7.\overline{)283.8.6} \rightarrow \begin{array}{r} 49.8 \\ 57\overline{)2838.6} \\ \underline{228} \\ 558 \\ \underline{513} \\ 45\ 6 \\ \underline{45\ 6} \\ 0 \end{array}$$

If answers do not have a remainder of 0, you can add 0's after the last digit and continue dividing.

Divide.

1. 72 ÷ 6
2. 6,000 ÷ 20
3. 26,568 ÷ 8
4. 5.6 ÷ 7
5. 120 ÷ 0.4
6. 936 ÷ 12
7. 3.28 ÷ 4
8. 0.1960 ÷ 5
9. 1968 ÷ 0.08
10. 16 ÷ 0.04
11. 1525 ÷ 0.05
12. 109.94 ÷ 0.23
13. 0.6 ÷ 24
14. 7.924 ÷ 0.28
15. 32.6417 ÷ 9.1
16. 24 ÷ 0.6
17. 1,784.75 ÷ 29.5
18. 0.01998 ÷ 0.37
19. 7.8 ÷ 0.3
20. 12,000 ÷ 0.04
21. 820.94 ÷ 0.02

MULTIPLYING AND DIVIDING FRACTIONS

To multiply fractions, multiply the numerators and then multiply the denominators. Write the answer in simplest form.

EXAMPLE Multiply $\frac{2}{5}$ and $\frac{7}{8}$.

SOLUTION $\frac{2}{5} \times \frac{7}{8} = \frac{2 \times 7}{5 \times 8} = \frac{14}{40} = \frac{7}{20}$

To divide by a fraction, multiply by the reciprocal of that fraction. To find the reciprocal of a fraction, invert (turn upside down) the fraction. The product of a fraction and its reciprocal is 1. Since $\frac{2}{3} \times \frac{3}{2} = \frac{6}{6}$ or 1, $\frac{2}{3}$ and $\frac{3}{2}$ are reciprocals of each other.

EXAMPLE Divide $1\frac{1}{5}$ by $\frac{2}{3}$.

SOLUTION $1\frac{1}{5} \div \frac{2}{3} = \frac{6}{5} \div \frac{2}{3}$ Write the mixed number as a fraction.

$\frac{6}{5} \div \frac{2}{3} = \frac{6}{5} \times \frac{3}{2}$ Rewrite division as multiplication by the inverse of the divisor.

$= \frac{6 \times 3}{5 \times 2} = \frac{18}{10}$ or $1\frac{4}{5}$ Multiply and simplify.

Multiply or divide. Write each answer in simplest form.

1. $\frac{2}{3} \div \frac{5}{6}$
2. $\frac{3}{5} \times \frac{10}{12}$
3. $\frac{5}{8} \div \frac{1}{4}$
4. $\frac{1}{2} \times \frac{2}{3}$
5. $\frac{2}{3} \times \frac{1}{2}$
6. $\frac{3}{4} \times \frac{5}{8}$
7. $\frac{1}{2} \div \frac{2}{3}$
8. $\frac{2}{3} \div \frac{1}{2}$
9. $\frac{3}{4} \div \frac{5}{8}$
10. $2\frac{2}{3} \div 1\frac{3}{5}$
11. $1\frac{1}{5} \times 2\frac{1}{4}$
12. $3\frac{1}{3} \times 1\frac{1}{10}$
13. $5\frac{2}{5} \div 2\frac{4}{7}$
14. $2\frac{4}{7} \div 5\frac{2}{5}$
15. $2\frac{4}{7} \times 5\frac{2}{5}$
16. $1\frac{7}{8} \div 1\frac{7}{8}$
17. $\frac{3}{4} \times \frac{2}{3} \times 1\frac{5}{8} \times 2\frac{2}{3}$

SKILLS WORKSHOP 6

ADDING AND SUBTRACTING FRACTIONS

To add and subtract fractions, you need to find a common denominator and then add or subtract, renaming as necessary.

EXAMPLE Add $\frac{3}{4}$ and $\frac{5}{6}$.

SOLUTION

$$\begin{array}{rcl} \frac{3}{4} = \frac{3}{4} \times \frac{3}{3} = & & \frac{9}{12} \\ + \frac{5}{6} = \frac{5}{6} \times \frac{2}{2} = & + & \frac{10}{12} \\ \hline & & \frac{19}{12} \end{array}$$

19 ← Add the numerators.
12 ← Use the common denominator.

Then simplify. $\frac{19}{12} = 1\frac{7}{12}$

EXAMPLE Subtract $1\frac{3}{5}$ from $5\frac{1}{2}$.

SOLUTION

$$\begin{array}{rcrcr} 5\frac{1}{2} & = & 5\frac{5}{10} & = & 4\frac{15}{10} \\ -1\frac{3}{5} & = & -1\frac{6}{10} & = & -1\frac{6}{10} \\ \hline & & \uparrow & & 3\frac{9}{10} \end{array}$$

You can not subtract $\frac{6}{10}$ from $\frac{5}{10}$, so rename again.

Add or subtract.

1. $\frac{1}{5} + \frac{1}{10}$

2. $\frac{2}{3} + \frac{1}{3}$

3. $\frac{5}{8} + \frac{3}{4}$

4. $\frac{6}{7} - \frac{2}{7}$

5. $\frac{3}{4} - \frac{1}{3}$

6. $\frac{5}{8} - \frac{1}{4}$

7. $2\frac{1}{2} + 3\frac{1}{2}$

8. $6\frac{5}{8} + 3\frac{7}{8}$

9. $3\frac{2}{3} + 4\frac{1}{2}$

10. $2\frac{3}{4} - 1\frac{1}{4}$

11. $5\frac{1}{8} - 3\frac{7}{8}$

12. $1\frac{1}{3} - \frac{2}{3}$

13. $6\frac{1}{2} + 5\frac{7}{9}$

14. $9\frac{2}{5} - 1\frac{1}{8}$

15. $7\frac{2}{3} + 6\frac{1}{5}$

16. $8\frac{1}{10} - 5\frac{2}{3}$

17. $6\frac{1}{2} - 5\frac{3}{5}$

18. $10\frac{5}{8} - 9\frac{3}{4}$

19. $1\frac{1}{5} + 2\frac{1}{3} + 5\frac{1}{4}$

20. $10\frac{7}{8} + 3\frac{3}{4} + 6\frac{1}{2} + 2\frac{5}{8}$

FRACTIONS, DECIMALS, AND PERCENTS

Percent means *per hundred.* Thus, 35% means 35 out of 100. Percents can be written as equivalent decimals and fractions.

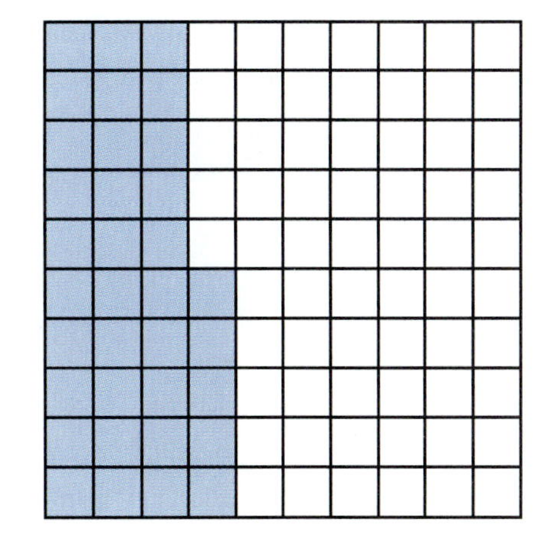

$35\% = 0.35$ Move the decimal point two places to the left.

$35\% = \frac{35}{100}$ Write the fraction with a denominator of 100.

$= \frac{7}{20}$ Then simplify.

EXAMPLES Write $\frac{3}{8}$ as a decimal and as a percent.

SOLUTIONS $\frac{3}{8} = 0.375$

$0.375 = 37.5\%$

To change a fraction to a percent, first divide and write the answer as a decimal. Then change the decimal to a percent by moving the decimal point two places to the right and annexing a percent symbol.

Percents greater than 100% represent whole numbers or mixed numbers.

$200\% = 2$ or 2.00 $\qquad 350\% = 3.5$ or $3\frac{1}{2}$

Complete each table. Write all fractions in simplest form.

	Fraction	Decimal	Percent
1.	$\frac{1}{2}$		
2.		0.63	
3.			10%
4.	$\frac{1}{4}$		
5.		0.15	
6.			12%
7.			100%

	Fraction	Decimal	Percent
8.	$\frac{3}{4}$		
9.		0.4	
10.			150%
11.		2.35	
12.	$3\frac{7}{8}$		
13.			160%
14.		10.125	

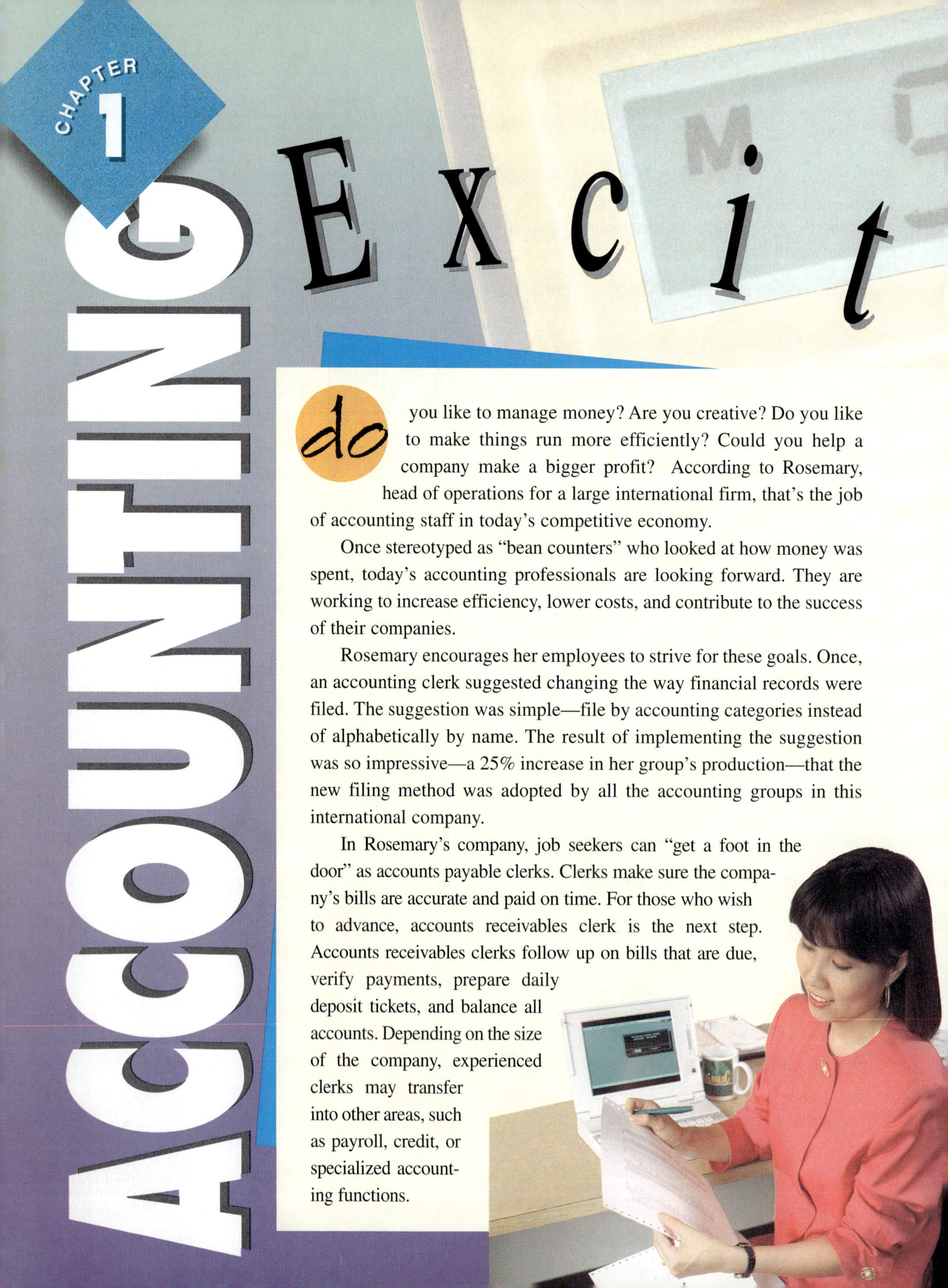

CHAPTER 1

ACCOUNTING Excit

do you like to manage money? Are you creative? Do you like to make things run more efficiently? Could you help a company make a bigger profit? According to Rosemary, head of operations for a large international firm, that's the job of accounting staff in today's competitive economy.

Once stereotyped as "bean counters" who looked at how money was spent, today's accounting professionals are looking forward. They are working to increase efficiency, lower costs, and contribute to the success of their companies.

Rosemary encourages her employees to strive for these goals. Once, an accounting clerk suggested changing the way financial records were filed. The suggestion was simple—file by accounting categories instead of alphabetically by name. The result of implementing the suggestion was so impressive—a 25% increase in her group's production—that the new filing method was adopted by all the accounting groups in this international company.

In Rosemary's company, job seekers can "get a foot in the door" as accounts payable clerks. Clerks make sure the company's bills are accurate and paid on time. For those who wish to advance, accounts receivables clerk is the next step. Accounts receivables clerks follow up on bills that are due, verify payments, prepare daily deposit tickets, and balance all accounts. Depending on the size of the company, experienced clerks may transfer into other areas, such as payroll, credit, or specialized accounting functions.

Marybelle has been in accounting for more than 40 years and advanced in her small company from accounts payable clerk to credit manager. She attributes her success to:

- **Good Memory and Eye for Detail** — to remember figures and customers
- **Problem-Solving Skills** — to make the dollars stretch when money is tight
- **Personal Skills** — to deal tactfully, yet forcefully, with bankers and clients
- **Estimating and Mental Math Skills** — to identify discrepancies at a glance
- **Understanding of Decimals, Percents, and Equations**

When hiring, Rosemary looks for these additional qualifications:

- ✔ data entry and spreadsheet skills
- ✔ skill using a 10-key calculator
- ✔ familiarity with an office environment

YOU BET.

Accounting clerks are needed in all industries. If considering a career in accounting, you may want to gain experience working for a temporary agency or a company that does accounting tasks for other companies. Most of all, remember Rosemary's advice: "Employees who can recognize areas for improvement and make them happen are the most valued."

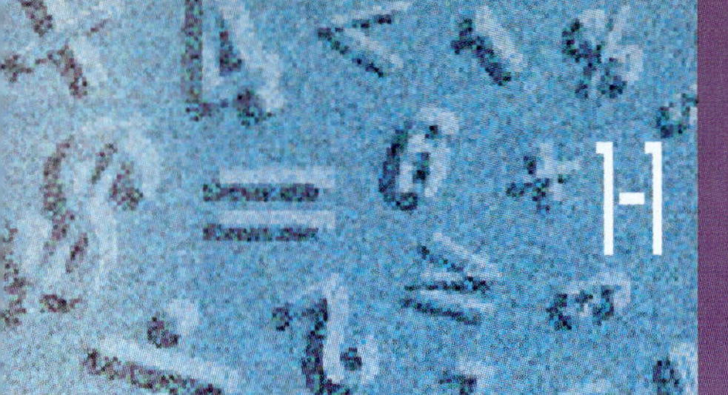

1-1 CASH RECEIPTS RECORDS

OBJECTIVES

In this lesson, you will learn to

- *use mental math and number patterns to simplify addition;*
- *add vertically; and*
- *maintain a cash receipts record.*

Cash receipts records, written records of money received, can be a great help to individuals and businesses when planning how to use money. Cash records may be kept manually in a notebook or on a computer spreadsheet or database. It is important to keep cash records up-to-date and accurate.

◆ Make a chart showing money you have received and spent during the past month. Be sure to include from whom you received the money and for what you spent the money.

WARM UP

Add.

a.	b.	c.	d.	e.	f.
23	802	768	45	90	230
398	9	56	309	230	71
107	23	134	96	16	524
9	512	9	749	225	85
+ 312	+ 229	+ 529	+ 190	+ 843	+ 43

THE CASH RECEIPTS RECORD

Laverne and Ramon Lewis receive weekly pay from their full-time jobs. Ramon also receives pay from a part-time job. Laverne and Ramon Lewis keep track of all their cash receipts manually using a form called a **cash receipts record.** Their cash receipts record for the week of June 6 is shown in Illustration 1-1.1

Cash Receipts Record				
NAME	*The Lewis Family*		ACCOUNT NO.	
ADDRESS				
DATE		EXPLANATION	AMOUNT	
19– – *June*	*6*	*Ramon's part-time pay for week*	*$85*	*50*
	7	*Refund on state income tax*	*87*	*20*
	8	*Refund on returned purchase*	*29*	*75*
	8	*Insurance dividend*	*44*	*25*
	10	*Ramon's full-time pay for week*	*480*	*75*
	10	*Laverne's pay for week*	*609*	*75*
		Total	*$1,337*	*20*

Illustration 1-1.1. Cash Receipts Record for the Lewis Family

At the end of the week, the Lewis' add the amount column and find the total of their cash receipts. Their total receipts for the week of June 6 were $1,337.20.

EXERCISE YOUR SKILLS

Prepare a cash receipts record; then find the total receipts.

1. VideoRama, a video store, had these cash receipts for a week: Monday, $548.66; Tuesday, $296.78; Wednesday, $361.93; Thursday, $500.24; Friday, $958.12; Saturday, $986.97; Sunday, $732.99.
2. Alice Buffom uses a computer to keep her cash receipts. Her computer screen shows the following receipts for the week of March 10: March 10, $50 for a birthday gift from her father and a $25 repayment of a personal loan from a friend; March 11, refund on aluminum cans recycled, $2.65; March 12, interest on savings, $115.15; March 13, refund on a cancelled insurance policy, $37.18; and March 14, her weekly pay, $382.40.
3. The senior class of Westport High School uses a computer to keep records of their cash receipts. For the month of October, their receipts were as follows: October 4, senior class car wash fundraiser, $124; October 6, senior class bake sale, $75; October 6, class dance night, $264.75; October 14, interest on savings, $10.34; October 28, food sales at football game, $165.85; and October 29, sales of class hats, $389.25.
4. An auto service station had these cash receipts for a week: auto parts, $367.45; repairs, $1,824.76; gasoline, $19,832.89; car wash, $215.00; other sales, $189.88.

ADDITION

Addition is a process of combining two or more numbers to get one number. The result is called the **sum,** or **total.** You have used addition to find total cash receipts. Of course, you can and will use addition in many other ways.

Checking Addition

Addition must be done accurately. Accuracy requires that you check your work. One way to check a total is by **reverse addition.** To do this, add the column in the opposite direction from the way you added it the first time. Reverse addition usually gives you new combinations, so you avoid making the same mistake twice.

Adding Faster by Finding Patterns

You can add faster by mentally combining two or three numbers that total 10, then adding that 10 to the other numbers. This method is shown in the example at the right. When you add from the top down, you think "10, 20, 30, 40, 48." When you check your work by adding from the bottom up, you think "8, 18, 28, 38, 48."

	5	10
	5	
	2	10
	3	
	5	
	9	10
10	4	
	1	
	6	
	+ 8	
	48	

EXERCISE YOUR SKILLS

Mental Math Add from the top down. Check your work by adding the columns from the bottom up.

5.	**6.**	**7.**	**8.**	**9.**	**10.**
5	6	9	5	4	6
5	4	1	5	6	1
6	5	9	6	8	8
4	2	1	2	3	1
2	1	8	3	7	5
8	9	1	5	6	3
4	4	5	2	3	6
1	6	3	1	2	2
+ 5	+ 7	+ 2	+ 7	+ 4	+ 7

Mental Math Discovering number patterns can be very helpful. Find the missing numbers below by finding the pattern in the numbers.

11. 10, 20, 30, 40, ?

12. 4, 9, 14, 19, ?

13. 1, 3, 5, ?, 9

14. 4, 14, 22, ?, 32

15. 3, 6, 12, ?, 48

16. 12, 14, 17, ?, 26

17. 82, 72, 64, ?, 54

18. 43, 46, 48, 51, 53, ?

19. $31, $72, $?, $318

20. 8, 21, 60, ?, 528

Solve. Check by reverse addition.

21.	**22.**	**23.**	**24.**	**25.**
$18.79	$ 7.45	$98.45	$55.38	$ 8.32
2.11	23.45	10.63	31.02	11.71
1.92	74.17	2.82	94.46	91.07
17.28	76.52	30.26	6.74	87.65
1.55	1.48	81.74	10.23	11.32
+ 21.05	+ 39.40	+ 26.36	+ 84.17	+ 10.13

26. Allnite Qwik Market's seven cash registers showed the following cash receipts for Saturday: cash register #1 $3,497.22; #2, $2,259.86; #3, $4,178.40; #4, $2,863.17; #5, $4,888.33; #6, $3,892.21; #7, $2,003.62. What were the total cash receipts for Saturday?

27. A neighborhood garage sale had these cash receipts for 5 days: $136.97, $256.21, $366.10, $218.06, $89.42. What were the total cash receipts?

28. Isabelle Santiago's home computer screen shows the following cash receipts for May. May 4, weekly pay, $584.04 and interest on savings, $26.86; May 7, reimbursement check for expenses, $782.33; May 11, weekly pay, $548.24; May 18, weekly pay, $548.23; May 25, weekly pay, $548.23, and May 30, stock dividend, $53.40. What were her total receipts?

29. The Little League in Scoville held 6 car washes to raise money for equipment. The cash receipts from the car washes were:

Week		Week	
1	$204.50	4	$ 77.75
2	137.25	5	120.50
3	93.15	6	137.25

What were the total cash receipts from the car washes?

30. A computer store's cash receipts for a week were:

Computers	$37,027.78	Computer books	$134.56
Computer software	8,107.39	Training	672.00
Printers	4,527.19	Repairs	879.42
Other equipment	2,365.08		

What were the store's total cash receipts for the week?

31. ◆ Write a short summary of why a cash receipts record is important. Consider including an example that shows the necessity of adding when working with cash receipts records. Keep a cash receipts journal for two weeks. You will refer to it later in this chapter.

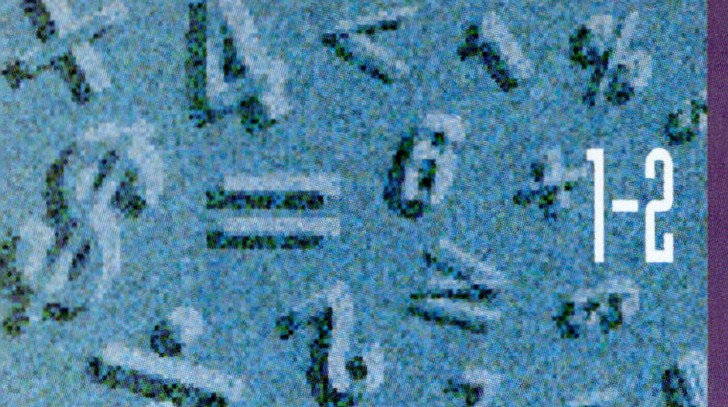

1-2 CASH PAYMENTS RECORDS

OBJECTIVES

In this lesson, you will learn to

- *add horizontally;*
- *use mental math to solve simple equations;*
- *read, interpret, and write numbers in the decimal system; and*
- *maintain a cash payments record.*

Heather could not understand it. She just got her paycheck last week, but now she is broke again! She is always broke! Where did the money go? Heather needs to keep track of where she spends her money. Then she may be able to find ways to spend more wisely and even to save money.

◆ What do you spend your weekly allowance and other income for? List the amounts you spend on each type of item during a typical week.

WARM UP

Add.

a.	b.	c.	d.
4	53	17.48	\$533.71
6	72	48.46	393.67
3	28	91.10	416.83
+ 9	+ 17	+ 32.74	+ 862.42

THE CASH PAYMENT RECORD

Laverne and Ramon Lewis also keep a record of the cash they spend. That record is called a **cash payments record.** The cash payments record for the Lewis family for the week of June 6 is shown in Illustration 1-2.1, on the following page.

At the end of each week the Lewises find the total of their payments by adding the amount column. Their total payments for the week of June 6 were \$1,380.75.

Calculator You can use a calculator to find the total for cash payments or cash receipts records. Be sure you know how to use your calculator properly. It is also a good idea to do each problem twice. It is easy to make errors when entering numbers on a calculator.

Cash Payments Record

NAME *The Lewis Family* ACCOUNT NO.

ADDRESS

DATE		EXPLANATION	AMOUNT	
19– – *June*	*6*	*Power bill for May*	*$92*	*41*
	7	*Home mortgage payment*	*772*	*33*
	8	*Auto payment*	*165*	*78*
	9	*Savings deposit*	*130*	*00*
	10	*Auto repair*	*85*	*00*
	10	*Groceries*	*135*	*23*
		Total	*$1,380*	*75*

Illustration 1-2.1. Cash Payments Record for the Lewis Family

EXERCISE YOUR SKILLS

For each of the following exercises, prepare a cash payments chart. Then find the total amount of the cash payments.

1. The cash payments for Ina Stein for the week of September 14 are: September 14, dinner out $24.88; September 15, groceries $65.75; September 15, movie rental $6.50; September 17, telephone bill $53.81; September 18, clothing purchase $145.29; September 19, lawn mowing service $15; September 20, gift $37.59. Find her total cash payments for the week.
2. Charlie O'Hara uses a spreadsheet program to keep his cash payments records. At the end of the month his spreadsheet shows these expenses: food, $319.53; clothing, $102.45; rent, $550; transportation, $61.36; miscellaneous, $38.76; savings, $100. Find Charlie's total payments for the month.
3. A small business spent these amounts in one month: salaries and wages, $8,129.33; delivery expenses, $134.89; rent, $1,230.00; supplies, $214.64; postage, $422.13; power, $692.61; telephone, $385.93. What were the total cash payments of the business for the month?
4. The Edgemont Data Processing Managers Association had the following cash payments for the month: national dues, $120.36; postage, $315.78; award plaques, $75.89; printing costs, $702.34; telephone, $27.34. How much did the Association spend in the month?
5. Sonia Perez uses fuel oil to heat her house. From September to June she spent these amounts on fuel oil: $67.63, $101.94, $197.55, $256.29, $177.92, $82.45, $73.12. What were her total cash payments for fuel during that period?

NUMERALS AND NUMBER SENTENCES

When you keep cash records or other records about numbers, you make marks to represent the numbers. These marks or symbols are called numerals. For example, the marks 2, 54, and 135 are numerals.

A **numeral** is a symbol or name for a number. A number may be represented by many different numerals or names. For example, the number 10 is usually represented by the numeral 10. The number ten may also be represented by the numerals $4 + 6$, $12 - 2$, 2×5, $30 \div 3$, $\frac{20}{2}$, X, and others.

To show that two numerals are names for the same number, you may write an equals sign (=) between them. For example, you may write $3 + 7 = 2 \times 5$. This statement is called **a number sentence** or an **equation.** It says that the number named by $3 + 7$ is equal to the number named by 2×5.

A number sentence may be either true or false. The sentence $3 + 7 = 2 \times 5$ is true because $3 + 7$ and 2×5 are names for the same number, 10. The sentence $3 + 7 = 3 \times 5$ is false because $3 + 7$ is a name for 10 but 3×5 is a name for 15.

A sentence like $2 + 7 = 3 + N$ has a numeral missing, so it is incomplete. These sentences are sometimes called **open sentences.** As it stands, the sentence is neither true nor false. To make it complete you must replace the N with a numeral that makes the sentence true. The numeral 6 makes the sentence true, so $N = 6$. You may write $2 + 7 = 3 + 6$. You now have a complete, true statement. Use this method to make complete and true sentences when solving many problems in this book.

EXERCISE YOUR SKILLS

Mental Math Tell whether each sentence is true (T) or false (F).

6. $28 = 11 + 1$
7. $23 - 8 = 14$
8. $9 + 16 = 19$
9. $34 - 6 = 32$
10. $2 + 7 = 7 + 2$
11. $8 - 5 = 5 - 2$
12. $4 + 7 = 15 - 3$
13. $8 - 2 = 3 + 3$
14. $6 - 2 = 3 + 1$
15. $9 - 4 = 15 - 9$
16. $8 + 7 = 2 + 7$
17. $3 - 1 + 8 = 5 + 5$
18. $6 + 7 - 9 = 4 + 8 - 8$
19. $4 + 5 - 3 = 6 - 2 + 3$
20. $9 - 8 + 5 = 2 + 9 - 5$
21. $7 + 7 - 6 = 5 - 2 + 6$

Find the value of N that makes each sentence true.

22. $5 + 4 = 4 + N$
23. $8 + 7 = N + 8$
24. $N - 4 = 6 + 9$
25. $18 - 9 = 4 + N$
26. $N + 3 = 9$
27. $4 + 5 = 7 + N$
28. $8 - 3 = N + 3$
29. $8 + N = 7 + 6$
30. $9 - 6 = 7 - N$
31. $3 + N = 15$
32. $N + 6 = 9 - 2$
33. $3 + 9 = N - 2$

READING AND WRITING NUMBERS

The symbols that represent the numbers from zero through nine are called digits. These ten digits are: 0, 1, 2, 3, 4, 5, 6, 7, 8, 9. Because this system uses ten digits, it is called the decimal system. The word "decimal" comes from the Latin word deca, which means ten.

Place Value

Using the digits of the decimal system, we can write larger numbers. We can do this because the value of a digit depends on its position in the number. We call this the **place value** of the digit. Notice the place value for each digit in 72,345.

Ten-thousands	Thousands		Hundreds	Tens	Ones
7	2	,	3	4	5

The 5 at the far right has a value of 5 *ones* or 5. The 4 to the left has a value of 4 *tens* or 40. The 3 has a value of 3 *hundreds* or 300. The 2 has a value of 2 *thousands* or 2,000. The 7 has a value of 7 *ten-thousands* or 70,000.

Look at the number 5,555. Notice that moving a digit one place to the left multiplies its value by ten. So 5,555 can be written as 5,000 + 500 + 50 + 5.

EXERCISE YOUR SKILLS

Look at this number: 34,967.

34. How many ten thousands are there?

35. How many thousands?

36. How many hundreds?

37. How many tens?

38. How many ones?

39. Write the number in words.

In the number 6,185 what digit is in the

40. thousands place?

41. hundreds place?

42. tens place?

43. ones place?

In the number 2,571 what is the value of the digit

44. 2?

45. 5?

46. 7?

47. 1?

48. Write the number in words.

The value of any number is the sum of the values of the digits in the number. For example, 8,358 is equal to 8,000 + 300 + 50 + 8. This can be written in **expanded form** as shown at the right.

$$\begin{array}{r} 8{,}000 \\ 300 \\ 50 \\ +\quad 8 \\ \hline 8{,}358 \end{array}$$

Write each number in expanded form. Then write the number in words.

49. 987 **50.** 1,111 **51.** 7,605 **52.** 25,000 **53.** 56,783

The numbers 243, 324, 432, 234, 423, and 342 are all made up of the same digits. Which number represents

54. the largest number? **55.** the smallest number?

To find the number that 6 thousands, 2 hundreds, 9 tens, and 5 ones equals, you find the sum of the values of the digits, as shown at the right. The sum is written in **standard form.**

```
  6,000
    200
     90
+     5
  6,295
```

Write each number in standard form and then in words.

56. 4 thousands, 9 hundreds, 3 tens, 5 ones

57. 6 ten-thousands, 1 thousands, 8 hundreds, 9 tens, 0 ones

58. 24 thousands, 6 hundreds, 2 tens, 9 ones

59. ◆ Write a short summary of why a cash payments record is important. Consider including an example that shows the necessity for adding when working with a cash payments record. Keep a record of your cash payments for two weeks. You will refer to it later in this chapter.

MIXED REVIEW

Add quickly. Check your work by reverse addition.

1.	**2.**	**3.**	**4.**
3	13	67.23	$357.31
8	75	28.58	923.17
7	50	32.10	515.53
2	47	41.32	682.62
4	25	72.17	108.97
+ 7	+ 68	+ 53.79	+ 397.64

5. Eu Hook Ti received these cash amounts during one week: wages, $531.93; returned bottles, $3.25; interest on savings, $51.77; gift, $50; refund on returned purchase, $28.56; tax refund, $167.32. What were his total cash receipts?

6. Lisa Orr's paychecks for June were: $267.91, $248.09, $264.33, and $275.26. Alan Orr's paychecks were: $235.55, $287.64, $253.68, and $277.83. Find their total cash receipts.

7. The Oli Co. spent these amounts last month: rent, $2,600; electricity, $467.29; natural gas, $251.72; salaries, $9,207.78; supplies, $1,508.09; travel expenses, $2,078.21; computer services, $2,145. How much did the company spend last month?

8. A canoe club spent these amounts during June, July, and August: $684.24 for canoe rentals, $187.23 for refreshments, $32.35 for copying, $27 for award certificates, and $45.32 for postage. How much did the club spend?

1-3 COLUMNAR CASH PAYMENTS RECORDS

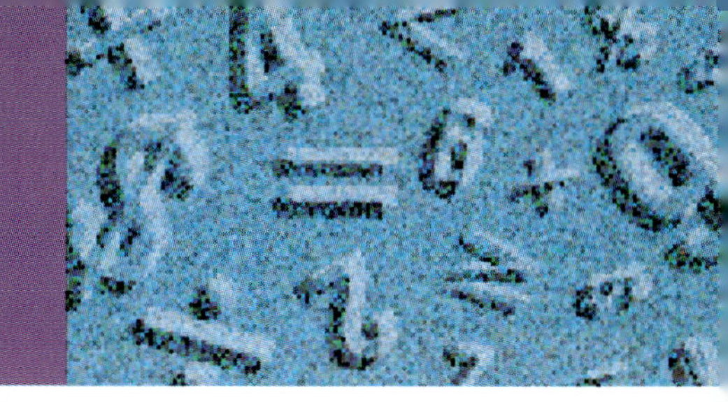

Most schools have many different clubs and organizations. Most of them require some kind of a treasury to operate. Think about the typical payment categories for a school organization, club, or team.

◆ Choose one school club or organization from your school. List the payments that you think are typical for that club or organization. Plan to contact a member or faculty sponsor to find out what their actual expenses and payments are. Compare the categories you listed with those used in Illustration 1-3.1.

OBJECTIVES

In this lesson, you will learn to

- *add horizontally and vertically;*
- *check addition in tables by finding the grand total; and*
- *complete a columnar cash payment record.*

WARM *UP*

Find the sum of each column below.

a.	**b.**	**c.**	**d.**
5	29	23	802
6	28	398	9
2	24	107	23
3	70	9	512
+ 1	+ 16	+ 312	+ 229

COLUMNAR CASH PAYMENTS RECORDS

To keep their cash payments, the Lewises use a cash payments record with a single column. Many people and businesses use records with multiple columns to keep similar payments together. An example is Ralph Kowalski's columnar cash payments record, shown in Illustration 1-3.1 on the following page.

At the end of each week or part week, and at the end of the month, Ralph Kowalski put the total amounts for each type of payment on one line. He added the amounts across to find the total for the week and put that amount in the Total Payments column. At the end of the month he added down each payment column to find how much he had spent for each type of payment.

CASH PAYMENTS RECORD

NAME *Ralph Kowalski* ACCOUNT NO.

ADDRESS

	WEEK		TYPES OF PAYMENTS					
			HOUSING	FOOD	PERSONAL	AUTO	SAVINGS	TOTAL PAYMENTS
1	19-- Dec.	1–7	$500.00	$124.45	$45.80	$23.56	$25.00	$718 81
2		8–14		52.12	25.84	17.87		95 83
3		15–21	67.54	107.22	17.45	37.60	25.00	254 81
4		22–28		37.23	12.45	33.33		83 01
5		29–31	124.78	72.34	65.12	18.98	25.00	306 22
		Total	$692.32	$393.36	$166.66	$131.34	$75.00	$1,458 68

Illustration 1-3.1. Cash Payments Record with Special Columns

Horizontal and Vertical Addition

Ralph Kowalski found his weekly totals by adding across each row without rewriting the figures. Adding across is called **horizontal addition.** When you add across you save the time and bother of recopying the figures. Adding up and down, as you do when you arrange the numbers in a column, is called **vertical addition.**

Ralph checked the accuracy of his cash payments journal by adding the totals of the Types of Payments columns ($1,458.68) and comparing that sum with the sum of the Total Payments columns ($1,458.68). These two sums should be equal. Ralph entered the sum in the lower right-hand corner of the cash payments record. This corner total is called a **grand total**. Grand totals are frequently used in business to check the accuracy of work.

Critical Thinking Why must the grand total obtained from horizontal addition be the same as the grand total obtained from vertical addition?

If your calculator has a memory, you can find grand totals easily. Start by finding the sum of the first column or row on the calculator. Press the memory plus [M+] key. Clear the screen (but not the memory) and find the sum of the next group of numbers. Press [M+]. Repeat for each group. After the last group has been entered into the memory, press the memory recall [MR] key. This total should be your grand total. Check the manual for your calculator if you have difficulty, or refer to the Technology Appendix.

EXERCISE YOUR SKILLS

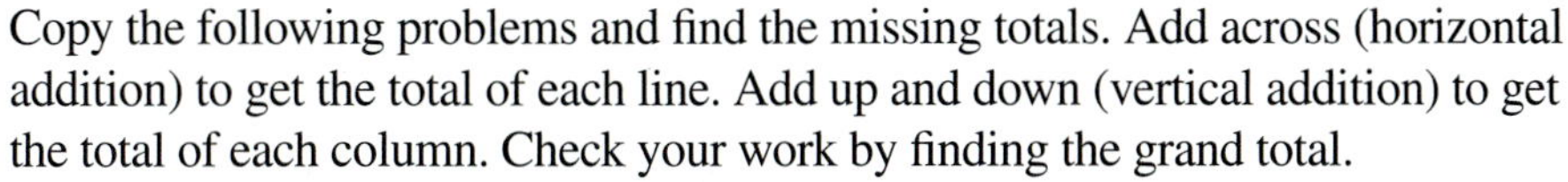

Copy the following problems and find the missing totals. Add across (horizontal addition) to get the total of each line. Add up and down (vertical addition) to get the total of each column. Check your work by finding the grand total.

1. $6 + 4 + 5 =$
$3 + 6 + 7 =$
$+ 8 + 5 + 2 =$

2. $34 + 56 =$
$21 + 39 =$
$16 + 85 =$

3. $24 + 33 + 53 =$
$51 + 27 + 22 =$
$33 + 96 + 56 =$

4. $7 + 9 + 3 + 1 =$
$5 + 5 + 6 + 8 =$
$3 + 8 + 2 + 5 =$
$4 + 4 + 2 + 7 =$

5. $93 + 122 =$
$237 + 154 =$
$593 + 47 =$
$134 + 89 =$

6. $123 + 39 + 221 =$
$140 + 66 + 84 =$
$155 + 31 + 219 =$
$79 + 51 + 106 =$

Olga Petroff is a college student who lives with her parents. Her cash payments for April are shown in the following chart.

CASH PAYMENTS RECORD

NAME *Olga Petroff* ACCOUNT NO.

ADDRESS

	WEEK		FOOD	CLOTHING	AUTO	SAVINGS	OTHER	TOTAL PAYMENTS
			TYPES OF PAYMENTS					
1	*19– – April*	*1–7*	*$15.76*	*$25.99*	*$15.23*	*$10.00*	*$3.25*	
2		*8–14*	*14.27*		*12.77*		*5.09*	
3		*15–21*	*22.88*	*23.79*	*12.87*	*10.00*	*17.49*	
4		*22–28*	*12.67*		*15.67*		*12.34*	
5		*29–30*	*6.45*		*3.78*		*5.56*	
		Total						

7. Complete Olga's cash payments record for April. Place a sheet of paper under the columns, add them, and write the totals on the paper. Check your work by adding the five Types of Payments totals horizontally (across). The sum should equal the total of the Total Payments column.

The Stroker Company had these expenses one month: October 1–7, rent $1,500, salaries $7,000, utilities $200, auto expense $338, miscellaneous $532; October 8–14, utilities $124, auto expense $205, postage $35, miscellaneous $320; October 15–21, salaries $1,500, auto expense $338, postage $27, miscellaneous $294; October 22–28, utilities $180, auto expense $536, postage $22, miscellaneous $534; October 29–31, auto expense $120, miscellaneous $184.

8. Make a columnar cash payments record for the Stroker Company for October. Be sure to check that both the horizontal and vertical totals are the same.

Metzger and Associates is a small legal firm. The firm's cash payments for February are shown below. Find the totals.

			CASH PAYMENTS RECORD				
NAME		*Metzger and Associates*				ACCOUNT NO.	
ADDRESS							
			TYPES OF PAYMENTS				
	WEEK		RENT	PAYROLL	OFFICE EXPENSES	OTHER EXPENSES	TOTAL PAYMENTS
1	*19-- Feb.*	*1–7*	*$975.00*	*$7,311.92*	*$345.61*	*$241.08*	**9.**
2		*8–14*		*617.25*	*207.67*	*97.12*	**10.**
3		*15–21*		*617.25*	*89.56*	*178.31*	**11.**
4		*22–28*		*617.25*	*306.45*	*81.02*	**12.**
5		*Total*	**13.**	**14.**	**15.**	**16.**	**17.**

18. Explain how the answer to problem 17 is used to check your work for problems 9–16.

19. Review the cash receipts and cash payment records you created in Lessons 1 and 2. Can you group similar items together? Redesign both records in columnar form.

20. ♦ Make up a columnar cash payments record for the school club or organization that you chose at the beginning of the lesson. You may either use real data from the club or make up your own data.

MIXED REVIEW

Find the value of N in each problem.

1. $15 + N = 19$ **2.** $N + 16 = 20$ **3.** $0 + N = 4$

4. $22 - N = 22$ **5.** $N - 7 = 18$ **6.** $14 + N = 7 + 22$

7. A roadside vegetable stand had these cash receipts for 5 weeks: $470.89, $592.93, $604.31, $556.82, $498.67. What were the total cash receipts for the 5 weeks?

8. Olga Svenson spent these amounts on a business trip: airfare, $478; taxis, $27; food, $145.23; hotel room, $245.98; gifts, $25.98; other expenses, $45.29. What total amount did Olga spend on the trip?

9. The Service League of Evansville ran a dart game at the town's 3-day homecoming event. The League collected $250.50 on Friday, $412.75 on Saturday, and $324.25 on Sunday. What was the total amount collected by the League?

1-4 CHECK REGISTER RECORDS

A small child asked for an expensive toy. His mother said she was out of money. The child said, "No, you're not out of money, you still have checks in your checkbook!"

◆ Have a small group discussion about your knowledge of checking accounts. Some questions to consider include: How do you know how much money you have left in your account? How do you keep track of money you put in the account or take out of the account?

OBJECTIVES

In this lesson, you will learn to

- *subtract whole numbers and decimals vertically and horizontally;*
- *check subtraction;*
- *maintain a check register; and*
- *use terms* gross *and* net *correctly.*

WARM UP

Find the value of N.

a. $25 + N = 39$

b. $N - 35 = 50$

c. $3 + 21 = N$

d. $N - 30 = 15$

e. $31 - N = 12$

f. $N - 7 = 0$

g. $82 - N = 82$

h. $21 + N = 8 + 31$

i. $32 - 6 = 9 + N$

j. $17 - N + 5 = 20$

CHECKBOOK RECORDS

Many people deposit their cash receipts in a checking account at a bank and make their payments by check. A checking account is safe and easy to use, and the canceled checks provide you with a record of your payments. You use addition and subtraction often when you have a checking account.

Deposit Slips

When Laverne and Ramon Lewis deposit their paychecks into their checking account, they fill out a **deposit slip** like the one shown in Illustration 1-4.1 on the following page.

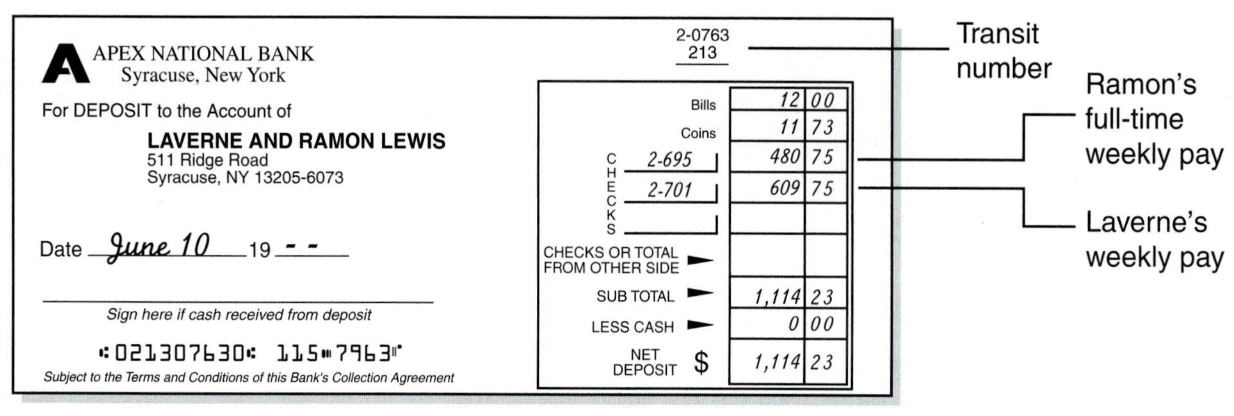

A APEX NATIONAL BANK
Syracuse, New York

2-0763
213

For DEPOSIT to the Account of

LAVERNE AND RAMON LEWIS
511 Ridge Road
Syracuse, NY 13205-6073

Date *June 10* 19 - -

Sign here if cash received from deposit

⑆021307630⑆ 115⑈7963⑈

Subject to the Terms and Conditions of this Bank's Collection Agreement

		Dollars	Cents
Bills		12	00
Coins		11	73
CHECKS	2-695	480	75
	2-701	609	75
CHECKS OR TOTAL FROM OTHER SIDE			
SUB TOTAL		1,114	23
LESS CASH		0	00
NET DEPOSIT $		1,114	23

Illustration 1-4.1. Deposit Slip for Laverne and Ramon Lewis

The Lewises list cash deposits as bills or coins. They list each check separately, using the check's **transit number.** This number is the top part of the group of bank numbers on the check shown in Illustration 1-4.2. It identifies the bank where you have your checking account.

If the Lewises' had more checks to deposit than lines provided for on the front of the deposit slip, they would list the additional checks (and transit numbers) separately on the back of the deposit slip. The total of those checks would be entered as the last item on the front of the deposit slip. Instead of writing a transit number, they would note "Total of checks on back." Then, all the amounts are added to find the sum, which is written on the TOTAL line.

If the Lewises want some cash back, that amount is written on the line below TOTAL.

Finally, the NET DEPOSIT is found by subtracting the cash back from the TOTAL. On most deposit slips, you only sign the deposit slip if you are receiving cash back.

Subtraction

You use **subtraction** to deduct cash back on the deposit slip. Subtraction is the reverse of addition. For example, look at these two problems.

$$\begin{array}{r} 38 \\ +\ 24 \\ \hline 62 \end{array} \qquad \begin{array}{r} 62 \\ -\ 24 \\ \hline 38 \end{array}$$

Notice that all of the numbers in the subtraction problem are the same as the numbers in the addition problem. The sum in the addition problem is what is subtracted from in the subtraction problem.

You could write an open sentence: $24 + N = 62$. To find the missing number, you reverse the addition by subtracting 24 from 62. The difference, 38, is the missing number.

Checking Subtraction Every subtraction should be checked as soon as it is finished. The best way to check subtraction is to add the amount subtracted and the difference. The sum should equal the top number.

EXERCISE YOUR SKILLS

Complete a deposit slip for each of the following.

1. Edna Schwartz deposited these items in a bank: (bills) 12 twenties, (coins) 12 quarters, 3 dimes; (checks) $218.49, $1.45. She received 100 one-dollar bills in cash back.
2. Brad Nolan, treasurer of Eaton's Microcomputer Managers Association, made this deposit: (bills) 12 twenties, 23 tens, 14 fives, 36 ones; (coins) 53 quarters, 32 dimes; (checks) $28.75, $2.07. He received no cash back.
3. The Craft Corner made this deposit: (bills) 72 twenties, 126 tens, 57 fives, 235 ones; (coins) 287 quarters, 312 dimes, 48 nickels, 347 pennies; (checks) $124.68, $132.08, $1.29, $5.79. They received no cash back.
4. Bruns' Bowling Alley made this deposit: (bills) 3 one-hundreds, 4 fifties; (checks) $15, $18.50, $12, $15, $47.50, $35, $17, $23.25, $44. They got back 50 one-dollar bills, 8 five-dollar bills, and 4 ten-dollar bills.

Find the differences. Show how you checked each problem.

5. 29 − 7
6. 25 − 13
7. 42 − 12
8. 105 − 75
9. 256.82 − 137.93
10. 964.64 − 0.55
11. $1,370.52 − 583.64
12. $500.00 − 84.34
13. $5,824 − 3,285
14. $2,000 − 1,432
15. $10,342.34 − 9,342.35
16. $99,099.43 − 98,549.43

CHECKS AND CHECK REGISTERS

Both Laverne and Ramon Lewis write **checks** on their checking accounts. These checks direct their bank to make payments. The checks you write are numbered in order. This makes it easy to keep track of them. A check completed by Laverne Lewis is shown in Illustration 1-4.2.

LAVERNE AND RAMON LEWIS
511 Ridge Road
Syracuse, NY 13205-6073

Date July 6 19 - -   1341

2-0763
213

PAY TO THE ORDER OF TV Time Magazine, Inc. $ 25.75

Twenty-five and 75/100 Dollars

APEX NATIONAL BANK
Syracuse, New York

For Classroom Use Only

Memo Subscription   Laverne Lewis

⑆021307630⑆ 115⑈7963⑈ 1341

Illustration 1-4.2. Completed Check

Critical Thinking Where is the transit number on the check? What is the check number? Why do you think these numbers are on checks?

The Lewises record each deposit made and check written in their **check register,** shown in Illustration 1-4.3. The check register is that part of the checkbook in which deposits and checks are recorded. Each deposit is added to their last **balance,** which is the amount of money left in their account. Each check is subtracted from their last balance. In this way they always know how much money they have in their checking account.

RECORD ALL CHARGES OR CREDITS THAT AFFECT YOUR ACCOUNT

NUMBER	DATE	DESCRIPTION OF TRANSACTION	PAYMENT/DEBIT (-)	√ T	FEE (IF ANY) (-)	DEPOSIT/CREDIT (+)	BALANCE
							$ 1500.00
1341	7/6	TV Time Magazine	$ 25.75		$	$	1474.25
Dep	7/6	Ramon's Part Time Check				100.00	1574.25
1342	7/7	Syracuse Electric	98.32				1475.93
1343	7/7	APEX Mortgage	772.33				703.60
1344	7/8	APEX Bank	165.98				537.82
Dep	7/10	Ramon's Pay Check				480.75	1018.57

RECORD ALL CHARGES OR CREDITS THAT AFFECT YOUR ACCOUNT

NUMBER	DATE	DESCRIPTION OF TRANSACTION	PAYMENT/DEBIT (-)	√ T	FEE (IF ANY) (-)	DEPOSIT/CREDIT (+)	BALANCE
							$ 1500.00
1341	7/6	TV Time Magazine (subscription)	$ 25.75		$	$	−25.75 1474.25
Dep	7/6	Ramon's Part Time Check (chk #3289)				100.00	100.00 1574.25
1342	7/7	Syracuse Electric (June bill)	98.32				−98.32 1475.93
1343	7/7	APEX Mortgage (July payment)	772.33				−772.33 703.60
1344	7/8	APEX Bank (auto payment - July)	165.78				−165.78 537.82
Dep	7/10	Ramon's Pay Check (full time #507934)				480.75	480.75 1018.57

Illustration 1-4.3. Check Register Pages Showing One-line and Two-line Entries

EXERCISE YOUR SKILLS

Eve Dent's check register shows her balance on March 1 as $501.32. Find the correct balance after each deposit and each check.

	No.	Date	Description	Payment	Deposit	Balance
		3/1	Current Balance			$501.32
17.	478	3/3	Mandy's Clothing Store	34.66		
18.	479	3/9	Evergreen Power Company	98.62		
19.	—	3/21	Paycheck		298.43	
20.	480	3/25	Keller Auto Company	108.45		
21.	—	3/30	Interest on savings		23.78	

Create a check register like the one in Illustration 1-4.3. Enter the transactions and determine the final balance by calculating the balance after each deposit and each check.

	Previous Balance	Deposits	Checks Issued	Final Balance
22.	$278.98	$125	$230.12, $1.50, $20.72	
23.	$67.54	$35.71	$3.67, $43.81, $9.33, $15.10	
24.	$807.22	$97.12, $50	$145.90, $6.08, $62.44, $35.19	

25. The Avalon Ladies Volleyball team's checking account balance was $482.63 on Monday. On Tuesday, a deposit of $178.55 from their bake sale was made and checks for $35, $40.19, and $56.09 were written. What was the club's new balance?

26. York, Inc.'s checking account balance was $4,070.65. A deposit of $1,006.54 was made and checks for $25.22, $1,980.30, $6.89, and $75 were written. What was the new balance of their account?

GROSS AND NET: APPLYING SUBTRACTION

Gross and net are very common business terms. For example, gross profit and net profit, gross weight and net weight, and gross wages and net wages are used often. A **gross** amount is the larger amount—the amount before anything is deducted. A **net** amount is the smaller amount that you get when you deduct or subtract something from a larger, gross amount. Looked at another way, a **gross** amount is the total of a net amount plus deductions.

Gross − Deductions = Net
Net + Deductions = Gross

For example, the gross cost of a camera is $375. When you subtract the rebate of $50 you get from the camera's manufacturer, the net cost of the camera is only $325 ($375 − $50 = $325).

PROBLEM SOLVING

When trying to solve word problems, having a plan is essential. Here is one plan for successful problem solving.

READ Carefully read the entire problem, being sure you understand what is given and what you are to find.

PLAN Decide which strategy to use. For example, write a number sentence.

SOLVE Carry out your plan, do the needed computations. Be sure to answer the question that is asked and be sure to properly label your answers.

CHECK See whether your answer is reasonable. Be sure to go back to the original problem to check your work.

EXERCISE YOUR SKILLS

Find the net amount in each problem by subtracting the deductions from the gross amount. Check your work by adding the net amount and the deductions.

27. The gross weight of a box and its contents was 14.2 pounds. The weight of the box alone was 1.8 pounds. What was the net weight of the contents?

28. Ann Ackerman's gross gas and electric bill for June was $137.98. If she pays early, the discount is $2.76. What will be the net amount she has to pay if she pays early?

29. A little league baseball club sold candy bars at a gross price of $2.00 each. Each bar cost the class 73 cents. What net amount did the club make per candy bar?

30. Sol Levine's gross pay for a week was $416.29. A total of $98.67 was deducted from his gross pay. What was Sol's net pay for the week?

31. A department store had gross sales of $287,806.22 last week. However, customers returned $11,512.24 of their purchases. What were the net sales of the store last week?

On Monday morning Sergio had $214.78 in his checking account. During the week he earned $45.00 for cutting lawns and $35.75 for yard work and deposited that money in his checking account.

32. What were Sergio's total earnings for the week?

33. If Sergio had $227.67 left in his checking account at the end of the week, how much did he spend during the week? (Assume all his expenses were paid by check.)

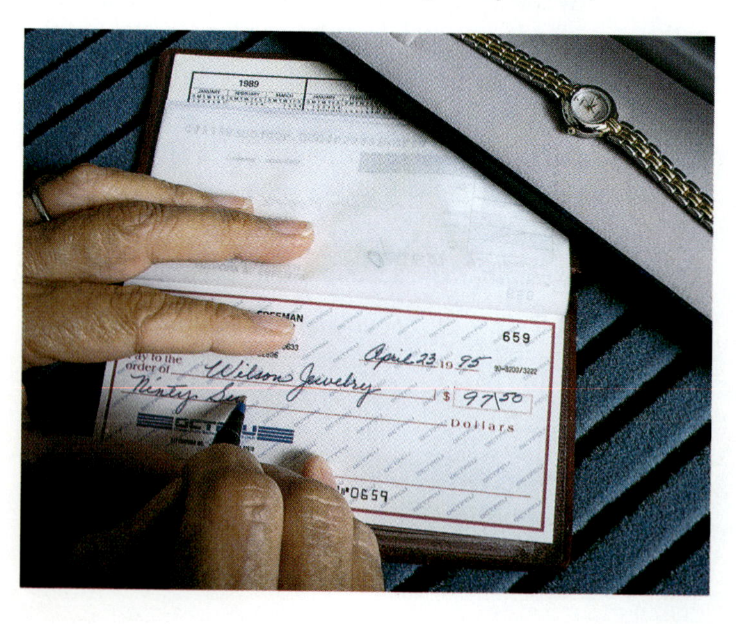

On October 1, Oki Kimura had on hand heating oil that had a gross cost of $178.56. During the winter Oki bought more oil. The gross costs were: $269.92; $326; $296.19; and $208. If Oki spent more than $1,000 on oil in 4 months, he received a discount of $100.

34. Find the net cost of the oil bought during those 4 months.

35. If Oki had $108.37 worth of oil left at the end of February, what was the net cost of the heating oil used between October 1 and the end of February?

PUT IT ALL TOGETHER

You have learned to find the total amount received, the total amount spent, and the amount left or balance. There are many different situations in which the same skills are used.

Patricia Jones earned a gross salary of $400 one week. Taxes and other deductions of $103.35 were subtracted from her gross salary. She also received a $3.50 rebate check and a $50 check from a friend in repayment of a debt.

36. What was the net amount of her paycheck?

37. Patricia deposited all the checks and got $35 cash back. What was her net deposit?

38. She wrote a check to Ari's Clothiers for $123.32. How should she write that amount in words on the check?

39. What was her checking account balance after writing the check to Ari's Clothiers?

40. ♦ Write a paragraph describing your understanding of how a checking account works and why one is useful to businesses and individuals.

MIXED REVIEW

Find each sum or difference.

1. 29 − 16 **2.** 125 − 75 **3.** 250 − 125

4. 18 + 83 **5.** 234 + 234 **6.** 823 + 91

7. 23,978.65 − 14,087.96 **8.** 23,978.65 + 14,087.96

9. $78,308.23 − 25,611.90 **10.** $68.02 + $0.98 + $126.01

Solve each problem.

11. A train station's four turnstiles collected these amounts in one day: $2,078.25, $1,467.40, $2,076.50, $1,208.20. What was the total amount of cash collected?

12. Alfredo Ramos spent these amounts landscaping his home: patio stones, $767.24; stone bed cover, $689.34; edging, $14.56; shrubs, $239.78; plastic sheeting, $6.89; and mulch, $34.56. How much did Alfredo spend on the landscaping?

13. The gross weight of a truck and its contents was 18,295 pounds. The weight of the truck alone was 15,207 pounds. What was the net weight of the contents?

14. Lee's check register showed a balance of $380.48. During the day he made a $108.56 deposit and wrote checks for $4.12, $1.69, $28.27, $100.87, and $0.75. Find his final balance.

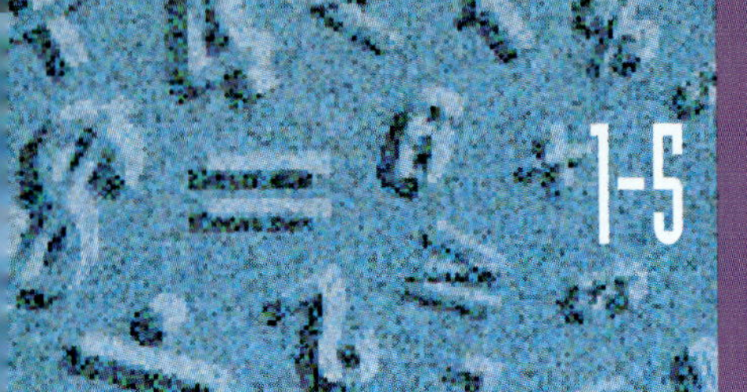

1-5 ELECTRONIC BANKING

OBJECTIVES

In this lesson, you will learn to

- *identify how electronic banking is used; and*
- *find the balance in an account after electronic banking transactions have been completed.*

You see someone approach a machine, slip a plastic card into a slot, push some numbers on a key pad, and presto, out comes cash. That person has used a **debit card.** One popular type of debit card is an **Automatic Teller Machine (ATM) card.**

◆ How many different places can you name in which you have used a debit card, or have seen someone else use one? Make a list of the places you remember.

WARM UP

Find each sum or difference.

a. $45 - 22 =$

b. $375 + 55 =$

c. $120 - 50 =$

d. $5{,}179.55 - 4{,}017.48$

e. $\$37{,}106.43 + 23{,}912.50$

f. $\$5{,}000.00 - 4{,}993.07$

ELECTRONIC BANKING

Computers are common in banking today. Many banking transactions are done electronically. No checks are actually printed. No deposit slips are made out. Many customers pay their bills electronically, and many companies pay each other and their employees electronically.

Electronic Funds Transfer Banks increasingly transfer deposits and checks, or "funds," from person to person and bank to bank electronically, using computers. This process is called **Electronic Funds Transfer,** or **EFT.**

Banks print the account numbers of customers and their banks in **Magnetic Ink Character Recognition,** or **MICR,** form at the bottom of deposit slips and checks. This allows special computer equipment to read the MICR data on the deposit slips and checks and then update bank accounts and transfer funds from account to account electronically.

For example, the MICR numbers written at the bottom of the deposit slip in Illustration 1-4.1 and the check in Illustration 1-4.2 are the Lewises' account number and the number of the Apex National Bank (Illustration 1-5.1).

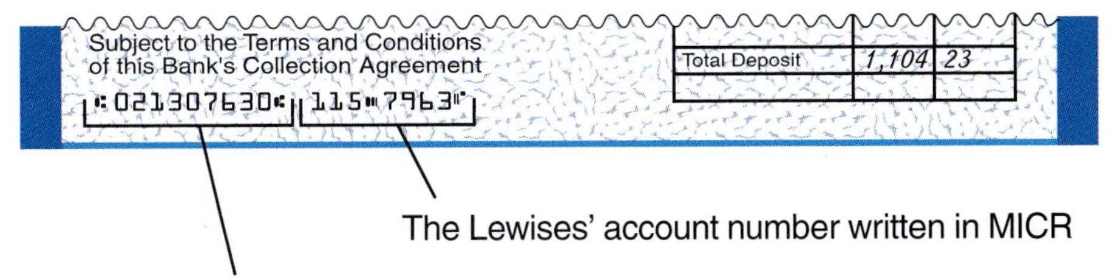

Illustration 1-5.1. Customer and Bank Numbers Printed in MICR Form

Individuals can also transfer funds electronically. Electronic funds transfers occur when you use **automatic teller machines.** Using a special card issued by your bank, you can withdraw or deposit money or make transfers from your savings account to your checking account from automatic teller machines at any time of the day. Automatic teller machines can be found at shopping malls, grocery stores, airports, and other places convenient to you.

Electronic funds transfers may also be used to pay your monthly bills, such as power and telephone bills. You can use the keypad of your telephone or computer. Or you can instruct your bank to transfer funds automatically each month from your bank account to the bank accounts of your utility company and telephone company. Either way, no checks are written or mailed. In fact, many companies pay their employees by transferring funds directly from their banks to their employees' banks without writing any checks to the employees. This is called **direct deposit.**

Debit Cards Many people today are able to use a special form of EFT called a **debit card.** Debit cards allow you to access your bank account through special computer terminals at cooperating stores and businesses. You pay for your purchases without cash by inserting your debit card into a special terminal. The computer system subtracts the amount of each purchase automatically from your bank account and adds the same amount to the store's bank account.

For many banks today, the ATM card also serves as the debit card. That is, the customer is given an ATM card that can be used for deposits and withdrawals at automatic teller machines and also used as a debit card at businesses. Fees may be charged for using these cards.

Recording ATM and Debit Card Payments

When you use a debit card or ATM card to make a payment, you do not complete a check. The payment you make is automatically deducted from your checking account. You receive a receipt for the payment from the store or ATM terminal. It is important to remember to save the receipt and immediately record the payment or withdrawal in your check register. In the place where you would note the check number, you may want to write ATM-WD—for withdrawal.

EXERCISE YOUR SKILLS

1. On Tuesday Terry Kowolski used a debit card to pay for: groceries, $74.89; clothing, $245.64; and garden plants, $35.77. If Terry's bank balance was $845.19 at the start of the day, what is her new balance?
2. After work Agnes O'Day used the ATM to deposit her paycheck for $568.33 and to withdraw $200 in cash. If her starting bank balance was $307.92, what is her new balance?
3. On Saturday Eiko Yoshino deposited a tax refund check for $328.66 in her bank's ATM at a shopping center. She then used her debit card to make these purchases: sweater, $56.89; wrench set, $78.99; audio tapes, $27.79. If her starting bank balance was $690.06, what is her new balance?
4. Ed Kosner started the day with a bank balance of $103.45. He used the ATM to deposit a check for $125.09 and his debit card to make these purchases: $56.08, $23.12, and $24.99. What will be the balance in his account after these transactions are processed by his bank?
5. ♦ Write a brief paragraph explaining the difference between writing a check and using debit cards and ATM cards.

MIXED REVIEW

Find the difference in each problem

1.		**2.**		**3.**	
	23,978.65		$78,308.23		$90,934.73
	− 14,087.96		− 25,611.90		− 89,239.61

4. On Friday, Anne Knowlton uses a debit card to pay for: auto fuel, $58.29; groceries, $63.12; and flowers, $25.46. If Anne's bank balance was $385.49 at the start of the day, what was her new balance at the end of the day?
5. Four city buses collected these fare amounts on one day: $709.15, $573.05, $297.40, $409.15. What were the total fares collected?
6. The gross weight of a crate and its contents was 1,295 pounds. The weight of the crate alone was 127 pounds. What was the net weight of the contents?
7. Chris Will's check register showed a balance of $250.87. During the day, a deposit was made for $158.67 and checks were written for $17.23, $2.98, $32.76, $90.71, and $0.88. What was the final balance?
8. John Rinaldi deposited these items: 21 twenties, 15 tens, 24 fives, 27 ones, 35 halves, 29 quarters, 183 dimes, 392 pennies, and checks for $42.58, $270.16, and $37.02. What was the total deposit?

1-6 RECONCILING THE CHECK REGISTER BALANCE

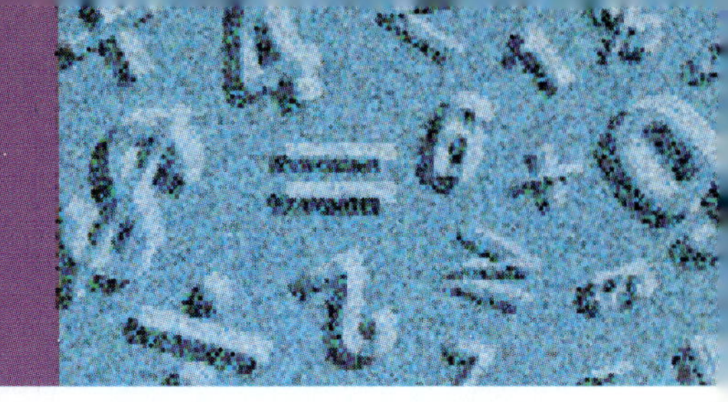

When you use a checking account, the bank periodically sends you a statement showing their records of your transactions. It is important that your records agree with the bank. Checking whether the records agree is called reconciling the account.

◆ Do you know what it means to bounce a check? Write your understanding of the term. How do you think you can avoid "bounced" checks?

OBJECTIVES

In this lesson, you will learn to

- ▸ *read and interpret a bank statement; and*
- ▸ *reconcile a checking account.*

WARM *UP*

Find each sum or difference.

a. 34 + 12

b. 105 − 25

c. 220 + 50

d. 3,209.25 − 118.58

e. 156 + 167 + 53

f. 23,567.34 − 3,567.34

g. \$172,306.63 − 141,947.76

h. \$89,342.75 + 74,814.03

THE BANK STATEMENT

The monthly report the bank sends to each depositor is called the **bank statement.** The bank statement that Karl Olson received from the Drury National Bank for the month of September is shown in Illustration 1-6.1, on the following page.

The bank statement lists (1) nine checks paid by the bank, (2) three deposits received by the bank, (3) a running balance, (4) a service charge of \$4.50, and (5) \$2.11 in interest earned. A **service charge** is a deduction charged by the bank for handling the checking account. Many banks now pay interest on checking accounts. Interest is money paid for the use of money. You will learn more about interest later in this course.

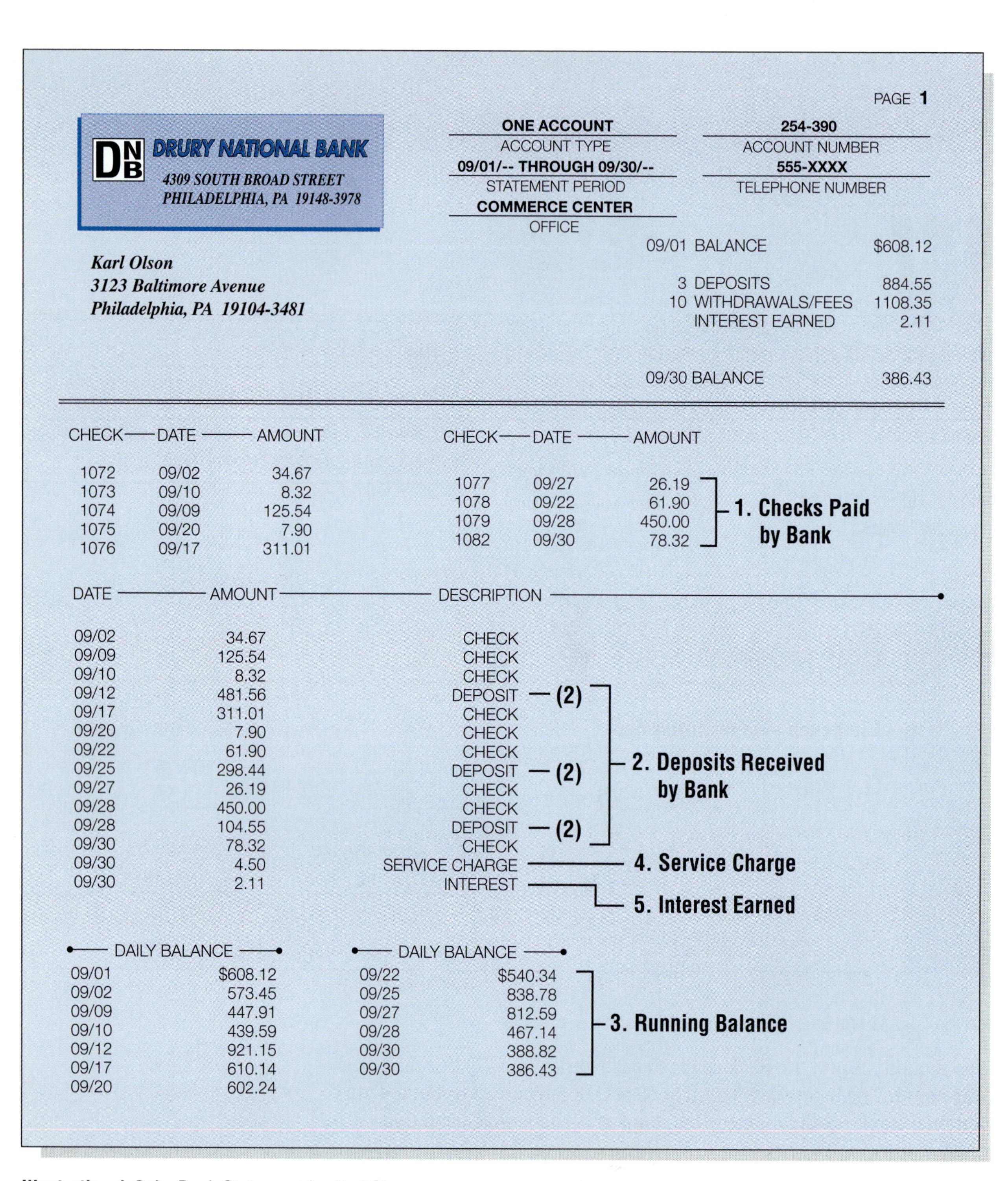

PAGE 1

DRURY NATIONAL BANK
4309 SOUTH BROAD STREET
PHILADELPHIA, PA 19148-3978

ONE ACCOUNT	254-390
ACCOUNT TYPE	ACCOUNT NUMBER
09/01/-- THROUGH 09/30/--	555-XXXX
STATEMENT PERIOD	TELEPHONE NUMBER
COMMERCE CENTER	
OFFICE	

Karl Olson
3123 Baltimore Avenue
Philadelphia, PA 19104-3481

09/01	BALANCE	$608.12
3	DEPOSITS	884.55
10	WITHDRAWALS/FEES	1108.35
	INTEREST EARNED	2.11
09/30	BALANCE	386.43

CHECK	DATE	AMOUNT	CHECK	DATE	AMOUNT
1072	09/02	34.67	1077	09/27	26.19
1073	09/10	8.32	1078	09/22	61.90
1074	09/09	125.54	1079	09/28	450.00
1075	09/20	7.90	1082	09/30	78.32
1076	09/17	311.01			

1. Checks Paid by Bank

DATE	AMOUNT	DESCRIPTION
09/02	34.67	CHECK
09/09	125.54	CHECK
09/10	8.32	CHECK
09/12	481.56	DEPOSIT — (2)
09/17	311.01	CHECK
09/20	7.90	CHECK
09/22	61.90	CHECK
09/25	298.44	DEPOSIT — (2)
09/27	26.19	CHECK
09/28	450.00	CHECK
09/28	104.55	DEPOSIT — (2)
09/30	78.32	CHECK
09/30	4.50	SERVICE CHARGE — 4. Service Charge
09/30	2.11	INTEREST — 5. Interest Earned

2. Deposits Received by Bank

DAILY BALANCE		DAILY BALANCE	
09/01	$608.12	09/22	$540.34
09/02	573.45	09/25	838.78
09/09	447.91	09/27	812.59
09/10	439.59	09/28	467.14
09/12	921.15	09/30	388.82
09/17	610.14	09/30	386.43
09/20	602.24		

3. Running Balance

Illustration 1-6.1. Bank Statement for Karl Olson

Outstanding Checks With his bank statement Karl Olson received several canceled checks. A **canceled check** is a check that the bank has paid and then marked so it can't be used again. (See Illustration 1-6.2.)

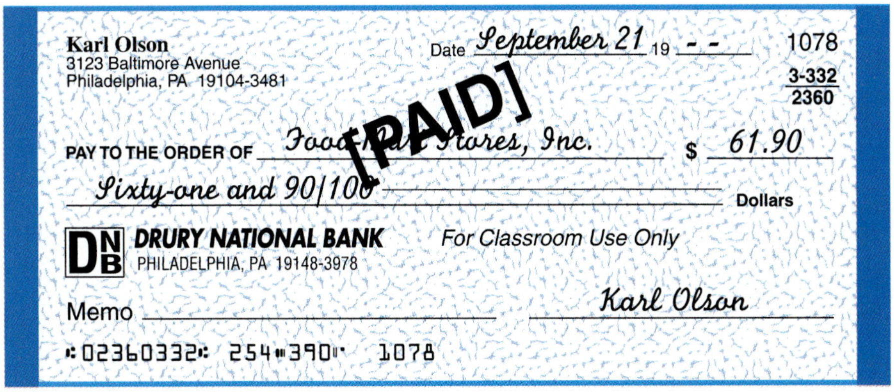

Karl Olson
3123 Baltimore Avenue
Philadelphia, PA 19104-3481

Date September 21 19 - - 1078

3-332/2360

[PAID]

PAY TO THE ORDER OF Food-Mart Stores, Inc. $ 61.90

Sixty-one and 90/100 Dollars

DNB DRURY NATIONAL BANK PHILADELPHIA, PA 19148-3978

For Classroom Use Only

Memo Karl Olson

⑆023603332⑆ 254⑈390⑈ 1078

Illustration 1-6.2. Canceled Check

When Karl Olson compared his bank statement with his check register he found that some checks (#1080 and #1081) that he had written had not been returned by the bank. These checks had not yet been received by the bank for payment. That is why they were not listed on the bank statement. These checks are called **outstanding checks**.

Reconciling a Bank Statement

The final balance on Karl Olson's bank statement was $386.43. The balance in his check register for September was $246.89. The difference between these records was caused by the service charge, the outstanding checks, and the interest earned.

Bringing the checkbook balance and bank balance into agreement with each other is called **reconciling** the account. The form showing the calculations is called the **reconciliation statement.**

To bring the two balances into agreement and to make sure that his records and the bank's records were correct, Karl Olson prepared the reconciliation statement shown in Illustration 1-6.3.

Karl uses a calculator as he reconciles his account. When he looked at his bank statement, Karl Olson found the $4.50 service charge that had been subtracted on the bank statement. Since he had not subtracted the service charge in his check register, he subtracted $4.50 from the checkbook balance on the left side of the reconciliation statement. He also added the interest that his account had earned to the left side of the statement to show the corrected checkbook balance, $244.50.

Next he matched the checks listed on the statement with his check register and found that check #1080 for $48.65 and #1081 for $93.28 were outstanding. He subtracted the sum, $141.93, from the bank balance on the right side of the statement to show the available bank balance, $244.50.

Since the available bank balance, $244.50, was the same as the corrected checkbook balance, $244.50, Karl Olson was satisfied that his records and the bank's records were correct. If the balances had not agreed he would have had to find out why.

Karl Olson

Bank Reconciliation Statement

September 30, 19– –

Checkbook balance	2	4	6	89	*Bank statement balance*	3	8	6	43
Deduct:					*Deduct:*				
Service charge			4	50	*Outstanding checks:*				
	2	4	2	39	*#1080* *$48.65*				
Add:					*#1081* *93.28*	1	4	1	93
Interest earned			2	11					
Correct checkbook balance	2	4	4	50	*Available bank balance*	2	4	4	50

Illustration 1-6.3. Reconciliation Statement for Karl Olson

When Karl Olson finishes his reconciliation statement he must bring his check register up to date. To do this, he records the service charge and interest earned. His updated check register is shown in Illustration 1-6.4.

RECORD ALL CHARGES OR CREDITS THAT AFFECT YOUR ACCOUNT

NUMBER	DATE	DESCRIPTION OF TRANSACTION	PAYMENT/DEBIT (-)		√ T	FEE (IF ANY) (-)	DEPOSIT/CREDIT (+)		BALANCE	
									$ 246	89
– –	9/30	*Service charge*	$ 4	50		$	$		– 4	50
									242	39
– –	9/30	*Interest earned*					2	11	+ 2	11
									244	50

Illustration 1-6.4. Updated Check Register for Karl Olson

EXERCISE YOUR SKILLS

Prepare a reconciliation statement for 1-6. Use your own name and the current date in the headings in 3-6. Watch for extra or missing information.

1. Anne Oster's checkbook balance on May 31 was $873.09. Her May 31 bank statement showed a balance of $1,378.49. Checks outstanding were #806 for $326.11, #808 for $97.56, and #809 for $82.03. The bank service charge for May was $5.75 and the interest earned was $5.45.
2. Ed Acker's bank statement balance on June 30 was $396.78. On that same date his checkbook balance was $277.91. The bank statement showed a service charge of $7.75 and interest earned of $1.65. The outstanding checks were #51 for $73.09, #52 for $46.28, and #54 for $5.60.
3. Service Charge: $5.70; Interest Earned: $4.13
   Checkbook Balance: $803.12
   Statement Balance: $978.44
   Outstanding Checks: #88, $167.41; #90, $9.48
4. Statement Balance: $678.12
   Service Charge: $10.12
   Interest Earned: $2.81
   Outstanding Checks: #23, $21.57; #25, $40.87
5. Checkbook Balance: $231.19
   Statement Balance: $268.56
   Number of checks: 23
   Service Charge: $5.10
   Interest Earned: $1.15
   Outstanding Checks: #21, $5.79; #23, $15.46;
   #25, $20.07
6. Checkbook Balance: $133.79
   Statement Balance: $237.70
   Number of checks: 45
   Average Balance: $394.43
   Service Charge: $9.50
   Interest Earned: $0.86
   Outstanding Checks: #63, $9.56; #64, $3.78; #65, $63.14; #66, $36.07
7. Elena Morales has had an account with Arcola Bank for five years. When she got her bank statement on October 31 it showed a balance of $907.46, a $12.75 service charge, and a $10 credit. Also included was an advertisement for savings certificates at 7.5%. Outstanding checks were: #451, $101.33; #452, $89.56; #455, $12.72; #456, $3.16. Her checkbook register showed a balance of $703.44 on October 31.

PROBLEM SOLVING

Too Much Information; Too Little Information

In the problems you have solved so far, you have been given only the information needed to solve the problems, and nothing more.

In real life, however, you may be given more information than you need. In this case, reread the problem, decide on a plan, and determine what information is needed to solve the problem. The rest of the information is not needed and can be ignored.

You may also be given problems with too little information to solve them. In those cases, the first step is to identify the missing information. Throughout this book, you may be given a problem that does not contain enough information to solve the problem. In those cases, write "Cannot be solved. Needed information is . . . ."

Artist hand-painting a piano.

8. Vera Tokheim, an artist, opened a checking account on July 1. On October 31 her bank statement balance was $1,207.88 and her check register balance was $978.34. Her statement showed twelve checks, two deposits, a service charge of $6.20, and interest earned of $5.78. A comparison of her check register and bank statement showed these outstanding checks: #2078 for $90.21, #2084 for $110.23, and #2085 for $29.52.

9. On June 30, Al Olm's check register balance was $603.49 and his bank statement balance was $870.15. The bank statement showed a service charge of $4.75. With the bank statement were 2 canceled checks: #453, $35.78 to A-1 Pet Shop and #454, $25 to the Town Fund. Outstanding checks were: #451, $101.33 to Benton Power Co.; #452, $89.56 to Al's TV Repair; #455, $12.72 to Arno Co.; #456, $67.80 to Route 54 Garage.

10. ♦ Write a paragraph explaining why reconciling your account is important.

MIXED REVIEW

Find the sum of each column and row. Then find the grand total.

1.	2.	3.
$4 + 5 + 8$	$34 + 12 + 89$	$45 + 11 + 54 + 23$
$7 + 3 + 2$	$59 + 51 + 37$	$63 + 17 + 78 + 22$
$9 + 2 + 1$	$28 + 17 + 32$	$64 + 97 + 36 + 13$

4. Sean Orr's gross power bill for July was $189.76. By paying early he earned a discount of $3.79. What was the net amount of his bill?

5. Dina Maglio started the day with $106.76 in cash. During the day she cashed a check for $175 and spent $92.78 in a store. How much money did she have left at the end of the day?

6. On April 30, Eve Kato's check register balance was $248.65 and her bank statement balance was $366.13. The bank statement showed a service charge of $10.75 and an $8 credit. These checks were outstanding: #302, $12.35; #304, $107.88. Prepare a reconciliation statement.

1-7 SPECIAL RECONCILIATION PROBLEMS

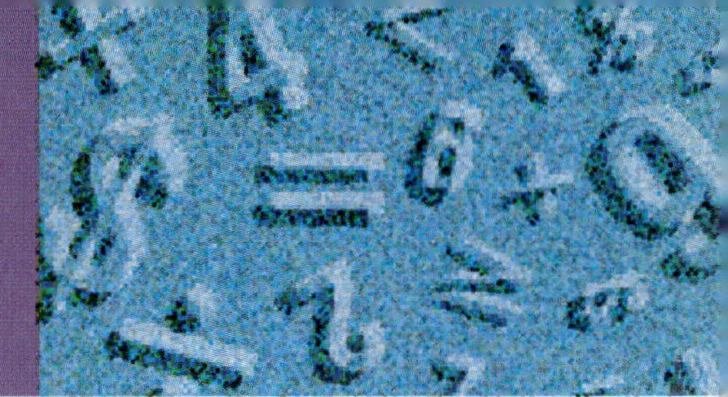

In Lesson 1-6 you learned to reconcile a checking account. All but two of the accounts in those problems "balanced" or came out correctly. In real life also the bank balance and checkbook balance do not always agree after you reconcile your account.

◆ List 5 reasons why the balances might not agree.

OBJECTIVES

In this lesson, you will learn to

- *identify needed corrections when reconciling a checking account;*
- *prepare a reconciliation statement; and*
- *find the corrected check register balance.*

WARM *UP*

Find each horizontal and vertical sum. Then find the grand total.

a. 34 + 16 + 29 + 19
21 + 7 + 72 + 45
53 + 82 + 10 + 15
28 + 31 + 9 + 62

b. 95 + 29 + 125 + 76
18 + 134 + 21 + 76
7 + 73 + 63 + 58
78 + 48 + 94 + 15

CORRECTING ERRORS

You have learned to reconcile bank statement and checkbook balances when service charges, interest earned, and outstanding checks occur. Now you will learn how to reconcile bank statement and checkbook balances when deposits and checks are not recorded and other errors are made.

On April 30, Ella LaGarce's bank statement balance was $99.35 and her check register balance was $58.20. When Ella compared the bank statement and canceled checks with her check register, she found that:

- A service charge of $7.75 had been subtracted from the bank statement balance but not from the check register balance.
- Check #98 for $9.50, which she wrote and cashed on April 5, had not been recorded and subtracted in the check register.
- Check #103 for $15 was recorded twice in the check register.
- A deposit of $76 made on April 21 had not been recorded in the check register.

- Interest of $2.48 was earned on the account.
- Check #106 for $7.50 had been recorded in the register as $5.70.
- Check #107 for $42 and #108 for $26 were outstanding.
- A deposit of $101.28 made on April 30 was recorded in the check register but deposited too late to appear on that month's bank statement.

Knowing these facts Ella prepared the reconciliation statement shown in Illustration 1-7.1.

The service charge for $10.75 and check #98 for $9.50 had not been recorded in the register. This left the check register balance higher than it should have been. Ella deducted these amounts from the register balance.

The unrecorded deposit of $76 and interest earned of $2.48 made the register balance lower than it should have been. Ella added these amounts to the register balance.

Since Ella had subtracted check #103 for $15 twice, she added $15 back once to the register balance.

To correct the error in recording check #106, Ella deducted the correct check amount, $7.50. She then added the incorrect amount, $5.70. This had the effect of wiping out the incorrect check amount.

Some people correct errors by finding the difference between the two checks. For example, Ella could subtract $5.70 from $7.50. The difference, $1.80, is the extra amount that would have been subtracted if she had recorded the check correctly. Ella could then subtract $1.80 in the Deduct section of the statement on the checkbook side.

Illustration 1-7.1.
Reconciliation Statement for Ella LaGarce

April 30, 19– –

Checkbook balance		58	20	*Bank statement balance*		99	35
Deduct:				*Deduct:*			
Service charge	*$7.75*			*Outstanding checks:*			
Check #98 not recorded	*9.50*			*#107*	*$42.00*	68	00
Check #106:				*#108*	*26.00*	31	35
Correct amount	*7.50*	24	75				
		33	45				
Add:				*Add:*			
Check #103				*Late deposit*		101	28
recorded twice	*$15.00*						
Deposit not recorded	*76.00*						
Interest earned	*2.48*						
Check #106:							
Incorrect amount	*5.70*	99	18				
Correct checkbook balance		132	63	*Available bank balance*		132	63

If the $7.50 check had been recorded for $9.30 instead of $5.70 the difference would still have been $1.80. But this extra amount would not have been subtracted. Ella would then have added $1.80 to the Add section of the statement on the check register side.

The outstanding checks left the bank statement balance $68 more than it should have been. The late deposit made the balance $101.28 less. Ella deducted $68 and added $101.28 to the bank statement balance to get the available bank balance of $138.63.

When reconciling your checking account, another error may be created if you incorrectly use a calculator. Be sure to double-check all calculations.

EXERCISE YOUR SKILLS

Find the available bank balances in Problems 1–3. Create a form like the right half of the reconciliation statement shown in Illustration 1-7.1.

	Date	Bank Statement Balance	Outstanding Checks	Late Deposit Not Recorded on Bank Statement
1.	Dec. 31	$381.36	$45.71; $9.45; $24.90	$85
2.	Mar. 31	$906.88	$89.16; $65.12; $1.78	$305
3.	Nov. 30	$580.04	$20.01; $9.41; $51.78; $3.35	$176

Find the correct check register balances in Problems 4–9. Use a form like the left half of Illustration 1-7.1.

	Check Register Balance	Not Recorded in Check Register			
		Service Charge	Check for	Other	Other Adjustments
4.	$829.76	$5.67	$55.78	Deposit of $135	Check #23 for $28.98 entered in register as $28.89
5.	$280.66	$2.25	$5.78	Interest of $1.14	Check #903 for $7.56 entered in register as $6.75
6.	$429.86	$4.10	None	Deposit of $110	Check #75 for $19.99 entered in register as $19.19
7.	$750.03	$5.01	$63.36	Interest of $2.78	Deposit of $86 entered twice in register

8. On March 31, your check register balance is $107.87. Your bank statement shows a service charge of $0.76 and interest earned of $0.85 which were not recorded in the register. You also find that check #307 for $35.29 had been entered in the register as $32.95.
9. At the end of May, Tomas Orroyo's check register balance was $812.45. An examination of his bank statement showed that a deposit of $135.90 had not been entered in the register and that check #201 for $92.49 had been entered in the register as $29.49.

Prepare reconciliation statements for Problems 10–13. Use Illustration 1-7.1 as a guide.

10. On May 31, Haru Gihei's checkbook balance was $200.89 and her bank statement balance was $278.99. Checks outstanding were: #407, $92.48; #410, $28.12; #154, $17.42. A deposit for $58.50 was not shown on the statement. The statement showed an ATM-use service charge of $2.67 and interest earned of $1.25.

11. On March 31, Melissa Sisson's check register balance was $92.45 and her bank statement balance was $167.23. The statement showed an ATM-use service charge of $1.45. Check #85 for $3.89 and #90 for $83.09 were outstanding. Melissa also found that she had recorded a deposit of $10.75 twice in the check register.

12. On September 30, Bud Benson's check register showed a balance of $700.61. His bank statement balance on that date was $856.90. With the statement, there was a slip showing that Bud's account had been charged $75 for his payroll savings plan. Checks outstanding were: #834 for $48.33, #837 for $21.19, #838 for $161.77.

13. Timothy Knowlton's check register balance on June 30 was $375.61. His June bank statement showed a balance of $498.34 on the same date. When Timothy examined the statement he found an ATM service charge of $2.75 and interest earned of $3.69. He also found that check #467 for $1.98 had been recorded in the check register as $1.89. Checks outstanding were: #477 for $64 and #478 for $57.88.

14. ◆ Look at the list you made at the start of this lesson. Add to or revise your list of reasons why balances might not agree. Explain how you would correct or avoid each error or omission.

MIXED REVIEW

1. 23,432 − 3,762

2. 347,437 + 372,653

3. Connie Bergman's check register balance on July 31 was $598.27. Her bank statement showed a balance of $674.21 on July 31. On the statement was a service charge of $4.73 and interest earned of $3.76. Outstanding checks were: #713 for $20.88, #718 for $32.10, and #714 for $23.93. Prepare a reconciliation statement.

4. On a business trip Yi Wang spent these amounts: hotels, $240.67; food, $148.89; gasoline, $57.12; tolls, $9.15; entertainment, $86.45; gifts, $125.89. Create a columnar cash payments record showing his trip expenses. What total amount did she spend?

5. Jack Smith's checking account had a balance of $507.87 on Monday. On the same day he deposited $165.23 and wrote checks for $15.67, $75.98, $267.11, and $75.75. What was his final balance?

What You Should Know Before You Begin

The Explore Technology lessons teach you how to work with and create spreadsheet and database applications. The ability to create software solutions to increase productivity is a valuable job skill.

A diskette that contains each of the spreadsheet and database files presented in the lessons is available for *Applied Business Mathematics.* The technology lessons are written assuming you have access to the *Applied Business Mathematics* template diskette.

You can use these lessons without the diskette by creating your own spreadsheets. Copy the column and row headings on the screens shown in the book and enter in appropriate formulas. If you do not have access to a computer and software, you can plan your spreadsheets and databases on paper.

Spreadsheets Most of the applications in the book are spreadsheet applications. Spreadsheet software is commonly used in business offices to prepare reports, such as budgets and payroll registers. Spreadsheets use a ledger format similar to the one you created when you made a multiple-column cash payments record in Lesson 3. Basically spreadsheets do the same mathematics an individual does only more quickly.

Databases You will also create and use some simple database files. Databases hold information for easy storage and retrieval. Frequently, the information in database files is linked to spreadsheet software to prepare financial reports.

Before You Begin working with these exercises, familiarize yourself with the basic functions and keystrokes of the software you will be using. We assume you know how to open, move around in, name, close, and save files. Refer to your software documentation if you need assistance. In the earlier lessons start-up instructions and illustrations are more detailed than in later lessons. Do not hesitate to consult the Help menu or software documentation when questions arise about procedures. You can also check with your school or community library for tutorials and other reference material.

As You Work Do not eat or drink while working at a computer. Spilling fluid can result in an electrical short that can damage your machine or keyboard. You do not need to be concerned, however, with experimenting with software packages. Experimenting is an excellent way to learn and will not harm the machine.

Save your files regularly, every 10–15 minutes and before stepping away from your computer station. Always give your files meaningful names so that you can retrieve them when you need them.

Print Outs Whenever you create a document that you may print, include a header or footer that identifies you as the creator. You may also want to include other identifying information, such as lesson number and date.

Getting the Most from the Lessons When you create your own spreadsheets and databases, write down the purpose of your application. Specify the mathematical skills necessary to generate the information you want. Then explore the software available to you to see what commands and functions may be available to simplify your work.

For More Information See the Technology Reference at the end of this book.

EXPLORE TECHNOLOGY

TEMPLATE: Using a Spreadsheet for a Cash Payments Record

Electronic spreadsheets are a commonly used computer application. A spreadsheet program lets you prepare many types of tables, reports, and forms quickly. They are very helpful for doing financial calculations like those learned in this chapter.

A spreadsheet containing structural information that will be used repeatedly can be saved in the form of a **template.** Each time the template is opened, the structure (row and column headings, and formulas) will appear and can be used to enter new data in.

Load ET1 (Explore Technology, Chapter 1) from your applications diskette. ET1 is a spreadsheet template that can be reused. After you have loaded the template, your computer screen should look like the illustration below.

	A	B	C	D	E	F	G
1				Curtis Jackson			
2				Cash Payments Record			
3							
4				Type of Payments			
5							
6							Total
7	Week	Housing	Food	Auto	Other	Savings	Payments
8							
9							
10							
11							
12							
13	Totals	0.00	0.00	0.00	0.00	0.00	0.00

Complete the cash payments record for Curtis Jackson by entering the data below into the designated cell of the spreadsheet template on your screen.

Cell	Data	Cell	Data	Cell	Data
A8	**Sept. 1–7**	B8	**350.00**	C8	**45.66**
A9	**8–14**	B9	**25.78**	C9	**102.34**
A10	**15–21**	B10	**179.98**	C10	**89.79**
A11	**22–28**			C11	**60.41**
A12	**29–30**			C12	**4.98**

Cell	Data	Cell	Data	Cell	Data
D8	**205.00**	E8	**19.23**	F8	**75.00**
D9	**91.88**	E9	**49.82**		
D10	**23.23**	E10	**9.20**		
D11	**82.10**	E11	**67.44**		
D12	**7.45**	E12	**69.50**		

Your completed spreadsheet should look like this.

	A	B	C	D	E	F	G
1				Curtis Jackson			
2			Cash Payments Record				
3							
4			Type of Payments				
5							
6							Total
7	Week	Housing	Food	Auto	Other	Savings	Payments
8	Sept 1–7	350.00	45.66	205.00	19.23	75.00	694.89
9	8–14	25.78	102.34	91.88	49.82		269.82
10	15–21	179.98	89.79	23.23	9.20		302.20
11	22–28		60.41	82.10	67.44		209.95
12	29–30		4.98	7.45	69.50		81.93
13	**Totals**	**555.76**	**303.18**	**409.66**	**215.19**	**75.00**	**1558.79**

Formulas have been included in the template which calculate the totals in Columns B, C, D, E, F, and G, as well as Row 13.

Answer these questions about your completed cash payments record.

1. What type of payment had the largest total for the month? the smallest total?
2. What was the total spent during the first week of the month?
3. What was the total spent during the last two days of the month?
4. What was the total spent on "Other" payments?
5. For what types of payment were no payments made in some weeks of the month?
6. Was the total spent on food expenses for the month larger or smaller than the total amount spent on auto expenses?
7. What is the spreadsheet doing arithmetically to get the amount in cell B13? in cell G8?
8. What spreadsheet function was used in cells B13 and G8 to get the totals?

Design Your Own

9. Although spreadsheets have many math functions built-in (one of the most commonly used is SUM), you should first learn to write formulas on your own that do the mathematics for you. In this way, you'll get the most out of your software.

First, use paper and pencil (and a ruler), to make a spreadsheet for your family or yourself that shows your cash payments for a month. Label your columns and rows in a way that is appropriate for your payments. Shade data-only cells to distinguish them from cells that require formulas. Then write the formula you would use to find the information needed in each of those cells.

Next create your spreadsheet using your software. Use the SUM function in your spreadsheet software to check your manual row and column totals. What does the SUM function do?

Explore the formatting options you have. For example, you may want to boldface your totals, or format them as dollars and cents.

CHAPTER 1 SUMMARY AND REVIEW

VOCABULARY REVIEW

ATM card *1-5*
Automatic Teller Machines ATM *1-5*
balance *1-4*
bank statement *1-6*
canceled check *1-6*
cash payments record *1-2*
cash receipts record *1-1*
check *1-4*
check register *1-4*
debit card *1-5*
deduction *1-4*
deposit slip *1-4*
Electronic Funds Transfer (EFT) *1-5*
grand total *1-3*
gross *1-4*
Magnetic Ink Character Recognition (MICR) *1-5*
net *1-4*
outstanding checks *1-6*
reconciliation statement *1-6*
service charge *1-6*
transit number *1-4*

MATH SKILLS REVIEW

1-1 ♦ To add faster, find combinations of ten.
♦ Check addition by reverse addition.

1. 14 + 8 + 12

2. 456.22 + 41.98 + 104.81

3. \$ 98.15 + 122.73 + 300.92

4. \$12,432.94 + 93,930.06 + 53,328.00

1-2 ♦ To make a number or open statement correct, replace missing or incorrect values with the numeral that makes the sentence true.

Find the missing numeral that makes the sentence true.

5. $23 + 6 = 12 + N$

6. $23 - 8 = N + 14$

1-3 ♦ Check columnar addition by verifying the grand total.

7.
23 + 123 + 3
287 + 7 + 209
4 + 83 + 18
509 + 75 + 821

8.
139 + 322 + 98
93 + 320 + 250
194 + 122 + 32

1-4 ♦ Check subtraction by adding the amount subtracted and the difference.

Subtract. Show your check.

9. 105 − 75 =

10. 220 − 80 =

11. 475 − 125 =

12. 964.64 − 0.55

13. \$5,008.37 − 3,905.88

APPLICATIONS

1-1 ◆ Add daily cash payments to find total weekly payments.

1-2 ◆ Add daily cash receipts to find weekly income.

1-3 ◆ When amounts are organized in columns, find the total for a day or week by adding horizontally; find the total for a type of payment by adding vertically.

◆ To check the grand total, add the sums for each type of payment, add the sums for each day or week, and compare.

14. Find the missing totals below. Verify by finding the grand total.

	WEEK		TYPES OF PAYMENTS: FOOD	CLOTHING	AUTO	SAVINGS	OTHER	TOTAL PAYMENTS
1	19– – Feb.	1–7	$89.15	$178.69	$25.83	$20.00	$6.95	
2		8–14	72.53		31.62		16.39	
3		15–21	82.30	15.92	24.67	20.00	22.89	
4		22–28	96.27	23.17	32.84		152.28	
		Total						

1-4 ◆ To maintain a check register, add deposits and subtract checks and other deductions from previous balance. Update after each transaction.

1-5 ◆ Be sure to record, and add or subtract, all electronic funds transactions in check register.

15. After work, Agnes O'Day used the automatic teller machine to deposit her paycheck for $568.33 and to withdraw $200 in cash for shopping. If her starting bank balance was $307.92, what is her new balance?

1-6 ◆ To reconcile a bank statement, add unrecorded deposits and subtract outstanding checks from statement balance. To the check register balance, subtract service charges and add interest earned.

16. Ed Acker's bank statement balance on June 30 was $400.78. On that same date, the checkbook balance was $277.91. The bank statement showed a service charge of $3.75 and interest earned of $1.65. The outstanding checks were No. 51 for $73.09, No. 52 for $46.28, and No. 54 for $5.60. Prepare a reconciliation statement.

1-7 ◆ To find the reason why a bank statement and check register do not reconcile, add checks recorded twice; subtract deposits recorded twice; and correct incorrectly recorded checks by adding back the wrong check and subtracting the correct check.

17. On May 31, Lance Carter's check register balance was $573.57 and his bank statement balance was $669.89. Checks outstanding were: #301, $115.87; #304, $13.15. Lance found that he had failed to record check #292 for $33.62 and that he had recorded check #298 for $6.30 as $5.30 in the check register. The statement showed a service charge for $1.25 and interest earned of $3.17. Prepare a reconciliation statement.

18. On June 30, Nolanda Brown's check register balance was $426.68 and her bank statement balance was $618.59. Checks outstanding were: #411, $37.73; #414, $143.06. Nolanda found that she had failed to record a deposit for $83.00. Another deposit for $71.20 was made too late to appear on the bank statement. The statement showed a service charge for $1.15 and interest earned of $0.47. Prepare a reconciliation statement.

WORKPLACE KNOWHOW: **HUMAN RESOURCE MANAGEMENT**

Assessment Skills include the ability to evaluate someone's knowledge and skill, distribute work accordingly, and evaluate and provide feedback to workers on performance. The ability to assess your own skills can be a powerful tool in building your career. You can match your skills to those needed by potential employers. You can identify skills you want to learn or improve. Many businesses include self-appraisals as part of the system they use to evaluate employees, establish performance goals, and determine raises and promotions.

Assignment On a scale of 1–5, rate your understanding of the following skills used in this chapter. Write a paragraph explaining why you gave yourself the rating. You will be asked to assess your skills at the end of each chapter of *Applied Business Mathematics.*

Addition Skills
Assessment Skills
Checkbook Reconciliation and Maintenance
Subtraction Skills
Cash Records Management
Spreadsheet Skills

Work Skills Portfolio Begin your work skills portfolio by including your assessment of your skills. In addition to your self-assessment, you may wish to include the following items in your work skills portfolio: writing samples from exercises, completed forms from the Integrated Project in the Student Workbook (check, deposit slip, check register, and reconciliation statement); and examples of spreadsheets you have designed. Include an explanation of why you have included the samples in your portfolio.

CHAPTER 1 TEST

Math Skills

Add or subtract. Use mental calculations when you can.

1. $\begin{array}{r} \$1{,}487.62 \\ 84.95 \\ 15.75 \\ +\ 2{,}092.75 \\ \hline \end{array}$

2. $\begin{array}{r} \$6{,}109.27 \\ -\ 3{,}727.89 \\ \hline \end{array}$

3. $\begin{array}{r} 50{,}035 \\ -\ 4{,}019 \\ \hline \end{array}$

4. $\begin{array}{r} 23{,}432 \\ 16 \\ 1{,}243 \\ +\ 9{,}209 \\ \hline \end{array}$

Find each missing number.

5. $423 + 36 + 199 + 23 = N$
6. $13{,}230 - 9{,}235 = N$
7. $5 + 22 = N + 16$
8. $N - 6 = 13 - 4$

Applications

Solve each problem. If a problem cannot be solved, tell what information is missing.

9. A cash-and-carry store's cash registers showed these amounts from sales on Friday: Register #1, $5,893.23; Register #2, $6,709.44; Register #3, $1,078.98; and Register #4, $9,101.45. What were the total sales on Friday?

10. Mario Torres deposited these items: 5 twenties, 6 tens, 5 fives, 61 ones, 55 quarters, 37 nickels, 19 pennies, checks for $309.78 and $7.24. What was the total deposit?

11. Jay Volmer sold bird houses. His gross sales were $106.50. The houses cost him $37.50. Other expenses were $24.63. What net amount did he make on the sale?

12. During one week, Diana Farley spent the following amounts on her car. gasoline, $54.57; oil, $14.79; windshield wiper blades, $15.78; city sticker for car, $25. How much did Diana spend for the week?

13. On April 30, Samantha had a check register balance of $344.05 and a bank statement balance of $504.25. A service charge of $1.95 and interest earned of $1.60 were shown on the bank statement. The outstanding checks were #21, $24.15; #23, $38.90; #24, $46.50; and #25, $51. Make a reconciliation statement.

14. Brad's bank statement on January 31 showed a balance of $1,165.75. His check register balance was $1,319.05. Checks were outstanding for $16.60, No. 89; $23.10, No. 91; and $34.60, No. 93. A deposit of $170.50, mailed on January 31, was not recorded on the bank statement. A canceled check for $57.10 had not been recorded in his check register. Prepare a reconciliation statement.

15. Claire Wieman started the day with a bank balance of $145.67. During the day she deposited a check for $50 in the ATM and used her debit card for these purchases: $13.78, $5.50, $26.89, and $59.98. What was her bank balance at the end of the day?

CHAPTER 2

Get it in the Armed Employees are Looking

HUT! HUT! HUT!

Running miles. Wearing fatigues. Sporting a close-cropped haircut. Are those the images you connect with the armed services?

Job training. Teamwork. Personal development. Computer skills. Careers. Those are the images connected with the armed services by Dawn, who spent 3 years in the Air Force; Robert, who has been in the U.S. Army for 15 years; and by many civilian employers.

While still in high school, Dawn joined under a delayed enlistment program. She saw the service as an opportunity to learn about people and the world while gaining independence. After the service, Dawn discovered potential employers saw her military experience as a professional asset.

Robert, a recruiter for the National Guard, joined as an alternative to college and now expects to retire from the service. He can then begin a second career using the sales and computer skills he learned in the service.

So what's it like to be a member of the U. S. Armed Forces?

To join, your first stop is likely to be the local recruitment office where you will take aptitude and skills tests. You'll receive an assignment, and then it is off to basic training, weeks of concentrated training aimed at teaching both personal and professional skills while also developing good physical health.

for Experience...

Forces

Studying and testing are integral parts of the service. For example, Dawn had to take Career Development Curriculum courses to move from one level in her job to the next. Robert reads a lot to stay current with policies. All military personnel undergo testing for physical fitness/readiness and weapon usage each year.

How important are math skills in the service? Tests that include basic math competencies are given both at the time of recruitment and promotion. On-the-job skills requiring more specific math vary. Dawn and Robert highlighted the following as frequently used.

- **Basic Mathematics**
- **Ratios and Percents**
- **Computer Skills**
- **Problem-Solving Strategies**

Dawn and Robert think service experience is valuable. Dawn says, "Keep in mind service life means structure and rules. It may not be for everyone." Robert says success "boils down to self-discipline."

If you are interested in the military, Dawn recommends that you "Find out about each branch's career training options. Talk to more than one recruiter. Talk to people in the service. Learn as much as you can before you sign up."

2-1 MULTIPLICATION AND ESTIMATION

OBJECTIVES

In this lesson, you will learn to

- *check multiplication;*
- *round numbers;*
- *estimate products;*
- *multiply by 10, 100, or 1,000;*
- *multiply by a multiple of 10, 100, or 1,000;*
- *multiply numbers that have end zeros; and*
- *multiply by 1¢ and 10¢.*

If someone asked you how much you earned at your part-time job last year, you might say "around $2,000" instead of giving the exact figure of $1,982.76. Why would you use a rounded amount? How might you have decided on the $2,000 amount?

◆ In a small group, take turns discussing each person's plans for after school or evening. Listen to people talk. Keep a count of how many times each person in your group uses rounded or estimated references. Compare your total count with the others in your group. What kinds of rounding and estimating references did you identify?

WARM UP

Find the value of N in each number sentence.

a. $18 + N = 34 + 12$	**c.** $16 + 19 = N - 33$	**e.** $48 - 16 = 32 - N$
b. $N + 46 = 23 + 34$	**d.** $N - 25 = 50 + 22$	**f.** $53 - N = 27 - 13$

Round to the nearest:

Tenth:	Hundredth:	Cent:	Thousand:
a. 7.3162	**c.** 362.3691	**e.** $5.8639	**g.** 8,288
b. 1.9821	**d.** 58.7806	**f.** $0.699	**h.** 13,612

FACTS ABOUT MULTIPLICATION

Multiplication is a short way of adding two or more equal numbers. So 3×5 is the same as $5 + 5 + 5$. Knowing the multiplication tables will eliminate having to add the numbers. You will find that doing multiplication quickly and accurately is an important business skill.

When you multiply, you use at least two numbers. Each number being multiplied is called a **factor**. The result is the **product**.

EXAMPLE What is the product of 15 and 26?

	SOLUTION		CHECK	
Factor	26	(Factor)	15	(Factor)
× Factor	× 15	(Factor)	× 26	(Factor)
Product	130		90	
	26		30	
	390	(Product)	390	(Product)

Checking Multiplication

One way to check multiplication is to reverse the factors and multiply again. In the example above, the answers (products) are the same even though the numbers (factors) that are multiplied are reversed.

Placing the Decimal Point in the Product

When multiplying decimal numbers, count the number of decimal places in each factor. The sum of the numbers is the number of decimal places in the product. To find where to place the decimal point in the product, start at the right of the number and move left the desired number of decimal places. Then place the decimal point.

EXAMPLE What gross pay does an employee earn by working 14.5 hours at $7.24 an hour?

SOLUTION

```
   $7.24
  × 14.5
   3 620
   28 96
   72 4
$104.980
```

EXPLANATION There are 2 decimal places in the first factor and 1 decimal place in the second factor, for a total of 3 places. Count left 3 places from the right-hand side of the product and place the decimal point.
Remember to label your answers with the dollar sign and decimal point.

CHECK

```
    14.5
  × 7.24
     580
    2 90
   101 5
$104.980
```

ANSWER The gross pay is $104.98.

Critical Thinking Why was the answer not written as $104.980?

EXERCISE YOUR SKILLS

1. Create and complete a multiplication table for the numerals 0–12.

Mental Math Place the decimal point in the products of the problems.

2. $2.4 \times 7.1 = 1704$
3. $3.12 \times 8.5 = 26520$
4. $1.11 \times 345 = 38295$
5. $7.1 \times 8.23 = 58433$
6. $42.5 \times 68.9 = 292825$
7. $9.92 \times 0.57 = 56544$
8. $0.7 \times 0.6 = 42$
9. $100 \times 8.7 = 8700$
10. $10 \times 0.03 = 30$
11. $10 \times 0.006 = 60$

ROUNDING NUMBERS

The results of many multiplication problems in business have to be **rounded.** Rounded numbers are easier to work with and often more useful.

Rounding Money Amounts

Money amounts are usually rounded to the cent, the smallest coin in our money system. The kind of rounding done in business depends on the business. Some businesses treat any part of a cent as a whole cent. Others drop the part of the cent if it is less than half a cent.

Note When you solve money problems in this text, round the final results to the nearest cent unless directed otherwise. This means ignore amounts less than one-half cent ($0.005). Round amounts of $0.005 and higher to a whole penny.

When writing money amounts, 58¢ is the same as $0.58. In both cases, the digit 8 is in the cents position. Also, 87.5¢ is the same as $0.875.

```
       8 7 . 5¢
  $0 . 8 7   5
cent ──┘     └── tenth of a cent
```

EXAMPLE Round 76.28¢ and $8.4281 to the nearest cent.

SOLUTION

	76.28¢	$8.4281
1. Identify cents place.	7<u>6</u>.28¢	$8.4<u>2</u>81
2. Check digit to right of cents place.	76.<u>2</u>8¢	$8.42<u>8</u>1
3. If the digit is less than 5 (< 5), leave the cents amount as shown. This is called *rounding down.* If the digit is 5 or greater (> 5), add 1 to the cents place. This is called *rounding up.*	2 < 5	8 > 5
4. Drop all digits to right of cents place. Show the rounded answer.	7<u>6</u>¢	$8.4<u>3</u>

Rounding Decimals and Other Numbers

Decimals can be rounded to tenths, hundredths, thousandths, and so on. The same rules for rounding money are used with decimals. For example, 3.4918 rounded to the nearest tenth is 3.5; to the nearest hundredth, 3.49; to the nearest thousandth, 3.492.

Whole numbers can also be rounded. For example, 62,299 rounded to the nearest thousand is 62,000; to the nearest hundred, 62,300; and so on.

Rounded numbers are often used in business reports in place of exact numbers. For example, assume that a business had as few as 188 and as many as 210 employees last year. By rounding the exact numbers to the nearest hundred, you could say that the company had about 200 employees.

EXERCISE YOUR SKILLS

Mental Math Round each to the nearest cent.

12. 3.56¢ **13.** 2.97¢ **14.** 50.449¢

15. $7.2894 **16.** $4.9548 **17.** $60.9961

As the billing clerk for the Delmar Company, you are to find the costs of several orders. Round these costs to the nearest cent.

18. $345.925 **19.** $40.006 **20.** $400.995

21. $67.999 **22.** $840.009 **23.** $999.996

Mental Math Round each decimal to the nearest tenth, to the nearest hundredth, and to the nearest thousandth.

24. 5.2379 **25.** 8.2741

26. 3.1875 **27.** 62.8684

28. 92.6597 **29.** 10.1999

Round each number to the nearest hundred thousand, to the nearest ten thousand, to the nearest thousand, and to the nearest hundred.

30. 435,715

31. 934,979

32. 3,050,049

33. A retail store had 2,106 customers on Monday and 1,890 customers on Tuesday. How many total customers did the store have for the two days, rounded to the nearest hundred?

34. While writing a report, you found that a certain state had a population of 2,369,872. What is the state's population rounded to the nearest hundred thousand?

ESTIMATION

You can use estimation to help check the accuracy of your multiplication. An estimate can help you determine if your actual product is reasonable. There are also short cuts in multiplication that you can use to save time.

Estimating the Product

Before you multiply, round both factors to simple numbers that you can multiply easily for an estimated product. Compare the exact product with the estimated product to see if the exact product is a reasonable answer.

EXAMPLE

Estimate $28.5 \times \$1.04$.

Round 28.5 to 30 and $1.04 to $1.00 to make the factors easier to multiply.

$30 \times 1 = 30$

The estimated product is $30 ($30 \times \$1$).

$28.5 \times \$1.04 \approx \$30$

Note $\approx$ is the symbol for "is approximately equal to." So this statement means $28.5 \times \$1.04$ is approximately equal to $30.

If you calculate an exact product of $29.64 (which is close to $30), it is a reasonable answer. If you calculate an exact product of $3.99 or $296.40, your answer would not be reasonable. If the exact product is not reasonable, check your estimate. If that seems correct, then recalculate the exact product. You probably made an error when you multiplied or placed the decimal point.

Round-Down Method One of the ways to estimate products is to **round down** both factors to the tens or hundreds. For example, to estimate the product of 23×48, round down both factors to the tens position. So 23 becomes 20 and 48 becomes 40. The estimated product is 800 (20×40). When the round-down method is used, the estimated product will *always be less* than the actual product.

When you estimate the product for smaller numbers, round down the factors to the ones place or even to the tenths place. For example, the estimated product of 2.1×0.87 may be found by rounding 2.1 to 2 and 0.87 to 0.8. The estimated product is 1.6 (2×0.8).

Round-Up Method Another way to estimate the product is to **round up** both factors to the tens or hundreds. To estimate the product of 37×128 by this method, round 37 to 40 and 128 to 130. The estimated product is 5,200 (40×130). Smaller factors may be rounded up to the ones place or to the tenths place to find the estimated product. When the round-up method is used, the estimated product will *always be greater* than the actual product.

Critical Thinking How can you use rounding to find a reasonable range for the product of two numbers?

EXERCISE YOUR SKILLS

Mental Math What rounded factors were used to find each estimated product?

35. $6.78 \times \$8.95$ \$63
36. $5.62 \times \$7.78$ \$48
37. 56.7×12.3 720
38. 93×27 2,700
39. $57¢ \times 876$ \$540
40. $1.874 \times \$0.48$ \$1

Mental Math Use the **round-down method** to identify factors used to estimate the product. Then estimate each product.

41. 56×72
42. $512 \times \$67$
43. 57.1×44.9
44. $4.29 \times \$5.76$
45. 0.87×0.56
46. 1.4×3.497
47. $2.45 \times 87¢$

Mental Math Use the **round-up method** to identify the factors used to estimate the product. Then estimate each product.

48. 49×78
49. 31×16
50. $\$15.80 \times 24$
51. 27.3×40.7
52. 5.87×0.56
53. 0.476×0.23

Rule-of-Five Method The **rule-of-five** method of estimating products will usually give you an estimated product that is closer to the exact product than other methods. In this method, you find the place to which you wish to round and look at the digit to the right. If it is 5 or more, round up. If the digit to the right is less than 5, round down. For example, if you round 484 to the hundreds position the answer is 500, and if you round the same number to the tens position the answer is 480.

EXAMPLE

Estimate 264×9.03	Use the rule-of-five method to round each factor.
$300 \times 9 \approx 2{,}700$	Multiply the rounded factors. Remember to use $\approx$ to show the product is approximate, not exact.

EXERCISE YOUR SKILLS

Use the rule-of-five method to identify the factors that will be used to estimate the product. Then estimate each product.

54. 56 × 78

55. 72 × 91

56. $25 × 850

57. 0.84 × 0.55¢

58. 473 × 65.5

59. $0.34 × 6.58

60. 0.004 × 0.98

61. 7,840 × 50¢

For each problem, estimate the product; find the exact product; check the exact product against the estimate; and check the exact product by reverse multiplication.

62. 52 × 4.8

63. 1.6 × 9.2

64. $0.85 × 5.6

65. $605.65 × 12

66. $737.20 × 9.1

67. $650.76 × 19

For each question, identify the method of rounding you would use to estimate the product. Justify your answers.

68. You are at a restaurant and know that 18 people each had meals that cost $9.25. You need to know if you have enough money.

69. You need to know the minimum you will earn in one week if you work 35 hours at $10.50 per hour.

70. You received a bill for $507.40 for 11.8 hours of work at $4.30 per hour. Is the bill correct?

MULTIPLICATION BY 10, 100, OR 1,000

To multiply by 10 or a multiple of 10 (such as 100, or 1,000), move the decimal point in the factor that is not a multiple of 10 one place to the right for each zero in the factor that is a multiple of 10. Drop the decimal point if the product is a whole number. Attach zeros to the product if needed.

EXAMPLES

10 × 9.42 = 94.2	10 × $9.42 = $94.20
100 × 9.42 = 942	100 × $9.42 = $942
1,000 × 9.42 = 9,420	1,000 × $9.42 = $9,420

Multiplication by a Multiple of 10, 100, or 1,000 To multiply by a number such as 40, 400, or 4,000, first multiply by 4; then multiply by either 10, 100, or 1,000. A number such as 400 is a multiple of 100, since 4 times 100 equals 400.

EXAMPLE Multiply $2.20 by 400.

SOLUTION

$2.20 × 4 = $8.80	Multiply $2.20 by 4.
$8.80 × 100 = $880	Now multiply by 100 by moving the decimal two places to the right in the first product, $8.80.
So, $2.20 × 400 = $880	

Multiplying Numbers That Have End Zeros When either one of the factors or both have end zeros, arrange and multiply only the numbers to the left of the end zeros. Then attach to the product as many zeros as there are in both factors combined.

EXAMPLES **A.** $140 \times 2{,}000 = ?$ **B.** $2{,}100 \times \$1.80 = ?$

SOLUTIONS

A. $14 \times 2 = 28$
140 has 1 zero
2,000 has 3 zeros
Product has 4 zeros
So $140 \times 2{,}000 = 280{,}000$

B. $21 \times 1.8 = 37.8$
2,100 has 2 zeros
1.80 has 0 zeros*
Product has 2 zeros
So $2{,}100 \times \$1.80 = \$3{,}780$

Note* 1.8 = 1.80—The zero is **not significant. Therefore, it is not counted when moving the decimal point in the answer. In a decimal, zeros at the end (far right) **after** the decimal point are not significant.

EXERCISE YOUR SKILLS

Mental Math Multiply each number by (a) 10, (b) 100, (c) 1,000.

71. 29	**72.** 0.568	**73.** 4.9	**74.** 0.08
75. \$3.88	**76.** \$0.425	**77.** \$13.12	**78.** \$0.08
79. 4¢	**80.** 70¢	**81.** 43.5¢	**82.** 8¢

Mental Math
Write the product only for each problem. Multiply the numbers mentally.

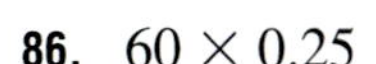

83. 20×0.37	**84.** 70×0.4	**85.** 40×0.95
86. 60×0.25	**87.** $60 \times 25¢$	**88.** $60 \times \$0.25$
89. $80 \times \$0.75$	**90.** $200 \times \$1.25$	**91.** $4{,}000 \times \$17$
92. 150×300	**93.** 230×80	**94.** 270×50
95. $200 \times \$7.40$	**96.** $50 \times \$2.50$	**97.** $70 \times \$9.20$

MULTIPLICATION BY 1¢ AND 10¢

To multiply by 1¢, or by \$0.01, move the decimal point in the other factor *two places to the left* and attach a dollar sign. *Note:* The @ is read as "at" and indicates multiplication.

EXAMPLE 186 sheets of paper @ \$0.01 → $186 \times 0.01 = \$1.86$

To multiply by 10¢, or \$0.10, move the decimal point in the other factor *one place to the left* and attach a dollar sign. For example:

EXAMPLE 38 folders @ \$0.10 → $38 \times 0.10 = \$3.80$

38 folders @ \$0.10 = \$3.80

Multiplication by a Multiple of 1¢ and 10¢ To multiply by a price such as \$0.05 or \$0.50, multiply by 5 or 50 and write the product. Then move the decimal point two places to the left in the product and attach a dollar sign.

EXAMPLE Find the cost of 135 items at \$0.50.

$135 \times 50 = 6{,}750$ ← Multiply 135 by 50.
$135 \times 0.50 = 67.50$ ← Move the decimal point *two places to the left.*
$135 \times \$0.50 = \$67.50$ ← Add the dollar sign.

EXERCISE YOUR SKILLS

Mental Math Find the cost.

98. 450 @ \$0.01 **99.** 156 @ \$0.10 **100.** 635 @ \$0.01
101. 218 @ \$0.10 **102.** 719 @ \$0.01 **103.** 73 @ 10¢
104. 245 @ 1¢ **105.** 1,300 @ 1¢ **106.** 82.6 @ 10¢
107. 240 @ \$0.10 **108.** 153 @ \$0.01 **109.** 2,730 @ 1¢
110. 24 @ \$0.04 **111.** 102 @ \$0.07 **112.** 80 @ \$0.05
113. 250 @ 3¢ **114.** 110 @ 60¢ **115.** 130 @ 30¢
116. 65 @ \$0.40 **117.** 73 @ \$0.70 **118.** 93 @ \$0.50

119. ♦ Write a report describing situations in which you have used rounded numbers in the last month.

MIXED REVIEW

1. Find the grand total. Show the answers for each row and each column.

$54 + 27 + 34 =$
$13 + 75 + \ \ 9 =$
$83 + 11 + 17 =$

Find the value of N.

2. $87 - 9 = N$
3. $1.35 \times 62 = N$
4. $24 - 15 + 7 = N$
5. $2 \times 2 \times 0 = N$
6. $17 \times N = 17$
7. $7 \times 9 = N$

Round 236.05 to the nearest:

8. hundred
9. tenth
10. ten

Round $53.831 to the nearest:

11. dollar
12. ten-dollars
13. cent

Write an estimate for each problem. Then find the exact product.

14. $4{,}900 \times 8.2$
15. 3.8×489
16. 505×12.3
17. 7.2×312
18. 281×4.1
19. 4.50×197

Solve each problem.

20. A store bought a carton of crackers that weighed a total of 276 ounces. The carton contained 24 boxes of crackers. If the empty carton weighed 12 ounces and each empty box weighed 1 ounce, what was the net weight of the crackers in ounces?

21. On April 1, Jean Monier's check register balance was $410. The bank statement balance for that date was $588.96. A comparison of the check register and bank statement showed a service charge of $5, interest earned of $3.20, and these outstanding checks: #523 for $105.40 and #542 for $75.36. Prepare a reconciliation statement.

22. Louise Trent makes this deposit at her bank: (coins) 36 pennies, 24 nickels, 36 dimes, 15 quarters, 9 halves; (bills) 19 ones, 8 fives, 6 tens, 13 twenties; (checks) $53.47, $83, and $541.27. What is the total amount of Louise's deposit?

23. In July, Edna, an editor, made these cash payments. Organize them into a multiple column payment chart. July 3: Copyediting $350; Writing $500. Typesetting $422. July 10: Answer Checking $225; Photo Scans $200. July 17: Copyediting $500; Answer Checking $175; Photo Scans $175. July 24: Writing $700. July 31: Photo Scans $90.

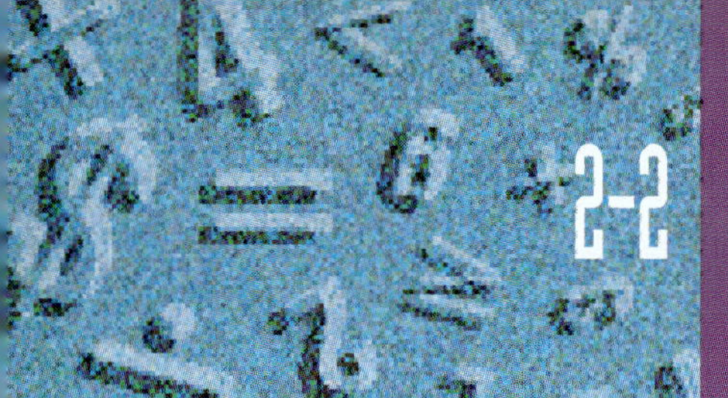

2-2 FINDING GROSS PAY

OBJECTIVES

In this lesson, you will learn to

- *calculate gross pay for hourly-rate, salaried, and piece-rate employees; and*
- *use a spreadsheet to calculate payrolls.*

When you apply for a job, your salary offer may be expressed as a certain amount of pay per hour, per week, per month, or per year. You may also be offered other payment plans. You need to be able to figure how the offer compares to your own needs. Most likely you think of these on a weekly or monthly basis.

◆ At the end of a job interview, you are told that you would start at $7.00 per hour for a 40-hour week. At the end of six months with a positive evaluation, your salary will increase to $7.50 per hour. If there are 4.3 work weeks in a month, how much would your raise be per week and per month? Make a chart showing your basic monthly expenses. Find the total, and determine if this job offer is enough to meet your needs.

WARM *UP*

a. $\begin{array}{r} 16{,}832 \\ -\ 9{,}705 \\ \hline \end{array}$ **b.** $\begin{array}{r} 761.36 \\ -\ 387.94 \\ \hline \end{array}$ **c.** $\begin{array}{r} 23{,}673.08 \\ -\ 12{,}079.44 \\ \hline \end{array}$ **d.** $\begin{array}{r} \$8{,}460.34 \\ -\ 7{,}726.85 \\ \hline \end{array}$

Find the sum of each row and each column. Then find the grand total.

e. $17 + 36 + 47 + 7 =$
$26 + 82 + 19 + 5 =$
$92 + 13 + 56 + 3 =$
$34 + 75 + 26 + 21 =$

f. $18 + 72 + 94 + 5 =$
$29 + 59 + 37 + 8 =$
$70 + 68 + 48 + 13 =$
$32 + 46 + 24 + 17 =$

GROSS PAY

Most people earn money by working for others. Those who work for others are called **employees.** The person or company an employee works for is called an **employer.**

An employee who is paid by the hour works for an **hourly rate,** which is a certain amount for each hour worked. An hourly rate employee is usually paid each week. An employee may be paid by the hour, day, week, month, or year. The total amount of money that an employee is paid is called **gross pay** or *gross wages*. Gross pay, total earnings, and total pay all mean the same thing.

Some employees are paid a **salary,** which is a fixed amount of money for a day, week, month, or year of work. Others, such as salespeople, may earn all or part of their gross pay from a percentage of the amount they sell. This type of pay is called **commission.** You will learn more about commissions in Chapter 4.

Gross Pay for Hourly-Rate Employees

The gross pay earned by employees who are paid by the hour is found by multiplying the pay per hour by the hours worked.

EXAMPLE Sandra Almond works as an auto mechanic and is paid $12 an hour. Find her gross pay for last week if she worked 34 hours.

SOLUTION Use your calculator.

12 [×] 34 [=]

Her gross pay was $408.

EXERCISE YOUR SKILLS

Solve each problem.

1. Paul Hunt earns $5 an hour at his part-time job. Last week he worked 16 hours. What was his gross pay for the week?

Kevin O'Leary worked the following schedule this week: Monday, 7 hours; Tuesday, 8 hours; Wednesday, 8 hours; Thursday, 6 hours; Friday, 5 hours. He was paid $9 an hour.

2. How many hours did Kevin work during the week?
3. What was his gross pay for the week?

Rachel Solero is paid $7 an hour. Last week she worked 8 hours a day for 5 days.

4. How many hours did Rachel work last week?
5. Find her gross pay for the week.

The Breslon company has five employees. The chart below shows the hourly pay of each employee and the hours each employee worked last week. Find last week's gross pay for each employee. Then find the total gross pay earned by all the employees last week.

	Employees	Hourly Pay	Hours Worked	Gross Pay
6.	Ed Alvarez	$8	40	
7.	Mark Dalton	$8	32	
8.	Vicky Mazur	$7	37	
9.	Tina McDowell	$9	40	
10.	Tom Dane	$7	35	
11.	Total	xx	xx	

Gross Pay for Salaried Employees

Salaried employees are paid a fixed amount of money for each time period worked, such as a day, week, or month. To find their gross pay, multiply the pay for a time period by the number of time periods worked.

EXAMPLE Tom Chin is paid a salary of $435 a week. How much gross pay will Tom receive for 4 weeks of work?

SOLUTION

$435	pay for one week
× 4	weeks worked
$1,740	gross pay

EXERCISE YOUR SKILLS

12. Art Hall is a part-time computer operator. When he works, he is paid a salary of $76 a day by his employer. What is Art's gross pay for 5 days of work?

13. Roy Lampley is a temporary employee. He earns $52 for each full day he works. What is Roy's gross pay if he works 7 full days during a 2-week period?

14. The Video-Tek Company pays its security guards a salary of $1,400 a month. What is each guard's gross pay per year?

15. Mildred Tompkins earns a salary of $389 a week. If Mildred is paid every 2 weeks, what gross pay does she receive each payday?

William Stolitz is paid a weekly salary of $375.

16. How much would William earn in 4 weeks of work?

17. How much would he earn by working for a year, assuming that he works 50 weeks and is paid for 2 weeks of vacation?

Find the yearly salary for each employee. Assume each is paid for 52 weeks a year.

	Employee	Weekly Salary	Yearly Salary
18.	Gene Clark	$300	
19.	Bertha Rowland	420	
20.	Lloyd Weisberg	495	

21. An employee earns $2,400 each pay period. He is paid on the first and fifteenth of each month. How much does he earn in one year?

Gross Pay for Piece-Rate Employees

Some employees are paid for each item or **piece** they produce. Their wages are paid on a **piece-rate** basis. To figure their gross pay, you must multiply their pay per piece by the number of pieces produced. If employees are paid only for usable pieces produced, they get no pay for the pieces that are rejected.

EXAMPLE Susan Gerby is paid $1.70 for each usable electronic timer she produces. Last week Susan produced the following quantities of usable timers: Monday, 48; Tuesday, 44; Wednesday, 47; Thursday, 50; Friday, 46. What was Susan's gross pay for the week?

SOLUTION $48 + 44 + 47 + 50 + 46 = 235$ timers produced
$235 \times \$1.70 = \$399.50$ gross pay

Calculator If you use a calculator, the decimal point will be automatically put in the right place. This feature is called the *floating decimal point.* Note your calculator will show $399.5. Add the 0 when you write your answer.

EXERCISE YOUR SKILLS

22. Lu Ying works at the Randor Bike Shop and is paid $2.75 for each bike he assembles. During the five working days of one week, he assembled these numbers of bikes: 27, 33, 29, 27, 31. What was Lu's gross pay that week?

For each of these piece-rate employees at Lockler Products, find the total pieces produced and the gross pay for the week.

	Name	Number of Pieces					Total Pieces	Rate per Piece	Gross Pay
		M	T	W	TH	F			
23.	Block, B.	54	55	59	62	60		$1.60	
24.	Campagna, T.	24	28	30	31	27		2.80	
25.	Schweib, V.	63	69	59	62	50		1.55	
26.	Zullo, Y.	68	65	72	74	75		1.18	

Melanie Crane is paid $1.60 for each of the first 50 usable lamps she assembles in a day and $1.76 for each usable lamp over 50. Find Melanie's gross pay for a day in which she assembles

27. 65 lamps, 11 of which are unusable.

28. 52 lamps, 7 of which are unusable.

Employees at the Graf Company are paid on this piece-rate schedule:

First 75 or fewer pieces, $0.95
Next 25 pieces, $1.05
All pieces over 100, $1.20

On Monday, nine Graf Company employees produced the number of pieces shown below. What was each employee's gross pay for the day?

29. Hart 65

30. Hassed 72

31. Cook 80

32 Nowak 84

33. Jenke 108

34. Wong 102

35. Dubois 118

36. Flynn 89

37. ◆ List some possible advantages and disadvantages of being paid on a per piece, hourly, and salaried basis for an employee. List the advantages and disadvantages of the three pay options for an employer.

MIXED REVIEW

Find the value of N in each number sentence.

1. $27 + N = 19 + 16$

2. $N - 30 = 17 + 14$

3. Arlon Hamsa worked these hours last week: Monday, 8 hours; Tuesday, 4 hours; Wednesday, 7 hours; Thursday, 6 hours; Friday, 8 hours. If he was paid $9 an hour, what was Arlon's gross pay last week?

Find the gross pay for the time worked by each employee.

	Employee	Pay Rate	Time Worked	Gross Pay
4.	Nick Marsa	$1,510 per month	12 months	
5.	Carter Woo	$380 per week	52 weeks	
6.	Karla Wright	$12 per hour	37 hours	

7. Cary Hunt is paid $2.08 for every usable machine part she makes. During one week, she made 220 parts, 14 of which were unusable. What was Cary's gross pay for the week?

8. Fred Lowery started the day with $35.67. He returned a pair of shoes to the store and received a cash refund of $41.60. Later in the day he paid cash for items costing $5.20, $14.56, $7.98, and $3.56. How much cash did he have left at the end of the day?

2-3 DIVISION AND ESTIMATION

You are at a restaurant with four friends. The bill comes. Everyone agrees to paying in equal shares. What arithmetic operation will you use to find how much each person should pay?

◆ In a small group, plan an activity such as eating out or going to the movies. Create a list of the costs you will encounter. Show how to split the total bill so that each person pays the same amount. Simulate the payment, and check to see if enough money was collected to cover the costs.

OBJECTIVES

In this lesson, you will learn to

- *identify factors and products and relate them to division;*
- *place the decimal point in a quotient;*
- *round quotients to a desired number of places;*
- *estimate quotients; and*
- *use division shortcuts.*

WARM UP

Find the value of N.

a. $48 \times N = 48$

b. $14 \times 6 = N \times 3$

c. $14 \times 3 = 6 \times N$

d. $9 \times 8 = N \times 3$

Multiply.

e. 320×70

f. $144 \times 10¢$

g. 2.1×1.000

h. $\$0.01 \times 800$

DIVISION

Many business problems are solved using division. In future lessons, you will use division to solve problems dealing with averages, investments, buying decisions, and tax rates.

Division is the opposite process of multiplication. For example, if you start with 6 and multiply by 3, you get 18. If you divide 18 by 3, you get 6, which is the number you started with. Dividing by 3 will undo what you did by multiplying by 3.

When you multiply, you use at least two factors to get a product. For example, when you multiply the factor, 6, by the factor, 3, you get the product, 18. The following shows how the factors and product are related:

$$\textbf{Factor} \times \textbf{Factor} = \textbf{Product}$$

$$\textbf{or} \quad \textbf{F} \times \textbf{F} = \textbf{P}$$

$$\textbf{3} \times \textbf{6} = \textbf{18}$$

Division is the opposite of multiplication. Division separates the product into the two factors. So when you divide the product, 18, by the factor, 3, you get the other factor, 6. Another way in which the product and the factors are related is shown in the following:

$$\textbf{Product} \div \textbf{Factor} = \textbf{Factor}$$

$$\textbf{or} \quad \textbf{P} \div \textbf{F} = \textbf{F}$$

$$\textbf{18} \div \textbf{3} = \textbf{6}$$

You can also think of division as a way of finding the unknown factor when you know one factor and the product. In solving $18 \div 3$, ask yourself what number multiplied by 3 gives 18 as a product. Since the answer is 6, we can say that $18 \div 3 = 6$.

Ways of Writing Division There are several ways to write division problems. For example, to show that 18 divided by 3 is 6, you may use any of these forms:

$$18 \div 3 = 6 \qquad \frac{18}{3} = 6 \qquad 3\overline{)18}^{\,6}$$

Computers and some calculators use a slash to indicate division: $18/3 = 6$.

In each case, 18 is the dividend, 3 is the divisor, and 6 is the quotient. A **dividend** is any number that is to be divided. A **divisor** shows the size or number of groups into which the dividend is to be split. A **quotient** shows how many times the divisor is included in the dividend.

Study the following illustration that shows how the terms used in division and multiplication are related.

$$3\overline{)18}^{\,6} \quad \text{or} \quad \text{Divisor}\overline{)\text{Dividend}}^{\,\text{Quotient}} \quad \text{or} \quad \text{Factor}\overline{)\text{Product}}^{\,\text{Factor}}$$

$$\frac{18}{3} = 6 \quad \text{or} \quad \frac{\text{Dividend}}{\text{Divisor}} = \text{Quotient} \quad \text{or} \quad \frac{\text{Product}}{\text{Factor}} = \text{Factor}$$

Using Factors and Products

While you work with number sentences in multiplication or division, you will find it helpful to identify mentally the factors and the product. You may also want to write the letter *F* or *P* above each numeral or letter in the number sentence. For example:

F F P	P F F	P F F	F P F
3 × 6 = 18	**18 = N × 6**	**18 ÷ 3 = 6**	**6 = N ÷ 3**

In the sentences $3 \times 6 = N$ and $N \div 3 = 6$, the two factors are known; and the product, *N*, is unknown. You can find the value of the unknown product, *N*, by multiplying 3×6. The product, 18, is the value of *N*.

In the sentences $3 \times N = 18$ and $18 \div N = 3$, the product, 18, and one factor, 3, are known. The other factor, *N*, is unknown. To find the value of *N*, divide the known product, 18, by the known factor, 3. The quotient, 6, is the value of *N*, the unknown factor.

Remember, in any division problem, the dividend is a known product, and the divisor is a known factor. The quotient is an unknown factor.

EXERCISE YOUR SKILLS

Find the quotients.

1. $27 \div 1$ **2.** $42 \div 6$ **3.** $24 \div 3$

4. $28 \div 7$ **5.** $12 \div 12$ **6.** $36 \div 12$

Copy each number sentence. Above each numeral or letter, write F or P to show whether it is a factor or product.

7. $5 \times 6 = 30$ **8.** $8 \times 9 = 72$ **9.** $42 \div 7 = 6$

10. $8 \times N = 104$ **11.** $N \div 4 = 14$ **12.** $N = 12 \times 8$

Find the value of *N*.

13. $35 = 5 \times N$ **14.** $12 \times N = 96$ **15.** $N \div 7 = 4$

16. $N = 352 \div 32$ **17.** $10 = N \div 5$ **18.** $19 \times N = 513$

19. $N = 735 \div 105$ **20.** $13 = 169 \div \text{N}$ **21.** $25 \times N = 625$

Placing the Decimal Point in a Quotient

Whole Number Divisors When the divisor is a whole number and the dividend has a decimal point in it, put the decimal point in the quotient above the decimal point in the dividend.

EXAMPLE $7\overline{)28.14}$ **SOLUTION** $\begin{array}{r} 4.02 \\ 7\overline{)28.14} \end{array}$ **CHECK** $4.02 \times 7 = 28.14$

EXPLANATION

When dividing 28.14 by 7, the decimal point in the quotient, 4.02, is placed directly above the decimal point in the dividend, 28.14.

Decimal Divisors When the divisor has a decimal, change the divisor to a whole number by multiplying both the divisor and the dividend by a multiple of 10 (10, 100, or 1,000, and so on). Then divide.

EXAMPLE Divide 54.6 by 3.25 **SOLUTION** $\begin{array}{r} 16.8 \\ 3.25\overline{)54.60.0} \end{array}$

EXPLANATION

The decimal points in both the dividend and divisor are moved to the right two places by multiplying both numbers by 100. This makes the divisor a whole number. The decimal point in the quotient is placed directly above the new position of the decimal point in the dividend.

CHECK $16.8 \times 3.25 = 54.6$

NOTE Multiplying both numbers by 100 does not change the value of the quotient because $\frac{100}{100} = 1$.

EXERCISE YOUR SKILLS

Solve and check.

22. $45.72 \div 6$ **23.** $5.88 \div 8$ **24.** $304.2 \div 9$

25. $7.24 \div 8$ **26.** $1{,}347.1 \div 5$ **27.** $8.533 \div 7$

28. $91.08 \div 22$ **29.** $18.025 \div 35$ **30.** $0.192 \div 24$

31. $1{,}640 \div 8.2$ **32.** $16.4 \div 0.082$ **33.** $0.82 \div 0.82$

Dividing to a Stated Number of Decimal Places

Problems involving the division of decimals might require that answers be found to a certain number of decimal places. In such cases, the division must be carried out one place beyond the specified decimal place. The number in that place is then used to round the quotient.

EXAMPLE Find the average wage per hour, to the nearest cent, for an employee who worked 36 hours and earned $312.42.

SOLUTION $36\overline{)\$312.42}$ = $8.678 **ANSWER** $8.68

EXPLANATION
The answer must be given correct to two decimal places. The quotient must be carried to three places. Since the last number in the quotient is an 8, the number to the left must be increased by 1. Rounded to two decimal places, the quotient is $8.68.

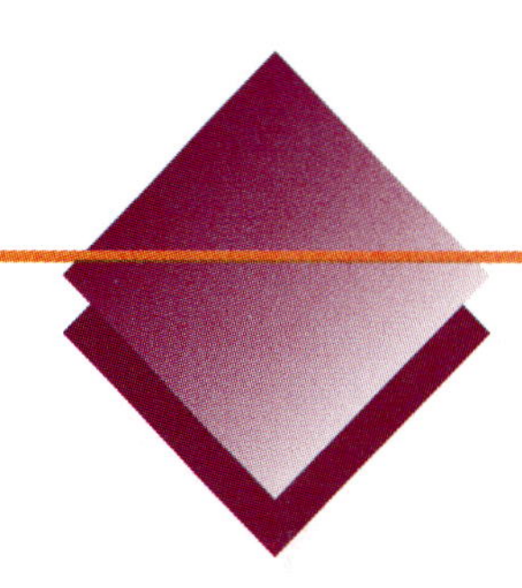

EXERCISE YOUR SKILLS

Find each quotient, rounded to the number of decimal places asked for.

To 3 decimal places

34. 34.1 ÷ 2.17 **35.** 5.34 ÷ 0.621 **36.** 45 ÷ 0.1312

To the nearest cent

37. $17.54 ÷ 504 **38.** $39.05 ÷ 304 **39.** $88.78 ÷ 235

Solve.

40. A store sold $851 worth of goods to 165 customers. To the nearest cent, how much did the average customer spend?

41. A basketball team scored 1,830 points in 23 games. Find the average points scored per game, to the nearest tenth.

ESTIMATING QUOTIENTS

In many business problems, you will divide by whole numbers such as 50, 100, or 2,000. For such problems, you can estimate answers and use shortcuts to help you get the correct answer and save time. Many of your answers will have to be rounded.

To avoid errors in division, estimate the quotient before you divide. Use your estimated quotient to check whether the exact quotient is a reasonable answer.

Compatible numbers are numbers that make calculations easier. In using compatible numbers with estimating quotients, round the divisor to a multiple of 10, 100, etc. Then round the dividend to a multiple of the divisor. Here are more examples of estimating the quotient using compatible numbers.

	Rounded Divisor	Rounded Dividend	Estimated Quotient	Exact Quotient
2,337 ÷ 61.5	60	2,400	40	38
7,644 ÷ 42	40	8,000	200	182
4,608 ÷ 36	40	4,800	120	128
4,257 ÷ 25.8	30	4,200	140	165

EXAMPLE Find the estimated and exact quotients of 4,590 ÷ 204.

SOLUTIONS Estimate: Round the divisor, 204, to 200.
Round the dividend, 4,590, to 4,600, the nearest easy multiple of 2.

4,600 ÷ 200 = 23 Estimated Quotient

$$204\overline{)4590} = 22.5$$ Exact Quotient

In the above example, 23 is close to 22.5, so 22.5 is a reasonable answer. If the exact quotient was 225 or 0.225, the answer would not be reasonable. If this happens, first check your estimate; then check your exact quotient.

Estimating with Divisors Less Than One When the divisor is a number less than 1, it must be changed to a whole number before being rounded. To do this, move the decimal point in the divisor to the right as many places as necessary to arrive at a whole number. Then move the decimal point in the dividend the same number of places. Now estimate the quotient as you did before. Study this example.

$$29 \div 0.025 = 0.025\overline{)29} = .025.\overline{)29.000.} \approx 30\overline{)30000} = 1000$$ Estimated Quotient

29 ÷ 0.025 = 1,160 Exact Quotient

EXERCISE YOUR SKILLS

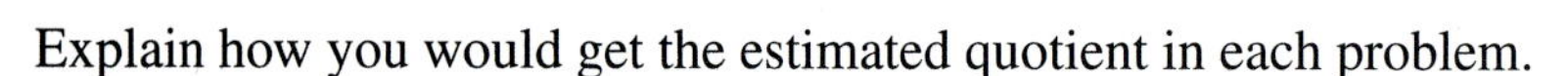

Explain how you would get the estimated quotient in each problem.

42. 1,623 ÷ 389 is about 4
43. 775 ÷ 205 is about 4
44. 1,020 ÷ 105 is about 10
45. 960 ÷ 279 is about 3
46. 537 ÷ 18 is about 30
47. 213.6 ÷ 9.8 is about 20
48. 482.6 ÷ 78.5 is about 6
49. $7,120 ÷ 55 is about $120
50. $7,762.40 ÷ 44 is about $200
51. $570.42 ÷ 11.8 is about $50

Show how to estimate each quotient. Then find the exact quotient.

52. $2,112 ÷ 32
53. $3,268 ÷ 43
54. $26.16 ÷ 2.18
55. 3.605 ÷ 10.3
56. 72 ÷ 0.18
57. 22 ÷ 0.04
58. 54 ÷ 0.75
59. 189 ÷ 0.35

DIVIDING BY 10, 100, OR 1,000

To divide by 10, 100, or 1,000, move the decimal point in the dividend to the left as many places as there are zeros in the divisor. Study the examples below.

5,700 ÷ 10 = 570
0.46 ÷ 100 = 0.0046
$16.40 ÷ 10 = $1.64
9,300 ÷ 1,000 = 9.3
$398 ÷ 100 = $3.98
$4,580 ÷ 1,000 = $4.58

Dividing by Multiples of 10, 100, or 1,000 Sometimes the divisor is a whole number with zeros at the end, but it is not 10, 100, or 1,000. In such cases, move the decimal point in the divisor to the left so you can rewrite the divisor without ending zeros. Then move the decimal point in the dividend as many places to the left as there were zeros in the divisor. Finally, divide in the usual way.

EXAMPLE What was the average number of gallons of paint used by a painting crew per day, to the nearest tenth of a gallon, if 231.5 gallons of paint were used in 20 days?

SOLUTION $\underset{\leftarrow}{231.5} \div \underset{\leftarrow}{20} = 23.15 \div 2 = 11.575$

EXPLANATION

The end zero in the divisor was eliminated by moving the decimal point one place to the left in the divisor and the dividend. This divides both the divisor and the dividend by 10. You would get the same answer if you did not divide by 10, but doing so makes the work easier.

EXERCISE YOUR SKILLS

Mental Math Write the quotient only.

60. $4,200 ÷ 10

61. 75,000 ÷ 100

62. $29,000 ÷ 1,000

63. 6.8 ÷ 10

64. $5.20 ÷ 10

65. $80 ÷ 100

66. 500 ÷ 1,000

67. 7 ÷ 100

68. 489.4 ÷ 100

69. $30 ÷ 1,000

70. 0.02 ÷ 10

71. 7.04 ÷ 100

Estimate each quotient. Then find the exact quotient, rounded to the given number of places.

Round quotients to the nearest hundredth.

72. 113 ÷ 40

73. 37.2 ÷ 420

74. 340 ÷ 3,500

Round quotients to 4 decimal places.

75. 125.6 ÷ 1,200

76. 2,100 ÷ 8,200

77. 880 ÷ 2,400

78. Arejay & Sons ordered 298,440 machine parts. The cost of the parts was $7,460. Find the cost of each part to the nearest tenth of a cent. Explain how you solved this problem.

79. ♦ Keep track of the amount you spend on entertainment and food for one week. Divide the total amount spent by seven days. The result is the average amount you spend on each in one day.

MIXED REVIEW

1. Multiply \$6.20 by 1,600
2. Multiply 360 by 58¢
3. Divide 834 by 0.1
4. Divide 92.5 by 0.01
5. Round to the nearest hundredth: 408.10284
6. Round to the nearest thousand: 4,492

Copy each problem, and place an F or P above each numeral or letter to show if it is a factor or a product. Then find the value of the unknown number.

7. $N \div 14 = 18$
8. $35 \times N = 245$

Estimate each quotient. Then find the quotient, correct to the nearest tenth.

9. $387 \div 42$
10. $1{,}815 \div 61$
11. $31{,}960 \div 77$
12. $38{,}522 \div 2200$

Solve.

13. Sal Furtado is paid \$13 an hour. Last week he worked these hours: Monday, 8 hours; Tuesday, 7 hours; Wednesday, 3 hours; Thursday, 5 hours; Friday, 8 hours. What were Sal's total earnings for the week?
14. Glenda Marcione earns \$1,620 a month, and Dominic Marcione earns \$370 a week. What is the total year's income for both?
15. On November 30, Nancy Wordlaw's bank statement showed a balance of \$670.42, a service charge of \$5.60, and interest earned of \$2.41. Her check register balance on that same date was \$205.81. When she compared her statement to her register, she found that a canceled check for \$96.82 had not been recorded in the check register and that a deposit of \$65.81 had been made too late to be recorded on the bank statement. There were two outstanding checks: #175, \$412; #182, \$218.43. Prepare a reconciliation statement.

2-4 FINDING AVERAGE PAY

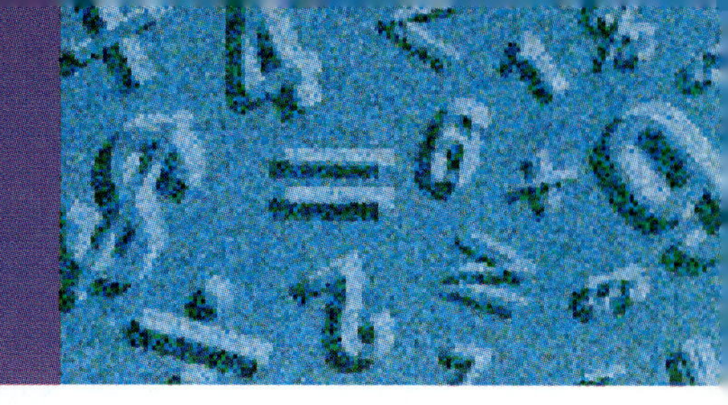

What does the word *average* mean to you? Think about the ways in which you have seen or heard the word *average* used, and write them on a sheet of paper.

◆ Compare your list with others in your class. Does the word have the same meaning everywhere it is used? As a class, determine five main categories under which most list items could fit. An example of a category is sports.

OBJECTIVES

In this lesson, you will learn to

- *find averages; and*
- *find an unknown item in a set of data.*

WARM *UP*

Multiply

a. $0.01 × 365

b. 10¢ × 34.6

c. 400 × 13

d. $387.9 × 100

Write an estimate. Then find the exact product.

e. 41 × 9.6

f. 18.8 × 47

g. $1.80 × 12

h. 372 × 8.2

AVERAGES

An **average** is a single number used to represent a group of numbers. The most commonly used average is the simple average.

Finding Simple Averages

A **simple average** is found by adding several numbers and dividing the sum by the number of items added. Another name for a simple average is the **mean.**

EXAMPLE 1 Maxine Hutchins earned these amounts for the 5 days she worked last week: Monday, $82; Tuesday, $91; Wednesday, $96; Thursday, $80; Friday, $86. What was her average pay for the 5 days?

SOLUTION

$82 + $91 + $96 + $80 + $86 = $435	Add to find total pay.
$435 ÷ 5 = $87	Divide the sum by the number of days to find the average pay per day.
Her average pay per day is $87.	Answer the question.

EXAMPLE 2 Dorothy Bruns earns $18,000 per year as assistant manager at a local store. Find her average pay per hour (to the nearest cent) if she works 37.5 hours for 50 weeks per year and gets 2 weeks paid vacation.

SOLUTION

37.5 × 52 = 1,950	Find total hours worked in 1 year.
18,000 ÷ 1,950 ≈ 9.2308	Divide the salary by the number of hours.
≈ 9.23	Round to the nearest cent.

Her average pay is $9.23 an hour.

EXERCISE YOUR SKILLS

1. Jack earned $210 for 35 hours of work. What was his average pay per hour?
2. Ellen hand carves wooden figures. Her wages depend on the number of figures she makes. Ellen's daily earnings for work she did from Monday through Saturday were $73, $90, $103, $108, $67, $81. What was her average daily pay for the days she worked?
3. Four employees are paid a monthly salary as follows: Bob, $1,300; Eleanor, $1,415; Howard, $1,177; Martha, $1,260. What average salary per month are these employees paid?

Two years ago you earned $17,200. Last year you earned $18,500. Suppose you earn $20,400 this year.

4. What total amount will you have earned for the three years?
5. What will be your average earnings per year?

A software trainer has been offered a job that pays $34,944 for working a 52-week year. What average amount does the job pay:

6. Per month?
7. Per week?

Farouk earned $435 by working five days a week at his full-time job. He earned $84 for 12 hours of work at his part-time job.

8. What average pay per day did he earn from full-time work?
9. What average amount per hour did he earn from part-time work?

Three weeks ago you worked 44 hours and earned $396. Two weeks ago you earned $342 by working 38 hours. Last week you worked 41 hours and were paid $369. For the three weeks

10. How many hours did you work?

11. What were your total earnings?

12. What were your average hourly earnings?

13. What were your average weekly earnings?

GROUPING DATA

A number or rate can occur more than once in a set of data. When that happens, you can find the sum quickly by grouping the common numbers. Look at the example shown below.

EXAMPLE Parker Brile works part time at a roadside stand selling fruits and vegetables. During the first month the stand was open, he earned $30. In each of the next three months he earned $150 per month. In the last month, Parker earned $60. What were his average earnings per month for those five months?

SOLUTION

1 month @ $30	=	$ 30
3 months @ $150	=	450
1 month @ $60	=	60
5 total months	=	$540 total earnings

$540 ÷ 5 = $108

His average earnings per month were $108.

To use grouped data with a calculator, press the [M+] key (instead of the Equal or Enter key) to complete the multiplication and store the product. Press [MR/C] to recall the total. Then divide by the number of items.

EXERCISE YOUR SKILLS

14. The Wilmark Company has 8 employees. Five of the employees earn $9 an hour, two earn $7 an hour, and 1 earns $13 an hour. What is the average amount per hour that these employees are paid?

15. The Hy-Gloss Paint Company gave its employees bonuses. Four employees received a bonus of $820 each; 5 employees received a bonus of $754 each; 6 employees were paid a bonus of $1,100 each. What was the average bonus paid to these employees?

16. At her job of repairing fences, LaShaundra worked 8 hours a day on Monday and Tuesday and earned $63 each day. On Wednesday, she earned $68. On Thursday and Friday, she earned $73 a day. What was LaShaundra's average daily pay for the 5 days she worked?

PROBLEM SOLVING

Using Estimation You have learned to round numbers and make estimates. When you solve problems, you should always estimate what you think the answer will be before solving the problem. Then compare your exact answer with your estimate to check whether your answer is reasonable.

Midori Saga was offered a new job. For the first 3 months she works, she will be paid a monthly salary of $1,600. Her monthly pay for the next 3 months will be $1,760. For the next 6 months after that, Midori's monthly salary will be $1,936.

17. How much will Midori earn during a full year?

18. What average pay per month will she receive if she works a full year?

Finding an Unknown Item

If one item in a group or set of data is unknown, you may have to find the unknown item. Averages are used often in finding the value of the unknown, or missing, item.

EXAMPLE The weekly pay of four glass blowers in a company averages $433 per employee. The weekly pay amounts of three of the four employees are $400, $410, and $460. What is the weekly pay of the fourth employee?

SOLUTION

4 × $433 = $1,732	total pay of 4 employees
$400 + $410 + $460 = $1,270	total pay of 3 employees
$1,732 − $1,270 = $462	weekly pay of fourth employee

EXPLANATION

Calculate the total amount using information on average amounts. Subtract the known amounts from the total amount to find the unknown amount.

EXERCISE YOUR SKILLS

19. Your daily pay for the first 4 days of the week was $68, $80, $75, and $79. How much do you have to earn on the fifth day to average $77 a day in earnings for the 5-day week?
20. Roy Bly was paid $422 for 5 days work. For 4 of those days, his average pay was $86. Find his pay for the fifth day.
21. The owner of Mid-Town Delivery has set a limit of $680 a day for employees' wages. The owner now has 7 employees who earn an average of $82 a day. What is the most a new employee can be paid without spending more money than planned?
22. Alma Thorpe plans to sell 200 computers this year. She earns $100 for every computer sold. For the first 8 months of this year, she sold an average of 14 computers a month and earned total gross pay of $11,200. How many computers must she sell, on average, in each of the remaining 4 months to reach her goal?
23. ♦ Think of how you have used "averages" in your life. Make up a word problem that requires finding an average.

PUT IT ALL TOGETHER

John must make a decision about keeping his current job, which he dislikes, or accepting an offer for a new job, which he thinks he would enjoy. His current job pays an hourly rate of $12. John works 40 hours a week.

At the new job, John would earn $0.75 for each item he produces up to 125 pieces per day. For each piece over 125 produced in a day, John would receive $0.80. The average production rate is 15 pieces an hour. Because of his experience and skill, John believes that he can produce 18 an hour. At the new job, John would work 8 hours a day, 5 days a week.

24. What is the average weekly pay received by employees at the new job?
25. How much does John expect to make each week at the new job?
26. What is John's weekly pay at his current job?
27. Create a chart, like the one below, that will allow you to compare both jobs by estimated daily, weekly, and annual earnings based on John's predictions about his productivity.

	Current Job	Production Job @ 18 pieces an hour
Daily Earnings		
Weekly Earnings		
Annual Earnings		

Comparison of Current and Production Job Earnings Potential

MIXED REVIEW

1. Multiply: 260 by 140
2. Multiply: 2,550 by \$0.01
3. Multiply: 8.3 by 1,000
4. Multiply: \$3.05 by 10,000
5. Round \$72.408 to the nearest cent.
6. Round 7.4291 to the nearest tenth.

Find the value of N.

7. $36 \times N = 36$
8. $N \times 2.4 = 4.8$
9. $28 \times N = 7 \times 12$
10. $0 \times 53 = N \times 34$
11. $10 \times 62 = N \times 6.2$
12. $14 \times 6 = 3 \times N$
13. Beverly's scores on seven tests were 77, 78, 88, 90, 85, 91, and 88. What was her average score on the tests, rounded to the nearest whole number?
14. The owner of a small business bought 5 boxes of computer disks for \$14 each, 8 boxes for \$22, and 11 boxes for \$36 each. There are 10 disks in each box. What average price did the owner pay for each disk, to the nearest cent?
15. In four days of work, Arnold produced these numbers of glass bowls: 52, 57, 55, 50. How many glass bowls must he make on the fifth day to average 53 glass bowls per day for 5 days?

TEMPLATE: Using a Spreadsheet for Payrolls

The Wilson Company pays its employees weekly. To complete a weekly payroll record for Department 16's employees, load ET2 from your applications disk. Enter the data shown in blue (in cells A7-G12) in the spreadsheet below. The spreadsheet will automatically calculate the totals in Columns H and I and Row 14.

	A	B	C	D	E	F	G	H	I
1	Wilson Company								
2	Weekly Payroll Record - Department 16								
3									
4								Total	
5		Pay						Hours	Gross
6	Employee	Rate	M	T	W	T	F	Worked	Pay
7	Bardo, N	6.75	8	8	8	8	8	40	270.00
8	Gruen, C	7.84	8	8	8	8	4	36	282.24
9	Ilsin, T	7.08	0	7	7	7	4	36	177.00
10	Trent, M	8.76	8	8	4	7	8	35	306.60
11	Wong, L	9.46	8	8	7	8	7	38	359.48
12	Woods, K	8.32	7	6	8	0	5	26	216.32
13									
14	Totals		39	45	42	38	36	200	1611.64

Answer these questions about your completed payroll record.

1. Which employee earned the highest gross pay for the week?
2. What was the total gross pay for the department?
3. The least total hours worked occurred on what day?
4. Which employee worked the greatest total number of hours during the week?

Answer these questions about the spreadsheet's design.

5. Write the formula used to calculate the Totals column on Monday (cell C14). What mathematical function is being used?
6. Write the formula used to calculate Cell H7 and I7. Was the same mathematical function used for each? Why or why not?

Design Your Own

The formulas in the template spreadsheet use both basic operations and the SUM function. It is important that you know how to write both types of formulas. Since you practiced writing formulas using SUM in the last chapter, for this lesson you will write formulas with simple operations.

7. Create a spreadsheet for the Put It All Together activity on page 71. Use simple math operations for the formulas.
8. Write a sentence explaining the format used for writing formulas that include a function.
9. Write a sentence explaining the format used for writing formulas that do not include a function.

CHAPTER 2 SUMMARY AND REVIEW

VOCABULARY REVIEW

average pay *2-4*
dividend *2-3*
division *2-3*
divisor *2-3*
employees *2-1*
employer *2-1*
estimate *2-3*
gross pay *2-1*
hourly rate *2-1*
mean *2-4*
payroll *2-1*
piece-rate *2-1*
quotient *2-3*
salaried employees *2-1*
salary *2-1*

MATH SKILLS REVIEW

2-1
- Multiply factors to get a product. Check products by reverse factors and multiply again.
- To locate decimal point in the product, use the total number of decimal places in both factors.
- To round numbers, use the Rule of 5 Rounding Method.
- Round money amounts to the nearest cent.

Round each number to the place indicated:

1. nearest cent, $1.475

2. nearest thousand, 108,581

- Estimate products by rounding both factors.
- To multiply by factors 10, 100, or 1,000, move the decimal point to the right in the other factor as many places as there are zeros.
- Move the decimal point in the other factor to the left *two* places when multiplying by 1¢ or $0.01 and *one* place when multiplying by 10¢ or $0.10, then add a dollar sign.

Use the rule-of-five method to write an estimate for each problem. Then find the exact product.

3. 81 × 48

4. 1.8 × 32.61

5. $0.72 × 39

2-3
- A quotient shows how many times a divisor is included in a dividend.
- Quotients and divisors are factors; a dividend is a product.
- When the divisor is a whole number and the dividend has a decimal point, put the decimal point in the quotient above the decimal point in the dividend.
- When the divisor has a decimal, change the divisor to a whole number by multiplying the divisor and dividend by a multiple of 10. Then divide.

- To divide by 10, 100, or 1,000, move the decimal point to the left in the dividend as many places as there are zeros in 10, 100, or 1,000.
- To divide by multiples of 10, 100, or 1,000, cross out the zeros in the divisor and move the decimal point in the dividend to the left as many places as there were zeros in the divisor. Then divide with the remaining numbers.

Find the value of the unknown number.

6. $84 = N \times 12$

7. $99.375 \div 75 = N$

8. $54.528 \div 12.8 = N$

9. $N \div 3.2 = 1.65$

10. Find the answer correct to 2 decimal places: $478 \div 13.5$

2-4
- Add several numbers and divide their sum by the number of items to find a simple average.
- To find an unknown item in a set of data, calculate the total from the average. Then subtract the known amounts from the total to find the unknown.

11. Find the average of 354, 222, 901, and 943.

APPLICATIONS

2-1
- Rounded numbers are used in estimating answers.

12. A company has 452 employees in Division A and 236 employees in Division B. Estimate the total number of employees in both divisions.

2-2
- Hourly Rate × Hours Worked = Gross Pay for Hourly Workers
- Salary Pay Rate × Time Period = Gross Pay for Salaried Workers
- Rate Per Piece × Number of Pieces Produced and Accepted = Gross Pay for Per Piece Workers

Find gross pay for each of the following employees.

13. John Green, 37 hours at $8 hour.

14. Alma Zeigler, 4 days at a daily rate of $96.

15. Phil Mckinley, 89 pieces produced; 5 pieces rejected. Per accepted piece rate of $1.15.

2-4
- Averages are used in many ways in business.

16. A study showed that 4,050 defects were reported by buyers of 1,870 customized vans. To the nearest tenth, what was the average number of defects per van?

17. Ida's part–time pay for five days was \$26.20, \$28.50, \$21.40, \$27.25, \$32.70. What was her average daily pay?

18. Last month, Tim bought seven dozen eggs as follows: 2 dozen @ 88¢, 1 dozen at 96¢, and 4 dozen at 93¢. What average price per dozen did Tim pay for last month's egg purchases?

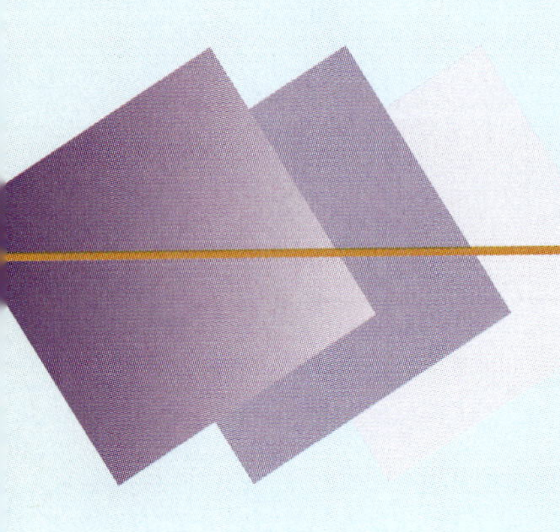

WORKPLACE KNOWHOW: **ACQUIRES AND EVALUATES INFORMATION**

Acquiring and Evaluating Information These skills include the ability to see when information is needed; to gather information from a variety of sources; and to evaluate the usefulness of the information.

Assignment Identify 3-5 jobs or industries that interest you. Look at job ads, contact companies, or do library research to identify job titles, the skills they require, and potential employers. Match your skills to the ads and other information you acquired. Group skills as (1) Have; (2) Need to Develop. Create an action plan (short or long term) showing steps you would take to develop skills for one of the jobs/industries that interest you.

Skills Assessment On a scale of 1-5, rate your understanding of the following skills used in this chapter. Write a paragraph explaining why you gave yourself this rating.

Multiplication
Estimation
Grouping Data
Calculating Gross and Average Pay
Division
Rounding
Writing Formulas for Spreadsheets

Work Skills Portfolio In addition to your assessment of your skills, you may wish to include the completed wage summary report from the Integrated Project. You may also include material from other classes or activities outside the classroom that reflect your skills and interests.

CHAPTER 2 TEST

Math Skills

1. Round $34.897 to the nearest cent.
2. Round 10.7201¢ to the nearest tenth of a cent.
3. Multiply 80 by 0.4.
4. Multiply 0.0963 by 1,000.
5. Divide 27.75 by 1.5.
6. Multiply 358 by $0.10.
7. Divide 6,450 by 1,000
8. Divide 392 by 2.5

Applications

9. Marva Hiller earns $8 an hour. What is her gross pay for a 40–hour week?
10. If she earned the same weekly pay for a year, what would be her gross pay for the year?
11. Viktor Miller earns a monthly salary of $2,056. What is his gross pay for one year?
12. Samantha works in a factory and is paid $2.40 for every usable chair she produces. Last week she produced these numbers of chairs in five days of work: 40, 37, 35, 41, 44. When the chairs were checked, 5 were found to be unusable. What was Samantha's gross pay for the 5 days of work?
13. Piece rate employees at a factory are paid in this way for the units they make in one work day: $0.70 each for the first 60 pieces, $0.80 each for the next 15 pieces, and $0.95 each for all pieces completed over 75 pieces. On Tuesday, Oscar completed 86 pieces. What were his gross earnings for that day?
14. Catherine's average sales for the last 9 weeks of last year were $1,138. For the first 4 weeks of this year, her sales were $1,228, $1,875, $1,290, and $1,329. What were Catherine's average sales for the 13 weeks?
15. A college student earned $246, $288, $234, and $276 over a 4-week period. How much must he earn during the fifth week to have average earnings of $255 for the five weeks?
16. Mattye Sparks worked 38 hours last week and had gross earnings of $296.40. What was her hourly rate of pay?
17. During the first 2 weeks in January, Jeri Melvin earned $250 each week clearing snow from parking lots. In the third week she earned $320, and in the fourth week, $400. Find the average amount earned per week for those four weeks.
18. A student allowed himself an average of $16 per day for meals, Monday through Friday. For the first four days of last week, he spent $18.80, $14.20, $15.75, and $17.15. What was the most he could spend on Friday and stay within his allowance?

CHAPTER 3

BIG CHANGES on the ASSEMBLY

Assembly-line workers are the backbone of factory-based industry. Until recently assembly-line workers only did the labor required to produce the line's product. Today line workers are being asked to participate in decisions about product development, equipment, and training.

Saren supervises an assembly line of 6 machines and 32 people. Every work day they transform 30,000 pounds of frozen meat logs into 6-ounce stacks of sliced, packaged, and boxed deli-meat.

Saren has many duties. He is responsible for making sure the assembly line runs smoothly and safely. Saren is responsible for the safety of his workers. He must make certain all workers understand their job and follow safety regulations. If a machine malfunctions, Saren decides if the power must be shut off to prevent injury. Saren also enforces health regulations, which require workers to wear hair nets, aprons, and plastic gloves.

Saren relies on his line workers to make decisions and handle many on-site problems. For example, when workers decided they wanted to be able to work on more than one machine, they helped develop a training plan and rotation schedule to achieve their goal.

Creativity and teamwork have been used to solve mechanical problems as well. Packaged meat was jamming up inside the packaging machines. As a result, 3,500 cartons were being wasted each day.

After analyzing the problem, Saren and his line workers determined that wider funnels were needed to drop the packaged meat into cartons. Saren helped design larger funnels. The new funnels have cut the number of cartons destroyed to fewer than 300 each day. The innovation will save the factory close to $50,000 a year in wasted cartons.

Math skills necessary for factory work varies. Saren uses basic math skills to determine how many packs of meat can be sliced from one log and how much packaging is needed each day.

Employee promotion opportunities are also different among companies. Saren recommends promotions based on intelligence, enthusiasm, communication skills, and efficiency. Other companies may promote employees who go to school to improve their work skills.

The competitive global market is changing assembly-line work. More and more companies are looking for and rewarding line workers who help improve product quality.

3-1 MULTIPLYING, SIMPLIFYING, AND DIVIDING FRACTIONS

OBJECTIVES

In this lesson, you will learn to

- *multiply fractions;*
- *simplify fractions; and*
- *divide fractions.*

Much of your work in this book has been with whole numbers. Some examples of **whole numbers** are 0, 1, 2, 3, 4, and so on. In some of the gross pay problems, though, you used parts of an hour such as $= \frac{1}{4}$, $\frac{1}{2}$, or $\frac{3}{4}$. Parts of whole numbers are called **fractional numbers**. A symbol for a fractional number, such as $\frac{1}{4}$, $\frac{1}{2}$, or $\frac{3}{4}$, is a **fractional numeral.** Fractional numerals are also called **fractions.**

◆ If you were to multiply $\frac{2}{3} \times \frac{1}{2}$, would your answer be greater than or less than $\frac{2}{3}$? Write your prediction in your notebook.

WARM UP

Multiply.		Divide.	
a. 4×5	**b.** 9×6	**i.** $27 \div 3$	**j.** $54 \div 6$
c. 8×2	**d.** 7×7	**k.** $64 \div 8$	**l.** $25 \div 5$
e. 3×1	**f.** 6×4	**m.** $81 \div 9$	**n.** $42 \div 7$
g. 8×7	**h.** 6×8	**o.** $24 \div 3$	**p.** $9 \div 9$

FRACTIONS

Fractions are written with a numeral above and a numeral below a line. The numeral above the line in a fraction is called the numerator. The numeral below the line is called the denominator. The denominator shows the number of equal parts into which a whole is divided. The numerator shows the number of parts with which you are working.

EXAMPLE In the fraction, $\frac{3}{4}$:

3	→	**Numerator**	→	**Three parts are being used.**
4	→	**Denominator**	→	**The whole is divided into 4 equal parts.**

$\frac{3}{4}$ **Three of four equal parts are shaded.**

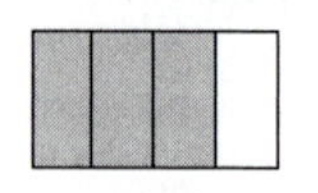

OR

Fractions may name either fractional numbers or whole numbers. For example, fractions such as $\frac{2}{3}$ and $\frac{5}{6}$ name fractional numbers. Fractions such as $\frac{3}{3}$ and $\frac{8}{2}$ name whole numbers. A fraction having the same numerator and denominator, such as $\frac{5}{5}$ or $\frac{7}{7}$, always names the number one.

A fraction can be read in different ways. For example, the fraction $\frac{3}{4}$ may be read as "three fourths," "three divided by four," or "three over four." To name a whole number as a fraction, use the number as the numerator and 1 as the denominator. For example, 3 named as a fraction is $\frac{3}{1}$.

EXERCISE YOUR SKILLS

Read each fraction in three ways. Then tell whether it represents a whole number or a fractional number.

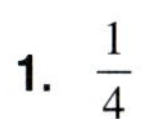

1. $\frac{1}{4}$ **2.** $\frac{1}{2}$ **3.** $\frac{2}{3}$ **4.** $\frac{4}{4}$ **5.** $\frac{3}{8}$

Write each as a fraction.

6. $12 \div 2$ **7.** $5 \div 8$ **8.** $4 \div 2$ **9.** $24 \div 4$

Write each with a division sign (÷).

10. $\frac{1}{7}$ **11.** $\frac{36}{72}$ **12.** $\frac{83}{75}$ **13.** $\frac{9}{3}$ **14.** $\frac{12}{12}$

MULTIPLYING FRACTIONS

To multiply two or more fractions, first multiply the numerators to find the numerator of the product. Then, multiply the denominators to find the denominator of the product.

EXAMPLE $\frac{2}{3} \times \frac{2}{5} = \frac{(2 \times 2)}{(3 \times 5)} = \frac{4}{15}$

A problem such as $\frac{2}{9}$ of 4 means the same as $\frac{2}{9} \times 4$. So, to find $\frac{2}{9}$ of 4, multiply the two fractions.

EXAMPLE $\frac{2}{9} \times 4 = \frac{2}{9} \times \frac{4}{1} = \frac{(2 \times 4)}{(9 \times 1)} = \frac{8}{9}$

Finding Equivalent Fractions

An **equivalent fraction** is a fraction that names the same number as another fraction. For example, $\frac{1}{3}$, $\frac{2}{6}$, and $\frac{4}{12}$ are equivalent fractions. When you add and subtract fractions, you may need to use equivalent fractions. To find an equivalent fraction, multiply or divide the numerator and denominator of a fraction by the same number.

EXAMPLE $\frac{1}{3} = \frac{1}{3} \times \frac{2}{2} = \frac{2}{6}$ $\quad \frac{8}{12} = \frac{8 \div 4}{12 \div 4} = \frac{2}{3}$

Critical Thinking What is another name for $\frac{2}{2}$? What about $\frac{3}{3}$? When you multiply a number by the number one, do you change the value of the original number? Explain your answer.

EXERCISE YOUR SKILLS

Find each product.

15. $\frac{1}{2} \times \frac{1}{3}$ **16.** $\frac{1}{3}$ of $\frac{1}{4}$ **17.** $\frac{1}{4} \times \frac{1}{5}$

18. $\frac{2}{3}$ of 3 **19.** $\frac{1}{2}$ of $\frac{5}{8}$ **20.** $\frac{1}{3} \times \frac{2}{1}$

21. $\frac{2}{5} \times \frac{4}{7}$ **22.** $\frac{3}{4} \times \frac{3}{8}$ **23.** $\frac{4}{5} \times \frac{2}{3}$

For each fraction, find the equivalent fraction you get when you multiply it by $\frac{2}{2}$, $\frac{3}{3}$, $\frac{4}{4}$, and $\frac{5}{5}$.

24. $\frac{1}{2}$ **25.** $\frac{3}{4}$ **26.** $\frac{5}{8}$ **27.** $\frac{2}{2}$

SIMPLIFYING FRACTIONS

You may find it easier to work with fractions if you simplify them. To simplify a fraction you reduce the terms of a fraction to their lowest form. The **terms** of a fraction are the numerator and the denominator. A fraction is in **simplest form** or **lowest terms** when 1 is the only number that will divide evenly into both terms.

For example, in the fraction $\frac{3}{6}$, the 3 and the 6 are the terms. To simplify $\frac{3}{6}$, divide the numerator and the denominator by the largest number that will divide both of them exactly (without a remainder). That number is called the ***greatest common factor*** or the ***greatest common divisor.*** For the fraction $\frac{3}{6}$, the greatest common divisor is 3 ($\frac{3}{6} \div \frac{3}{3} = \frac{1}{2}$). The fraction $\frac{3}{6}$ is in lowest terms when it is simplified to the equivalent fraction $\frac{1}{2}$.

Another way to simplify a fraction is to write the fraction showing all of the factors of both the numerator and denominator. Then divide the numerator and denominator by common factors. This is often referred to as ***cancelling*** or ***cancellation.***

EXAMPLE $\frac{12}{30} = \frac{2 \times 2 \times 3}{2 \times 3 \times 5}$ 2 and 3 are common factors.

$\frac{12}{30} = \frac{12 \div 2}{30 \div 2} = \frac{6}{15} \rightarrow \frac{6 \div 3}{15 \div 3} = \frac{2}{5}$ Divide the numerator and the denominator by those factors.

Critical Thinking In the example above, the numerator and denominator were first divided by 2 and then divided by 3. Could you get the same result by dividing the numerator and denominator by 6 to start with? Why?

Simplifying the Product of Fractions

You can simplify the product of two or more fractions using cancellation. Study these examples.

$$\frac{1}{9} \times 6 = \frac{1}{\cancel{9}_{3}} \times \frac{\cancel{6}^{2}}{1} = \frac{2}{3} \qquad \frac{5}{12} \times \frac{18}{35} = \frac{\cancel{5}^{1}}{\cancel{12}_{2}} \times \frac{\cancel{18}^{3}}{\cancel{35}_{7}} = \frac{3}{14}$$

Changing Improper Fractions to Mixed Numbers

A **proper fraction** has a numerator that is smaller than the denominator. For example, $\frac{3}{4}$ and $\frac{5}{9}$ are proper fractions. An **improper fraction** has a numerator that is equal to or greater than the denominator. For example, $\frac{2}{2}$ and $\frac{5}{3}$ are improper fractions.

When you multiply fractions the product may be an improper fraction. For example, $\frac{2}{1} \times \frac{3}{5} = \frac{6}{5}$. Improper fractions may be changed to mixed numbers. A **mixed number,** such as $1\frac{5}{6}$ or $3\frac{1}{8}$, consists of both a whole number and a fraction. To change an improper fraction to a mixed number divide the numerator by the denominator.

EXAMPLE $\frac{6}{5}$ means $6 \div 5$.

$$\begin{array}{r} 1 \\ 5\overline{)6} \\ \underline{5} \\ 1 \end{array} \leftarrow \text{remainder}$$

Write the remainder as a fraction $\frac{1}{5}$. The remainder is the numerator, the divisor is the denominator. The answer is $1\frac{1}{5}$.

Critical Thinking How would you change $2\frac{2}{3}$ to an improper fraction?

EXERCISE YOUR SKILLS

Simplify each fraction.

28. $\frac{6}{8}$ **29.** $\frac{5}{10}$ **30.** $\frac{8}{20}$ **31.** $\frac{15}{18}$ **32.** $\frac{18}{24}$

Change each improper fraction to a mixed number or a whole number.

33. $\frac{3}{2}$ **34.** $\frac{5}{3}$ **35.** $\frac{9}{4}$ **36.** $\frac{3}{3}$ **37.** $\frac{8}{8}$

Find each product in simplest form.

38. $\frac{1}{2} \times \frac{3}{8}$ **39.** $\frac{3}{5} \times \frac{5}{6}$ **40.** $\frac{3}{8}$ of $\frac{2}{3}$ **41.** $\frac{5}{8}$ of $\frac{8}{9}$

42. $\frac{1}{2} \times \frac{6}{7}$ **43.** $\frac{3}{5} \times \frac{5}{3}$ **44.** $\frac{1}{9}$ of 9 **45.** $\frac{3}{4} \times \frac{2}{5}$

DIVIDING FRACTIONS

To divide by a fraction, you must multiply the first factor by the **reciprocal** of the other factor. The reciprocal of any number is the number that gives a product of 1 when it is multiplied by the original number. For example, $\frac{3}{2}$ is the reciprocal of $\frac{2}{3}$, because $\frac{3}{2} \times \frac{2}{3} = 1$. In the same way, $\frac{1}{5}$ is the reciprocal of $\frac{5}{1}$, because $\frac{1}{5} \times \frac{5}{1} = 1$. So, the easy way to find the reciprocal of any number is to turn it upside down, or ***invert*** it.

EXAMPLE $\frac{7}{4} \div \frac{5}{6} = \frac{7}{4} \times \frac{6}{5}$ Rewrite the division as multiplying by the reciprocal.

$= \frac{7}{\not{4}_{2}} \times \frac{\not{6}^{3}}{5}$ Simplify.

$= \frac{21}{10}$ or $2\frac{1}{10}$ Multiply. Rewrite as a mixed number.

Multiplying or Dividing a Decimal and a Fraction

When working with decimals and fractions together, you can treat the decimal as a fraction with a denominator of 1. For example, \$4.24 can be written as $\frac{\$4.24}{1}$.

EXAMPLE Find the pay for $\frac{3}{4}$ hour of work at \$8 an hour and $\frac{2}{3}$ hour of work at \$9.60 an hour.

SOLUTION $\frac{3}{4} \times \frac{\$8}{1} = \frac{\$24}{4} = \$6$ $\qquad \frac{2}{3} \times \frac{\$9.60}{1} = \frac{\$19.20}{3} = \$6.40$

Critical Thinking When you multiply an amount by a fraction, is the answer larger or smaller than the amount you started with? What about when you divide by a fraction? Explain how this information can be used to check that an answer is reasonable.

EXERCISE YOUR SKILLS

Write the reciprocal of each number.

46. $\frac{3}{4}$ **47.** $\frac{4}{5}$ **48.** $\frac{7}{5}$ **49.** 14 **50.** $\frac{1}{9}$

Find each quotient in simplest terms.

51. $\frac{3}{5} \div 3$ **52.** $\frac{3}{8} \div \frac{5}{8}$ **53.** $\frac{1}{2} \div \frac{1}{4}$ **54.** $\frac{5}{6} \div 15$

55. $\frac{1}{3} \div \frac{1}{3}$ **56.** $\frac{8}{15} \div \frac{4}{5}$ **57.** $6 \div \frac{2}{3}$ **58.** $36 \div \frac{4}{5}$

59. $24 \div \frac{3}{8}$

60. $2 \div \frac{7}{9}$

61. $\frac{3}{8} \div \frac{5}{6}$

62. $\frac{2}{3} \div \frac{4}{5}$

63. $\frac{1}{2} \times \$36$

64. $\$7.20 \times \frac{1}{4}$

65. $\$7.20 \div \frac{1}{4}$

66. $\$6 \times \frac{2}{3}$

67. $\$9.50 \times \frac{4}{5}$

68. $\$6.90 \div \frac{1}{3}$

PUT IT ALL TOGETHER

Solve each problem, stating any fraction in lowest terms.

69. The school pep team has 8 members, including 6 girls and 2 boys. What fractional part of the pep team is boys?

70. While taking a test, Arnold Briggs guessed the answers to $\frac{1}{4}$ of the questions. He found out later that $\frac{2}{3}$ of his guesses were correct. For what part of the total test did Arnold correctly guess the right answers?

71. Hilda is paid \$8.80 an hour. What is her pay for $\frac{1}{4}$ hour?

72. Celeste usually earns \$360 a week. This week she earned only $\frac{3}{4}$ of her usual earnings. Find her pay this week?

73. The average weekly pay at the RJ Co. is \$480.72. Gordon's weekly pay is only $\frac{5}{6}$ that amount. Find his weekly pay.

74. ♦ Make a list of ways in which you use fractions in daily life.

MIXED REVIEW

Change each fraction to a mixed number.

1. $\frac{9}{2}$

2. $\frac{17}{4}$

3. $\frac{22}{5}$

Multiply. Write each answer in simplest form.

4. $\frac{1}{2} \times \frac{5}{6}$

5. $\frac{3}{4} \times \frac{2}{15}$

6. $\frac{1}{3} \times \frac{3}{7}$

Solve. Write each answer in simplest form.

7. Rolf Lada worked $38\frac{1}{4}$ hours regular hours and 8 hours overtime. Rolf's regular rate is \$12 per hour and overtime rate is \$18 per hour. Find Rolf's gross pay for the week.

Florence McHale earns \$1,860 a month.

8. What estimated amount does Florence earn in a year?

9. What are Florence's actual earnings in a year?

10. What is Florence's pay for a week, to the nearest cent?

3-2 ADDING AND SUBTRACTING FRACTIONS

OBJECTIVES

In this lesson, you will learn to

- *find common and least common denominators;*
- *add and subtract fractions with common denominators; and*
- *add and subtract fractions that do not have common denominators.*

You ordered two medium pizzas that had six slices each. After lunch you were told that $\frac{5}{6}$ of one pizza was eaten and $\frac{2}{3}$ of the second pizza was eaten.

◆ Draw a diagram to show how many pieces of pizza were left in each pie. How many pieces of pizza are left in all? On a sheet of paper, write an explanation of how you determined the number of pieces of pie to show in your diagram.

WARM UP

Change each improper fraction to a mixed number or a whole number.

a. $\frac{8}{3}$ b. $\frac{13}{4}$ c. $\frac{21}{3}$ d. $\frac{26}{4}$

Replace N with a numeral to make the sentences true.

e. $\frac{1}{3} = \frac{N}{9}$ f. $\frac{1}{2} = \frac{N}{6}$ g. $\frac{3}{4} = \frac{N}{12}$ h. $\frac{2}{3} = \frac{N}{6}$

COMMON DENOMINATORS

Fractions that have the same denominator are said to have a **common denominator.** The fractions $\frac{2}{5}$, $\frac{3}{5}$, and $\frac{4}{5}$ have common denominators. Fractions like $\frac{3}{4}$, $\frac{4}{5}$, and $\frac{3}{10}$ do not have common denominators.

Adding and Subtracting Like Fractions

Fractions with common denominators are called **like fractions.** In adding or subtracting like fractions, you add or subtract the numerators only and write the result over the common denominator.

EXAMPLES $\frac{3}{8} - \frac{1}{8} = \frac{2}{8}$ or $\frac{1}{4}$ $\frac{3}{4} + \frac{3}{4} = \frac{6}{4}$ or $1\frac{1}{2}$

EXERCISE YOUR SKILLS

Rewrite each fraction with the specified denominator.

1. $\frac{1}{2}, \frac{3}{4}$ to 8ths

2. $\frac{1}{2}, \frac{1}{5}, \frac{3}{5}$ to 10ths

3. $\frac{1}{4}, \frac{1}{3}, \frac{1}{6}$ to 12ths

4. $\frac{1}{2}, \frac{3}{4}, \frac{3}{8}$ to 16ths

Add or subtract. Write each answer in simplest form.

5. $\frac{1}{5} + \frac{2}{5}$

6. $\frac{3}{7} + \frac{3}{7}$

7. $\frac{3}{8} + \frac{4}{8}$

8. $\frac{3}{6} + \frac{2}{6}$

9. $\frac{3}{8} + \frac{3}{8}$

10. $\frac{1}{9} + \frac{2}{9}$

11. $\frac{3}{7} - \frac{2}{7}$

12. $\frac{9}{10} - \frac{6}{10}$

13. $\frac{5}{6} - \frac{1}{6}$

14. $\frac{9}{10} - \frac{4}{10}$

15. $\frac{7}{8} - \frac{5}{8}$

16. $\frac{7}{12} - \frac{4}{12}$

17. $\frac{5}{12} + \frac{3}{12}$

18. $\frac{9}{16} - \frac{1}{16}$

19. $\frac{3}{5} + \frac{4}{5}$

20. $\frac{11}{16} - \frac{5}{16}$

21. $\frac{6}{7} + \frac{6}{7}$

22. $\frac{3}{4} - \frac{1}{4}$

Solve.

23. Janet Meeks rode for $\frac{3}{4}$ hour Monday, $\frac{3}{4}$ hour Tuesday, and $\frac{1}{4}$ hour Wednesday. How long did she ride altogether during those three days?

24. David Yo lives $\frac{8}{10}$ mile from the library. He is on his way to the library. He has walked $\frac{2}{10}$ mile. How much farther does he have to walk?

25. Anna Jungst had $\frac{7}{9}$ yard of ribbon. She used $\frac{4}{9}$ yard. How much ribbon does she have left?

LEAST COMMON DENOMINATOR

Fractions that do not have common denominators, such as $\frac{1}{3}$ and $\frac{1}{6}$, are called **unlike fractions.** To add or subtract unlike fractions, you must change one or both so they have a common denominator. The work is simpler if you use the **least common denominator (LCD).** The least common denominator is the smallest number that is a multiple of both denominators.

EXAMPLE Find the least common denominator of $\frac{1}{12}$ and $\frac{1}{8}$.

SOLUTION Think:

$12 \times 1 = 12$	$8 \times 1 = 8$
$12 \times 2 = 24$	$8 \times 2 = 16$
$12 \times 3 = 36$	$8 \times 3 = 24$
$12 \times 4 = 48$	$8 \times 4 = 32$

24 is the least common denominator of $\frac{1}{12}$ and $\frac{1}{8}$.

Hints for finding LCD

- If the denominators have no common factors, the product of the denominators is the least common denominator. For example, $\frac{2}{5}$ and $\frac{3}{8}$ have a common denominator of 40 (5×8).
- If one denominator is a multiple of the other, the larger denominator is the least common denominator. For example, $\frac{1}{5}$ and $\frac{1}{10}$ have a common denominator of 10 because 10 is a multiple of 5.

Critical Thinking How can you find the least common denominator of 3 or more fractions? Write down your answer. Compare answers with others in your class. Then use your method to find the LCD of $\frac{1}{2}$, $\frac{2}{3}$ and $\frac{4}{5}$.

Adding and Subtracting Unlike Fractions

EXAMPLES

$\frac{1}{2} + \frac{2}{3} = \frac{?}{6} + \frac{?}{6}$ Find the least common denominator. Rewrite each fraction, using the least common denominator.

$= \frac{3}{6} + \frac{4}{6}$

$= \frac{7}{6}$ or $1\frac{1}{6}$ Add and simplify, if needed.

$\frac{7}{8} - \frac{1}{4} = \frac{?}{8} - \frac{?}{8}$ Notice that 8 is a multiple of 4, so the least common denominator is 8.

$= \frac{7}{8} - \frac{2}{8}$

$= \frac{5}{8}$ Subtract and simplify, if needed.

EXERCISE YOUR SKILLS

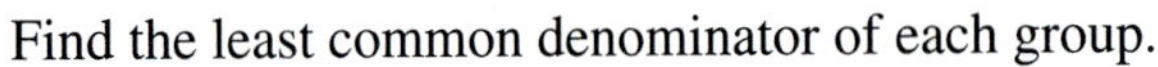

Find the least common denominator of each group.

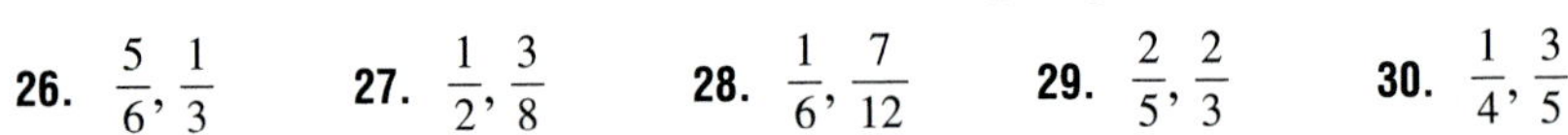

26. $\frac{5}{6}, \frac{1}{3}$ **27.** $\frac{1}{2}, \frac{3}{8}$ **28.** $\frac{1}{6}, \frac{7}{12}$ **29.** $\frac{2}{5}, \frac{2}{3}$ **30.** $\frac{1}{4}, \frac{3}{5}$

31. $\frac{7}{8}, \frac{3}{4}$ **32.** $\frac{1}{2}, \frac{3}{4}, \frac{7}{8}$ **33.** $\frac{2}{3}, \frac{1}{2}, \frac{3}{5}$ **34.** $\frac{5}{6}, \frac{7}{8}, \frac{1}{3}$

Find each sum or difference. Write your answers in simplest form.

35. $\frac{5}{6} + \frac{1}{3}$ **36.** $\frac{1}{2} - \frac{3}{8}$ **37.** $\frac{1}{6} + \frac{7}{12}$

38. $\frac{1}{3} + \frac{5}{8}$ **39.** $\frac{1}{4} + \frac{3}{8}$ **40.** $\frac{2}{3} + \frac{1}{2}$

41. $\frac{3}{4} - \frac{1}{8}$ **42.** $\frac{3}{8} - \frac{1}{16}$ **43.** $\frac{3}{4} - \frac{1}{16}$

44. $\frac{2}{3} - \frac{1}{6}$ **45.** $\frac{3}{4} - \frac{7}{12}$ **46.** $\frac{1}{2} - \frac{1}{4}$

47. $\frac{1}{2} + \frac{1}{3} + \frac{1}{4}$ **48.** $\frac{4}{5} + \frac{2}{3} + \frac{3}{4}$ **49.** $\frac{2}{3} + \frac{2}{3} + \frac{5}{6}$

50. Mr. Jennings jogged $\frac{1}{4}$ mile in the morning and $\frac{2}{5}$ mile in the afternoon. How far did he jog that day?

51. Jessica has two pieces of copper tubing. One piece is $\frac{3}{4}$ of a yard long, the other piece is $\frac{3}{8}$ of a yard long. What is the difference in length between the two pieces of tubing?

52. Jennifer, a nurse's aide, drives 20 miles to reach the hospital where she works. For $\frac{1}{10}$ of the trip Jennifer drove on a dirt road. She drove on a gravel road for $\frac{3}{8}$ of the trip. For what part of the trip did Jennifer drive on paved roads?

53. ♦ Look back at the diagram that you drew at the start of this lesson. How many pieces of pizza do you show remaining in each pie? Calculate the answer based on what you have reviewed in this lesson. Look at the explanation that you wrote for how you determined how many pieces of pizza you should show in your diagram. How does what you learned in this lesson and the previous lesson relate to the opening exercise?

MIXED REVIEW

Replace N to make each statement true.

1. $\$10 \times N = \$2.50$
2. $\$9 \div \frac{3}{7} = N$
3. $\frac{1}{3} + \frac{1}{6} = N$
4. $\frac{4}{5} \times \frac{15}{2} = N$
5. $\frac{7}{8} - \frac{1}{4} = N$
6. $\frac{2}{9} + \frac{7}{9} = N$

Round.

7. \$126,491 to the nearest hundred
8. 0.9987 to the nearest thousandth

Estimate.

9. \$37,745.34 − \$9,829.01
10. 357 ÷ 63.1
11. 5,800 × 22

Solve each problem.

12. The average weekly cash receipts of a bus company for the first 10 weeks of the year were \$45,600. The cash receipts for weeks eleven and twelve were \$48,230 and \$56,170. What were the average weekly cash receipts for all 12 weeks?
13. This year Sarah won a local election in which 10,800 votes were cast. In the previous election 12,500 people voted. If Sarah received $\frac{5}{8}$ of the total votes in this year's election, what number of votes did she get?
14. A store owner bought 240 cartons of eggs. The owner sold $\frac{5}{8}$ of the eggs and set aside 5 cartons which were not to be sold because of damage. How many cartons of eggs did the owner have left to sell?

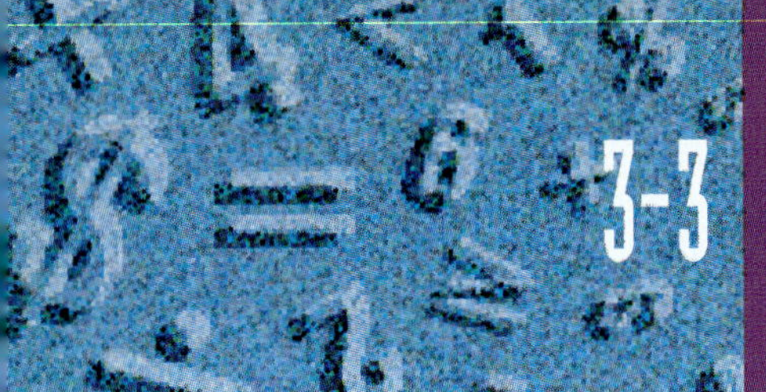

3-3 RATIO AND PROPORTION

OBJECTIVES

In this lesson, you will learn to

- *write ratios in simplest form;*
- *use cross products to solve proportions; and*
- *use proportions to solve problems.*

A recipe for punch requires two parts of soda to one part of fruit juice. If you have 10 two-liter bottles of soda, how many two-liter bottles of fruit juice do you need?

◆ Draw a picture to help solve this problem. Exchange your picture with a partner. Are the pictures helpful? Tell your partner why it is or is not helpful.

WARM *UP*

Find the value of N to make each statement true.

a. $\$8 \times N = \$1.50$ **b.** $\frac{2}{5} \times \frac{3}{8} = N$ **c.** $9 \div \frac{3}{5} = N$

d. $\frac{2}{3} - \frac{3}{8} = N$ **e.** $\frac{1}{2} + \frac{3}{5} = N$ **f.** $\frac{4}{9} + \frac{5}{9} = N$

RATIOS

A small food market has 12 loaves of bread, 4 of which are Brand A and 8 of which are Brand B. One way of comparing the numbers of each brand is to say that Brand A is to Brand B as 4 is to 8. That way of comparing two numbers is a **ratio.** The ratio can be written in 3 ways:

$\frac{4}{8}$ **4 ÷ 8** **4 : 8** **Read as "the ratio of 4 to 8."**

You can use the information to state different ratios:

- Ratio of Brand B to Brand A: 8 to 4
- Ratio of Brand B to total number of loaves: 8 to 12
- Ratio of total number of loaves to Brand A: 12 to 4

Ratios are fractions. Therefore, ratios can be simplified by dividing the terms by their greatest common divisor. For example, the ratio of Brand A to Brand B, $\frac{4}{8}$, can be simplified to $\frac{1}{2}$ by dividing each term by 4. The ratio of Brand B to Brand A, $\frac{8}{4}$, can be simplified to $\frac{2}{1}$ in the same way.

Using ratios, you may also say that for every two Brand B loaves there is one Brand A loaf. Or, for every Brand A loaf there are two Brand B loaves. The ratio of Brand A to Brand B is shown in Illustration 3-3.1.

Illustration 3-3.1. Ratio of Brand A to Brand B is 1:2

If you are comparing weights or measures in a ratio, you must show both terms in the same units. For example, to compare a measure in inches with a measure in feet, you must express both measures in either inches or feet. Example: 5 in. : 2 ft must be changed to 5 in. : 24 in.

Using Ratios

Suppose your gross pay for the year was earned from two jobs, a full-time job and a part-time job, in the ratio of 9 to 1. That ratio tells you that you earned $9 at your full-time job for every $1 you earned at your part-time job. In this ratio there are 10 equal parts, so other ratios can also be written, such as $\frac{9}{10}$ to $\frac{1}{10}$.

The ratio of 9 to 1 does not tell you how much pay you earned. But, for any amount of earnings, you can use the ratio to find how much pay you earned from full-time and part-time work.

For example, if you earned gross pay of $25,000 for the year, $\frac{9}{10}$ of $25,000, or $22,500, would be from full-time work and $\frac{1}{10}$, or $2,500, would be from part-time work. This use of ratios is shown in Illustration 3-3.2.

Illustration 3-3.2. Full and Part-Time Earnings as a Ratio

EXERCISE YOUR SKILLS

Of 24 cars, 16 are small cars and 8 are large cars. Without simplifying the terms, what is the ratio of

1. small cars to large cars?
2. large cars to small cars?
3. small cars to the total number of cars?
4. large cars to the total number of cars?

A baseball team played 60 games, winning 45 and losing 15. In lowest terms, what is the ratio of the

5. games won to games lost?
6. games lost to games played?
7. games played to games won?
8. games won to games played?

Find each ratio in simplest form. Remember to find common units of measure.

9. $250 to $1.25
10. $0.20 to $4
11. 3 yards to 2 feet
12. 4 feet to 12 inches
13. 2 pencils to 1 dozen pencils
14. 3 dimes to 2 nickels

The Quinn family has $12,600 to spend for food and housing. The ratio of their food expenses to their housing expenses is 5 to 7. How much of the $12,600 should they plan to spend for

15. food?
16. housing?
17. Wes and Jared are buying a rare coin for $9,500 and investing money in the ratio of 3 to 2. How much will each man invest?
18. An office buys and uses 1 box of small envelopes to every 5 boxes of large envelopes. Of 72 boxes of envelopes in stock, how many boxes of each type of envelope are there?
19. Phoebe Switzer earns income from regular pay and overtime pay in the ratio of 15 to 1. Last year she earned a total of $24,000. What amount did Phoebe earn from regular pay? overtime pay?
20. Alva Courville's gross earnings for a year were $26,000. Of that amount she saved $1,300. What was the ratio of Alva's savings to her earnings?

PROPORTIONS

Meaning of Proportion

The ratio $\frac{6}{9}$, in lowest terms, is $\frac{2}{3}$. The ratio $\frac{8}{12}$ when simplified, is also $\frac{2}{3}$. Therefore, $\frac{6}{9}$ and $\frac{8}{12}$ are equivalent fractions. You can show that equality as $\frac{6}{9} = \frac{8}{12}$. A statement such as that one, which shows that ratios are equal, is called a **proportion.** Read the proportion this way: 6 is to 9 as 8 is to 12.

Cross Products The product that results from multiplying the numerator of one fraction in a proportion with the denominator of the other fraction is called the **cross product.** *In all proportions, the cross products are equal.*

EXAMPLE

$$\frac{5}{6} = \frac{15}{18}$$

$6 \times 15 = 90$ $5 \times 18 = 90$

Cross products are equal.

Finding the Unknown Term in a Proportion

You can find an unknown term in a proportion by applying the fact that the cross products are equal.

EXAMPLE Chester Guzdal worked 8 hours and was paid \$64. At the same rate, how much would he earn in 24 hours?

SOLUTION

$$\frac{8}{24} = \frac{\$64}{N}$$

$8 \times N = 24 \times \$64$

$8 \times N = \$1{,}536$

$N = \$192$ He would earn \$192.

EXERCISE YOUR SKILLS

Solve each proportion.

21. $\frac{2}{3} = \frac{10}{N}$ **22.** $\frac{9}{10} = \frac{N}{50}$ **23.** $\frac{5}{20} = \frac{N}{8}$

24. A newspaper carrier received \$3.45 in tips from 12 customers. At the same rate, how much would she earn in tips from 32 customers?

25. In 6 weeks Faye Blum spent \$63 for gas for her car. At that rate, how much would she spend for gas in 52 weeks?

26. In 5 months Jose Arroyo earned \$9,205. At the same rate, how much would he earn in 12 months?

27. Eight packages of paper weigh a total of 38 pounds. What is the weight of 50 packages of the same kind of paper?

28. Adam Janson is planning a picnic and figures that 6 pounds of meat will serve 10 people. At the same rate, how many pounds of meat will he need for 250 people?

29. Ray Cates made a 235-mile car trip in 5 hours. How many miles will he travel in 7 hours driving at the same rate?

30. At the Delta Computer Center, a computer monitor regularly priced at $150 is on sale for $115. At that same rate of discount, what would be the sale price of a printer regularly priced at $450?

Silk cloth is on sale at a price of $19.80 for 5 yards. If 9 yards of the cloth are bought at the same rate, what is the

31. estimated cost of the cloth?

32. actual cost of the cloth?

Last year a business paid its employees a total of $600,000. Of that amount, $420,000 was paid to hourly rate employees and $180,000 to salaried employees. What was the ratio of

33. pay of hourly employees to total pay?

34. total pay to pay of salaried employees?

35. pay of salaried employees to pay of hourly employees?

36. ◆ Find a recipe for a cake in a cookbook. Write at least 3 ratios involving the ingredients in the cake. Compare your ratios with other students' results. Are there any general relationships that seem to be true of most cakes?

MIXED REVIEW

1. $1,400 × 400

2. 9,300 ÷ 100

3. $N \times \frac{3}{8} = \frac{3}{4}$

4. $\$295 \times \frac{3}{5}$

5. $\$540 \div \frac{3}{4}$

6. $\frac{18}{35} + \frac{12}{35}$

Estimate each answer. Then find the exact answer.

7. $1,299 × 42

8. $40,950 ÷ 78

9. $23,932 × 531

10. A salary of $312 per week is how much per year? Per month?

A computer store sells 8 boxes of $3\frac{1}{2}$ inch diskettes to every 3 boxes of $5\frac{1}{4}$ inch diskettes. Find the ratio of

11. sales of $5\frac{1}{4}$ inch diskettes to $3\frac{1}{2}$ inch diskettes.

12. sales of $3\frac{1}{2}$ inch diskettes to $5\frac{1}{4}$ inch diskettes.

13. sales of $3\frac{1}{2}$ inch diskettes to total diskette sales.

14. sales of $5\frac{1}{4}$ inch diskettes to total diskette sales.

3-4 REGULAR AND OVERTIME PAY

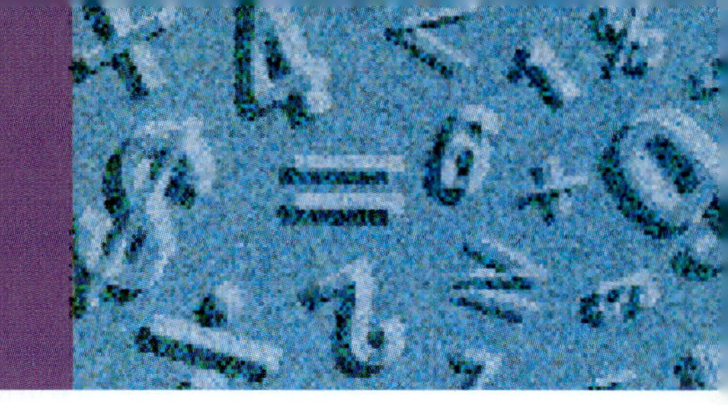

What happens if you arrive at school late or leave early without permission? What do you think happens to employees who come in late or leave early? Do you think rules that require employees to be at work on time and work the entire day are necessary? fair?

♦ Imagine you are the supervisor of a small group of workers. Write down 5 guidelines for your employees about arriving at work, leaving work, arriving late, being ill, etc. Be fair to your employees, to yourself, and to the company. As a class discuss everyone's guidelines. Make a list of the 5 rules the entire class agrees on.

OBJECTIVES

In this lesson, you will learn to calculate

- *time worked as shown on a time card;*
- *regular, overtime, and total wages for hourly employees;*
- *overtime pay rates;*
- *overtime pay; and*
- *gross pay.*

WARM UP

a. 5,200 × 230

b. 5,783 × 0.001

c. 536 × 100

d. 98.72 ÷ 0.01

e. $683 ÷ 100

f. 53 + 36 + 17 + 28

g. $56.81 − $37.56

h. Round 6,725 to the nearest hundred.

i. Round 82.037816 to the nearest thousandth.

WAGES

An employee is paid a certain amount of money for each hour worked. The employee's gross or total pay is called **wages,** and wages can be calculated according to an established pay schedule, such as daily or weekly.

Recording Time Worked

Many companies keep an exact record of the number of hours their employees work. They record the times their employees arrive, take breaks, and leave for the day. This may be done using bar-coded employee badges, hand-written time cards or **time cards** stamped in a **time clock.** The amount of time worked in a day or week is calculated from the time card. A sample time card is shown in Illustration 3-4.1.

Overtime

Some companies pay their employees for **overtime,** which is time worked beyond the regular working day or week. Daily overtime is based on a regular working day, such as an 8-hour day. So, an employee who works 10 hours in one day will be paid for 8 hours regular time and 2 hours overtime.

If the regular working week is 40 hours, then an employee who works 45 hours is paid for 40 regular hours and 5 overtime hours.

Finding Hours Worked from a Time Card

Paula Steele's time card for the week ending August 7 is shown in Illustration 3-4.1. Paula's regular working hours are from 8:00 A.M. to 12:00 noon and from 1:00 P.M. to 5:00 P.M. She is not paid for her lunch hour, which lasts from 12:00 noon to 1:00 P.M. For all time worked during her regular working hours, Paula is paid a **regular-time** rate of $8 an hour. For any time worked over those regular working hours, Paula is paid an overtime rate of $12 an hour.

No. ***63*** Pay Period Ending

Employee: ***Paula Steele*** ***August 7, 19–***

Days	In	Out	In	Out	In	Out	Regular Hours	Overtime Hours
1	M 8^{05}	M 12^{10}	M 12^{55}	M 5^{02}			$7\frac{3}{4}$	
2	T 7^{56}	T 11^{32}	T 1^{03}	T 4^{58}			$7\frac{1}{2}$	
3	W 8^{20}	W 12^{00}	W 1^{00}	W 4^{45}			$7\frac{1}{4}$	
4	Th 7^{59}	Th 11^{58}	Th 12^{59}	Th 5^{00}	Th 6^{00}	Th 8^{30}	*8*	$2\frac{1}{2}$
5	F 8^{01}	F 12^{02}	F 12^{55}	F 4^{30}			$7\frac{1}{2}$	
	In	Out	In	Out	In	Out	Total Regular	Total Overtime
	Morning		Afternoon		Overtime		*38*	$2\frac{1}{2}$

Arrived Late

Left Early

Overtime

Illustration 3-4.1. Time Card for Paula Steele

The times when Paula arrived (In) and left (Out) each morning and afternoon are stamped on the lower part of her time card. The total regular-time hours and overtime hours worked each day are recorded at the right of the times.

Many companies make rules for finding hours worked from a time card. **NOTE** *Use these rules to solve time-card problems in this text. Real employers may have other rules.*

- The smallest unit of time is $\frac{1}{4}$ hour (15 minutes).
- No credit is given for arriving before usual work hours.
- No credit is given for leaving after usual work hours unless it is for scheduled overtime work.
- A penalty in units of $\frac{1}{4}$ hour is applied for arriving more than 3 minutes late or leaving more than 3 minutes early.

These rules apply to Paula's time card in this way. On Monday, Paula's arrival at 8:05 A.M. was treated as 8:15 A.M. She left for lunch at 12:10 P.M., which was treated as 12:00 noon. Paula returned from lunch at 12:55 P.M., which was treated as 1:00 P.M. She left for the day at 5:02 P.M., which was treated as 5:00 P.M.

On Tuesday, 7:56 A.M. was treated as 8:00 A.M., 11:32 A.M. was treated as 11:30 A.M., 1:03 P.M. was treated as 1:00 P.M., and 4:58 P.M. was treated as 5:00 P.M. On Wednesday, 8:20 A.M. was treated as 8:30 A.M.

Critical Thinking Describe how Paula Steele's time is calculated for Thursday and Friday.

EXERCISE YOUR SKILLS

Arnold's regular working hours are from 8:00 A.M. to 12:00 noon and from 1:00 P.M. to 5:00 P.M. Calculate the number of hours Arnold would be paid in each of these problems. *Use the rules on page 96 to find the time worked.*

	In	Out		In	Out
1.	7:56	12:00	**2.**	8:00	12:06
3.	7:45	12:09	**4.**	8:07	12:00
5.	9:25	11:57	**6.**	12:56	4:10
7.	1:05	4:58	**8.**	1:02	5:02
9.	1:20	3:45	**10.**	12:45	5:02

Alma Pelgreen's and Osami Kinoshita's regular-time work hours are from 8:00 A.M. to 12:00 noon and from 1:00 P.M. to 5:00 P.M. Overtime is credited daily for time worked beyond 8 hours in a day. Their time cards for a week are shown below.

Employee: ***Alma Pelgreen***

	Morning		Afternoon		Evening	
	In	Out	In	Out	In	Out
M	8^{00}	12^{01}	12^{59}	5^{00}		
T	7^{58}	11^{55}	12^{52}	3^{50}		
W	8^{02}	12^{00}	1^{01}	4^{58}	5^{30}	9^{05}
Th	8^{27}	11^{58}	12^{54}	5^{02}		
F	7^{59}	11^{50}	1^{05}	4^{32}		

Employee: ***Osami Kinoshita***

	Morning		Afternoon		Evening	
	In	Out	In	Out	In	Out
M	7^{56}	12^{00}	1^{05}	4^{58}		
T	7^{45}	12^{09}	1^{20}	3^{45}		
W	9^{25}	11^{57}	12^{56}	4^{10}	5^{30}	9^{05}
Th	8^{00}	12^{06}	1^{02}	5^{02}		
F	8^{07}	12^{00}	12^{45}	5^{02}	6^{00}	9^{00}

11. How many regular-time hours did Alma work?

12. How many overtime hours did she work?

13. How many regular-time hours did Osami work?

14. How many overtime hours did he work?

15. If overtime were based on a 40-hour work week, would Osami have earned overtime pay?

FINDING REGULAR AND OVERTIME WAGES

For an employee who has worked both regular time and overtime, find gross pay using these three steps:

Step 1 Find the regular-time pay by multiplying the regular-time hourly rate by the number of regular-time hours worked.

Step 2 Find the overtime pay by multiplying the overtime hourly rate by the number of overtime hours worked.

Step 3 Add the regular-time pay and the overtime pay.

EXAMPLE Last week Brian worked 40 regular-time hours at \$6 an hour and 4 overtime hours at \$9 an hour. What was Brian's gross pay for the week?

SOLUTION

$40 \times \$6 = \$240$ regular-time pay

$4 \times \$9 = \$36$ overtime pay

$\$240 + \$36 = \$276$

His gross pay was \$276.

Critical Thinking Alma earns \$8 per hour for the first 8 hours worked in a day and \$12 per hour for overtime. Write a description of the relationship between her regular pay and her overtime pay. Then find her weekly wages for the week shown on the time sheet on page 97.

EXERCISE YOUR SKILLS

16. Last week Harold Perkins worked 38 regular-time hours and 3 overtime hours. He is paid \$10 an hour for regular-time work and \$15 an hour for overtime work. What was Harold's gross pay for the week?

17. Marta Gruen's average pay is \$277.50 a week. Last week Marta worked 40 hours at the regular-time rate of \$6.80 an hour and $6\frac{1}{2}$ hours at the overtime rate of \$10.20 an hour. What was Marta's gross pay for the week?

Ralph Humes is paid overtime for all time worked past 40 hours in a week. His regular-time pay rate is \$7 an hour, and his overtime pay rate is \$10.50 an hour. Last week Ralph worked 47 hours.

18. How many regular-time hours did Ralph work last week?

19. How many overtime hours did Ralph work last week?

20. What was Ralph's regular-time pay last week?

21. What was his overtime pay last week?

22. What was his gross or total pay last week?

Art Lantzy and Kurt Voss work for Lansco Electric. Their time cards for 1 week are shown below. Employees of Lansco Electric have regular working hours of 8:00 A.M. to 12:00 noon and 12:30 P.M. to 4:30 P.M. Art and Kurt are both paid \$16 an hour for regular-time work. An overtime rate of \$24 an hour is paid for time worked beyond 8 hours in a day.

Employee: ***Art Lantzy***

	Morning		Afternoon		Evening	
	In	Out	In	Out	In	Out
M	7^{58}	12^{04}	12^{33}	4^{29}		
T	8^{02}	11^{50}	12^{29}	4^{31}		
W	8^{57}	11^{32}	12^{30}	4^{15}		
Th	8^{30}	12^{00}	1^{00}	4^{32}		
F	8^{00}	11^{59}	12^{30}	4^{30}	5^{00}	7^{01}

Employee: ***Kurt Voss***

	Morning		Afternoon		Evening	
	In	Out	In	Out	In	Out
M	8^{25}	12^{01}	12^{31}	4^{30}		
T	7^{59}	11^{58}	12^{30}	4^{31}	6^{00}	10^{00}
W	8^{01}	12^{00}	1^{12}	4^{32}		
Th	9^{00}	11^{59}	12^{30}	3^{45}		
F	8^{03}	12^{02}	12^{29}	2^{30}		

Use the time card rules to answer these questions.

23. How many regular and overtime hours did Art work?

24. What was Art's gross pay for the week?

25. How many regular and overtime hours did Kurt work?

26. What was Kurt's gross pay for the week?

The time cards for Ruth Hirsch and Tod Groot are shown below. Their working hours are from 8:00 A.M. to 12:00 P.M. and 1:00 P.M. to 5:00 P.M. Ruth earns \$6.80 an hour regular pay and \$10.20 overtime pay. Tod's regular-time rate is \$6.60 an hour, and his overtime rate is \$9.90 an hour. Ruth and Tod are paid on an 8-hour-day basis, with daily overtime based on time beyond 8 hours.

Employee: ***Ruth Hirsch***

	Morning		Afternoon		Evening	
	In	Out	In	Out	In	Out
M	7^{58}	12^{04}	12^{55}	5^{02}		
T	8^{02}	11^{50}	12^{59}	5^{00}		
W	8^{00}	11^{32}	1^{08}	4^{58}		
Th	8^{01}	12^{00}	1^{01}	5^{00}		
F	7^{57}	11^{59}	12^{57}	4^{59}	5^{15}	7^{17}

Employee: ***Tod Groot***

	Morning		Afternoon		Evening	
	In	Out	In	Out	In	Out
M	8^{10}	12^{01}	12^{59}	4^{20}		
T	7^{59}	11^{58}	1^{00}	5^{04}	6^{00}	9^{05}
W	8^{01}	12^{00}	12^{57}	5^{00}	5^{45}	7^{50}
Th	9^{00}	11^{59}	1^{00}	5^{09}		
F	8^{03}	12^{02}	1^{01}	4^{08}		

27. What is Ruth's gross pay for the week?

28. What is Tod's gross pay for the week?

OVERTIME PAY RATES

Overtime pay is an extra amount of money paid for working more than the usual work hours in a day or a week. Overtime pay is often figured at one and a half times (1.5) the regular-time rate and is called **time-and-a-half pay.** Sometimes **double-time pay** is given for work over a certain number of hours, or for work on weekends and holidays. Double-time pay is twice the regular-time pay rate.

Calculating Overtime Pay

To find time-and-a-half pay, first multiply the regular pay rate by 1.5 to get the time-and-a-half rate. Then multiply the time-and-a-half rate by the number of time-and-a-half hours.

1.5 × Regular Pay Rate = Time-and-a-Half Rate (Do NOT round.)

Time-and-a-Half Hours × Time-and-a-Half Rate
= Time-and-a-Half Pay

Critical Thinking When finding the time-and-a-half overtime rate, why do you multiply by 1.5?

To find double-time pay, first multiply the regular pay rate by 2 to get the double-time rate. Then multiply the double-time rate by the number of double-time hours.

2 × Regular Pay Rate = Double-Time Rate

Double-Time Hours × Double-Time Rate = Double-Time Pay

EXAMPLE Last week Mary Vaughn worked 6 hours at time-and-a-half pay and 2 hours at double-time pay. Her regular pay rate was $9.60 per hour. What was her total overtime pay for the week?

SOLUTION Use a calculator.

$1.5 \times \$9.60 = \$14.40$ time-and-a-half rate
$2 \times \$9.60 = \$19.20$ double-time rate
$6 \times \$14.40 = \$86.40$ time-and-a-half pay
$2 \times \$19.20 = \$38.40$ double-time pay
$\$86.40 + \$38.40 = \$124.80$ total overtime pay

EXERCISE YOUR SKILLS

29. Felix Garcia worked 6 hours at time-and-a-half pay last week. His regular pay rate was $11 an hour. What was Felix's total overtime pay for the week?

The Grafix Shop pays double time for work done on weekends. An employee who earns a regular pay rate of $6.95 an hour works 8 hours on Saturday. What is the employee's

30. estimated pay for work on Saturday?

31. actual pay for Saturday work?

32. Joan McFarland worked 4 hours at time-and-a-half pay and 3 hours at double-time pay. Her regular pay rate was \$7.72 an hour. What was Joan's total overtime pay for the week?

You are paid \$9.95 an hour with time-and-a-half pay for all hours you work over 40 hours a week. Last week you worked 47 hours.

33. How many overtime hours did you work?

34. What was your overtime rate? Remember, do NOT round overtime pay rate.

35. Find your overtime pay. Round your answer to the nearest cent.

36. In 1 week Mei-yu Li worked 38 hours at her regular-time rate, 3 hours at time and a half, and 4 hours at double time. Her regular hourly pay rate was \$12. What was Mei-yu's gross pay that week?

Sheldon Berger worked 40 hours last week at his regular pay rate of \$13 an hour. He also worked 5 hours at time-and-a-half pay.

37. What was Sheldon's regular pay for the week?

38. What was his overtime pay for the week?

GROSS PAY

To find gross pay, including overtime hours, follow these steps:

Step 1 Find the number of hours worked at the regular and overtime rates.
Step 2 Find the overtime pay rate or rates.
Step 3 Multiply and combine regular pay and overtime pay.

When calculating gross pay, you can use the memory on your calculator.

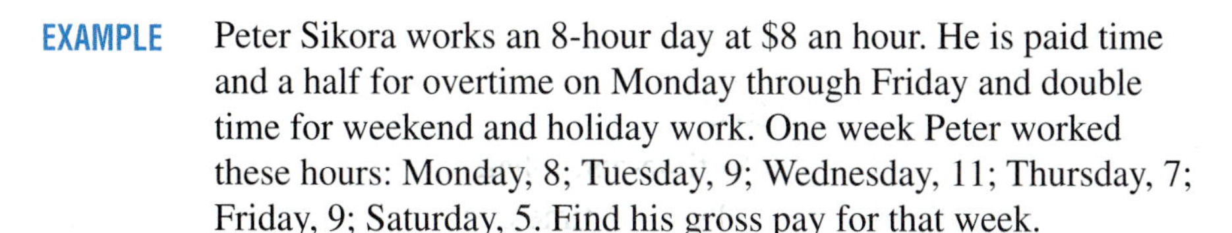

EXAMPLE Peter Sikora works an 8-hour day at \$8 an hour. He is paid time and a half for overtime on Monday through Friday and double time for weekend and holiday work. One week Peter worked these hours: Monday, 8; Tuesday, 9; Wednesday, 11; Thursday, 7; Friday, 9; Saturday, 5. Find his gross pay for that week.

SOLUTION

	Regular Time	Time and a Half	Double Time
Monday	8	0	0
Tuesday	8	1	0
Wednesday	8	3	0
Thursday	7	0	0
Friday	8	1	0
Saturday	0	0	5
Totals	39	5	5
	× 8	× 12	× 16
	\$312 +	\$60 +	\$80

EXPLANATION

$1.5 \times \$8 = \$12$ time-and-a-half rate
$2 \times \$8 = \$16$ double-time rate

$\$312 + \$60 + \$80 =$ **\$452**

His gross pay was \$452.

EXERCISE YOUR SKILLS

Your job pays $5 an hour with time and a half for overtime hours worked beyond 8 hours a day. This week you worked these hours: Monday, 6; Tuesday, 8; Wednesday, 10; Thursday, 9; Friday, 11.

39. How many regular hours did you work?

40. How many overtime hours did you work?

41. What is your overtime pay for the week?

42. What is your gross pay for the week?

43. During the week of November 8, Roseann Walker worked 9 hours a day, Monday through Friday. Her regular pay rate was $14 an hour, with time and a half for all work beyond 40 hours a week. What were Roseann's earnings for that week?

Elnora Kar's regular pay rate is $9 an hour. She is paid time and a half for overtime work beyond 40 hours a week. Last week Elnora worked 8 hours a day on Monday, Wednesday, Thursday, and Friday. She worked 10 hours on Tuesday.

44. How many regular hours did Elnora work?

45. How many overtime hours did she work?

46. What was Elnora's regular-time pay for the week?

47. What was her overtime pay for the week?

48. What was Elnora's gross pay for the week?

Jeff Irfan is paid on an 8-hour-day basis, with time and a half for all overtime. Jeff worked these hours in one week: Monday, 8; Tuesday, 6; Wednesday, 11; Thursday, 10; Friday, 8. Jeff is paid $10.30 an hour.

49. What was Jeff's regular-time pay for the week?

50. What was his overtime pay for the week?

51. What was Jeff's gross or total pay for the week?

52. Tim Bjorum worked these hours during one week: Monday, 9; Tuesday, 4; Wednesday, 8; Thursday, 10; Friday, 10. He was paid $10.67 an hour on the basis of an 8-hour working day, with time and a half for all overtime. What was Tim's gross pay for those five days?

53. The G.R. Price Tool Company pays Marsha Carmichael a regular hourly rate of $11.70 an hour, with time and a half for all hours worked over $37\frac{1}{2}$ hours per week. What is Marsha's gross pay for a week in which she worked $7\frac{1}{2}$ hours a day Monday through Thursday, and 10 hours on Friday?

54. Jason Swada is paid $8.15 an hour for an 8-hour work day, with time and a half for overtime Monday through Friday. Work on weekends and holidays is paid as double time. In one recent week, Jason worked these hours: Monday, 9; Tuesday, 10; Wednesday, 8; Thursday, 8; Friday, 9; Saturday, 7. What was Jason's gross pay for that week?

55. ◆ Write a paragraph comparing the benefits and disadvantages of working overtime.

MIXED REVIEW

1. $\$16 \times \frac{3}{4}$
2. $38 \div \frac{1}{4}$
3. $\frac{12}{35} \times \frac{7}{2}$
4. $\frac{3}{8} + \frac{11}{8}$
5. $\frac{2}{5} + \frac{2}{3}$
6. $\frac{3}{4} - \frac{1}{8}$

Find the value of N.

7. $N \times 13 = 82 - 43$
8. $0 \div 3 = N$
9. $64 \div N = 1\frac{1}{3} + 6\frac{2}{3}$
10. $16 \times 1 = 1 \times N$

Find the estimated and exact answers.

11. $1{,}023 \times 96$
12. $1{,}248 \div 78$
13. $32.1 \div 0.75$
14. 0.92×7.1

Solve.

15. Multiply each by 0.1: $95; $28.90; $576
16. Multiply each by 0.01: $24; $89.89; $540
17. Divide each by 0.01: 52; $1.70; $3
18. Divide each by 10: $18; $36.72; 7,000
19. Stella Lorida worked these hours last week: Monday, 10; Tuesday, 8; Wednesday, 9; Thursday, 6; Friday, 8; Saturday, 5. Stella is paid $11.40 per hour for regular hours, time and a half for overtime during the week, and double time for weekend hours. If Stella works on an 8-hour-day basis, what was her pay for the week?

An auto dealer sells two types of vehicles: cars and vans. The dealer now has 80 cars and 20 vans in stock. Find the ratio of

20. number of cars to total vehicles.
21. number of vans to total vehicles.
22. number of vans to cars.
23. number of cars to vans.
24. In one day a record store sold 20 albums at $16; 14 albums at $12; and 18 albums at $9. The store also sold 26 compact disks at $16 each. What was the average price of the albums sold that day?

3-5 MIXED NUMBERS

OBJECTIVES

In this lesson, you will learn to

- *change mixed numbers to improper fractions;*
- *multiply and divide mixed numbers; and*
- *add and subtract mixed numbers.*

To get to work, Kate Anders walks $\frac{1}{2}$ mile to the train depot. Her train ride is $24\frac{1}{4}$ miles long. She then takes a bus $3\frac{7}{10}$ miles farther. She walks the last $\frac{1}{10}$ mile to work. The entire trip takes $1\frac{1}{4}$ hours. She works $9\frac{1}{3}$ hours. To get home, she reverses the steps. The only difference is her husband picks her up at the train station and they drive home together. How many miles does Kate commute each day? Do NOT solve this problem yet.

◆ On a piece of paper draw a chart with three columns. In the first column list any unnecessary information; in the second column list the information needed to solve the problem, and in the third column, identify any information that is needed to solve the problem but not provided.

WARM *UP*

Write an estimate. Then find the exact product.

a. $2{,}009 \times 303$ **b.** $32.1 \div 0.75$

Find the value of *N*.

c. $\$12 \times \frac{3}{4} = N$ **d.** $N \times 18 = 82 - 28$ **e.** $52 \div \frac{1}{8} = N$

MIXED NUMBERS

In the previous lesson, you found an overtime rate by multiplying the regular pay rate by a mixed number, 1.5, which can also be written as the mixed number $1\frac{1}{2}$. You will use mixed numbers to solve business and consumer problems in other chapters. These problems include:

- Finding the amount of interest at a rate of $5\frac{1}{4}\%$.
- Finding the city income tax due when the rate is $1\frac{1}{2}\%$.
- Writing a stock price of $86\frac{5}{8}$ in dollar terms.
- Finding the selling price when the markup is $33\frac{1}{3}\%$.

Changing Mixed Numbers to Improper Fractions

In previous lessons, you changed improper fractions to mixed numbers. For example, $\frac{11}{3} = 3\frac{2}{3}$. As you solve problems involving fractions, you may also need to know how to change a mixed number to an improper fraction.

EXAMPLE Name $3\frac{1}{4}$ as an improper fraction.

SOLUTION $3\frac{1}{4} = \frac{(4 \times 3) + 1}{4}$

$3\frac{1}{4} = \frac{13}{4}$

$$3\frac{1}{4} \;(\times,\ +)\; = \frac{13}{4}$$

Use the pattern shown at the right as a memory aid.

Multiplying with Mixed Numbers

When solving problems you may have to multiply a whole number by a mixed number or multiply mixed numbers. Study the following examples to learn two ways to handle mixed numbers in multiplication.

EXAMPLE Multiply 8 by $6\frac{5}{7}$.

SOLUTION 1

Step 1 $6\frac{5}{7} = \frac{47}{7}$ Change the mixed number to an improper number.

Step 2 $8 \times \frac{47}{7} = \frac{(8 \times 47)}{7}$ Multiply.

Step 3 $= \frac{376}{7} = 53\frac{5}{7}$ Simplify the product.

SOLUTION 2

Step 1 $8 \times \frac{5}{7} = \frac{40}{7} = 5\frac{5}{7}$ Multiply the whole number by each part of the mixed number.

Step 2 $8 \times 6 = 48$

Step 3 $5\frac{5}{7} + 48 = 53\frac{5}{7}$ Add the results.

To multiply two mixed numbers, change each to an improper fraction.

EXAMPLE Multiply $4\frac{2}{3}$ by $3\frac{3}{4}$.

SOLUTION

Step 1 $4\frac{2}{3} \times 3\frac{3}{4} = \frac{14}{3} \times \frac{15}{4}$

Step 2 $\frac{\overset{7}{\cancel{14}}}{\underset{1}{\cancel{3}}} \times \frac{\overset{5}{\cancel{15}}}{\underset{2}{\cancel{4}}} = \frac{35}{2} = 17\frac{1}{2}$

EXERCISE YOUR SKILLS

Mental Math Change each mixed number to an improper fraction.

1. $2\frac{1}{5}$
2. $4\frac{1}{8}$
3. $5\frac{2}{3}$
4. $12\frac{1}{2}$

Find each product in simplest form.

5. $16 \times 3\frac{1}{2}$
6. $3\frac{1}{3} \times 1\frac{2}{5}$
7. $45 \times 6\frac{3}{5}$
8. $24 \times 4\frac{3}{4}$
9. $6\frac{1}{4} \times 3\frac{3}{10}$
10. $1\frac{7}{8} \times 3\frac{1}{6}$
11. $18\frac{3}{8} \times 24$
12. $27\frac{2}{5} \times 35$
13. $17\frac{5}{6} \times 72$
14. $16 \times 3\frac{5}{8}$
15. $3\frac{3}{4} \times 4\frac{2}{5}$
16. $27 \times 4\frac{2}{3}$
17. $60 \times 6\frac{4}{5}$
18. $6\frac{3}{10} \times 4\frac{2}{7}$
19. $9\frac{1}{3} \times 6\frac{3}{4}$

20. When Phoebe Milton started working 25 years ago, her pay rate was \$1.60 an hour. Her hourly pay has increased and is now $4\frac{1}{4}$ times what it used to be. What is Phoebe's current hourly pay rate?
21. Tom's Lawn Care Service uses an average of $3\frac{1}{2}$ gallons of gasoline each working day. How many gallons of gasoline will Tom use in a month that has 26 working days?

Dividing with Mixed Numbers

When dividing mixed numbers, first change each mixed number to an improper fraction. Then multiply by the reciprocal of the divisor.

EXAMPLE Divide $37\frac{1}{2}$ by $1\frac{1}{2}$.

SOLUTION

$37\frac{1}{2} \div 1\frac{1}{2} = \frac{75}{2} \div \frac{3}{2}$ Write as improper fractions.

$= \frac{75}{2} \times \frac{2}{3}$ Multiply by the reciprocal.

$= \frac{\overset{25}{\cancel{75}}}{\underset{1}{\cancel{2}}} \times \frac{\overset{1}{\cancel{2}}}{\underset{1}{\cancel{3}}}$ Cancel.

$= 25$ Multiply and simplify if necessary.

CHECK Multiply the answer by the divisor.

$$1\frac{1}{2} \times 25 = \frac{3}{2} \times 25 = \frac{75}{2} \text{ or } 37\frac{1}{2}$$

Critical Thinking Explain why $8\frac{1}{3} \div 5 \neq 5 \div 8\frac{1}{3}$. Justify your answer by finding each quotient.

EXERCISE YOUR SKILLS

Find each quotient in simplest form. Check each answer.

22. $3\frac{1}{9} \div 4$ **23.** $4\frac{4}{5} \div 6$ **24.** $5\frac{5}{6} \div 7$

25. $8 \div 1\frac{1}{4}$ **26.** $6 \div 4\frac{4}{5}$ **27.** $28 \div 1\frac{1}{6}$

28. $2\frac{3}{4} \div 3\frac{1}{2}$ **29.** $4\frac{2}{5} \div 2\frac{3}{4}$ **30.** $16\frac{1}{2} \div 3\frac{2}{3}$

31. $87\frac{1}{2} \div 3\frac{1}{2}$ **32.** $42\frac{1}{2} \div 8$ **33.** $16\frac{1}{5} \div 3\frac{1}{3}$

Adding and Subtracting Mixed Numbers

To add mixed numbers, change the fractions to equivalent fractions.

EXAMPLE Rhonda Clayton ran $4\frac{3}{4}$ miles on Monday and $3\frac{3}{8}$ miles on Wednesday. What total distance did she run in those two days?

SOLUTION

$$\begin{array}{r} 4\frac{3}{4} = 4\frac{6}{8} \\ +\,3\frac{3}{8} = +\,3\frac{3}{8} \\ \hline 7\frac{9}{8} \\ = 8\frac{1}{8} \end{array}$$

Think $\frac{3}{4} = \frac{?}{8}$

Add the fractions, then add the whole numbers.

Simplify.

Think $\frac{9}{8} = 1\frac{1}{8}$.

To subtract mixed numbers, use equivalent fractions.

EXAMPLE The height of a building designed to be $120\frac{1}{2}$ feet high was reduced by $6\frac{1}{3}$ feet to cut costs. What is its new height?

SOLUTION

$$\begin{array}{r} 120\frac{1}{2} = 120\frac{3}{6} \\ -\,6\frac{1}{3} = -\,6\frac{2}{6} \\ \hline 114\frac{1}{6} \end{array}$$

Think $\frac{1}{2} = \frac{?}{6}$

Think $\frac{1}{3} = \frac{?}{6}$

Subtract the fractions, then subtract the whole numbers.

The building's new height will be $114\frac{1}{6}$ feet.

Sometimes, you may have to rename, or borrow, to complete the subtraction.

EXAMPLE Mike Ito works $37\frac{1}{2}$ hours each week. He has already worked $21\frac{3}{4}$ hours. How many more hours does he have left to work?

SOLUTION

$$\begin{array}{r} 37\frac{1}{2} = \quad 37\frac{2}{4} \\ -\ 21\frac{3}{4} = -\ 21\frac{3}{4} \\ \hline \end{array}$$

Think $\frac{1}{2} = \frac{?}{4}$

You cannot subtract $\frac{3}{4}$ from $\frac{2}{4}$.

$$\begin{array}{r} 36\frac{6}{4} \\ \not{37}\frac{2}{4} \\ -\ 21\frac{3}{4} \\ \hline 15\frac{3}{4} \end{array}$$

Think $37\frac{2}{4} = 36 + 1 + \frac{2}{4}$ or

$36 + \frac{4}{4} + \frac{2}{4} = 36 + \frac{6}{4}$

Subtract the fractions, then subtract the whole numbers.

Critical Thinking How can you check addition? Subtraction?

EXERCISE YOUR SKILLS

Find each sum in simplest form.

34. $4\frac{3}{4} + 3\frac{1}{4}$

35. $5\frac{2}{3} + 9\frac{1}{4}$

36. $7\frac{7}{8} + 2\frac{1}{2}$

37. $6\frac{3}{4} + 3\frac{1}{2} + 2\frac{5}{8}$

38. $8\frac{7}{10} + 14\frac{2}{5} + 9\frac{1}{4}$

39. $10\frac{1}{4} + 9\frac{2}{3} + \frac{7}{8}$

Find each difference in simplest form. Check your work.

40. $8\frac{7}{8} - 4\frac{3}{4}$

41. $12\frac{5}{6} - 7\frac{2}{3}$

42. $7\frac{2}{5} - 4\frac{1}{4}$

43. $8\frac{1}{3} - 2\frac{5}{8}$

44. $21\frac{1}{5} - 16\frac{1}{2}$

45. $100\frac{4}{5} - 99\frac{9}{10}$

PROBLEM SOLVING

Choose an Operation

When trying to solve problems, one important step is deciding which operation to use.

Work in small groups. Discuss how you decide which operation to use when solving a problem. Use the problems in Put It All Together to answer each of these questions.

- In problem 46, you are to find a total. What operation will you use? Explain your choice.
- In problem 47, you are to calculate how much she earned given the hourly wage. What operation will you use? Explain your choice.
- For problems 48 and 49, multiple operations may be necessary. Explain what operation you will use in which order.

PUT IT ALL TOGETHER

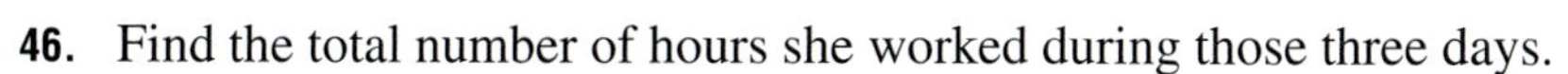

Jan Barnes worked $10\frac{1}{2}$ hours Sunday, $7\frac{1}{4}$ hours Monday, and $9\frac{3}{4}$ hours Tuesday.

46. Find the total number of hours she worked during those three days.

47. If she earns $8 per hour with no overtime pay, how much did she earn in those three days?

48. If she earns $8 per hour with time-and-a-half for time over 8 hours in a day, how much did she earn in those three days?

49. If she earns $8 per hour with double time for all time worked on Sunday and time-and-a-half for time over 8 hours on a weekday, how much did she earn in those three days?

50. ♦ Solve the problem found at the beginning of the lesson. Show all steps.

MIXED REVIEW

Find each answer in simplest form.

1. $\frac{2}{7} + \frac{4}{7} + \frac{6}{7}$ **2.** $\frac{4}{5} - \frac{2}{7}$ **3.** $\$78.40 \times \frac{3}{4}$

4. $1\frac{1}{4} + 3\frac{5}{6}$ **5.** $6\frac{3}{4} \div 3\frac{3}{8}$ **6.** $2\frac{1}{2} \times 4\frac{1}{4}$

Solve each problem.

7. On a recent trip Eric and Ron agreed to share gasoline expenses in the ratio of 3 to 2, with Eric paying the larger share. If they spent $90 on gasoline, how much did each pay?

8. Elmo Tork is paid 98¢ for each welding tip he produces that is accepted. Last week he produced these amounts of tips: Monday, 68; Tuesday, 74; Wednesday, 70; Thursday, 75; Friday, 78. Last week, 11 of the tips Elmo made were rejected; 18 items were rejected the week before. What was Elmo's gross pay last week?

9. On September 30, Alicia Alvero's check register balance was $534.04 and her bank statement balance was $945.63. A comparison of her register and the bank statement showed a service charge of $4.25 and earned interest of $3.60 on the statement but not recorded in the register. Check #346 for $210 and #352 for $202.24 were outstanding. Prepare a reconciliation statement.

During one year, Elsie Wyle worked 48 weeks and was on vacation for 4 weeks. Without simplifying the terms, find the ratio of her

10. work weeks to vacation weeks.

11. vacation weeks to work weeks.

12. work weeks to total weeks in year.

13. vacation weeks to total weeks in year.

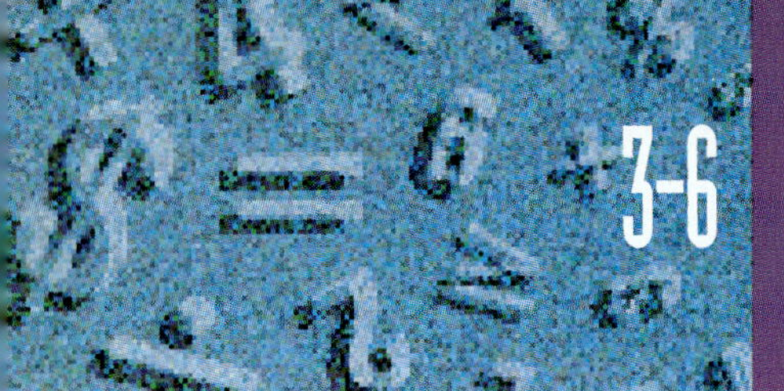

3-6 FRACTIONAL RELATIONSHIPS

OBJECTIVES

In this lesson, you will learn to find

- *a part of a number;*
- *a number greater than or smaller than another number;*
- *what part one number is of another; and*
- *what part one number is greater than or smaller than another number.*

Assume you work part-time. A co-worker tells you that another company will pay you $\frac{1}{4}$ more than your current gross salary of $100. How much will the competitor pay you?

◆ Write a sentence or a formula that shows how you calculated the salary paid by the competitor.

WARM UP

Find the value of *N*.

a. $\frac{2}{9} + \frac{4}{9} + \frac{5}{9} = N$

b. $\frac{6}{7} - \frac{2}{3} = N$

c. $\$96.50 \times \frac{2}{5} = N$

d. $1\frac{5}{6} + 3\frac{1}{3} = N$

e. $6\frac{1}{8} - 1\frac{2}{5} = N$

f. $2\frac{1}{3} \times 3\frac{3}{4} = N$

g. $\$100 \times 1\frac{4}{5} = N$

h. $6\frac{1}{8} \div \frac{1}{6} = N$

USING FRACTIONAL RELATIONSHIPS

You have already seen many ways in which fractions and mixed numerals are used to solve problems. In this lesson, you will learn how to solve four more types of problems using fractions. They are

- finding a part of a number,
- finding a number greater than or smaller than another number,
- finding what part one number is of another; and
- finding what part one number is greater than or smaller than another number

Finding a Part of a Number

To find a fractional part of a number, multiply the number by the fraction.

EXAMPLE Bev Keegan's gross pay last week was \$376. One eighth ($\frac{1}{8}$) of that amount was overtime pay. How much overtime pay did Bev earn?

SOLUTION $\frac{1}{8} \times \$376 = \$47$ overtime pay

$F \times F = P$

EXPLANATION

To find $\frac{1}{8}$ of \$376, multiply \$376 by $\frac{1}{8}$.

You will see other phrases such as "$\frac{1}{4}$ as much as . . . ," "$\frac{5}{8}$ as great as . . . ," "$\frac{2}{3}$ as large as . . . ," and "$\frac{3}{5}$ as many as . . . ." All of those phrases mean the same as "of" or "times." Find the product by multiplying the number by the fraction ($F \times F = P$).

Finding a Number That is Greater Than or Smaller Than Another

Suppose you are to find a number that is $\frac{1}{4}$ greater than another number, or $\frac{1}{4}$ more than another number. To do so, *add* $\frac{1}{4}$ of the number to the given number. For example, the hourly pay rate that is $\frac{1}{4}$ more than \$8 an hour is \$10 an hour.

$$\frac{1}{4} \times \$8 = \$2 \qquad \$8 + \$2 = \$10$$

On the other hand, suppose you are to find a number that is $\frac{1}{4}$ less than another number, or $\frac{1}{4}$ smaller than another number. To do that, *subtract* $\frac{1}{4}$ of the number from the given number. For example, the pay rate that is $\frac{1}{4}$ less than \$12 an hour is \$9 an hour.

$$\frac{1}{4} \times \$12 = \$3 \qquad \$12 - \$3 = \$9$$

EXERCISE YOUR SKILLS

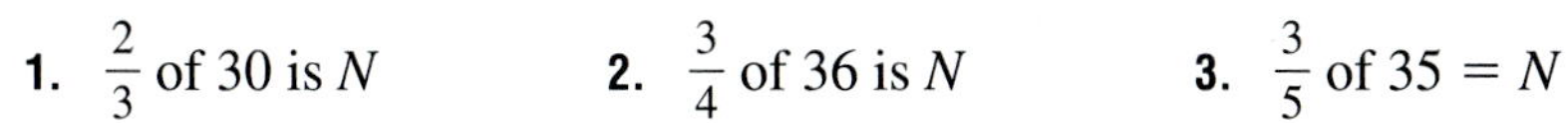

Mental Math Find the value of N.

1. $\frac{2}{3}$ of 30 is N

2. $\frac{3}{4}$ of 36 is N

3. $\frac{3}{5}$ of $35 = N$

4. $\frac{1}{3} \times 18 = N$

5. $42 \times \frac{5}{6} = N$

6. $64 \times \frac{3}{8}$ is N

7. $\frac{7}{8}$ as large as $56 = N$

8. $\frac{4}{5}$ as many as $60 = N$

9. $\frac{3}{10}$ as much as $50 = N$

10. $\frac{9}{20}$ of $150 = N$

Solve each problem.

11. In 1 week Marc Kipple earned $360.42, $\frac{5}{6}$ of which was in regular-time pay. What was Marc's regular-time pay?

12. Kim Wintell's annual overtime pay was $\frac{1}{10}$ as much as her regular pay of $22,910. What was her overtime pay?

13. Last year, Pablo Arellano's gross earnings were $26,780. This year, his earnings were $\frac{4}{5}$ of last year's because he was unemployed part of this year. Find his earnings this year.

14. The Car Security Company sold 15,816 burglar alarms the year before last. Last year, they sold only $\frac{7}{8}$ as many alarms. How many alarms did they sell last year?

Find the number that is

15. $\frac{1}{5}$ more than 20

16. $\frac{1}{6}$ greater than $30

17. $\frac{1}{8}$ more than $16

18. $\frac{1}{5}$ larger than $40

19. $12 plus $\frac{1}{4}$ of itself

20. $\frac{1}{6}$ smaller than $30

21. $\frac{1}{3}$ less than $60

22. $\frac{1}{4}$ less than $24

23. $\frac{1}{7}$ smaller than $42

24. $24 minus $\frac{1}{8}$ of itself

Solve.

25. Phil's salary is $24,300 a year, or $2,075 a month. Norma's yearly pay is $\frac{3}{10}$ more than Phil's. What is Norma's yearly pay?

26. Last March, Gene Booth worked overtime and earned $\frac{1}{8}$ more than his usual pay of $1,712 a month. How much did Gene earn last March?

27. Two years ago a cable company's total sales were $246,750. Last year the cable company increased its sales by $\frac{4}{7}$ over the year before. What were the company's total sales last year?

28. Ella Stearns lost a job that paid $13.20 an hour. Her new job pays $\frac{1}{5}$ less. How much does her new job pay per hour?

29. The Reese Company has 70 employees and plans to cut its labor costs by $\frac{1}{12}$. The company's total labor cost is now $1,800,000 a year. What will be the company's labor cost per year after the cut?

30. Fred's average weekly earnings of $486 dropped by $\frac{1}{9}$ after he stopped working overtime. What was his new average pay per week?

Finding What Part a Number is of Another

To find what fractional part one number is of another number, write the numbers as a fraction. The number that is the part should be written as the numerator. The number that is the whole with which the part is being compared should be written as the denominator. Simplify the fraction.

EXAMPLE Kim Logan's total pay for a week was \$400. Of that amount, \$50 was overtime pay. What fractional part of Kim's total pay was the overtime pay?

SOLUTION $\frac{\$50}{\$400} = \frac{1}{8}$ \$50 is the part.
\$400 is the whole.

CHECK $\frac{1}{8} \times \$400 = \$50$

Using Estimation — Is Your Answer Reasonable? When finding what part one number is of another, use these estimation tips as a quick check of your answer.

A number compared with itself equals 1. For example, 6 compared to $6 = \frac{6}{6} = 1$.
When a number is compared with a larger number, the result is less than 1. For example, 4 compared to $7 = \frac{4}{7}$.
When a number is compared with a smaller number, the result is greater than 1. For example, 3 compared to 2 is $\frac{3}{2}$, or $1\frac{1}{2}$.

Finding What Part a Number is Greater or Smaller Than Another

Suppose that a pay rate has been increased from \$7.20 to \$9 per hour. If you want to find what part the new rate of \$9 is greater than the old rate of \$7.20, follow these steps:

EXAMPLE 1
\$9 is what part greater than \$7.20?

SOLUTION $\$9.00 - \$7.20 = \$1.80$

$\frac{\$1.80}{\$7.20}$ ← the part / ← the whole

$\frac{\$1.80}{\$7.20} = \frac{1}{4}$

EXAMPLE 2
\$7.20 is what part smaller than \$9?

$\$9.00 - \$7.20 = \$1.80$

$\frac{\$1.80}{\$9.00}$ ← the part / ← the whole

$\frac{\$1.80}{\$9.00} = \frac{1}{5}$

EXPLANATION

Step 1 Find the number representing the part by subtracting the smaller number from the larger number. That number will be called the part.

Step 2 Write the part as the numerator of a fraction. As the denominator of the fraction, write the whole number with which the part is being compared.

Step 3 Simplify the fraction.

EXERCISE YOUR SKILLS

Mental Math 5 is what part of

31. 12? **32.** 9? **33.** 8?

34. 6? **35.** 4? **36.** 2?

$12 is what part of

37. $12 **38.** $36? **39.** $24?

40. $16? **41.** $14? **42.** $7?

What part of $12 is

43. $3? **44.** $4? **45.** $11?

46. $15? **47.** $16? **48.** $24?

Solve.

49. Last year Ruth Delisle's salary was $24,000. This year her salary was increased by $1,500. What fractional part of Ruth's old salary was the increase?

50. Leonard Chin got a 90¢ raise in his hourly pay rate. His old rate was $8.10 per hour. What part of the old rate was the increase?

Mental Math 20 is what part less than

51. 30? **52.** 24? **53.** 22? **54.** 50? **55.** 100?

What part more than 12 is

56. 13? **57.** 19? **58.** 14? **59.** 16? **60.** 18?

Solve.

61. 14 equals 10 increased by what part of itself?

62. 18 equals 24 decreased by what part of itself?

63. In her first month on her new job as a city spokesperson, Annie Campo earned $1,400. In her second month, Annie earned $1,600. What part more were her second month's earnings than her first month's earnings?

64. In his first year on his new job, Mick Grote earned $18,300. In his second year, Mick earned $20,130. What part greater were his earnings in the second year than in the first year?

65. In March, Ileana Herrera's regular and overtime pay totaled $2,025. In April, Ileana worked no overtime hours, so her pay dropped to $1,800. What part smaller than her March pay was Ileana's April pay?

Last year the Yamaguchi family's expenses totaled $18,000. This year their expenses totaled $19,500.

66. What part greater were this year's expenses than last year's?
67. What part less were last year's expenses than this year's?

In July, Alan Rojas' earnings were $600. In August his earnings totaled $720.

68. What part of July's earnings were Alan's August earnings?
69. What part of August's earnings were Alan's July earnings?
70. ♦ Write an advertisement comparing the price of a product at Store A and Store B using the information learned in this lesson.

MIXED REVIEW

Find each of the following.

1. $\frac{1}{8}$ of 72
2. $\frac{2}{5}$ as much as 40
3. $\frac{1}{7}$ less than 35
4. $\frac{1}{6}$ more than 24
5. 8 is what part of 72?
6. The Soto-Weil Software Company sold $\frac{2}{7}$ more software programs this year than last year. If they sold 742 programs last year, how many programs did the company sell this year?
7. Last month the Merdex Company paid wages of $134,000. Of that amount, $26,800 was paid for work on weekends. What part of the month's wages were paid for weekend work?

Trek Products has four workers in the painting department. Last month the workers received these hourly pay raises: first worker, $0.75; second worker, $1.10; third worker, $0.90; fourth worker, $0.62.

8. What average hourly pay raise did these four workers earn, to the nearest cent?
9. Based on the average pay raise, what estimated pay increase will a worker earn for a 40-hour week?
10. A company spent $13,200 for temporary help this year. Last year a total of $14,850 was spent for the same services. What part less did the company spend on temporary help this year than it spent last year?
11. Herb Austin's pay is $\frac{5}{6}$ as large as Elaine Frost's pay. If Elaine's pay is $1,230 a month, what is Herb's pay per month?

EXPLORE TECHNOLOGY

TEMPLATE: A Database File to Hold Names and Addresses

A **database** is a collection of related information. Your city's telephone book is an example of a database. It is organized alphabetically by name and used to find an address and/or phone number. A database template for this lesson is on the applications diskette.

When information in a database is stored electronically, you can easily update, retrieve, and report on that information in a variety of ways. For example, for a phone book stored in an electronic database you can search for a specific name, a specific phone number, and for all telephone numbers beginning with a certain area code.

In database terminology, each listing in the phone book is a record. A record is made up of fields. For example, Last Name is a field on the blank record shown on the screen below. When you search for an item in a database, you search on one or more fields. For example, you could search our address database Last Name and ZIP.

Before you create a database, think about how you will want to retrieve information. Then create field names to meet your objectives. The fields in a local phone directory usually are last name, first name, street address, city, and phone number. There are different kinds of fields; text fields, numeric fields, logic (true/false or yes/no), and date fields are the most common. In this lesson, we use only text and numeric fields. You will use other field types in later lessons.

In this lesson, you will learn to create a simple address database. Your database will contain the 7 fields shown on the screen below. To begin, load ET3 from your applications disk. The fields Last Name and First Name have already been set up. Add the remaining five fields to the file on your computer screen. Check you software HELP screen or documentation to see how fields are entered. The colon is used by Microsoft Works to show when a field is being created. You will also be asked to identify the field type and the number of characters for each field. Allow at least 20 characters for street addresses. You only need 2 characters for state abbreviations. After you have created the five fields, your computer screen should look like this:

ADDRESS.WDB

Last Name:

First Name: ....

Address: ....

City: ....

State: ....

ZIP: ....

Phone Number: ....

Next enter data about five people into your database. Use the **TAB** key and arrow keys to move from field to field and record to record. When you have finished entering the records, save your file.

See what happens as you move about the menu and use the commands available to you. Start by viewing your database in LIST or Standard view. (LIST, FORM, Column and Standard are terms used by many software programs. Check your documentation for the views provided by your software.) Your screen should look like the one shown.

ADDRESS1.WDB

	Last Name	First Name	Address	City	State	Zip	Phone Number
1	Velez	Sonia	123 Parkway Dr	Detroit	MI	48219-2001	313-555-0100
2	Smith	Virginia	P.O. Box 456	Orlando	FL	32837	407-555-0132
3	Leon	Victor	Rural Route 7	Greenwood	IN	46142	317-555-0178
4							

1. How did the screen change?
2. View the file as a form or column. Write a sentence describing the difference between the LIST or Standard view and the FORM or Column view.
3. Sort on Last Name field and Phone Number in ascending order. What name is at the top of the order? What number is at the top of the order? What does ascending order mean?
4. What problems would you have encountered if you had not separated the first and last names?
5. How would you change your record to include an individual name and organization name?

Design Your Own

Database files can be used to track many things, such as baseball card or CD collections. Think of something you would like to keep track of using an electronic database. Follow these steps.

6. How would you like to look things up in your database file? Make a list of possible fields.
7. Create a database file using at least 3, but not more than 8 fields.
8. Enter data and experiment with the SORT command.
9. Save your file.
10. Would you add fields or change the structure of your database file in any way? How? Why?

CHAPTER 3 SUMMARY AND REVIEW

VOCABULARY REVIEW

cross product *3-3*
double-time pay *3-5*
greatest common factor *3-1*
gross pay *3-5*
least common denominator *3-2*
overtime *3-1*
proportion *3-3*
ratio *3-3*
reciprocal of a fraction *3-1*
regular pay *3-1*
regular time *3-1*
time-and-a-half pay *3-5*
time cards *3-1*
time clock *3-1*

MATH SKILLS REVIEW

3-1
- To multiply fractions, multiply the numerators and the denominators.
- To find an equivalent fraction, multiply or divide the numerator and denominator by the same number.
- Simplify fractions by changing them to lowest terms.
- To change an improper fraction to a mixed or whole number, divide the numerator by the denominator.
- To divide by a fraction, multiply by the reciprocal of the divisor.

Find each product or quotient in simplest form.

1. $\frac{1}{5} \times \frac{1}{4}$ **2.** $\frac{2}{7} \times \frac{3}{4}$ **3.** $\frac{1}{3} \div \frac{5}{6}$

4. Of the 72 animals at the local animal shelter, $\frac{1}{8}$ were cats. How many cats were in the shelter?

3-2
- To add or subtract like fractions, add or subtract the numerators and write the result over the common denominator.
- To add or subtract unlike fractions, find a common denominator, then add or subtract as usual.

5. $\frac{7}{12} - \frac{5}{12}$ **6.** $\frac{2}{3} - \frac{1}{4}$ **7.** $\frac{2}{9} + \frac{5}{12}$

3-3
- To compare two numbers, write them as a ratio.
- Write a proportion to show that two ratios are equal.
- To test a proportion, calculate the cross products to see if they are equal.
- To find an unknown term in a proportion, calculate the cross products and solve for the unknown term.

A survey found that 3 out of 5 workers drive to work. There are 480 workers who drive to work.

8. How many workers are there in all?

9. What is the ratio of workers who drive to workers who do not drive to work?

3-4 ◆ Multiply whole numbers.

10. 42×78 **11.** $37 \times \$8$ **12.** 23×56

3-5 ◆ Add or subtract mixed numbers by first adding or subtracting the fractions, renaming when necessary, then add or subtract the whole number part.

◆ Change mixed numbers to improper fractions before multiplying.

◆ When dividing mixed numbers, change the mixed numbers to improper fractions then multiply by the reciprocal of the divisor.

13. $12\frac{1}{4} \times 16$ **14.** $2\frac{1}{7} \times 4\frac{2}{3}$

15. $6\frac{1}{2} + 7\frac{5}{6}$ **16.** $6\frac{1}{4} - 3\frac{4}{7}$

17. $\frac{1}{8} \div 3\frac{2}{3}$ **18.** $18 \div 1\frac{1}{2}$

3-6 ◆ To find a part of a number, multiply the number by the fraction.

◆ To find a number that is a part greater or smaller than a known number, find the part and add or subtract it from the known number.

◆ To find what part a number is of another, write a fraction using the part as numerator and the whole as denominator.

◆ To find what part a number is greater or smaller than another, write the difference between the numbers (the part) as a numerator and the number with which it is being compared (the whole) as the denominator.

19. Kim's earned \$20,560, of which $\frac{1}{8}$ was from overtime pay. How much overtime pay did Kim earn last year?

20. Earl Wilby received a pay raise of \$0.60 an hour and now earns \$9.60 an hour. What part of Earl's former pay is the pay raise?

APPLICATIONS

3-4 ◆ Calculate time worked as shown on a time card.

◆ Find regular and overtime gross wages for hourly employees.

◆ To find time-and-a-half pay rate, multiply the regular rate by $1\frac{1}{2}$. Overtime pay rates are not rounded.

◆ To find double-time rate, multiply the regular rate by 2.

◆ To find gross pay, add the overtime pay and regular pay.

Find the gross pay for each employee.

Alan Kirbaum works on an 8-hour day basis and is paid \$8 an hour for regular time work and \$12 an hour for overtime work. Last week he worked these hours: Monday, 9; Tuesday, 6; Wednesday, 11; Thursday, 8; Friday, 9. Find Alan's:

21. regular time pay. **22.** overtime pay.

23. gross pay.

Stephen Woll's regular pay is $9.50 an hour. He is paid time-and-a-half for overtime work beyond 40 hours a week. He worked: Monday, 11; Tuesday, 10; Wednesday, 10; Friday, 12. Find Stephen Woll's

24. regular time pay.

25. overtime pay.

26. gross pay.

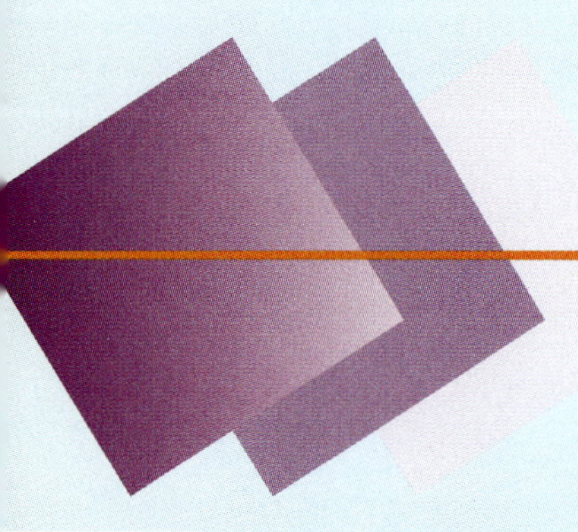

WORKPLACE KNOWHOW: **NEGOTIATION AND LEADERSHIP SKILLS**

Negotiation skills include the ability to work toward an agreement that may involve sharing or exchanging resources or resolving differences. Good negotiators understand the history of the circumstances, set realistic goals, and are willing to listen, ask questions, respond to new information and make reasonable compromises. Leadership skills include the ability to persuade, convince, and motivate with reasonable arguments for their position.

Assignment In Chapter 3 you negotiated attendance rules for a business. Break into small groups. One group should be observers, another actors. Keeping in mind the skills described in the first paragraph, the acting group should discuss how those rules should be enforced and identify penalties for breaking the rules. After the acting group has finished, the observers should critique the actors on their demonstration of negotiation and leadership skills. The class should then have a discussion about what they learned about negotiation and leadership from the exercise.

Work Skills Portfolio Include in your portfolio and assessment of your skills with fractions. You may want to include some comments on your experience with developing a database in the Explore Technology lesson. If you completed the integrated project in the student workbook, consider including a copy of the completed time cards, payroll register, and responses to payroll questions.

CHAPTER 3 TEST

Math Skills

1. $\frac{1}{5} \times \frac{1}{6}$
2. $\frac{7}{15} - \frac{2}{5}$
3. $\frac{7}{9} + \frac{1}{2}$
4. $8\frac{1}{2} + 3\frac{3}{4}$
5. $\frac{3}{4}$ of \$12
6. $3\frac{1}{8} \times 2\frac{2}{5}$
7. $3\frac{1}{2} \times 18$
8. $\$8.60 \div \frac{2}{3}$
9. $\frac{12}{15} \div \frac{3}{15}$
10. $6\frac{2}{7} - 3\frac{3}{7}$
11. $9\frac{5}{6} + 3\frac{3}{4}$
12. $16\frac{1}{4} - 15\frac{7}{8}$
13. Simplify $\frac{8}{40}$.
14. Name the reciprocal of $\frac{4}{7}$.
15. What part of \$1.80 is 36¢?
16. \$16 is what part less than \$56?
17. Find the amount that is $\frac{1}{2}$ greater than \$4.60.

Applications

In one firm there are 10 office workers, 70 factory workers, and 2 managers. Without simplifying the terms, find the ratio of:

18. Office to total workers
19. Factory workers to managers
20. Alphonse Scanetti is paid \$12.70 an hour for regular-time work with time-and-a-half paid for work beyond 40 hours a week. He worked: Monday, 7; Tuesday, 9; Wednesday, 9; Thursday, 10; Friday, 8. Find his gross pay for the week.
21. A worker spent $23\frac{1}{2}$ hours making a part. If the pay is \$12.40 an hour, what was the cost of the worker's time to make the part?
22. John Czarnota earned \$25 last week and $4\frac{1}{2}$ times that much this week. How much did he earn this week?
23. Zeline Thomas earned \$440 last week. Of that amount, $\frac{1}{8}$ came from a part-time job. What were Zeline's earnings from her part-time job?
24. The profit from a business owned by Misek and Brent is shared in the ratio of 5 to 3 with Misek getting the larger share. If last year's profit was \$96,000, what amount did Misek get?
25. Eve Marsten found a job closer to home that pays $\frac{1}{7}$ less than her old hourly pay rate of \$10.50. What is her new hourly pay rate?

CHAPTER 4

SALES

do the work

Are you motivated by seeing your hard work rewarded with dollars? If so, you may like to work as a sales professional. Sales — for services or products — is part of every business.

Gizelle sells advertising space for a technical magazine. Finding and keeping advertisers is Gizelle's primary job responsibility. She attends trade shows, makes cold calls (calls on customers without prior introduction), and pursues leads (potential advertisers) identified by her company.

Before contacting a lead, Gizelle learns as much as she can about them and their products. She reads their annual report and sales literature and looks for articles about them in community newspapers and industry publications. If she thinks they will be successful advertising in her magazine, she calls them.

Gizelle's goal is to present a proposal showing the advantages of advertising in her magazine. It may take months to convince someone to see her — much less to advertise. No wonder

AND THE DOLLARS

Gizelle says a good salesperson has to be persistent without being obnoxious, plus be able to take rejection.

Gizelle keeps the names of her customers and prospects, along with information on their products and ads, on a computer database. This allows her to generate personalized letters and flyers. She also uses a fax machine to send and receive orders and other information quickly. And when it comes to math, she uses:

- **Percents and Averages —** to put together proposals for clients.
- **Problem Solving —** to find the best deal for her customers.
- **Mental Math —** to help find and negotiate a fair advertising rate.
- **Personal Financial Management Skills —** to manage her finances through seasonal high and low cycles. (Gizelle is paid commission from her sales.)

In the two years she's been selling, Gizelle has learned never to assume she knows what a customer needs. She listens to what customers tell her.

Gizelle has also learned she can't focus on sales dollars. She believes the best way to be successful in sales is to focus on doing the job — making calls, visiting customers, mailing flyers — whatever it takes. She says, "If you do the job, the money will come."

If you are interested in a career in sales, Gizelle recommends starting as a sales assistant. You'll learn organizational skills and how to work with customers. Her advice: "Be positive. Be polite. Be professional. Know your product. You'll get your chance."

WILL COME !!! $

4-1 DECIMALS

OBJECTIVES

In this lesson, you will learn to

- *change fractions to decimals;*
- *change decimals to fractions; and*
- *multiply and divide by 0.1, 0.01, and 0.001.*

Try this experiment. Use a calculator. Choose a whole number between 1 and 9. Multiply by 259. Then multiply that result by 429. Divide by the number you started with. What is the result?

◆ Based on the result of the first experiment, write a prediction about what will happen if you do the same experiment with a second number. Then choose another number. Perform the same experiment. Write down your results. Did your results match your prediction?

WARM UP

a. $\frac{1}{5}$ less than 115 is ?

b. $\frac{1}{4}$ larger than 48 is ?

c. $\frac{3}{7} \times \$57.50 = ?$

d. $4\frac{1}{5} \div 2\frac{1}{3} = ?$

e. $10\frac{5}{9} - 7\frac{3}{4} = ?$

f. $\frac{3}{8} = \frac{?}{1{,}000}$

DECIMALS AND FRACTIONS

A decimal is just another way of naming a fractional number when the fraction has a denominator such as 10, 100, or 1,000. For example, one dollar has 100 cents. One cent is one one-hundredth of a dollar, or $1 \div 100 = 0.01$. One cent can be written as $\frac{1}{100}$ of a dollar, or \$0.01. Both $\frac{1}{100}$ and 0.01 are numerals that name the same fractional number. The zero is written to the left of the decimal point for clarity.

The following illustrates place value and will help you read numbers with decimal forms. The decimal point is read as "and."

hundreds	tens	ones		tenths	hundredths	thousandths	ten-thousandths
1	2	3	.	4	5	6	7

The number is read as one hundred twenty-three and four thousand five hundred sixty-seven ten-thousandths.

To change a fraction with a denominator of 10, 100, or 1,000 to a decimal, divide the numerator by the denominator. For example:

$\frac{1}{10} = 1 \div 10 = 0.1$ $\qquad$ $\frac{6}{10} = 6 \div 10 = 0.6$

$\frac{1}{100} = 1 \div 100 = 0.01$ $\qquad$ $\frac{67}{100} = 67 \div 100 = 0.67$

$\frac{1}{1{,}000} = 1 \div 1{,}000 = 0.001$ $\qquad$ $\frac{67}{1{,}000} = 67 \div 1{,}000 = 0.067$

As a shortcut, you can divide by denominators of 10 and multiples of 10 by simply moving the decimal point in the numerator to the *left* the same number of places as there are zeros in the denominator. Thus, $235 \div 10 = 23.5$.

Changing Fractions to Decimals

Any fraction can be changed to a decimal by dividing the numerator by the denominator. Divide to one digit beyond the needed decimal place to round.

EXAMPLE Find the decimal equivalent of $\frac{5}{8}$, correct to two decimal places.

SOLUTION

```
  0.625 ≈ 0.63
8)5.000
  4 8
    20
    16
     40
     40
```

EXPLANATION

The numerator, 5, is divided by the denominator, 8. The division is carried to three decimal places, then rounded back to 2 places.

EXERCISE YOUR SKILLS

Write each fraction as a decimal.

1. $\frac{21}{100}$ **2.** $\frac{592}{1{,}000}$ **3.** $\frac{8}{10}$ **4.** $\frac{5}{1{,}000}$ **5.** $\frac{74}{100}$

6. $\frac{3}{10{,}000}$ **7.** $\frac{11}{1{,}000}$ **8.** $\frac{3{,}000}{10}$ **9.** $\frac{639}{100}$

Find the decimal equivalent of each fraction to three decimal places.

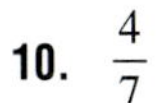

10. $\frac{4}{7}$ **11.** $\frac{5}{9}$ **12.** $\frac{11}{12}$ **13.** $\frac{8}{11}$

14. $\frac{3}{14}$ **15.** $\frac{5}{6}$ **16.** $\frac{7}{9}$ **17.** $\frac{4}{15}$

OTHER FACTS ABOUT DECIMALS

- The fractions $\frac{7}{10}$, $\frac{70}{100}$, and $\frac{700}{1,000}$ are equivalent fractions. Since $\frac{7}{10} = \frac{70}{100} = \frac{700}{1,000}$, 0.7 = 0.70 = 0.700. In other words, we may attach zeros to a decimal without changing the number it represents.
- Any whole number can be shown in decimal form by attaching a decimal point and as many zeros as desired. For example, 75 = 75.0 = 75.00.
- Decimals can be used instead of common fractions to make calculations easier. For example, it is easier to add 5.65 and 1.75 than $5\frac{13}{20}$ and $1\frac{3}{4}$.
- It is often easier to compare fractions when they are shown as decimals. For example, it is easier to see that $\frac{5}{9}$ is a larger number than $\frac{7}{13}$ if $\frac{5}{9}$ is shown as 0.556 and $\frac{7}{13}$ as 0.538.

Multiplication by 0.1, 0.01, or 0.001

To multiply by 0.1, 0.01, and 0.001, move the decimal point in the number to be multiplied to the *left* as many places as there are decimal places in the number by which you are multiplying. Add zeros at the beginning, if needed.

EXAMPLES $39.6 \times 0.01 = 0.396$ $\quad 7 \times 0.001 = 0.007$
$\$9.70 \times 0.1 = \$0.97$ $\quad \$362 \times 0.01 = \$3.62$

Changing a Decimal to a Fraction

To change a decimal to an equivalent fraction, drop the decimal point and write the number as the numerator of a fraction. For the denominator, write 1 followed by as many zeros as there are decimal places in the decimal. Then, simplify the fraction.

EXAMPLES $0.25 = \frac{25}{100} = \frac{1}{4}$ $\quad 0.025 = \frac{25}{1,000} = \frac{1}{40}$

Division by 0.1, 0.01, or 0.001

When you divide by 0.1, 0.01, or 0.001, move the decimal point in the number to be divided to the *right* as many places as there are decimal places in the number by which you are dividing. Attach zeros if needed.

EXAMPLES $2.87 \div 0.1 = 28.7$ $\quad 527 \div 0.1 = 5,270$
$2.87 \div 0.01 = 287$ $\quad 339 \div 0.01 = 33,900$
$2.87 \div 0.001 = 2,870$ $\quad 924 \div 0.001 = 924,000$

Critical Thinking When you divide by a decimal, is the quotient larger or smaller than the dividend? Why?

EXERCISE YOUR SKILLS

Mental Math Multiply each problem mentally. Write the products only.

18. 98.2×0.1
19. 14.7×0.01
20. 845×0.001
21. 906×0.001
22. 3.1×0.001
23. 0.46×0.1
24. $\$107 \times 0.01$
25. $\$57.40 \times 0.1$

Change each decimal to an equivalent fraction in lowest terms.

26. 0.56
27. 0.45
28. 0.98
29. 0.025
30. 0.12
31. 0.625
32. 0.375
33. 0.85

Mental Math Divide each problem mentally. Write the quotients only.

34. $6.28 \div 0.1$
35. $0.085 \div 0.01$
36. $0.0065 \div 0.001$
37. $0.208 \div 0.01$
38. $7.23 \div 0.01$
39. $33 \div 0.01$
40. $0.2 \div 0.01$
41. $2.76 \div 0.001$
42. $2.2 \div 0.01$
43. $\$0.18 \div 0.01$
44. $\$3.00 \div 0.1$
45. $\$5.31 \div 0.001$
46. $\$7.34 \div 0.001$
47. $\$28.78 \div 0.01$

48. ◆ Look at the calculations you did at the beginning of the lesson. Explain why it works.

MIXED REVIEW

1. $3\frac{3}{5} \div 4\frac{2}{3}$
2. $\frac{5}{8} + \frac{3}{4}$
3. 23.7×0.001
4. $\$8.24 \div 0.1$
5. $\frac{3}{4} \div \frac{1}{2}$
6. $\frac{3}{4}$ of $\frac{1}{2}$

7. Change $\frac{5}{7}$ to a decimal correct to three places.

8. Change $\frac{2}{11}$ to a decimal, rounded to the nearest hundredth.

9. Change 0.065 to a fraction in lowest terms.

10. Nathan Donnely works on a piece-rate basis. He produced these numbers of items last week: Monday, 73; Tuesday, 84; Wednesday, 79; Thursday, 82. How many items did Nathan produce Friday so as to average 80 items per day?

11. Novak and Oakes are partners in a lumber yard. They sell \$637,109 and \$475,224 worth of lumber during the year, respectively. They also share profits in the ratio of 3 to 5, respectively. The profit for the lumber yard for the year is \$180,512. What is Novak's share of the profit?

4-2 PERCENTS

OBJECTIVES

In this lesson, you will learn to

- *change decimals and fractions to percents; and*
- *change percents to decimals and fractions.*

A part of a number can be represented as a fraction, percent, or decimal.

◆ Write a paragraph that compares fractions, percents, and decimals. For each form, identify a situation in which it is commonly used. For example, batting averages are typically expressed in decimal form. Give a reason why one form is used instead of the others.

WARM *UP*

Find each answer in simplest form.

a. $\frac{2}{5} - \frac{1}{7}$

b. $2\frac{2}{18} + 3\frac{1}{6}$

c. $6\frac{2}{3} - 4\frac{1}{7}$

d. $\frac{2}{5}$ more than 340

e. $\frac{1}{7}$ less than 280

f. 35×150

g. 6 is what part of 36?

h. 15 is what part more than 10?

i. Write $\frac{2}{9}$ as a decimal, rounded to the nearest hundredth.

PERCENT

Percent means "per hundred," "by the hundred," or "out of a hundred." A percent shows the comparison or ratio of any number to 100. For example, if 20 video tapes out of 100 in a video store are comedy films, you can say that 20% of the tapes are comedies.

You have already learned that a ratio is another way of showing a fraction. For example, the ratio 20 to 100 (20:100) is the same as $\frac{20}{100}$. Therefore, the ratio 20:100, the fraction $\frac{20}{100}$, the decimal 0.20, and the percent 20% are different numerals that stand for the same number.

Changing Decimals and Whole Numbers to Percents

To change a decimal or whole number to a percent, the denominator must become 100. Multiply the number by 1 expressed as $\frac{100}{100}$. Then, change the fraction to a percent by dropping the denominator. The numerator becomes the percent value. For example:

$$0.47 \times \frac{100}{100} = \frac{47}{100} = 47\% \qquad 6 \times \frac{100}{100} = \frac{600}{100} = 600\%$$

$$0.01 \times \frac{100}{100} = \frac{1}{100} = 1\%$$

In short, to change a decimal or whole number to a percent, move the decimal point two places to the right and attach a percent sign:

0.04 = 4% **0.18 = 18%** **2 = 200%** **0.009 = 0.9%**

Changing Fractions to Percents

You can change any fraction with a denominator of 100 to a percent by dropping the 100 and attaching a percent sign. For example, $\frac{5}{100} = 5\%$.

Any fraction that has a denominator that is a factor of 100, such as 1, 2, 4, 5, 10, 20, 25, and 50, can also be changed to a percent easily. All you have to do is change the fraction to an equivalent fraction with a denominator of 100.

EXAMPLES $\frac{1}{2} \times \frac{50}{50} = \frac{50}{100} = 50\%$ $\qquad \frac{2}{5} \times \frac{20}{20} = \frac{40}{100} = 40\%$

$\frac{1}{4} \times \frac{25}{25} = \frac{25}{100} = 25\%$

For fractions that cannot be so easily changed, change the fraction to a decimal and round the results to a stated number of places.

EXAMPLE Express $\frac{4}{7}$ as a percent to the nearest tenth of a percent.

SOLUTION $7\overline{)4.0000}$ = 0.5714 $\rightarrow \frac{4}{7} \approx 0.5714 \approx 57.1\%$

EXPLANATION

Carry the division to four places in order to round to three. Since the fourth place is less than 5, the answer is rounded to 0.571. So, $\frac{4}{7} = 0.571 \approx 57.1\%$, to the nearest tenth of a percent.

Saving Time with Calculators If your calculator has a percent key, you can get the answer as a percent in one step. Press the percent key instead of the equals key on your calculator:

Press: [4] [÷] [7] [%] and [57.142857] appears in the display.

Now, round the answer to the nearest tenth of a percent, and attach a percent sign (57.1%). Notice that by pressing the percent key, the decimal point was *automatically* placed in the correct position.

EXERCISE YOUR SKILLS

Show each ratio as a fraction, a decimal, and a percent. Follow this format: 26 to 100 = $\frac{26}{100}$ = 0.26 = 26%.

1. 100 to 100
2. 110 to 100
3. 500 to 100
4. 375 to 100

Change each decimal or whole number to a percent.

5. 0.58
6. 1.0
7. 0.03
8. 2.6
9. 0.2812
10. 0.7514
11. 4
12. 0.0049

Change each fraction to an equivalent fraction with a denominator of 100. Then show the new fraction as a percent.

13. $\frac{1}{2}$
14. $\frac{3}{10}$
15. $\frac{12}{25}$
16. $\frac{3}{1}$

Change each fraction to a percent, rounded to the nearest hundredth of a percent.

17. $\frac{3}{16}$
18. $\frac{2}{15}$
19. $\frac{15}{16}$
20. $\frac{5}{32}$

Change each fraction to a percent. Round each answer to the nearest tenth of a percent.

21. $\frac{1}{14}$
22. $\frac{4}{15}$
23. $\frac{15}{18}$
24. $\frac{5}{32}$

Changing Percents to Decimals

To change a percent to a decimal, drop the percent sign and move the decimal point two places to the left.

EXAMPLES $7\% = 0.07$ $37\% = 0.37$ $31\frac{1}{2}\% = 0.31\frac{1}{2}$, or 0.315

$7.8\% = 0.078$ $409\% = 4.09$ $\frac{1}{2}\% = 0.00\frac{1}{2}$, or 0.005

Changing Percents to Fractions

To change a percent to a fraction, drop the percent sign and write the number as the numerator over the denominator 100. Simplify the fraction.

EXAMPLES $8\% = \frac{8}{100} = \frac{2}{25}$ $140\% = \frac{140}{100} = \frac{7}{5}$, or $1\frac{2}{5}$

Changing Percents with Decimals to Fractions

If the percent has a decimal, change the percent to decimal form, before changing the decimal to a fraction.

EXAMPLE $7.5\% = 0.075 = \frac{75}{1{,}000} = \frac{3}{40}$

EXERCISE YOUR SKILLS

Change each percent to a decimal or a whole number.

25. 178% **26.** 2% **27.** 6.1% **28.** 100%

29. 0.8% **30.** 625% **31.** $15\frac{3}{4}\%$ **32.** 1,000%

Change each percent to a fraction or a mixed number and simplify.

33. 12% **34.** 3% **35.** 5% **36.** 90%

37. 60% **38.** 175% **39.** 125% **40.** 250%

41. 0.3% **42.** 1.3% **43.** $37\frac{1}{2}\%$ **44.** 1.5%

45. ♦ Work in small groups. Make a list of places where percents are used. Organize the list from most often to least often used. Present your list to the class for discussion.

MIXED REVIEW

1. Write the ratio 15:100 as a fraction, decimal, and percent.
2. Change to percents: 2.75, 0.37, $\frac{4}{5}$.
3. Change to decimals: $\frac{3}{8}$, 67.4%, 0.4%, $\frac{1}{4}\%$.
4. Change to fractions or mixed numbers and simplify: 25%, 250%, 10%.
5. A cash register clerk started with $255 in cash. During the day, the clerk took in $2,072.47 and paid out $67.82. How much cash should remain in the drawer at the end of the day?
6. Six boxes of computer disks cost $168. At the same rate, how many boxes of disks could a company buy for $420?

Allen is paid $7.80 an hour for an 8-hour day, time and a half for time past 8 hours per day, and double time on weekends. Last week he worked these hours: Mon., 8; Tues., 8; Wed., 7; Thurs., 11; Fri., 8; Sat., 5. Deductions were: $37 in withholding and 7.65% in FICA. Find Allen's

7. regular pay.
8. overtime pay.
9. net pay.

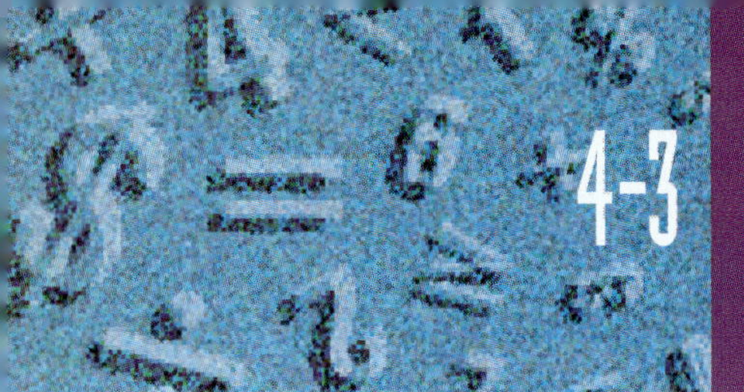

4-3 DEDUCTIONS AND TAKE-HOME PAY

OBJECTIVES

In this lesson, you will learn to

- *calculate common deductions from wages; and*
- *find net pay or take-home pay.*

When employees are paid, money is deducted from the total pay resulting in less pay. Typical deductions include federal and state taxes, life insurance, health insurance, retirement funds, and union dues.

◆ Make a chart with the headings Deduction; Type, Estimate. Estimate what percentage of an employee's paycheck goes to each deduction.

WARM UP

a. $34 + 9 + 102 + 6$

b. $92 - 18 + 24$

c. $24 \times \frac{3}{4}$

d. $16\frac{2}{3} \div 8\frac{1}{3}$

e. Divide 9 by 4, round to the nearest tenth.

f. What number is $\frac{5}{6}$ of \$384?

g. Find $\frac{4}{5}$ of \$135.

h. 20 is $\frac{2}{3}$ of what number?

DEDUCTIONS AND TAKE-HOME PAY

In this section you will learn to calculate some common deductions from a worker's pay and to find the net pay a worker takes home.

Deductions are subtractions from gross pay. An employer must make deductions from an employee's pay for federal and state income taxes and social security taxes. Deductions may also be subtracted for union dues, health and life insurance, government bonds, etc.

After all deductions are subtracted from total or gross wages, an amount remains that is called **net pay,** or **take-home pay.**

Gross Pay − Deductions = Net Pay or Take-home Pay

The gross pay, deductions and net pay for a typical employee are shown in Illustration 4-3.1.

Statement of Employee Earnings and Payroll Deductions											
	EARNINGS			DEDUCTIONS							
WEEK ENDED	REGULAR	OVER-TIME	TOTAL	FED. WITH.	FICA	LIFE INS.	HEALTH INS.	OTHER	TOTAL	NET PAY
2/4	*375.00*		*375.00*	*42.00*	*28.69*	*12.50*	*56.45*	*13.75*	*153.39*	*221.61*

NO. 4798

Illustration 4-3.1. Part of a Paycheck Showing Deductions

WITHHOLDING TAXES

The federal government, as well as many states and cities, require employers to deduct money from employee wages for income taxes. The deduction is called a **withholding tax.**

The amount of withholding tax depends on a worker's wages, marital status, and number of withholding allowances claimed. **A withholding allowance** is used to reduce the amount of tax withheld. Workers may claim one withholding allowance for themselves, one for a husband or a wife, and one for each child or dependent.

To find the amount withheld from a worker's wages, you can use an income tax withholding table prepared by the government. Part of a federal income tax withholding table for married employees paid on a weekly basis is shown in Illustration 4-3.2.

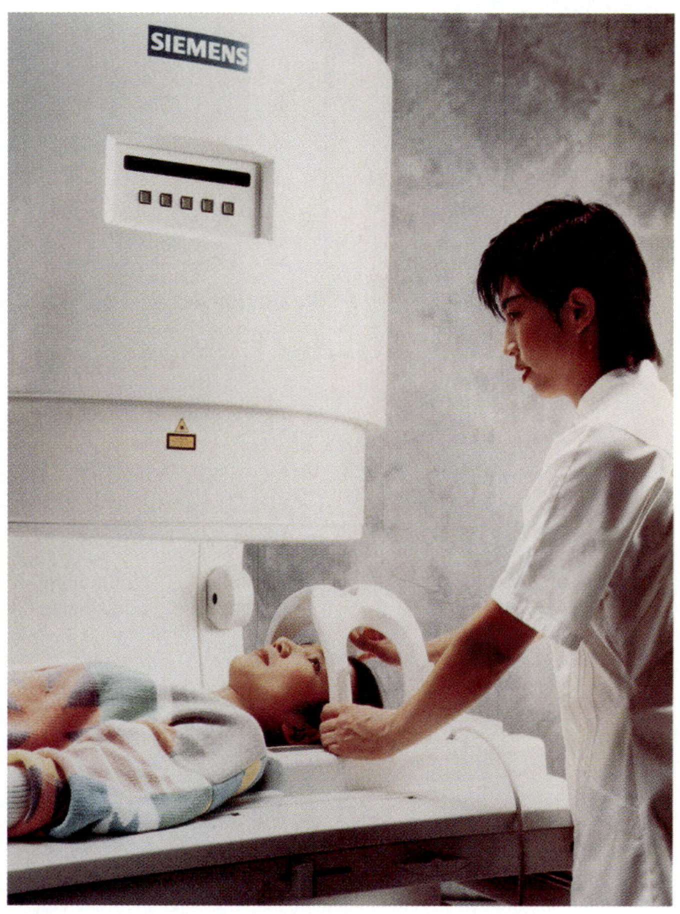

Using Withholding Tables

To find the amount of tax on an employee's wages, read down the *And the wages are—* column at the left until you reach the correct wage line. Read across to the column headed by the number of withholding allowances claimed by the employee.

EXAMPLE A medical technician's weekly wages are $455 with 2 withholding allowances.

SOLUTION The wages, $455, are on the last line of this part of the table. Read across to find the column under 2 withholding allowances. The amount of tax is $35.

Critical Thinking Look at the following withholding chart. What is the relationship between income, tax, and the number of withholding allowances? Why would less money be taken out when more allowances are claimed?

MARRIED PERSONS—WEEKLY PAYROLL PERIOD												
And the wages are—		And the number of withholding allowances is—										
At least	But less than	0	1	2	3	4	5	6	7	8	9	10
		The amount of income tax to be withhold shall be—										
310	320	29	22	14	7	0	0	0	0	0	0	0
320	330	30	23	16	9	1	0	0	0	0	0	0
330	340	32	25	17	10	3	0	0	0	0	0	0
340	350	33	26	19	12	4	0	0	0	0	0	0
350	360	35	28	20	13	6	0	0	0	0	0	0
360	370	36	29	22	15	7	0	0	0	0	0	0
370	380	38	31	23	16	9	2	0	0	0	0	0
380	390	39	32	25	18	10	3	0	0	0	0	0
390	400	41	34	26	19	12	5	0	0	0	0	0
400	410	42	35	28	21	13	6	0	0	0	0	0
410	420	44	37	29	22	15	8	1	0	0	0	0
420	430	45	38	31	24	16	9	2	0	0	0	0
430	440	47	40	32	25	18	11	4	0	0	0	0
440	450	48	41	34	27	19	12	5	0	0	0	0
450	460	50	43	35	28	21	14	7	0	0	0	0

Illustration 4-3.2. Part of an Income Tax Withholding Table

EXERCISE YOUR SKILLS

Use the tax table in Illustration 4-3.2 to find the withholding tax in each problem.

	Total Wages	Withholding Allowances		Total Wages	Withholding Allowances
1.	$390.00	1	**2.**	$399.00	2
3.	$411.00	0	**4.**	$444.00	4
5.	$328.97	5	**6.**	$407.81	3
7.	$457.07	6	**8.**	$438.88	9

FICA TAXES

The tax for social security is part of the Federal Insurance Contributions Act. It is called the **FICA tax** or *social security tax*. The tax rate and the maximum amount of wages on which it is based are changed by Congress from time to time. The tax rate has two portions: a social security tax and a Medicare tax. In this text, an overall FICA tax rate of 7.65% will be used; 6.2% for social security and 1.45% for Medicare. The social security portion is applied to a maximum wage of $70,000. The Medicare portion is applied to all wages.

To find the amount of tax, change 7.65% to the decimal 0.0765. Then multiply the employee's wages for the week by the FICA tax rate, or 0.0765.

EXAMPLE An employee earned $531.10 a week. Find the FICA tax.

SOLUTION $0.0765 \times \$531.10 = \$40.629$, or $40.63

EXERCISE YOUR SKILLS

Find the FICA tax (0.0765) on each weekly wage.

9. $475.00 **10.** $556.34 **11.** $249.40

12. $497.45 **13.** $749.23 **14.** $180.04

15. $289.48 **16.** $863.78 **17.** $2,000.00

USING FICA TABLES

In a business that completes payrolls manually, a worker usually uses a table to find the FICA tax. A partial FICA table is shown in Illustration 4-3.3, on the following page. In the complete FICA table, the FICA tax is shown for pay from $0.07 to $100, and for multiples of $100. To find the tax on pay of more than $100, add the tax for $100, or multiple of $100, to the tax for the amount less than $100.

EXAMPLE Find the FICA tax on a wage of $531.10.

SOLUTION Use the table.

$38.25	← Tax on $500
+ 2.38	← Tax on $31.10
$40.63	← Total FICA tax

Critical Thinking Find the tax by multiplying the wages by 7.65%. Compare your results with those found using the table. What is the purpose of having a chart if using a calculator will result in the same answer?

SOCIAL SECURITY EMPLOYEE TAX TABLE—7.65%										
Wages at least	But less than	Tax to be withheld	Wages at least	But less than	Tax to be withheld	Wages at least	But less than	Tax to be withheld		
31.05	31.18	2.38	32.88	33.01	2.52	34.71	34.84	2.66		
31.18	31.31	2.39	33.01	33.14	2.53	34.84	34.97	2.67		
31.31	31.44	2.40	33.14	33.27	2.54	34.97	35.10	2.68		
31.44	31.57	2.41	33.27	33.40	2.55	35.10	35.23	2.69	The FICA tax to be withheld on multiples of $100 is:	
31.57	31.70	2.42	33.40	33.53	2.56	35.23	35.36	2.70		
31.70	31.84	2.43	33.53	33.67	2.57	35.36	35.50	2.71		
31.84	31.97	2.44	33.67	33.80	2.58	35.50	35.63	2.72		
31.97	32.10	2.45	33.80	33.93	2.59	35.63	35.76	2.73	Wage	Tax to be withheld
32.10	32.23	2.46	33.93	34.06	2.60	35.76	35.89	2.74	100	7.65
32.23	32.36	2.47	34.06	34.19	2.61	35.89	36.02	2.75	200	15.30
32.36	32.49	2.48	34.19	34.32	2.62	36.02	36.15	2.76	300	22.95
32.49	32.62	2.49	34.32	34.45	2.63	36.15	36.28	2.77	400	30.60
32.62	32.75	2.50	34.45	34.58	2.64	36.28	36.41	2.78	500	38.25
32.75	32.88	2.51	34.58	34.71	2.65	36.41	36.54	2.79	600	45.90

Illustration 4-3.3. Part of a Social Security Tax Table for Weekly Wages

EXERCISE YOUR SKILLS

Use the table in Illustration 4-3.3 to find the taxes.

18. $235.78
19. $434.99
20. $631.05
21. $333.53
22. $536.53
23. $432.75
24. $436.24
25. $331.43

Emi Isobe's gross weekly wage is $436. She has 2 withholding allowances. Each week, her employer deducts $35.20 for payment to the credit union and $35.80 for insurance. For the week, use Illustrations 4-3.2 and 4-3.3 to find Emi's

26. Federal withholding tax.

27. FICA tax.

PUT IT ALL TOGETHER

Todd Pratt works a 40-hour week at $9.40 an hour with time and a half for overtime. Last week he worked 44 hours. From his gross pay, $30 is deducted for his savings plan and $46.89 for insurance. He has 1 withholding allowance. For last week, use Illustrations 4-3.2 and 4-3.3 to find Todd's

28. Gross regular pay

29. Gross overtime pay

30. Gross pay

31. Federal withholding tax

32. FICA tax

33. Net pay

Complete the table below.

	Name	Allowances	Gross Wages	FICA Tax	Income Tax	Other	Total	Net Wages
34.	Dean	1	$431.31			$56.81		
35.	Dent	0	335.74			38.93		
36.	Dern	4	436.42			65.39		
37.	Dest	2	333.07			49.23		
38.	Totals							

39. ◆ Work in groups. Interview at least four people who receive paychecks and have deductions taken out of them. Make a two-column report. The first column should show required deductions; the second column personal or optional deductions. The third column, percent of paycheck. Compare this list to the list you made in the beginning of this lesson.

MIXED REVIEW

1. $\$25.60 \times \frac{1}{8}$

2. $3\frac{1}{8} \div \frac{5}{8}$

3. $5\frac{1}{4} - 3\frac{5}{8}$

4. $\frac{2}{5}$ as large as $44.50

5. $\frac{1}{4}$ more than 40

6. Faye Sim is paid $276 a week. Her employer deducts $27 for the federal withholding tax, $11.39 for insurance, and 0.0765 for the FICA tax. What is Faye's net pay for 1 week?

7. Lana Hatton is paid $0.38 for each lever she produces. Last week she produced 627 levers. Her employer deducted $21 for the federal withholding tax, $25 for her credit union savings plan, and 0.0765 for the FICA tax. What was Lana's net pay for the week?

8. Tyrone's yearly pay is $18,400 a year. Lavonda's yearly pay is $\frac{2}{5}$ more than Tyrone's. What is Lavonda's yearly pay?

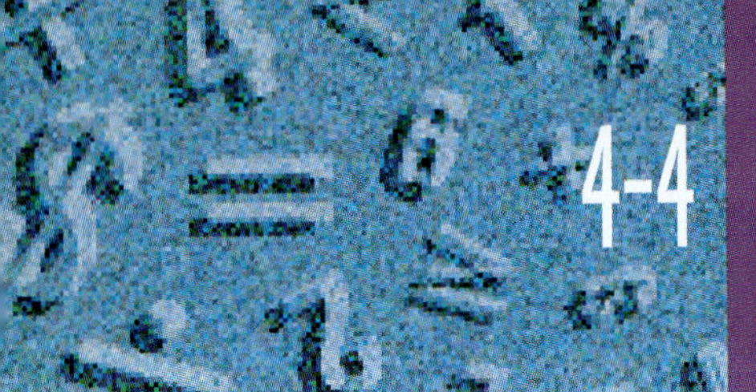

4-4 MORE PERCENTS

OBJECTIVES

In this lesson, you will learn to

- *find a percent of a number;*
- *find what percent one number is of another;*
- *use fractional equivalents of percents; and*
- *use fractional parts of 1%.*

Jana scored 75% on a test. Zenia scored 80%. Jana looked at their scores and complained, "But you only had one more right answer than I had!" How many questions were on the test that Jana and Zenia took?

◆ Work with a partner. Write down the steps you would take to determine the number of questions on the test.

WARM UP

Change to decimals.

a. $\frac{1}{8}$ **b.** 32.9% **c.** 0.25% **d.** $\frac{3}{4}\%$

Change to fractions or mixed numbers in simplest form.

e. 40% **f.** 325%

WORKING WITH PERCENTS

Finding a Percent of a Number

Suppose you earn $10,000 a year and your fringe benefits are 25% of your wages. Since percent means "hundredths," 25% means 25 hundredths. So, finding 25 percent of $10,000 is the same as multiplying 25 hundredths (0.25) times $10,000.

In other words, when you need to find a percent of a number, change the percent to a decimal and multiply. For example:

$$25\% \times \$10{,}000 = 0.25 \times \$10{,}000 = \$2{,}500$$

Mental Math Tip Multiply by the number in the percent, then put the decimal point in the answer. For example, first multiply 25 × $10,000 to find $250,000, then move the decimal point two places to the left to get $2,500.

Finding What Percent a Number is of Another

To find what percent a number is of another, divide the one number (the part) by the other number (the whole). Then show the result as a percent.

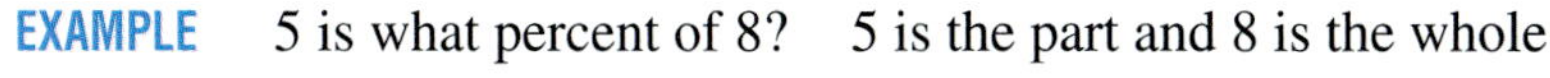

EXAMPLE 5 is what percent of 8? 5 is the part and 8 is the whole

SOLUTION $8\overline{)5.000}$ = 0.625 = 62.5% **CHECK:** 62.5% of 8 = 0.625 × 8 = 5

Hint Before you divide a fraction, simplify it to make the division easier.

EXAMPLE Write $\frac{27}{81}$ as a percent.

SOLUTION $\frac{27}{81} = \frac{1}{3} = 0.33\frac{1}{3} = 33\frac{1}{3}\%$

Any Number is 100% of Itself You can show that any number is 100% of itself by solving this problem: 26 is what percent of 26? Since the fraction $\frac{26}{26}$ shows the relationship of 26 to 26, then

$$\mathbf{\frac{26}{26} = 26 \div 26 = 1 = 100\%.}$$

When you compare one number to another number that is larger, the result is *less* than 100%. When you compare one number to itself, the result is *equal* to 100%. When you compare one number to another number that is smaller, the result is more than 100%.

EXAMPLES 24 compared to 30 is 80%, since $\frac{24}{30} = \frac{4}{5} = 80\%$

24 compared to 24 is 100%, since $\frac{24}{24} = 1 = 100\%$

24 compared to 20 is 120%, since $\frac{24}{20} = \frac{6}{5} = 120\%$

EXERCISE YOUR SKILLS

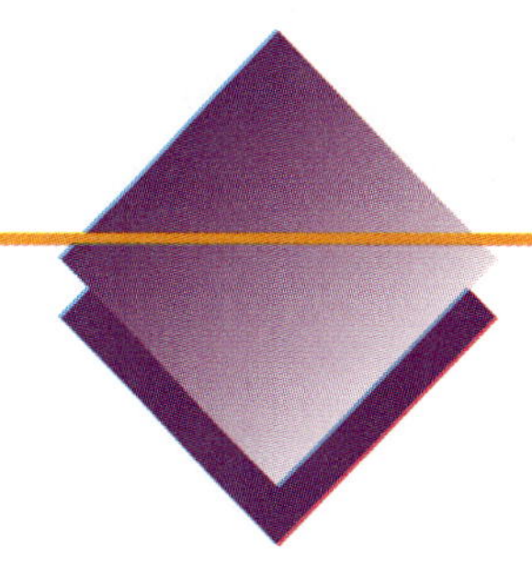

Multiply each problem mentally. Write the answers only.

1. 2% of \$600 **2.** 6% of \$110 **3.** 4% of \$120
4. 15% of \$40 **5.** 32% of \$500 **6.** 1% of \$480
7. 100% of \$68 **8.** 300% of \$3.10 **9.** 70% of \$31
10. 80% of \$75 **11.** 105% of \$600 **12.** 1% of \$1,856

Find the unknown percent in each problem.

13. 30 is ?% of 60 **14.** \$8 is ?% of \$32 **15.** \$10 is ?% of \$25
16. ?% of \$30 is \$6 **17.** 50 is ?% of 25 **18.** ?% of 7 is 28
19. \$24 is ?% of \$20 **20.** 45 is ?% of 180 **21.** ?% of \$1,200 is \$54
22. \$4.80 is ?% of \$60 **23.** $5\frac{1}{3}$ is ?% of 16 **24.** $10\frac{1}{8}$ is ?% of $60\frac{3}{4}$

TECHNIQUES FOR SIMPLIFYING COMPUTATIONS

Use Fractional Equivalents Some percents are contained in 100% an exact number of times. For example, 25% is contained exactly 4 times in 100%. Thus, 25% is $\frac{1}{4}$ of the base number, 100%.

$$\frac{\mathbf{25\%}}{\mathbf{100\%}} = \frac{\mathbf{25}}{\mathbf{100}} = \frac{\mathbf{1}}{\mathbf{4}}$$

When you multiply by a percent that is contained evenly within 100%, you can multiply mentally using the fractional equivalent. For example, let's say that your fringe benefits are 25% of your wages, which are $12,000. To find the dollar amount of your fringe benefits, multiply $12,000 by $\frac{1}{4}$ instead of 0.25. Remember that multiplying by $\frac{1}{4}$ is the same as dividing by 4.

EXAMPLE $25\% \text{ of } \$12{,}000 = \frac{1}{4} \times \$12{,}000 = \$3{,}000$

The same process can be used for numbers which are multiples of these fractional equivalents. For example, 75% is 3 times 25%, or $\frac{3}{4}$ of 100%; 125% is 5 times 25%, or $\frac{1}{4}$.

Common fractional equivalents of percents are shown below. Learn these percents and their fractional equivalents to simplify computation.

Common Fractional Equivalents			
$50\% = \frac{1}{2}$	$25\% = \frac{1}{4}$	$75\% = \frac{3}{4}$	$20\% = \frac{1}{5}$
$40\% = \frac{2}{5}$	$60\% = \frac{3}{5}$	$80\% = \frac{4}{5}$	$12\frac{1}{2}\% = \frac{1}{8}$
$37\frac{1}{2}\% = \frac{3}{8}$	$62\frac{1}{2}\% = \frac{5}{8}$	$87\frac{1}{2}\% = \frac{7}{8}$	$33\frac{1}{3}\% = \frac{1}{3}$
$66\frac{2}{3}\% = \frac{2}{3}$			

Use 1%, 10%, 100%, and 1,000% of a Number The decimals or whole numbers that are equivalent to 1%, 10%, 100%, and 1,000% are:

1% = 0.01 **10% = 0.1** **100% = 1** **1,000% = 10**

To multiply by 1%, 10%, etc., move the decimal point in the number as if you were multiplying by 0.01, 0.1, 1, or 10.

EXAMPLES

To Find	Multiply	By Moving the Decimal Point	Answer
1% of 375	0.01 × 375	2 places to left	3.75
10% of 375	0.1 × 375	1 place to left	37.5
100% of 375	1 × 375	no places	375.
1,000% of 375	10 × 375	1 place to right	3,750.

EXERCISE YOUR SKILLS

Mental Math Use fractional equivalents to find the following.

25. 50% of $28

26. 25% of $48

27. 75% of $36

28. $12\frac{1}{2}\%$ of $24

29. 20% of $20

30. 37% of $48

31. $66\frac{2}{3}\%$ of $39

32. $87\frac{1}{2}\%$ of $40

33. $62\frac{1}{2}\%$ of $80

34. $33\frac{1}{3}\%$ of $27

Solve. Use fractional equivalents when it is easier.

35. 25% of $1,600

36. 80% of $5,800

37. 30% of $2,900

38. 3% of $1,467

39. 5% of $1,840

40. 106% of $805

41. $12\frac{1}{2}\%$ of $1,200

42. 125% of $88

43. $66\frac{2}{3}\%$ of $270

44. $187\frac{1}{2}\%$ of $16.40

Mental Math Write the answers only. If the answer has a fraction of a cent, round it to the nearest cent.

45. 1% of 40

46. 1% of $7.48

47. 10% of $47.45

48. 100% of 284

49. 1,000% of $15

50. 1,000% of $7.50

PROBLEM SOLVING

Choose a Method

Any of these methods can be used to solve problems: paper and pencil, estimation, a calculator, mental math, or a computer.

Assuming you had all the needed tools available:

a. When would you use mental math to solve a problem?

b. When would you use estimation to solve a problem?

c. When would you use a calculator to solve a problem?

d. When would you use a computer to solve a problem?

e. When would you draw a diagram or use other visual aids to solve a problem?

FRACTIONAL PARTS OF 1%

The numeral $\frac{1}{2}\%$ means $\frac{1}{2}$ of 1%. There are several ways to find $\frac{1}{2}\%$ of a number.

EXAMPLE Find $\frac{1}{2}\%$ of $2,400.

SOLUTION 1 Use mental math. 1% of $2,400 = $24

$\frac{1}{2}$ of $24 = $12

SOLUTION 2 Use mental math. $\frac{1}{2}$ of $2,400 is $1,200

1% of $1,200 = $12

SOLUTION 3 Use a calculator. $\frac{1}{2}\% = 0.005$

$0.005 \times \$2,400 = \$12$

To show a fraction of a percent in decimal form, change the fraction to its decimal equivalent. Then, drop the percent sign and move the decimal point two places to the left.

EXAMPLES $\frac{3}{4}\% = 0.75\% = 0.0075$ $\frac{1}{4}\% = 0.25\% = 0.0025$

When the numerator is not exactly divisible by the denominator, drop the percent sign, move the decimal point two places to the left, and leave the fraction as a fraction.

EXAMPLES $\frac{2}{3}\% = 0.00\frac{2}{3}$ $\frac{4}{9}\% = 0.00\frac{4}{9}$

EXERCISE YOUR SKILLS

Find the indicated percent of each amount.

51. $\frac{1}{4}\%$ of $1,600; $400; $3,600; $2,800; $820

52. $\frac{1}{8}\%$ of $3,200; $5,600; $8,800; $480; $800

Mental Math Write only the product.

53. 25% of $3,200 **54.** $\frac{1}{4}$ of $3,200 **55.** $\frac{1}{4}\%$ of $3,200

56. $\frac{1}{2}\%$ of $32 **57.** $\frac{1}{5}\%$ of $500 **58.** $\frac{3}{4}\%$ of $400

Write each as a decimal.

59. $\frac{1}{2}$ of 1% **60.** $\frac{2}{5}$ of 1% **61.** $\frac{5}{8}\%$

Write each numeral as a percent, with the fractional part of 1% written as a fraction.

62. 0.0075 **63.** 0.005 **64.** 0.00375

65. $0.0033\frac{1}{3}$ **66.** 0.0875 **67.** 0.00875

68. ◆ Copy the following guess and check table on a piece of paper. Complete the table to answer the question at the beginning of the lesson about the number of questions on Zenia and Jana's test.

Number of Guesses	Jana	Zenia
10	*7.5*	*8*
20	*15*	*16*
30	*22.5*	*24*
40	*30*	*32*

MIXED REVIEW

Solve. Round all answers to the nearest cent.

1. 10% of $23.52

2. 1% of $9.33 is ?

3. $\frac{1}{5}$% of $745 is ?

4. $37\frac{1}{2}$% of $896 is ?

5. $350 is what percent of $500?

6. Write $87\frac{1}{2}$%, as a decimal and as a fraction in simplest form.

7. Ella Dent works on an 8-hour-day basis at $12.46 an hour with time and a half for overtime. During one week she worked these hours: Monday, 9; Tuesday, 8; Wednesday, $6\frac{1}{2}$; Thursday, 9; Friday, $8\frac{1}{2}$. What were Ella's gross earnings for the week?

8. A factory worker with 15 years experience is paid 72 cents for each clamp produced that passes inspection. In one week the worker produced these amounts: 137, 101, 186, 156, and 134. A total of 57 clamps did not pass inspection. What were the worker's gross earnings?

9. A fruit drink is made with 4 parts of orange to every 5 parts of grapefruit. If a company makes 1,278 gallons of the fruit drink, how many gallons of orange do they use?

10. Casandra Williams goes to college and works a part-time job. February 1 is on a Monday. In February, she earned $215 each week from her job and was paid each Friday. She also earned $20 baby sitting on February 13 and $15 on February 26. She received $100 from her parents on February 15. Reimbursement for book purchases of $183.50 came in the mail on the February 10. Prepare a cash receipts record for the month of February for Casandra.

4-5 FRINGE BENEFITS AND JOB EXPENSES

OBJECTIVES

In this lesson, you will learn to

- *find total fringe benefits; and*
- *find net job benefits; and*
- *compare jobs.*

Bill's uncle asked him to help with a remodeling job that will take 10 days to complete. He said he would pay Bill $60 a day or start with $1 for the first day and double his pay each day. Which wage will earn Bill the most money? How much more?

◆ Make a table or spreadsheet with three columns. Label the first column **Day Number,** the second column **$60 Day,** and the third column **Double.** Number 10 rows for each day and calculate the money earned that day in the first column, and the total money earned for each method under the appropriate column.

WARM UP

a. Change $\frac{2}{11}$ to a decimal correct to two places.

b. $4\frac{1}{4} \times 6\frac{2}{5} = ?$

c. $\frac{1}{3} + \frac{2}{7} = ?$

d. $309 \times 0.01 = ?$

e. $\$54.63 \div 0.01 = ?$

f. $\frac{6}{7}$ of $\$145 = ?$

NET JOB BENEFITS

In addition to wages, many employers provide their workers other things of value called **fringe benefits.** For example, employers may provide free or low-cost health and accident insurance, life insurance, and pensions. They may also provide paid holidays, paid sick leave, paid vacation time, the use of a car, a credit union, uniforms, parking, discounts for purchases of merchandise, recreational facilities, and education or training.

Finding Total Job Benefits Fringe benefits are an important part of a job's total value. Typically, fringe benefits are worth as much as 15% to 40% of the amount paid in wages. When you are considering a job offer, the value of the fringe benefits should be added to the amount of wages to find the **total job benefits.**

Gross Pay + Fringe Benefits = Total Job Benefits

Critical Thinking The value of some fringe benefits, such as paid holidays, can be figured very accurately. However, the value of other fringe benefits, such as free recreation facilities, can only be estimated. If you were offered a benefit package that included use of a free gymnasium, how would you estimate its dollar value?

EXAMPLE Sonia Burgoz is a systems analyst with Tri-State Paper Company. Last year Sonia calculated her total job benefits.

SOLUTION

Gross pay		$32,400
Fringe benefits:		
Paid pension	$2,592	
Health insurance	1,440	
Paid vacations, holidays	2,278	
Paid course tuition	925	
Use of gymnasium (estimated)	360	7,595
Total job benefits		$39,995

Job Expenses Almost every job has expenses. Some examples of job expenses are union or professional dues, commuting expenses, uniforms, licenses, and tools. There may also be funds for birthdays and other events. These expenses can add up.

To find net job benefits, subtract total job expenses from total job benefits.

Total Job Benefits − Job Expenses = Net Job Benefits

EXAMPLE Denelda Ogg had total job benefits of $27,320. Her job expenses were $624 for travel, $35 for a required license, $175 for professional dues, and $25 for the company birthday fund. Find her net job benefits.

SOLUTION

Total job benefits		$27,320
Job expenses:		
Travel	$624	
License	35	
Professional dues	175	
Birthday fund	25	859
Net job benefits		$26,461

EXERCISE YOUR SKILLS

Ray Novotney's gross pay for a year is $18,512. He estimates his yearly fringe benefits to be worth these amounts: paid pension, $1,226; health and life insurance, $1,534; paid vacations and holidays, $1,993; use of company car, $2,936; free parking (estimated), $690.

1. Find the total value of Ray's fringe benefits for the year.
2. Find his total job benefits for the year.

Jarvis Oertel has been offered a job that pays $12.45 an hour for a 40-hour week. He estimates that his weekly fringe benefits would be: pension, $30; health insurance, $15.65; parking, $14; uniforms, $5. Find Jarvis'

3. estimated weekly wage.
4. exact weekly wage.
5. total weekly fringe benefits.
6. total job benefits per week.

Karin Hedberg is paid $8.72 an hour for a 40-hour week. Her estimated fringe benefits are 37% (0.37) of her wages. Find her

7. estimated weekly wage.
8. exact annual wage.
9. total yearly fringe benefits.
10. total yearly job benefits.
11. Bob Dees is paid $1,128 per month. He estimates that his fringe benefits amount to 0.23 of his pay. Find his total job benefits for a year.

Kumar Godhwani estimates that his total job benefits are $17,220 per year. He estimates these yearly job expenses: uniforms, $276; transportation, $398; parking, $260; dues, $96; other, $74.

12. Find his yearly job expenses.
13. Find his yearly net job benefits.
14. Jay Orr's job pays $1,374 a month. His estimated monthly fringe benefits are worth $384 and his monthly job expenses are tools, $38; dues, $22; travel costs, $24; parking, $42. Find his monthly net job benefits.

Mary Farr earns $420 a week. Estimated fringe benefits are 35% (0.35) of her wages. Estimated yearly job expenses are licenses, $345; uniforms, $326; travel, $745; dues, $535. Find her

15. annual wage.
16. annual fringe benefits.
17. total annual job benefits.
18. annual job expenses.
19. annual net job benefits.

Comparing Jobs When you compare jobs you should consider many features about each job. For example, how much you like the job, the chances for raises and promotions, the chances of layoffs, or job security, and the net job benefits offered by each job.

PUT IT ALL TOGETHER

Ikuko Goro's job pays $23,650 in yearly wages and 19% of her wages in yearly fringe benefits. She estimates that yearly job expenses are $1,354. Another job which she is looking at pays $22,490 in yearly wages and has estimated yearly fringe benefits of 26%, with job expenses of $1,080.

20. Find the net job benefits of Ikuko's current job.

21. Find the net job benefits of the other job.

22. Which job offers the greater net job benefits, and how much greater?

23. Georgio Bakalis can work for TRI Co. for $360 per week or Ellis, Inc. for $1,505 per month. Fringe benefits average 19% of yearly wages at TRI Co. and 24% at Ellis, Inc. Job expenses are estimated to be $956 per year at TRI Co. and $237 per year at Ellis, Inc. Which job would give Georgio more net job benefits? How much more?

24. ◆ Make a list of five job benefits you would like to receive. Next to each item, estimate the cost of each benefit and give one reason why an employer might want to offer such a benefit to their employees.

MIXED REVIEW

Find each answer in simplest form.

1. $\frac{3}{7} - \frac{2}{5}$

2. $3\frac{4}{9} + 8\frac{2}{3}$

3. $8\frac{7}{8} - 3\frac{1}{6}$

4. $\frac{1}{3}$ less than 36 is ?

5. $\frac{2}{5}$ more than 45 is ?

You are paid $21,800 a year. You estimate the value of your fringe benefits to be 18% of your yearly income and your annual job expenses to be $1,230 for travel, $185 for dues, and $420 for other expenses. Find your

6. total job benefits per year.

7. net job benefits per year.

8. Sally has earned marks of 87 on four tests, 90 on three tests, 82 on three tests, and 98 on four tests. Find the average mark she has earned so far, to the nearest tenth.

9. Julio's July 31 check register balance was $951.21. His bank statement showed earned interest of $5.67 and a service charge of $1.34. He found that he had not recorded check No. 211 for $17.82 and that check No. 238 for $51.98 had been recorded as $51.89. What was his correct check register balance?

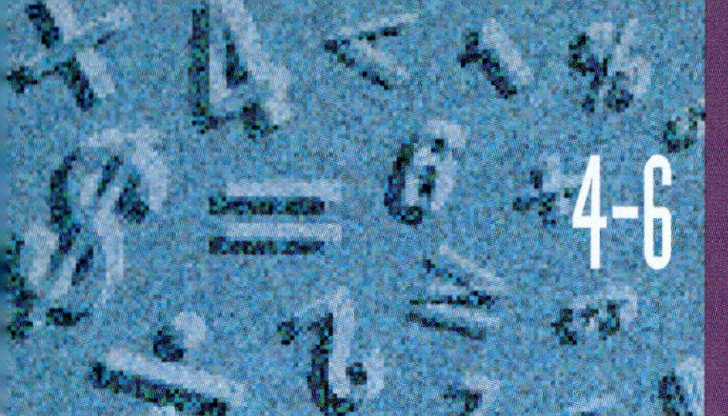

4-6 PERCENT RELATIONSHIPS

OBJECTIVES

In this lesson, you will learn to

- *find a number that is a percent greater or smaller than another;*
- *find what percent a number is of another; and*
- *find percent of increase or decrease.*

Angie said, "I earned 25% more than John." John said, "I earned 20% less than Angie."

◆ Discuss with a partner if it is possible that both people are telling the truth. Explain your reasoning.

WARM UP

a. $\frac{1}{4}$% of \$328 is ?

b. 1% of \$1,600 = ?

c. 1,000% of \$20 = ?

d. 480 is what percent of 600?

e. 40% of \$350 = ?

f. Write $62\frac{1}{2}$% as a decimal and as a fraction.

PERCENT RELATIONSHIPS

In Chapter 3 you studied relationships with fractions. You will study similar relationships with percent in this lesson.

Finding a Number That is a Percent Greater or Smaller Than Another

When we say "25% greater than 32" and "25% more than 32," we mean 25% added to 32, or 32 + 25% of 32. The result is 32 + 8, or 40.

When we say "25% smaller than 32" and "25% less than 32," we mean 25% subtracted from 32, or 32 − 25% of 32. The result is 32 − 8, or 24.

EXAMPLE 1 32 increased by 25% of itself equals what number?

SOLUTION

32	known number
25% of 32 = 8	increase
32 + 8 = 40	unknown number

CHECK: 32 + (25% × 32) = 32 + 8 = 40

EXAMPLE 2 32 decreased by 25% of itself equals what number?

SOLUTION

32	known number
25% of 32 = 8	increase
32 − 8 = 24	unknown number

CHECK: 32 − (25% × 32) = 32 − 8 = 24

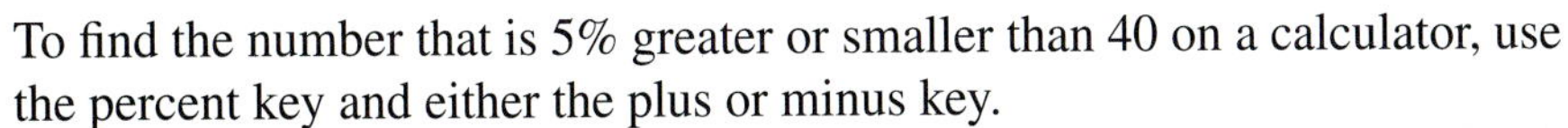

To find the number that is 5% greater or smaller than 40 on a calculator, use the percent key and either the plus or minus key.

EXAMPLE Press: 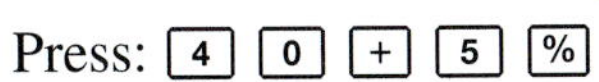 4 0 + 5 %

Display: 42.

EXERCISE YOUR SKILLS

Solve.

1. 10% more than $500
2. 10% greater than $200
3. 10% less than $900
4. 10% less than $600
5. 10% smaller than $700
6. 1% larger than $700
7. $500 decreased by 20%
8. $80 increased by 10%
9. $90 reduced by $33\frac{1}{3}$%
10. 30% larger than $50
11. 200% more than $20
12. $4 increased by 200%
13. A new job pays 20% more per hour than an old job. If the old job paid $8 an hour, what does the new job pay per hour?
14. Since your employer changed health insurance companies, your health insurance deduction decreased by 10%. If your old health insurance deduction was $60, find your new deduction.

Finding What Percent One Number is Greater or Smaller Than Another

If one job you are offered pays $400 a week while another pays $320, you may want to know by what percent the one wage is greater than the other wage.

EXAMPLE $400 is what percent greater than $320?

SOLUTION $400 − $320 = $80
$80 is the part.

Divide the part by the original number and change the result to a percent.

$80 divided by $320 = $\frac{1}{4}$ $\frac{1}{4} = 25\%$

$400 is 25% greater than $320.

To find the percent that one number is smaller or less than another, do the same thing.

EXAMPLE 24 is what percent smaller, or less, than 30?

SOLUTION $30 - 24 = 6$ **CHECK** $30 - 20\%$ of $30 = 30 - 6 = 24$

$\frac{6}{30} = \frac{1}{5} = 20\%$

To use a calculator to solve the problem:

Press: [3] [0] [−] [2] [4] [=] Display: 6.

Press: [÷] [3] [0] [%] [=] Display: 20.

Add a percent sign and write the answer as 20%.

EXERCISE YOUR SKILLS

Find the unknown percent in each problem.

15. $5 is ?% greater than $4

16. $4 is ?% smaller than $5

17. $3 is ?% smaller than $4

18. ?% less than $10 is $5

19. $24 is ?% less than $32

20. ?% more than $16 is $20

21. $44 is ?% more than $40

22. $7 is ?% less than $10

23. $5 is ?% more than $4

24. $6 is ?% more than $5

25. $15 is ?% more than $5

26. ?% more than $10 is $25

27. $10 is ?% less than $25

28. $12 is ?% more than $6

Percent Increase or Decrease

A practical use for finding what percent one number is greater or smaller than another is to compare amounts for two periods of time. You may want to compare this year's net job benefits with last year's. A business may want to compare this year's sales with last year's sales.

Sales This Year $600,000 **Sales Last Year $450,000**

THAT'S A $33\frac{1}{3}\%$ INCREASE

When you show the increase or decrease as a percent, use the amount for the *earlier* period of time as the base.

EXAMPLE A student earned a score of 84% on this week's test and 80% on last week's test. What was the percent increase or decrease in the score?

SOLUTION $84\% - 80\% = 4\%$ difference $\frac{4\%}{80\%} = \frac{1}{20} = 5\%$ increase

EXERCISE YOUR SKILLS

29. Isabel Muyo's salary is $20,832 this year. Last year her salary was $18,600. What was the percent of increase or decrease?

30. In Ed's new job his fringe benefits are $75.24 a week. In his old job the benefits were $66 a week. Find the percent of increase or decrease.

31. Garry's net pay on his new job is $205.77. His net pay on his old job was $180.50. What is the percent increase or decrease in his net pay?

32. This year Vera's monthly net job benefits total $1,255. Last year her benefits totaled $1,225. By what percent did Vera's job benefits increase or decrease, to the nearest tenth of a percent?

Since she started carpooling, Carole's job expenses dropped from $75 to $51 per month.

33. Find the estimated percent of decrease in job expenses.

34. Find the actual percent of decrease in job expenses.

35. ◆ Tell how to solve the problem at the beginning of the lesson. Then write a similar problem of your own.

MIXED REVIEW

1. Show $\frac{1}{4}\%$ as a decimal.
2. $\frac{1}{5}\%$ of \$380 is what amount?
3. \$45 increased by 24% of itself is?
4. \$7.00 decreased by 16% of itself is?
5. Owen Company pays its employees an average of \$9.10 an hour. The average paid by Barons is \$8.75 an hour. By what percent is Owen Company's average hourly pay greater than Barons?
6. Doris's net pay this week is \$210. This amount is only $\frac{5}{6}$ as much as her pay for last week. What was Doris's pay last week?
7. A cashier started with \$175 cash in bills and change. At the end of the day, the cash register totals showed \$1,864.26 taken in and \$58.38 paid out. How much cash should be in the cash register at the end of the day?
8. You write a check for \$4,200.25. Write that amount in words.
9. On May 31 Elmer's account balances were: bank, \$763.14; check register, \$717.08. While comparing his check register to the bank statement, Elmer found a service charge of \$1.65 and interest earned of \$3. Checks outstanding were: #877, \$19.78; #879, \$10; #880, \$14.93. Prepare a reconciliation statement for Elmer.
10. Sherry Downey can work for Southwest Auto Co. for \$5.25 an hour for a 40-hour week or take a similar job at Office Solutions, Inc. for \$6.10 an hour for a 40-hour week. Fringe benefits at Southwest Auto Co. would be 0.27 of yearly wages. Fringe benefits at Office Solutions, Inc. would be 0.17 of yearly wages. Sherry estimates her yearly job expenses would be \$568 at Southwest Auto Co., and \$785 at Office Solutions, Inc. Which job would give Sherry greater net job benefits per year? How much more?
11. 8 is what part of 64?
12. 9 is what part less than 12?
13. $\frac{3}{11}$ as a decimal to the nearest hundredth is ?
14. $\frac{5}{14}$ as a decimal rounded to the nearest thousandth is ?
15. Ed O'Donnel is paid \$8.20 an hour for a 40-hour week. Each week his employer deducts \$40 in withholding taxes, FICA tax at the rate of 7.65%, and \$12 for a savings plan. Find Ed's net pay for one week.

4-7 COMMISSION

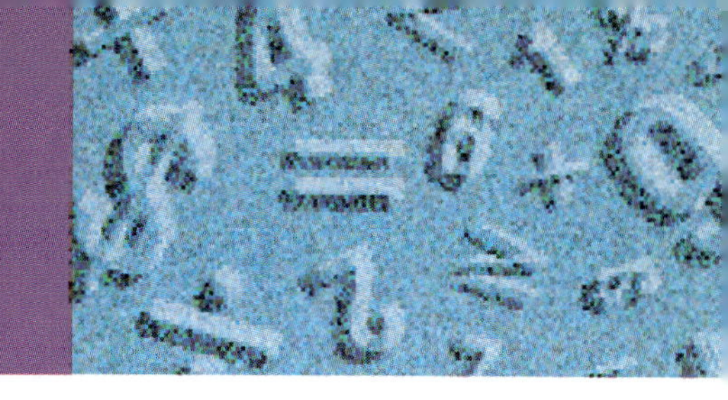

There are two ads in the newspaper for sales jobs. The first job pays a straight commission of 10% on all sales; the second job pays a commission of 5% on sales up to a certain amount, and 15% on sales over that amount.

◆ You are going to interview for both jobs. Make a list of questions you will ask each employer in order to determine which job offers the higher salary potential.

OBJECTIVES

In this lesson, you will learn to find

- *straight commissions;*
- *graduated commissions;*
- *the rate of commission; and*
- *net proceeds.*

WARM UP

a. $\frac{1}{3}\%$ of \$195 is what amount?

b. Write 0.087 as a percent.

c. \$8.75 increased by 36% of itself = ?

d. Write $1\frac{3}{4}\%$ as a decimal.

e. \$135 is what percent greater than \$120?

COMMISSION

Some salespeople earn a commission instead of a fixed salary. A **commission** is a fee or percentage of interest paid for sales made. A commission may be an amount for each item sold, or it may be a percent of the dollar value of sales. A higher commission may be earned for goods that are harder to sell than for goods that are easy to sell. Both a salary and a commission may be earned.

Straight Commission

Salespeople who earn only a commission work on a **straight commission** basis. When the rate of commission is an amount for each item sold, multiply the number of items by the rate to find the commission.

$$\overset{(F)}{\textit{Rate of Commission}} \times \overset{(F)}{\textit{Quantity Sold}} = \overset{(P)}{\textit{Commission}}$$

EXAMPLE Earl Brown sells greeting cards and is paid a straight commission of $0.75 on each box of cards he sells. During March, he sold 145 boxes. Find his commission.

SOLUTION $0.75 × 145 = $108.75 commission

When the rate of commission is a percent, multiply the amount of the sales by the rate to find the commission.

$$\overset{(F)}{\textit{Rate of Commission}} \times \overset{(F)}{\textit{Sales}} = \overset{(P)}{\textit{Commission}}$$

EXAMPLE Terri Ames is paid a straight commission of 5% on her sales. During September, her sales were $32,000. What was her commission?

SOLUTION 0.05 × $32,000 = $1,600 commission

Use a calculator to find commission for more complicated problems.

EXAMPLE Angie Barnes earns 12% commission on sales. She sends in orders once a week. During four weeks, her orders were: $1,050.30, $725.05, $1,305.55, and $925.85. Find her commission earned for those four weeks.

SOLUTION Use a calculator.

Press: 1050.30 [+] 725.05 [+] 1,305.55 [+] 925.85 [=]

Display: 4,006.75

Press: [×] 12 [%] [=] Display: 480.81

She would earn $480.81.

EXERCISE YOUR SKILLS

Find the commission in each problem.

	Item	Quantity	Commission on each	Total Commission
1.	Boxes of Cards	78	$2	
2.	Books	32	$0.55	
3.	Pennants	49	$1.20	

	Total Sales	Rate of Commission	Total Commission
4.	$35,986	4%	
5.	$12,435	6%	
6.	$5,750	$12\frac{1}{2}$%	

7. Dora Cache is paid a straight commission of $5.75 for each item she sells. Last month she sold 103 items. Find her estimated and exact commissions.
8. Cory Snyder earns a commission of $8\frac{1}{2}$% on sales. Last week he had sales of $534.03, $200.53, $2,340.23, $658.66, and $923.53. Find his commission to the nearest cent.

SALARY PLUS COMMISSION

Salespersons may be paid a salary plus a commission. The commission may be a percent of their total sales or of their sales above a fixed amount. This fixed amount is called a **quota.**

EXAMPLE Mae Barr is paid a salary of $250 a week plus 5% on her sales over $4,000. Last week her sales totaled $7,000. What were her total earnings for the week?

SOLUTION

Sales:	$7,000	Salary:	$250
Quota:	− 4,000	Commission:	+ 150
Sales over Quota:	$3,000	Total earnings:	$400

Commission: 5% of $3,000 or $150

Mae earned $400 for the week.

Graduated Commission

Some salespersons are paid a **graduated commission.** This means their rate of commission increases as their sales increase. For example, the rate may be 3% on the first $12,000 of sales; 4% on the next $6,000; and 5% on sales over $18,000. Graduated commissions may also be based on the number of units sold.

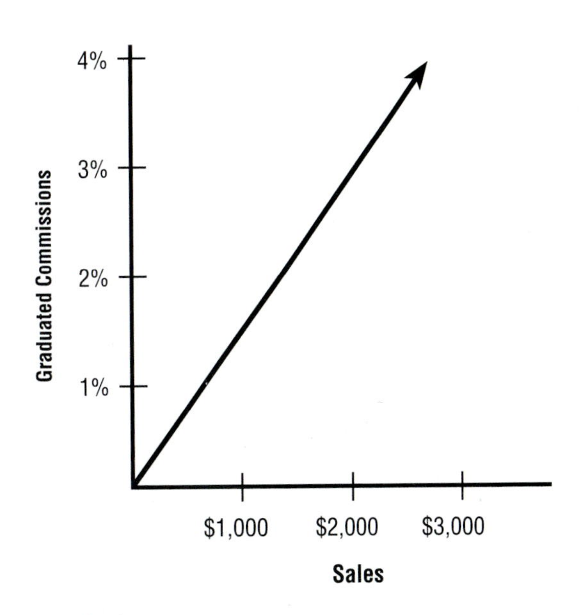

EXAMPLE Lori Wells is paid 5% commission on the first $10,000 of monthly sales and 10% on all sales over $10,000. Last month her sales were $34,000. What was her commission?

SOLUTION Sales over $10,000: $34,000 − $10,000 = $24,000

Commission on first $10,000: 0.05 × $10,000 = $500

Commission on sales over $10,000: 0.10 × $24,000 = $2,400

Total commission: $500 + $2,400 = $2,900

EXERCISE YOUR SKILLS

Find the total commission in each problem.

	Sales	Commission
9.	$9,000	10% on first $5,000; 15% on sales over $5,000
10.	$800	3% on first $400; 5% on excess over $400
11.	$2,600	6% on first $900; 9% on sales over $900
12.	$8,370	5% on first $3,500; 7% on next $6,500

13. Jo Ann White is paid a salary of $325 a week and a commission of 5% on all sales. Her sales last week were $5,780. Find her total earnings for the week.

14. Hollis Riggs earns a salary of $200 a week and a commission of 6% on all sales. If Hollis's sales for one week were $4,120, what were his total weekly earnings?

15. Toni Wild earns a salary of $1,200 a month and a commission of 7.5% on all sales over $4,000. This month her sales were $21,400. Find her total earnings for the month.

16. Kamil Wise receives a weekly salary of $400 plus $\frac{1}{2}$% commission on all sales in excess of $11,500 a week. Last week his sales were $43,870. What were his total earnings for the week?

17. Armando Lamas sells house siding. He is paid 4% commission on his first $15,000 of monthly sales and 6% commission on all sales over $15,000. In June his sales were $27,000; in July, his sales totaled $32,000. What was the total commissions earned for the 2 months?

18. Rita Gray sells cosmetic kits on a part-time basis. She is paid a weekly commission of $1.50 each on the first 50 kits she sells, $1.75 each on the next 100 kits, and $2 on any kits she sells over 150. Last week she sold 225 kits. What was her commission for the week?

FINDING THE RATE OF COMMISSION ON SALES

Suppose you know the amount of sales and how much commission was paid on the sales. To find the rate of commission, follow the same steps you used to find the percent one number is of another. In this case, the amount of sales is the base and the amount of commission is the part. The percent, or rate of commission, is found by dividing the commission by the sales.

(F) Rate of Commission = (P) Amount of commission ÷ (F) Sales

EXAMPLE A salesperson sold a computer for $3,000 and received a $150 commission. What percent commission did the salesperson receive?

SOLUTION 1 $150 ÷ $3,000 = 0.05 = 5%

SOLUTION 2 $\frac{\$150}{\$3{,}000} = \frac{1}{20} = 0.05 = 5\%$

EXERCISE YOUR SKILLS

Find the rate of commission in each problem.

	Sales	Commission Amount	Rate		Sales	Commission Amount	Rate
19.	$6,400	$448		**20.**	$625	$18.75	
21.	$480	$72		**22.**	$2,800	$350	
23.	$600	$33		**24.**	$5,400	$189	

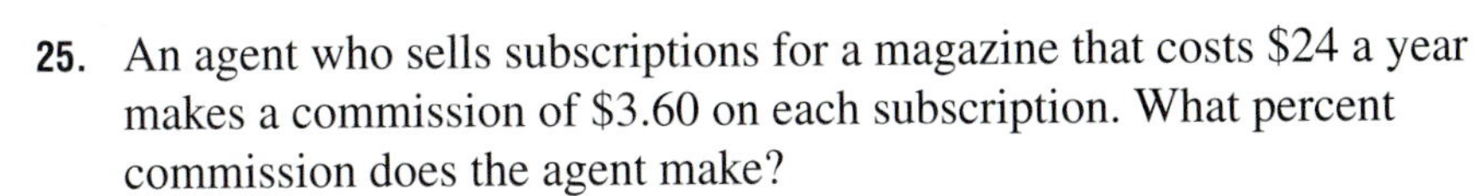

25. An agent who sells subscriptions for a magazine that costs $24 a year makes a commission of $3.60 on each subscription. What percent commission does the agent make?

Austin Orr sells motorcycles. He is paid a salary of $900 a month plus a commission on all sales. Each month his employer contributes 8% of his salary to his pension and $56 to his health insurance. Last month his sales were $35,000, and he earned a total salary and commission of $3,000.

26. How much commission was he paid?

27. What rate of commission was he paid?

28. Akemi Soga is paid a commission on all sales over $3,000 a week. Last week she earned a commission of $520 on sales of $16,000. What rate of commission was she paid?

AGENTS AND NET PROCEEDS

A person who legally acts for someone else is called an **agent.** An agent is usually paid a commission. The person for whom the agent acts is called the **principal.**

When a principal has a real estate agent sell property, the principal usually pays the agent a percent of the sale price as commission. Sometimes the principal will also pay the agent for unusual expenses caused by the sale of the property. The amount left for the principal, after deducting the agent's commission and expenses, is called the **net proceeds.**

Collection agents collect unpaid bills. They usually are paid a percent of the amount they collect. They may also be repaid for unusual expenses involved in making the collection.

Finding Commission and Net Proceeds

The rate of commission of a real estate agent is based on the sale price of the property sold. For example, a real estate agent sells a house for $80,000 and charges a 6% commission. His commission would be $80,000 × 0.06 = $4,800.

The commission of a collection agent is based on the amount collected. The net proceeds is the amount of money received by the principal after subtracting the agent's commission and expenses, if any.

Sale Price − (Commission + Expenses) = Net Proceeds

Amount Collected − (Commission + Expenses) = Net Proceeds

EXAMPLE Cal Kerr, a collection agent, collected a bill of $1,500 for B-Trix Co. He charged 35% commission and $75 for expenses. Find the net proceeds that B-Trix Co. will receive.

SOLUTION

Amount collected		$1,500
Deductions:		
Commission		
(35% × $1,500)	$525	
Expenses	+ 75	
Total Deductions	$600	− 600
Net proceeds		$ 900

EXERCISE YOUR SKILLS

Wright Corporation hired an agent to collect accounts totaling $7,700. The agent collected 78% of the accounts and charged $37\frac{1}{2}$% for collecting them.

29. What estimated amount was collected?

30. How much did the Wright Corporation receive?

Find the commission, total deductions, and net proceeds in each problem.

	Sale Price or or Amount Collected	Commission Rate Amount	Agent's Expenses	Total Deductions	Net Proceeds
31.	$45,000	6%	$187		
32.	$2,070	30%	None		
33.	$85,900	7%	None		
34.	$143,750	6%	$207.88		

35. Bi-State Real Estate Agency sold a house for Oscar Lavelle for $65,800. The agency charged 5% for commission and expenses of the sale were $320. What were the net proceeds?

36. ◆ In a small group discuss these two questions: What are the advantages and disadvantages of working on commission for an employee? for an employer?

MIXED REVIEW

1. $\frac{1}{4}$% of $36,000 is ?
2. $37\frac{1}{2}$% of $4,800 = ?
3. $92,078 × 1% = ?
4. 60 is 15% of ?
5. $108 = ?% of $360
6. 17% more than $27 is ?
7. $175 decreased by what percent of itself is $168?
8. Myrna Blair earns 10% commission on all sales and 3% more commission on sales over $15,000 per week. Her sales this week were $19,800. What was her commission for the week?
9. Luke Starr earns $425 a week in salary and 3% commission on all sales over $5,000 a week. Last week his sales were $8,089. What was Luke's gross income for the week?
10. Fatima Wyatt's sales last month were $43,200. Her total earnings for the month were $2,578, of which $850 was her monthly salary. What rate of commission was Fatima paid?
11. Zika, Inc., a computer software firm, sold 214,537 copies of a spreadsheet package this year. Last year they sold 175,850 copies. What was the percent of increase or decrease in the sales of the package?
12. Leon scored 80 on a test last week. This week his score was 15% more than last week's score. What was Leon's score this week?

EXPLORE TECHNOLOGY

TEMPLATE: Spreadsheet Application for Calculating FICA Taxes and Net Pay

Load ET4 from your applications diskette. You will see a shell of the completed spreadsheet shown below. Enter the data shown in blue (in cells D7–E16) into your spreadsheet. The spreadsheet will calculate FICA taxes and net pay for each employee in the Totals columns. Your computer screen should look like the one shown below when you are done.

	A	B	C	D	E	F	G	H	I
1				Nashua-Brinkley Corporation					
2				Payroll Sheet for Jan. 15, 19—					
3						Deductions			
4			Allow-	Gross	Income	FICA			Net
5	No.	Name	ances	Wages	Tax	Tax	Other	Total	Pay
6									
7	1	Allen	0	380.67	39.00	29.12	45.56	113.68	266.99
8	2	Baker	4	428.88	16.00	32.81	23.89	72.70	356.18
9	3	Cruz	1	395.35	34.00	30.24	93.12	157.36	237.99
10	4	Dolan	3	438.97	25.00	33.58	55.23	113.81	325.16
11	5	Elgin	2	450.00	35.00	34.43	37.56	106.99	343.01
12	6	Faber	0	458.23	50.00	35.05	34.87	119.92	338.31
13	7	Greb	0	354.56	35.00	27.12	29.98	92.10	262.46
14	8	Hirt	2	349.21	19.00	26.71	44.34	90.05	259.16
15	9	Isobe	5	438.39	11.00	33.54	37.44	81.98	356.41
16	10	Jain	1	317.44	22.00	24.28	64.55	110.83	206.61
17									
18		Totals		4,011.70	286.00	306.88	466.54	1,059.42	2,952.28
19									
20		FICA RATE	0.0765						

Answer these questions about your completed payroll sheet.

1. Which employee had the largest net pay for the period?
2. Which employee had the largest amount of deductions?
3. What was the total amount of Income Taxes withheld from wages for the week?

Now move the cursor to cell C20, labeled "FICA Rate." Enter the rate 0.0775. Notice how the FICA Tax, Total Deductions, and Net Pay amounts have all changed. These changes show what would happen to these workers if the FICA tax rate was changed from 7.65% to the higher 7.75%.

Answer these questions about your updated payroll sheet.

4. What is the formula used in Column F? What arithmetic is done?
5. What is the formula used in Column I? What arithmetic is done?
6. Why did changing cell C20 change the amounts throughout the rest of the spreadsheet?

Design Your Own

You received two job offers. The expenses and benefits for each job are shown below.

	Offer 1	Offer 2
Salary Information		
Hourly Rate	$5.15	$6.05
Annual Benefits		
Health Insurance	$2,400	$1,200
Life Insurance*	100%	150%
Health Club Membership	——	$350
401K (Retirement)*	8%	6%
Free Parking	$450	——
Expense Information		
Health Insurance	——	$1,200
Commuting Costs	$477	$755
Dues	$87.50	$175
Tools	——	$280
Uniforms	——	$315

*Benefit is stated as a percent of annual salary.

7. Create a spreadsheet that will allow you to compare the jobs based on net benefits. Assume that each job is for a 40-hour week; 52-week year. Your spreadsheet should contain a row for each of the items shown above. Also enter row or column labels and formulas to calculate annual pay, retirement benefit, total job benefits, total job expenses, and net benefits.

After completing your spreadsheet, answer these questions:

8. How did you calculate annual retirement benefits?
9. What were the total job benefits of Offer 1? Offer 2?
10. Which job offered the highest net benefits? How much higher?
11. Find what percent net job benefits for Job 1 are greater or less than net job benefits for Job 2. Write down the formula you entered into the spreadsheet to calculate the answer.
12. Which job benefits package do you think is better?
13. If you were to change the pension percentage rate for Job 1 to 4%, and change the hourly rate to $6.50 for Job 1, and eliminate life insurance as a benefit from both jobs, how do you think the difference in net benefits between the two offers will be affected? Enter the changes in your spreadsheet and answer the question in the last statement.

CHAPTER 4 SUMMARY AND REVIEW

VOCABULARY REVIEW

agent *4-8*
deductions *4-1*
FICA tax *4-1*
fringe benefits *4-3*
graduated commission *4-7*
gross pay *4-1*
job benefits *4-3*
job expenses *4-3*
net proceeds *4-8*
principal *4-8*
quota *4-7*
straight commission *4-7*
take-home pay *4-1*
withholding allowance *4-1*
withholding tax *4-1*

MATH SKILLS REVIEW

4-1
- Change fractions to decimals by dividing the numerator by the denominator, rounding as needed.
- Change decimals to fractions by dropping the decimal point and using the result as the numerator, using a denominator of 1 and as many zeros as there are decimal places, and simplify.
- Multiply by 0.1, 0.01, and 0.001 by moving the decimal point to the left in the number to be multiplied as many places as there are decimal places in the multiplier.
- Divide by 0.1, 0.01, and 0.001 by moving the decimal point to the right in the number to be divided as many places as there are decimal places in the divisor.

1. $4{,}523 \times 0.001$ **2.** $9.27 \div 0.1$

Write each as a decimal.

3. $\frac{5}{7}$ to the nearest thousandth **4.** $\frac{21}{1{,}000}$

Write each as a fraction in simplest form.

5. 0.45 **6.** 0.2 **7.** 0.008 **8.** 0.49

4-2
- Change decimals and whole numbers to percents by moving the decimal point two places to the right and adding a percent sign.
- Change fractions to percents by changing the fractions to decimals and then changing the decimals to percents.
- Change percents to decimals by dropping the percent sign and moving the decimal point two places to the left.
- Change percents to fractions by dropping the percent sign and writing the number as a numerator over a denominator of 100 and simplifying.

Complete the table. Write fractions or mixed numbers in simplest form. Round decimals to the nearest thousandth.

Fraction	Decimal	Percent	Fraction	Decimal	Percent
9. $\frac{7}{20}$			**10.**		140%

4-4 ♦ Find a percent of a number by changing the percent to a decimal and multiplying.

♦ Find what percent one number is of another by dividing the number (the part) by the other number (the whole) and showing the result as a percent.

♦ Multiply by fractional equivalents of parts of 100% by mentally changing the percent to a fraction and multiplying.

♦ Find 1%, 10%, 100%, and 1,000% of a number by changing the percent to a decimal and multiplying.

11. $37\frac{1}{2}\%$ of $560

12. 1,000% of $2.30

13. $0.47 is what percent of $235

4-6 ♦ Find a number that is a percent greater or smaller than another by multiplying by the percent and then either adding or subtracting the result to the number.

♦ Find what percent one number is greater or smaller than another by subtracting the smaller number from the larger to find the part and then dividing the part by the original number.

14. $8 is what percent greater than $6?

15. 1% less than $475

APPLICATIONS

4-3 ♦ Multiply gross pay by the FICA rate to find FICA taxes.

♦ Add deductions to find total deductions, then subtract deductions from gross pay to find net pay.

16. Luisa Medina is paid $356 a week. Her employer deducts $28 for federal withholding tax, $14.58 for insurance, and 0.0765 for FICA tax. For the week, what is her FICA tax, total deductions, and net pay for the week?

4-5 ♦ Add fringe benefits to find total fringe benefits, then add gross pay and fringe benefits to total job benefits.

♦ Add job expenses to find total job expenses. Subtract total job expenses from total job benefits to find net job benefits.

17. Karen Polk's job pays $45,650 per year plus 0.22 of wages in benefits. She estimates that yearly job expenses are $1,956. A job she has been offered pays $49,250 in yearly wages with estimated fringe benefits of 0.18 with job expenses of $2,122. What are her current job's net job benefits? What are the other job's net job benefits? Which job offers the greater job benefits? How much greater?

4-7 ♦ Find the amount of straight commission by multiplying sales beyond a quota, if there is one, by the rate of commission.

♦ Find total earnings from salary plus commission by finding the commission and adding it to the salary.

♦ Find the rate of commission on sales by dividing the amount of commission by sales.

♦ Find the net proceeds of the principal by subtracting commission and expenses from the amount collected or sales price.

18. Reito Akita is paid a commission of $3.25 on each of the first 100 perfume boxes sold; $4.00 on the next 100 boxes, and $5.00 on every box over 200. Last week Reito sold 215 boxes of perfume. What was her commission for the week?

19. Bendel Corporation hired a collection agency to collect $5,450. The agency collected 82% of the account and charged 32% commission and $74 for expenses. Find the net proceeds.

WORKPLACE KNOWHOW: ORGANIZE AND MAINTAIN INFORMATION

Organize and Maintain Information The ability to manage information in a way that makes it easy to find and use involves the ability to categorize and sort information. That information may be work paper, computer files, or an item, such as a test tube.

Assignment At the end of each chapter in this book, you will add your self assessments and work samples to your Work Skills Portfolio. You may also want to include material about skills needed for jobs that interest you. Write a plan for organizing the material you will keep in your portfolio. Your plan should answer these questions: How will I sort material? How will I find something specific? When will I remove something from my portfolio? Think about different ways of organizing your papers, such as in physical files, on disk, or some combination. A table of contents or an index may be helpful. Whenever you update your portfolio, ask yourself if the organization is working or needs to be modified. You may reorganize material as you become more aware of your requirements.

Skills Assessment On a scale of 1–5, rate your understanding of the following skills used in this chapter. Write a paragraph explaining why you gave yourself that rating.

Multiplication	Rounding	Information Management
Division	Estimation	Writing Formulas for Spreadsheets
Grouping Data		

Work Skills Portfolio If you completed the Integrated Project for this chapter, you may include information on the completed wage summary report. You may also include material from other classes or activities outside the classroom that reflect your skills and interests.

CHAPTER 4 TEST

Math Skills

Calculate mentally when you can.

1. Write $\frac{16}{1,000}$ as a decimal.
2. Find the decimal equivalent of $\frac{5}{16}$ to the nearest hundredth.
3. Change 0.36 to a fraction in lowest terms.
4. Change $\frac{7}{19}$ to a percent (round to nearest tenth of a percent).
5. Change 160% to a fraction or mixed number and simplify.
6. $18 is what percent greater than $10?
7. 208×0.1
8. $\frac{1}{4}\%$ of $36
9. $1,026 \times 0.001$
10. $51.7 \div 0.1$
11. $12.92 \div 0.01$
12. 30% of $820
13. $34 is what percent less than $40?

Applications

Peter Noeltner earns $7.50 an hour for a 40-hour week, with time and a half for overtime. Last week he worked 48 hours. His deductions were: $34 federal withholding tax, 0.0765 FICA tax, and $23.88 health insurance. Find each of the following for Peter:

14. FICA tax.
15. total deductions.
16. take-home pay.
17. Wilma Bollini earns an annual wage of $23,589. She estimates benefits at 0.32 of her wages and her job expenses at: insurance, 0.08 of her wages; transportation, $269; dues, $245; and birthday fund, $38. What are her annual net job benefits?
18. Phan Am Van makes a salary of $650 a month plus a 10% commission on sales over $5,000. Last month Phan's sales were $8,459. What were Phan's gross earnings?
19. Leta Fischer is paid a 5% commission on the first $10,000 of sales and 8% on sales over $10,000. Last week Leta had sales of $35,678. What was her gross pay?
20. A salesperson was paid a salary of $1,200 a month plus a commission on all sales. Last month her sales were $30,000, and the total amount she earned in salary and commission was $2,700. What rate of commission was she paid?

UNIT 1 TEST: CHAPTERS 1–4

◆ Choose the best definition for each term.

1-5	**1.** debit card	**a.** Allows access to your bank account through special computer terminals
1-6	**2.** reconciliation statement	**b.** Amount earned after any deductions are subtracted
2-2	**3.** gross pay	**c.** Amount earned as percentage of sales
4-3	**4.** net pay	**d.** Amount earned before any deductions are subtracted
4-3	**5.** FICA tax	**e.** Amount of pay withheld for federal income taxes
4-7	**6.** commission	**f.** Amount of pay withheld for social security and Medicare
4-7	**7.** net proceeds	**g.** Amount remaining after agents' fees and other expenses are deducted
		h. Shows calculations needed to balance a check register with bank statement

◆ Choose the best answer.

1-2 **8.** In the number 640,397, the value of the six is

a. sixty thousand **b.** six hundred
c. six thousand **d.** six hundred thousand

2-1 **9.** Choose the best estimate for 65.6×19.95.

a. 140 **b.** 12,000
c. 1,400 **d.** none of these

3-1 **10.** When any number is multiplied by a proper fraction, the answer

a. is larger than the number you started with
b. is smaller than the number you started with
c. is always smaller than the fraction
d. might be larger or smaller than the number

4-1 **11.** Which of the following is **not** another name for 0.8?

a. 0.08 **b.** $\frac{8}{10}$
c. 0.80 **d.** $\frac{80}{100}$

4-2 **12.** The percent equal to $\frac{5}{8}$ is

a. 160% **b.** 6.25%
c. 62.5% **d.** 0.625%

◆ Solve each problem.

1-1 **13.** Quintana's Service Shop had these cash receipts: parts sold, \$2,353.35; repairs, \$5,155.40; accounts receivable, \$525.93; and miscellaneous sales, \$235.35. Find the total receipts.

1-3 **14.** Make a columnar cash payments record for the week if Ms. Shin's cash payments while making sales calls were:

Monday: food, \$24.30; transportation, \$45.50
Tuesday: food, \$18.85; faxes sent, \$12.50; other, \$15
Thursday: food, \$17.64; faxes sent, \$19.54; other, \$9.53
Friday: food, \$15.95; transportation, \$15.50; other, \$5

1-4 **15.** Your check register balance was \$375.07. You deposited \$603.64 and wrote checks for \$25.53, \$39, \$157.45, and \$475. What is your new check register balance?

1-7 **16.** Your check register balance is \$96.46. Your bank statement shows a service charge of \$3.50 and interest earned of \$0.57. You also find that check #705 was entered as \$44.53 but should have been \$45.43. Find your corrected check register balance.

2-3 **17.** Jose ordered 14,000 parts that cost \$3,450. Find the cost per part to the nearest cent.

2-4 **18.** Zoa earned \$1,687.50 in four weeks. She worked 37.5 hours per week. Find her average pay per hour.

3-5 **19.** You had $4\frac{1}{6}$ yards of plastic tubing. You used $2\frac{1}{2}$ yards on a job. How much tubing do you have left?

3-3 **20.** Jake can make 150 parts in 2.5 hours. How many parts can he make in 7 hours?

3-4 **21.** Nicki earns \$8.50 an hour and time-and-a-half over 35 hours. How much does she earn if she works 42 hours?

3-6 **22.** Alexi earned \$18,000 last year. How much did he earn this year if he earned $1\frac{1}{5}$ as much as last year?

4-5 **23.** Sonya earns \$2,550 per month. Estimated fringe benefits are 35% of wages. Estimated job expenses total \$3,500 per year. Find her annual net job benefits.

4-6 **24.** Kamil earned \$13.40 per hour last year. Due to poor company earnings, he is now earning 23% less than last year. To the nearest cent, how much per hour is he earning now?

4-7 **25.** Hollis earns 8% commission on the first \$15,000 in sales and 10% on sales over \$15,000. How much commission will he earn on sales of \$26,500?

CHAPTER 5

the ACTING

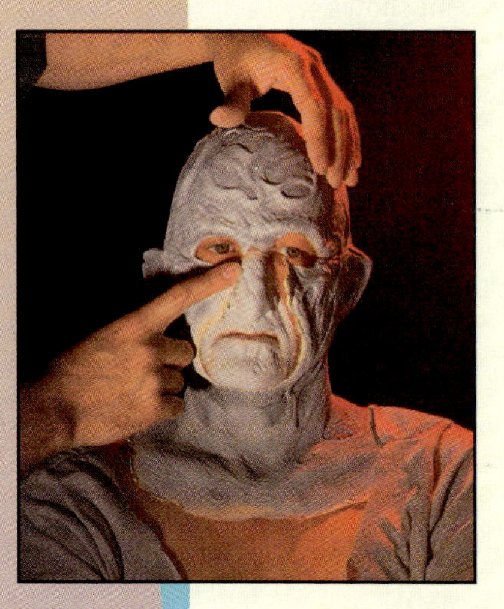

Do you think an acting career means movies, TV, Broadway? Living in New York City, California, or Chicago? While those are the major markets for actors, other cities offer rewarding acting opportunities.

Toya is an actress in the Ohio-Indiana-Kentucky area. She says work in smaller, less competitive markets can be very satisfying. Toya makes a comfortable living acting in industrial training videos and occasional commercials and doing voice-over work. Voice-over actors serve as off-screen narrators in commercials and videos.

If you want to pursue a career as an actor, Toya advises you learn:

- **technical skills** — to work with cameras, prompters, makeup, lighting, and microphones.
- **verbal skills** — for reading aloud and interpreting material.
- **memorization skills** — for scripts and speeches.
- **professionalism** — Toya says that means "be reliable, arrive on time, and be ready to go to work."
- **"a look"** — because "type" is everything.
- **personal skills** — such as patience, friendliness, and a good attitude, so people will want to work with you again.
- **self-promotion skills** — to tell prospective clients who you are and what you can do. "Pounding the pavement," sending out letters, resumes, and demo tapes (videos with samples of work), are all part of the work.

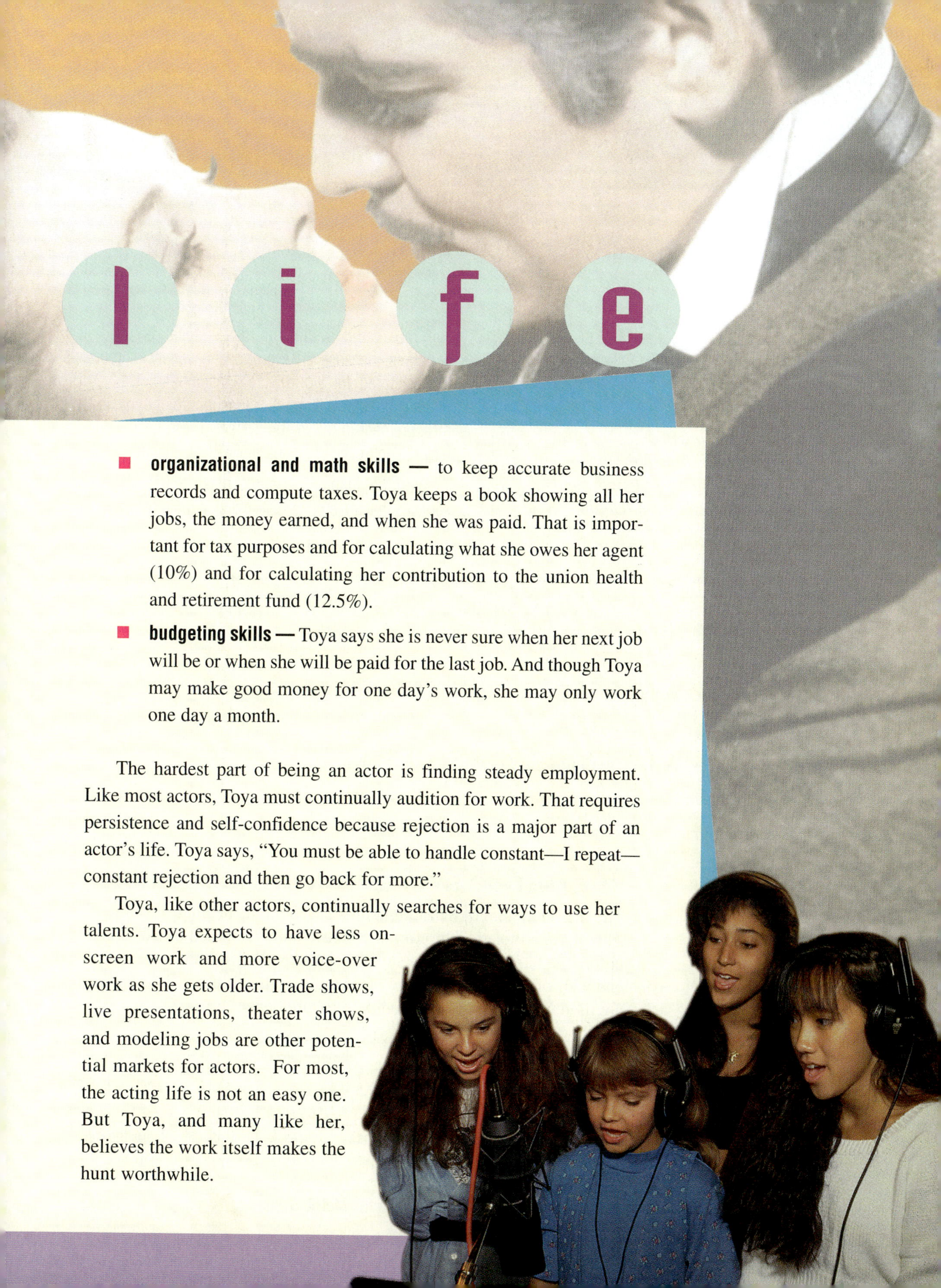

- **organizational and math skills** — to keep accurate business records and compute taxes. Toya keeps a book showing all her jobs, the money earned, and when she was paid. That is important for tax purposes and for calculating what she owes her agent (10%) and for calculating her contribution to the union health and retirement fund (12.5%).
- **budgeting skills** — Toya says she is never sure when her next job will be or when she will be paid for the last job. And though Toya may make good money for one day's work, she may only work one day a month.

The hardest part of being an actor is finding steady employment. Like most actors, Toya must continually audition for work. That requires persistence and self-confidence because rejection is a major part of an actor's life. Toya says, "You must be able to handle constant—I repeat—constant rejection and then go back for more."

Toya, like other actors, continually searches for ways to use her talents. Toya expects to have less on-screen work and more voice-over work as she gets older. Trade shows, live presentations, theater shows, and modeling jobs are other potential markets for actors. For most, the acting life is not an easy one. But Toya, and many like her, believes the work itself makes the hunt worthwhile.

5-1 LENGTH

OBJECTIVES

In this lesson, you will learn

- *the basic metric units;*
- *to use metric units of length;*
- *to convert metric units of length; and*
- *to compute metric units of length.*

The metric system of measurement is used in most countries. Because an ever-growing variety of products are manufactured for world-wide consumption, Unit 2, Buying Wisely, begins with an introduction to the basic units of measurement in the metric system. These basic units are the meter (length), the liter (capacity), and the gram (mass or weight).

◆ Working in small groups, make a list of products or services that are sold or distributed in units of metric measure.

WARM *UP*

Write the symbol for each unit of measure in the customary system.

a. feet **b.** inches **c.** yards **d.** miles

Complete. Write fractional answers in simplest form.

e. 48 inches = ? feet **f.** 2 miles = ? feet

THE METRIC SYSTEM: AN INTRODUCTION

The metric system of measurement is based on the decimal system. Therefore, all relationships among metric units are based on 10. Prefixes before the base unit indicate smaller and larger units. Memorize the prefixes and their order as shown in the place table in Illustration 5-1.1. The prefixes are used with all of the base metric measurements.

	Larger			← **Base Unit** → **(meter, liter, gram)**			**Smaller**
Prefix + Symbol	Kilo K	hecto h	deca da	m, L, or g	deci d	centi c	milli m
Meaning	thousands 1,000	hundreds 100	tens 10	one 1	tenth 0.1	hundredth 0.01	thousandth 0.001

Illustration 5-1.1. Prefixes Used in the Metric System

METRIC MEASURES OF LENGTH

Length can be shown in many ways in metric units. Highway signs may give distances in both kilometers and miles. Rulers often have both a centimeter and an inch scale. Olympic athletes run in 100-meter races. The basic unit of length in the metric system is the **meter.** One meter is equivalent to 3.28 feet, or slightly more than one yard in the customary system.

The parts (smaller units) and multiples (larger units) of the meter are named by the prefixes shown in the illustration below. The prefixes are used with all metric measurements.

	Unit	Symbol	Value in Meters
Parts of a meter	**millimeter**	**mm**	0.001 m (one-thousandth meter)
	centimeter	**cm**	0.01 m (one-hundredth meter)
	decimeter	dm	0.1 m (one-tenth meter)
Basic unit	**meter**	**m**	1 m (one meter)
Multiples of a meter	dekameter	dam	10 m (ten meters)
	hectometer	hm	100 m (one hundred meters)
	kilometer	**km**	1 000 m (one thousand meters)

(Commonly used units are in **bold** type.)

Illustration 5-1.2. Metric Units of Lengths

A meter can be divided into smaller parts such as tenths, called *deci*meters; hundredths, called *centi*meters; or thousandths, called *milli*meters. These parts are used to measure lengths of less than a meter.

The meter or its multiples can also be used to measure longer lengths. Multiples of the meter include the *deka*meter, or ten meters; the *hecto*meter, or one hundred meters; and the *kilo*meter, or one thousand meters.

The metric units of length, their symbols, and their values in meters are shown in Illustration 5-1.2.

Writing Metric Amounts

When writing metric amounts, follow these rules:

1. Write the symbol, not the unit name, in small letters.
   (18 millimeters = 18 mm)
2. Use the same symbol for singular and plural values. Do not add an "s."
   (3 centimeters = 3 cm   0.03 centimeters = 0.03 cm   1 centimeter = 1 cm)
3. To break up large numbers, use a space instead of a comma.
   (1200 meters = 1 200 m   4243 kilometers = 4 243 km)

EXERCISE YOUR SKILLS

1. Name the units of length in the metric system. Give the symbol for each.

How many meters are there in a

2. kilometer? 3. dekameter? 4. hectometer?

A meter can be divided into how many

5. millimeters? 6. centimeters? 7. decimeters?

Identify the metric unit you would use to describe each of the following items, then estimate the length in that unit.

8. A pencil? 9. A room?
10. The distance from your home to school?

CONVERTING FROM ONE METRIC UNIT TO ANOTHER

To compute with or compare metric units, you must work with common units.

It is easy to change or convert metric measurements because the metric system is a decimal system. Each position in the metric system is either 10 times more than, or one-tenth of, the next unit. To change units, multiply or divide by 10 as many times as needed to change to the unit you want.

Another way to do this is to move the decimal point. When changing from a larger to a smaller unit, move the decimal point to the right. When changing from a smaller to a larger unit, move the decimal point to the left. You can use the decimal place chart to be sure you move the correct number of decimal places.

EXAMPLE 1 Change 85 m to ? cm.

1. Use the place chart to count the number of places to the unit you want to convert to. To move from meter (the larger unit) to centimeter (the smaller unit), you must move the decimal point two places to the right of meter.

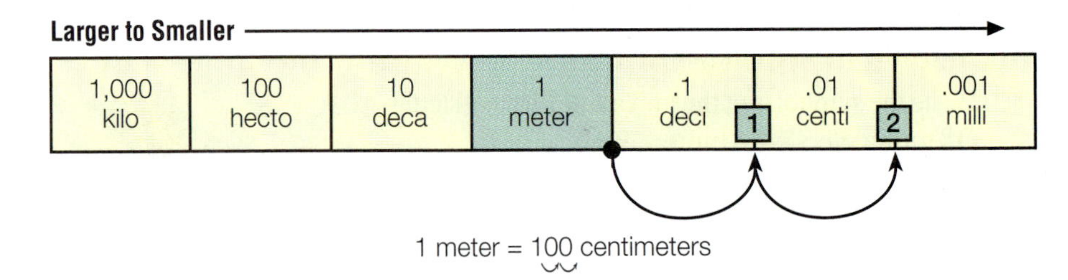

2. To change 85 meters to centimeters, move the decimal point two places to the right.

85 m = 8 500 cm

EXAMPLE 2: Change 8 500 cm to km.

1. Use the place chart to count the number of places to the unit you want to convert to. Since you are converting a smaller to a larger unit, you move the decimal point left. To move from centimeter to kilometer, you must move the decimal point 5 places to the left.

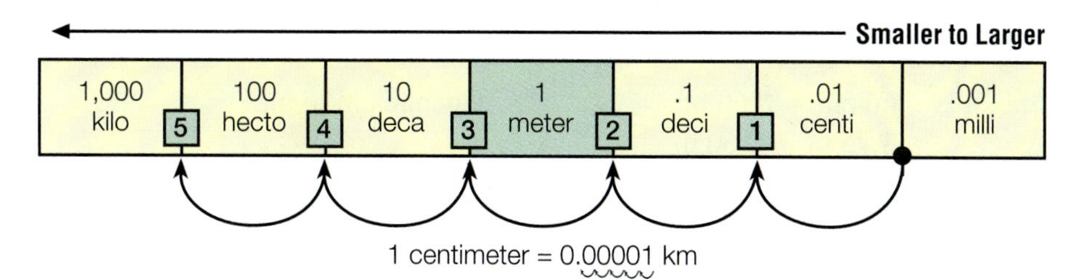

2. To change 8 500 cm to km, move the decimal point five places to the left.

   8 500 cm = 0.08500 km

Comparing Measurements

To compare measurements, change to common units if necessary. Then compare the numbers.

EXAMPLE Which is longer, 2 meters or 330 centimeters?

SOLUTION

2 m = 200 cm	Change 2 meters to centimeters.
200 cm is less than 330 cm.	Compare numbers.
330 cm is longer than 2 m.	Answer the question.

EXERCISE YOUR SKILLS

Write the missing metric values.

11. 1 m = ? dm **12.** 1 m = ? cm **13.** 1 m = ? mm

14. 1 km = ? m **15.** 1 cm = ? m **16.** 1 dm = ? m

17. 1 mm = ? m **18.** 1 cm = ? mm **19.** 1 mm = ? cm

20. 3 mm = ? cm **21.** 600 mm = ? m **22.** 548 m = ? km

23. 3 m = ? cm **24.** 6 m = ? mm **25.** 7 cm = ? mm

26. 4 km = ? m **27.** 700 m = ? km **28.** 340 cm = ? m

29. 41 mm = ? m **30.** 176 mm = ? cm **31.** 6.2 m = ? mm

Tell which is longer.

32. 0.2 km or 2 100 m **33.** 0.3 m or 3 000 cm

34. 42 mm or 42 dm **35.** 1 040 mm or 1.4 cm

36. 9 232 m or 92.32 cm **37.** 2 400 m or 2.04 km

COMPUTING WITH METRIC UNITS

Metric measurements are added, subtracted, multiplied, and divided in the same way as customary values. Before doing any operations, change measurements so that they all are the same unit. When units are the same, complete the computation in the same way you did for other decimals.

ADDITION EXAMPLE Add 3.75 meters and 60 centimeters.

SOLUTION 60 cm = 0.60 m — Change cm to m.
3.75 m + 0.60 m = 4.35 m — Add.

Critical Thinking Could you have changed 3.75 meters to centimeters? Would your answer have been the same? Explain your answer.

MULTIPLICATION EXAMPLE Multiply 1.7 cm by 200.

SOLUTION 1.7 cm × 200 = 340 cm

EXERCISE YOUR SKILLS

Find each sum or difference.

38. 3 cm + 2 cm + 1 cm
39. 2.8 m + 4.6 m + 1.5 m
40. 5 km + 7 km + 400 m
41. 7 m + 50 cm + 20 cm
42. 78 mm − 12.6 mm
43. 8.3 m − 4.7 m
44. 128 m − 430 cm
45. 33 cm − 124 mm
46. 40 cm + 55 cm + 157 mm
47. 1.5 m − 8 cm

Find each product or quotient.

48. 15 m × 3
49. 60 km × 7
50. 4.3 cm × 3
51. 21.4 m × 6
52. 80 km × 0.5
53. 200 mm × 5
54. 20 m ÷ 5
55. 72 cm ÷ 3
56. 5.2 mm ÷ 4
57. 2 100 km ÷ 7

Solve each problem.

58. Eric Feldman needed these lengths of cable: 4 m, 2.5 m, 3.7 m, 2.1 m. What total amount of cable, in meters, did Eric need?

59. A store sold 170 cm of gold chain from a roll that originally held 680 cm of chain. How much chain was left on the roll?

60. Max Rivard has set a goal of running 53 kilometers per week. In 6 days this week he ran 46.4 kilometers. How many kilometers must Max run on the seventh day to reach his goal?

61. Kellie Milon needs 6 500 m of wire fencing. She also needs a total of 250 m of wood fencing. If the kind of wire fencing she needs is in 500 m rolls, how many rolls will she need?

62. Yvette Wyzek is filling an order for copper tubing. The order is for these pieces and lengths of tubing: 10 pieces, 208 cm long; 24 pieces, 55 cm long; 115 pieces, 2 m long. What total length of tubing, in meters, does Yvette need to fill the order?

On a business trip, Enrique Gutierrez drove these distances in 3 days: 280 km, 244 km, 64 km.

63. How many km did he drive in those 3 days?

64. What estimated average number of km did he drive per day?

65. What actual average number of km did he drive per day?

CONVERTING METRIC AND CUSTOMARY MEASUREMENTS OF LENGTH

Once in a while, you may find it necessary to convert from the metric system to the customary system or vice versa. Illustration 5-1.3 lists some common equivalencies between the metric and customary systems of measurements.

Metric to Customary	Customary to Metric
1 cm ≈ 0.39 inches	1 inch = 2.54 cm
1 m ≈ 3.28 feet	1 foot ≈ 0.305 m
1 km ≈ 0.62 miles	1 mile ≈ 1.61 km

Illustration 5-1.3 Common Equivalencies for Metric and Customary Measurements

Critical Thinking If 1 ft ≈ 0.305 m, approximately how many meters equal a yard?

EXAMPLE A sailboat is 10 meters long. What is its length in feet?

SOLUTION Think: 1 m ≈ 3.28 feet, so 10 m ≈ 32.8 feet
The length of the sailboat is *about* 32.8 feet.

EXAMPLE Sam bicycled 5 miles. How many kilometers did he bicycle?

SOLUTION 1 mile ≈ 1.61 km, so 1.61 km × 5 ≈ 8.05 km
Sam bicycled *about* 8.05 km.

EXERCISE YOUR SKILLS

66. A sign states that the next highway exit is 56 km. To the nearest tenth of a mile, how many miles must you travel to reach the next exit?
67. You run from one end of a field to the other end four times. The field measures 105 m. To the nearest foot, how far did you run?
68. A circle cut in a pattern has a diameter of 16 cm. What is the diameter of the circle in inches?
69. The inside of a garage is 15 feet deep when the door is closed. Will a car that measures 4.7 m in length fit into the closed garage?
70. ◆ Make a reference chart for your own use that shows the relationship between inches and centimeters; feet and meters, yards and meters, and miles and kilometers.

MIXED REVIEW

Solve each problem.

1. \$168 is what part smaller than \$252?
2. Show the sum in centimeters. 70 cm + 95 cm + 240 mm
3. \$340 increased by $\frac{1}{5}$ of itself is?
4. Divide $3\frac{3}{4}$ by $2\frac{1}{2}$.

Rae Murphy needs these amounts of electrical wire: seven 4 m pieces; four 2.75 m pieces; six 1.5 m pieces.

5. How many meters of wire does she need?
6. At \$0.24 per meter, how much will Rae pay for the wire?
7. A salesperson in a fabric store sold three types of ribbon in these amounts: 8 m at \$0.84 per m; 12 m at \$1.40 per m; 5 m at \$1.12 per m. What was the average selling price of the ribbon per meter, to the nearest cent?
8. Ric Webb was paid \$8.50 per hour for 40 hours of work. From his pay, his employer deducted \$43 for withholding taxes, 7.65% for FICA tax, and \$32 for insurance. What was Ric's net pay?

5-2 AREA

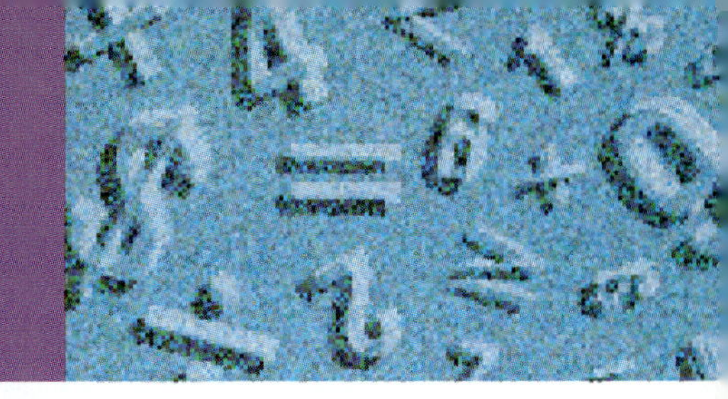

In the metric system, the basic unit of area is the square meter. It is equivalent to 1.2 square yards in the customary system.

◆ List four ways area is measured in the customary system. Estimate the metric equivalent of each of those four customary measurements.

OBJECTIVES

In this lesson, you will learn to

- *use metric units of area;*
- *convert metric units of area; and*
- *compute with metric units of length.*

WARM UP

Complete.

a. $1 \text{ ft}^2 = ? \text{ in.}^2$

b. $1 \text{ yd}^2 = ? \text{ ft}^2$

c. \$72 is what part smaller than \$120?

d. Divide $8\frac{1}{4}$ by $5\frac{1}{2}$.

e. Multiply 10,000 by 0.01.

f. Divide 340 by 10,000.

g. Multiply 0.1 by 0.001.

AREA

The amount of surface an item has is called its **area.** To find the area of any flat surface, such as a desk or floor, multiply its length by its width.

Knowing how to calculate area is a skill used in many businesses. For example, the owner of a paving company must find the area of a parking lot to estimate how much to charge for paving the lot. The parking lot shown on the next page is 120 feet long by 60 feet wide and has an area of 7,200 square feet. Area is found by multiplying the length by the width.

Area = length × width or $A = l \times w$

Area = 60 feet × 120 feet = 7,200 square feet.

In the customary system, area may be in square inches, square feet, square yards, or square miles, depending on what is being measured. For example, a tablecloth may cover a certain area measured in square inches. The area covered by a box of floor tile can be measured in square feet. Carpeting is sold by the square yard, while forests may be measured in square miles.

In the metric system, the basic unit of area is the **square meter.** The square meter is equivalent to approximately 1.2 square yards in the customary system, or about one square yard.

METRIC UNITS OF AREA

In the metric system, square meter is written as m^2. The word "square" is not written. It is named by the exponent "2" that appears to the right and above the symbol for meter, m.

The parts (smaller units) and multiples (larger units) of the square meter are named by the prefixes you have already studied: milli-, centi-, deci-, deka-, hecto-, and kilo-. Metric units of area are shown in Illustration 5-2.1.

	Unit	Symbol	Value in Square Meters
Parts of a square meter	**square millimeter**	**mm^2**	0.000 001 m^2 (one-millionth square meter)
	square centimeter	**cm^2**	0.000 1 m^2 (one-ten-thousandth square meter)
	square decimeter	dm^2	0.01 m^2 (one-hundredth square meter)
Basic unit	**square meter**	**m^2**	1 m^2 (square meter)
Multiples of a square meter	square dekameter	dam^2	100 m^2 (one hundred square meters)
	square hectometer or **hectare**	hm^2 or **ha**	10 000 m^2 (ten thousand square meters)
	square kilometer	**km^2**	1 000 000 m^2 (one million square meters)

(Commonly used units are in **bold** type.)

Illustration 5-2.1 Metric Units of Area

Square centimeter and square meter are the measures of area used most often. Large areas are measured in either hectares (ha) or square kilometers (km^2). Hectare is another name for square hectometer. Hectare is used because it is easier to say and write.

EXERCISE YOUR SKILLS

1. What customary unit is approximately equal to the square meter?
2. Explain how area of a flat surface is found.
3. Name the basic metric unit of area and its symbol.
4. Of the six prefixes, name the one that stands for the largest quantity; the smallest quantity.
5. What is a simpler name for the square hectometer?

CONVERTING FROM ONE METRIC AREA UNIT TO ANOTHER

Metric measures of area are changed in a similar way as metric measures of length. Each square unit is **100 times** the next smaller unit, or one-hundredth of the next larger unit, as illustrated in the diagram below. So, to change from a metric unit of area to the next smaller unit, move the decimal point **two** places to the *right*. To change to the next larger unit of area, move the decimal **two** places to the *left*.

Remember that to determine area, multiply length times width. In this case, 1 cm × 1 cm = 1 cm^2. Since 1 cm = 10 mm, the area can be expressed as 10 mm × 10 mm = 100 mm^2. So 1 cm^2 = 100 mm^2. Study these examples to see how the decimal point is moved.

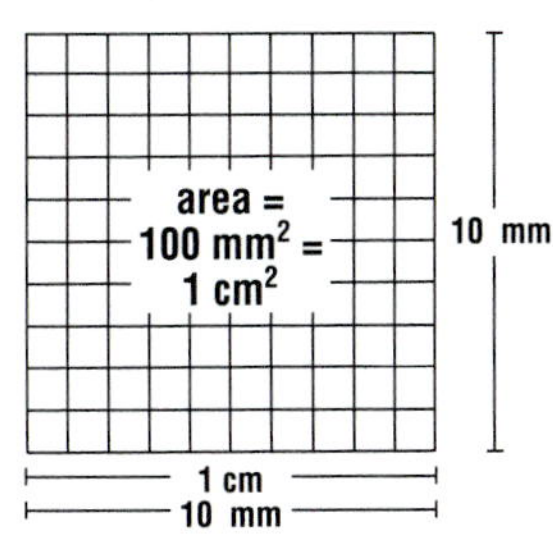

EXAMPLE Change to a smaller unit of area:

1 cm^2 = 100. mm^2 1 m^2 = 10 000. cm^2

1 ha = 10 000. m^2 1 km^2 = 1 000 000. m^2

EXAMPLE Change to a larger unit of area:

100 mm^2 = 1.00 cm^2 10 000 cm^2 = 1.0 000 m^2

10 000 m^2 = 1.0000 ha 1 000 000 m^2 = 1.000 000 km^2

COMPUTING WITH SQUARE UNITS

When you add or subtract square units, the unit in the measure remains the same. Remember, measurements must have the same units in order to be added or subtracted.

When two measurements that are in centimeters are multiplied, the answer is given in cm^2. When square units, such as km^2, are divided by a measurement, such as km, the quotient is in units, not square units.

EXERCISE YOUR SKILLS

Complete.

6. $1\ cm^2 = ?\ mm^2$
7. $1\ m^2 = ?\ cm^2$
8. $1\ ha = ?\ m^2$
9. $1\ km^2 = ?\ ha$
10. $4\ ha = ?\ km^2$
11. $0.5\ m^2 = ?\ cm^2$
12. $7.2\ cm^2 = ?\ mm^2$
13. $1.6\ ha = ?\ m^2$
14. $97\ km^2 = ?\ ha$

Find each sum or difference. Be sure to use like units.

15. $7\ km^2 + 5\ km^2$
16. $12\ cm^2 + 100\ mm^2$
17. $8\ ha + 16\ ha$
18. $10\ 000\ m^2 + 6\ ha$
19. $9\ m^2 + 7\ 000\ cm^2$
20. $25\ km^2 - 7\ km^2$
21. $5\ cm^2 - 100\ mm^2$
22. $8\ km^2 - 200\ ha$
23. $1\ ha - 6\ 000\ m^2$
24. $4\ km^2 - 1\ 000\ 000\ m^2$

Find each product or quotient. Be sure to label your answer correctly.

25. $35\ m \times 4\ m$
26. $0.4 \times 58\ cm^2$
27. $100\ m \times 100\ m$
28. $12\ mm^2 \times 3$
29. $3\ km^2 \times 0.5$
30. $64\ m^2 \div 8$
31. $0.75\ m^2 \div 3$
32. $300\ cm^2 \div 6$

Solve each problem.

33. The Crown Home Builders bought these amounts of land: 8 ha, 120 ha, 56 ha, 1 340 ha. Find the total amount of land they purchased.
34. Talva wants to buy carpeting for a room 4 m by 4.5 m. How much carpeting will she need?
35. Armando's yard is 28 meters long by 22 meters wide. He turned part of his yard into a garden 9 meters long by 4 meters wide. He also has a 3 m by 3 m shed on his lawn. He has to mow the rest of the lawn. How much lawn does he mow?

Mike cut 2 boards for shelving out of a $2.88\ m^2$ sheet of wood. One board was 20 cm by 120 cm; the other was 30 cm by 120 cm.

36. How much wood, in m^2, was left from the large sheet?

PROBLEM SOLVING

Make a Drawing

When trying to solve problems, sometimes it helps to make a drawing.

EXAMPLE A board is 1.2 m long and 50 cm wide. A section 20 cm by 30 cm is cut from one end of the board. Find the area of the board.

SOLUTION

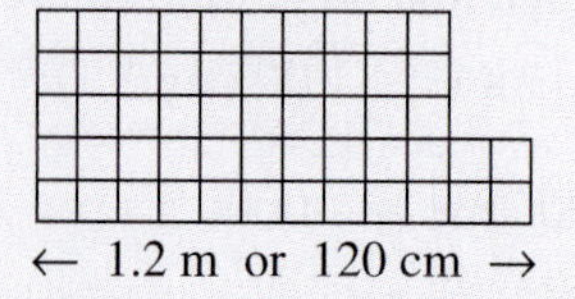

Each □ represents $100\ cm^2$.

$120\ cm \times 50\ cm = 6\ 000\ cm^2$

$20\ cm \times 30\ cm = 600\ cm^2$

$6\ 000\ cm^2 - 600\ cm^2 = 5\ 400\ cm^2$

CHECK The drawing shows 54 squares or $5\ 400\ cm^2$.

CONVERTING METRIC AND CUSTOMARY AREA MEASUREMENTS

You may occasionally find it necessary to convert from the metric system to the customary system or vice versa. Illustration 5-2.2 shows relationships of area measures.

1 $cm^2 \approx$ 0.16 $in.^2$	1 $in.^2 \approx$ 6.5 cm^2
1 $m^2 \approx$ 10.8 ft^2	1 $ft^2 \approx$ 0.09 m^2
1 ha $\approx$ 2.5 acres	1 acre = 0.4 ha

Illustration 5-2.2 Metric and Customary Equivalents for Area Measurements

EXAMPLE Trudy bought a roll containing 8 m^2 of gift wrap. How many square feet of gift wrap are in the roll?

SOLUTION Think: 1 $m^2 \approx$ 10.8 ft^2, so 8 $m^2 \approx 8 \times$ 10.8 or 86.4 ft^2.

EXAMPLE A stock clerk found a carpet remnant that covers 215 square feet of floor space. How many square meters will the remnant cover?

SOLUTION 1 $ft^2 \approx$ 0.09 m^2, so 215 $ft^2 \approx 215 \times$ 0.09 or 19.35 m^2
The remnant will cover 19.35 m^2.

EXERCISE YOUR SKILLS

37. Joan inherited a 350 ha farm in Belgium. What is the equivalent size of the farm in acres?

38. A paper mill produced 120 000 m^2 of coated paper last month. What was last month's coated paper production in square feet?

39. Meg plans to use pieces of fabric that each have an area of 1 ft^2 for an art project. However, the plans for the project are given in centimeters. To the nearest 100, how many cm^2 are equal to 1 ft^2?

40. Visitors from another country toured a farm and were told that it had 2,000 acres of land. The visitors want to know its size in hectares. What is the area in hectares?

A box holds 40 floor tiles. Each tile measures 30.5 cm by 30.5 cm.

41. What estimated area, in cm^2, is covered by a tile?
42. What estimated area, in ft^2, is covered by a tile?
43. What area, in cm^2, is covered by a box of tiles?
44. What area, in ft^2, is covered by a box of tiles?
45. What area, in m^2, will a box of tiles cover?
46. What area, in yd^2, will a box of tiles cover?
47. Helen Quincy has a front lawn that measures 66 ft by 23 ft and a backyard lawn that measures 66 ft by 46 ft. She bought a bag of lawn fertilizer that covers 980 square meters. How many times can she fertilize her entire lawn using the fertilizer she bought? Describe the plan you will use to solve this problem. Show any conversions that are necessary to carry out the plan. Solve the problem and show all your work.
48. ◆ Create a reference chart for your own use that shows the equivalents of square inches to square centimeters; square feet to square meters; square yards to square meters, and square miles to square kilometers.

MIXED REVIEW

1. Marcy Duvall's boat dock is 1.3 m wide by 14 m long. What is the area of the dock in square meters?
2. Ludwik Wadarski is building a fenced exercise area for his dog. Each of the sides will be 2.2 m wide and each of the long ends will be 8.1 m long. If he buys 28 m of fencing and wastes none of it, how many meters of fencing will he have left over?
3. Ned Stellif's sales for June were $27,540. In May, his sales were $25,500. What was the percent increase or decrease in Ned's June sales compared to his May sales?
4. The sales of the Hartwell Company for this year are only $\frac{5}{6}$ as much as last year's sales. If last year's sales were $2,448,000, what are this year's sales?

A real estate agent was paid a 7% commission by the owner of a property for selling it for $132,000.

5. What was the amount of the agent's commission?
6. What net proceeds did the owner get from the sale?

5-3 CAPACITY AND MASS

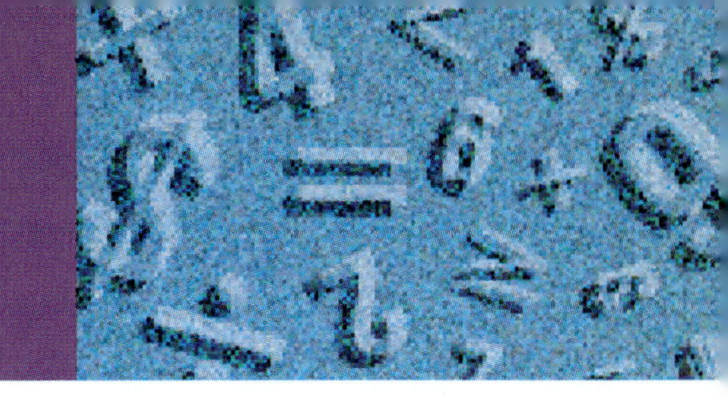

Fruit juice and cooking oil are sold by capacity. You may buy them in quarts, gallons, or liters. Goods such as rice, flour, apples, or tomatoes are sold by weight. Their weight may be given in customary pounds or in metric measures of grams or kilograms.

◆ Make a list of items that you are familiar with that are packaged in liters, grams, or kilograms.

OBJECTIVES

In this lesson, you will learn to

- *use metric units of capacity and mass;*
- *convert metric units of capacity and mass; and*
- *compute with metric units of capacity and mass.*

WARM UP

Complete.

a. 100 cm = ? mm

b. 50 m = ? km

c. 2.7 m = ? cm

d. Write 0.45 as a fraction in simplest form.

e. Write 0.075 as a percent.

f. Write $\frac{4}{5}$ as a decimal.

g. $42 is what percent of $120?

METRIC MEASURES OF CAPACITY

The basic metric measure of capacity is the **liter.** A liter is slightly larger than a quart. In this textbook, the symbol for the liter is a capital L. Another common symbol for the liter is l .

The prefixes milli-, centi-, deci-, deka-, hecto-, and kilo- are used with the liter. Notice in Illustration 5-3.1 that, as with all metric measurements, different parts and multiples of a liter are either one-tenth of the next larger unit or ten times the next smaller unit.

Most small capacity measures are shown in milliliters or as decimal parts of a liter. For example, a measure of 400 mL is equivalent to 0.4 L. Larger measures may be written in liters or kiloliters. For example, 4 000 L is the same as 4 kL.

	Unit	Symbol	Value in Liters
Parts of a liter	**milliliter**	**mL**	0.001 L (one-thousandth liter)
	centiliter	cL	0.01 L (one-hundredth liter)
	deciliter	dL	0.1 L (one-tenth liter)
Basic unit	**liter**	**L**	1 L (one liter)
Multiples of a liter	dekaliter	daL	10 L (ten liters)
	hectoliter	hL	100 L (one hundred liters)
	kiloliter	**kL**	1 000 L (one thousand liters)

Illustration 5-3.1 Metric Units of Capacity

Using Metric Measures of Capacity

You can add, subtract, multiply, and divide capacity measures in the same way as other numbers. When measures are in different units, convert them to the same unit. Metric units of capacity are changed to other units in the same way units of length and area were converted.

EXERCISE YOUR SKILLS

1. Name the basic unit of capacity in the metric system.
2. What customary measure is the liter approximately equal to?
3. Are units of capacity related to each other in the same way as units of length? In the same way as units of area?
4. What prefixes name units less than a liter?

How much larger than a liter is a:

5. dekaliter?
6. kiloliter?

Complete.

7. 1 L = ? mL
8. 1 L = ? kL
9. 1 mL = ? L
10. 1 kL = ? L
11. 1 000 mL = ? L
12. 800 mL = ? L
13. 320 L = ? kL

Solve. Be sure to correctly label your answer.

14. 37 L + 14 L
15. 34.68 L + 79.15 L
16. 400 mL + 26.1 mL
17. 14 L + 344 mL
18. 786 L − 554 L
19. 45.29 kL − 35.31 kL
20. 250 L × 300
21. 4.8 L × 5
22. 20 L × 1.9
23. 0.5 × 682 mL
24. 64 L ÷ 8
25. 180 kL ÷ 9
26. 1620 mL ÷ 5

Solve each problem.

27. Milfred Xeno is hosting a party for 12 members of the drama club. If Milfred buys 18 liters of juice, how many liters of juice will there be for each person at the party?

A 354 mL can of frozen concentrate makes 0.94 liter of lemonade.

28. How many liters of lemonade can you make from 6 cans?

29. How many cans of concentrate will you need to make 4.7 liters of lemonade?

30. A tank holds 130 kL of fuel oil when full. On Monday, the tank was 90% full. These amounts of fuel oil were pumped from the tank in the next 3 days: 20 500 L, 24 000 L, 21 750 L. How many kL of fuel oil were left in the tank at the end of the third day?

31. Natural Farm Products has 30 kiloliters of catsup ready to ship. It is packed in 48 one-liter bottles per case. How many cases will be shipped?

On a recent car trip, Frances Zaphir stopped and bought these amounts of gasoline: 56 L, 63.5 L, 47.3 L, and 59.1 L.

32. How many liters of gasoline did she buy in these 4 stops?

33. Find the average number of liters of gasoline bought per stop, to the nearest tenth liter?

Clarence Hanson saves 37¢ a liter by buying orange juice in bulk. His family uses 2.5 liters of orange juice a week.

34. How many liters of orange juice does the Hanson family use in a year?

35. How much will they save in a year by buying juice in bulk?

METRIC MEASURES OF MASS

In this text and in everyday conversation, **mass** and **weight** are used interchangeably. Technically, however, mass is a measurement of the quantity of matter. Weight is a force exerted by gravitational pull against a mass. That is why astronauts on the moon weigh approximately $\frac{1}{6}$ of their weight on earth. An astronaut's mass, however, is the same on earth and in space.

In the metric system, the **gram** is the standard unit of *mass*. One kilogram, or 1 000 grams, is equivalent to approximately 2.2 customary pounds. A **gram** is equivalent to about 0.04 of a customary ounce.

Prefixes are used with the gram to show the parts and multiples of the gram. These are shown in Illustration 5-3.2. Notice that each metric unit of mass is one-tenth or ten times the next unit.

Usually grams or milligrams are used to measure small mass. Kilograms or **metric tons** are used to measure large mass.

You can complete the addition, subtraction, multiplication, and division of metric mass measures as you would other numbers. Remember, to do these computations, all measures must be the same unit.

Unit	Symbol	Value in Grams
milligram	**mg**	0.001 g (one-thousandth gram)
centigram	cg	0.01 g (one-hundredth gram)
decigram	dg	0.1 g (one-tenth gram)
gram	**g**	1 g (one gram)
dekagram	dag	10 g (ten grams)
hectogram	hg	100 g (one hundred grams)
kilogram	**kg**	1 000 g (one thousand grams)
metric ton	**t**	1 000 kg (one thousand kilograms)

(Commonly used units are in **bold** type.)

Table 5-3.2. Metric Measurements of Mass

Critical Thinking What is an advantage of using common prefixes in the metric system?

EXERCISE YOUR SKILLS

36. What is the basic unit of weight in the metric system? What is its symbol?

37. How many grams make a kilogram?

38. What part of a gram is a milligram?

39. How many kilograms are there in a metric ton?

40. What part of a metric ton is a kilogram?

41. What metric weight measures are generally used with large amounts? with small amounts?

42. Is a gram larger or smaller than an ounce in the customary measurement system?

Complete.

43. 1 kg = ? g

44. 1 g = ? kg

45. 1 g = ? mg

46. 1 mg = ? g

47. 1 t = ? kg

48. 1 kg = ? t

49. 900 mg = ? g

50. 250 g = ? kg

51. 0.75 kg = ? g

52. 1 000 mg = ? g

53. 400 kg = ? t

54. 75 mg = ? g

55. 0.7 kg + 0.6 kg = ? kg

56. 0.65 g − 0.25 g = ? g

57. 2 g − 500 mg = ? mg

58. 12 kg × 9 = ? kg

59. 600 g × 7 = ? kg

60. 560 kg ÷ 28 = ? kg

Solve each problem. Be sure to correctly label your answers.

61. Millie Sorensen bought these items at the market: a 6.2 kg turkey, a 5 kg bag of potatoes, a 4.5 kg bag of flour, and 1.7 kg apples. These items were placed in one grocery bag. What was the total weight in kilograms of the items that were placed in the bag?

62. At the beginning of the week, the Wisteg Company had 350 metric tons of coal on hand. During the week the Wisteg Company received a shipment of 52 metric tons of coal. These amounts of coal were used during the week: 24 t, 28 t, 17 t, 21 t, 23 t. How many metric tons of coal should have been on hand at the end of the week?

63. A box contains 240 steno pads. The pads alone weigh 31.2 kg and the box weighs 3 kg. How many grams does each steno pad weigh?

64. A full box of crackers weighed 0.46 kg. The box is now $\frac{1}{2}$ full. What is the weight in grams of the crackers left?

65. Clarke LeBlanc bought a box of soap powder that held 4.2 kilograms of soap. He figures that he uses 210 grams of the soap for each load of clothes he washes. How many loads of wash can Clarke do with one box of soap powder?

66. A generic hand soap is sold in 8-bar packs. Each pack weighs 1.136 kg. Hand soap made by Prose is sold in 3-bar packs with a total weight of 405 g. What is the average weight in grams of one bar of each brand of soap?

The Chelsay Company is shipping 800 computer keyboards. The weight of one keyboard and its shipping box is 1.8 kg.

67. What is the total estimated weight in kilograms of this shipment?

68. What is the actual total weight in kilograms of this shipment?

A certain model of a car used to weigh 1 300 kg. The car has been redesigned so that its total weight is now 1 196 kg.

69. By how many kilograms was the weight of the car reduced?

70. What percent of the car's original weight is this reduction?

CONVERTING METRIC AND CUSTOMARY MEASUREMENTS OF CAPACITY AND MASS

Use Illustration 5-3.3 when converting from the metric system to the customary system or vice versa.

Metric to Customary	Customary to Metric
1 L ≈ 1.06 quarts	1 quart ≈ 0.95 L
1 L ≈ 0.26 gallon	1 gallon ≈ 3.79 L
1 g ≈ 0.035 ounce	1 ounce ≈ 28.3 g
1 kg ≈ 2.2 pounds	1 pound ≈ 0.45 kg

Illustration 5-3.3

EXAMPLE Amelia has a 2-liter container. How many quarts will it hold?

SOLUTION 1 liter ≈ 1.06 quarts $2 \times 1.06 \approx 2.12$ quarts.

EXAMPLE A boat anchor weighs 300 pounds. What is its equivalent weight in kilograms?

SOLUTION 1 pound ≈ 0.45 kg $300 \times 0.45 \approx 135$ kilograms.

EXERCISE YOUR SKILLS

71. A painter used 55 gallons of paint on his last job. How much paint was used in liters?

72. Because of the heat and the heavy exercise you are doing, you feel that you need to drink 2 L of water daily. To the nearest half glass, how many 8 oz. glasses of water must you drink to reach your goal?

73. A recipe for canning pickles that you received from Australia via the Internet directed you to use 80 grams of allspice. You do not have a metric scale. How many ounces of allspice should you use?

74. Until Lauren's back injury heals completely, she is not allowed to pick up anything heavier than 45 lbs. The weight of a box is printed as 21 kg. May Lauren pick up the box? Justify your answer.

Your car can hold 15 gallons of gas, weighs 2,000 pounds, and averages 35 miles to a gallon on the highway. The cost of gasoline in the United States is $1.28 per gallon.

75. Find the cost of a liter of gasoline at American prices.

76. On a trip to Canada, your car breaks down and you are towed 100 kilometers to the American border. The towing service charged you $1.50 per kilometer plus $0.15 for each kilogram of your car's weight. Calculate the total towing charge.

77. ◆ Create a capacity reference chart for your own use that shows the equivalents of ounces to liters; cups to liters; and quarts to liters. Create a mass reference chart for your own use that shows the equivalents of ounces to grams, and pounds to kilograms.

PUT IT ALL TOGETHER

78. Choose something to describe (for example, your classroom, an aisle in the supermarket, a garden, or a product.) Write a paragraph using only metric units to describe the length, area, capacity, and mass.

MIXED REVIEW

Complete.

1. 23.8 L + 14 L = ? L
2. 7000 mg + 15 g = ? g
3. 40 cm + 82 cm + 145 mm = ? cm
4. 38 L − 40 mL = ? L
5. 9 kg − 3 500 g = ? kg
6. 24.6 m − 450 mm = ? m
7. 25.6 t @ $8.40 = ?
8. 4 cm × 4 cm = ? cm^2
9. 832 L ÷ 14, to the nearest liter = ?

Solve.

John Matthews is foreman for a goldsmith shop. Each month, the shop receives 1 000 grams of gold to make gold bracelets. Each bracelet requires 30 grams of gold. Three jewelers are assigned to make the bracelets.

10. How many gold bracelets can be made each month out of the 1000 grams of gold ordered?
11. If gold is currently selling for $32.22 per gram, how much is the cost of gold in one bracelet?
12. If each bracelet takes 8 hours to make, and that time is billed at $25 per hour, what is the value of the labor involved in making the bracelet?
13. Paul needs to find the area of two storage sheds on his farm. One shed is 6 m long by 5 m wide. The other shed is 12 m long by 7.5 m wide. What is the total area of the two storage sheds in square meters?

Fern Wolters needs these lengths of plastic pipe: 1.5 m, 3 m, 0.4 m, 2.6 m, and 1.9 m. The pipe costs $4.80 a meter.

14. How many meters of plastic pipe does Fern need?
15. What will be the total cost of the pipe Fern needs?

The Clair Transit Company started the week with 3 560 L of gasoline in its storage tank. During the week it used these amounts of gasoline for its trucks: 340 L, 180 L, 248 L, 427 L, and 205 L.

16. How many liters of gasoline were used?
17. How many liters of gasoline were left in the tank?

TEMPLATE: Customer Service

You may recall working with an electronic database to store and retrieve address information in Chapter 3.

In this activity, you will see how a company can use a product database to assist customers and keep track of inventory. Your job will be to create a database that helps counter clerks at Do-It-Yourself Auto Parts answer customer questions. The database should also let clerks know when to place a new order so that no store part is ever out of stock.

Load ET5 from your Spreadsheet Applications diskette. The Item Name field has been set up. Set up the following fields: Part #, Bin #, Quantity, and Price. Then enter the data shown below.

Item Name	Part #	Bin #	Quantity	Price
Wiper Blades	44879	123	147	12.95
Oil Filter	77899	477	25	3.15
Air Filter	11447	014	5	6.75
Windshield Wash Fluid	99988	111	112	1.19
Spark Plug	55774	568	14	1.12
Fan Belt	24561	825	30	7.89
Battery	33478	679	3	39.95
Floor Mats	04897	033	12	59.99

You may also remember from Chapter 3 using SORT and LIST to help you locate and retrieve information. As a warm-up, view the Part # field in LIST format. Try to perform a numerical SORT with the same list.

Another way to retrieve information from a database is with a query. A **query** is a question you ask a computer about the data. You must ask the question in a way that your software will understand. The computer finds the records that match your query and displays only the data you want in the order you want. When you select query from the menu, your screen may look similar to the one shown below.

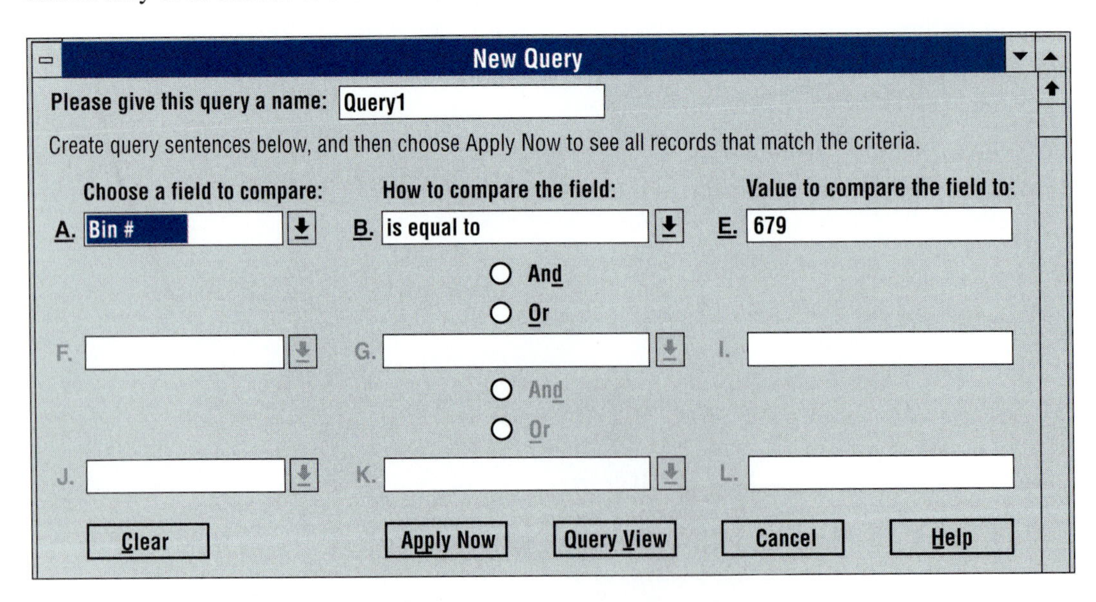

Imagine you are working the counter at Do-It-Yourself Auto Parts and two phone calls come in from customers. Complete the steps and answer the questions below.

1. A customer wants to know the cost of wiper blades. Write a query to search for that information.
2. The second customer asks about the availability and cost of Part #04897. Search for the part number in the computer. List the item, how many there are in stock, and the price.
3. Reorders are made when an item count drops below 20 in stock. Write a query to find all items with less than 20 on hand. Name the items needing to be ordered.

Design Your Own

4. What other information do you think a parts counter clerk at Do-It-Yourself might need? Add at least two fields to the database just created that would hold that information. Then write three queries of your own that find different ways to sort and match records.
5. Many companies keep customer records that show customer purchases and other useful information. If you were to create a customer database for Do-It-Yourself, what fields might you include? Make a list of three fields (other than address information).

CHAPTER 5 SUMMARY AND REVIEW

VOCABULARY REVIEW

area *5-2*
centimeter *5-1*
Customary System *5-1*
gram *5-3*
hectare *5-2*
Kilogram *5-3*
liter *5-3*
meter *5-1*
metric system *5-1*
metric ton *5-3*
milliliter *5-3*
square meter *5-2*

MATH SKILLS REVIEW

5-1
- A meter is the basic unit of length in the metric system.
- A meter can be divided into smaller parts such as a decimeter, centimeter, or millimeter.
- Multiples of a meter used to measure longer lengths are the dekameter, hectometer, and kilometer.
- Change from one metric unit to a smaller unit by moving the decimal point to the right one place for every unit of change.
- Change from one metric unit to a larger unit by moving the decimal point to the left one place for every unit of change.
- Change customary and metric measurements by finding their equivalent value and multiplying.

1. Write the metric symbol for meter.

2. Write 1 kilometer in meter terms.

3. Write 1 meter in millimeter terms.

4. Write the metric symbol for kilometer.

5. 18 dm = ? cm

6. 340 mm = ? m

7. 30 m – 1 800 mm = ? m

8. 80 cm × 6 = ? cm

9. 729 mm ÷ 9 = ? mm

10. 0.8 km + 2 100 m = ? km

5-2
- Smaller parts of a square meter include the square decimeter, square centimeter, or square millimeter.
- Common multiples of a square meter used to show larger areas are square hectometer (hectare), and square kilometer.
- Change from one metric unit to a smaller unit by moving the decimal point to the right two places for every unit of change.
- Change from one metric unit to a larger unit by moving the decimal point to the left two places for every unit of change.
- Change customary and metric measurements by finding their equivalent value and multiplying.

11. Write the metric symbol for square meter.

12. Write 1 hectare in square meter terms.

13. What is the process for finding area?

14. Is a mm^2 a part or a multiple of a square meter?

15. $2.8 \text{ ha} = ? \text{ m}^2$

16. $569 \text{ mm}^2 = ? \text{ cm}^2$

17. $8 \text{ m}^2 + 2\ 500 \text{ cm}^2 = ? \text{ m}^2$

18. $3 \text{ ha} - 8\ 000 \text{ m}^2 = ? \text{ ha}$

19. $0.6 \times 340 \text{ mm}^2 = ? \text{ cm}^2$

20. $120\ 000 \text{ cm}^2 \div 4 = ? \text{ m}^2$

5-3
- A liter is the basic metric unit of measure for capacity.
- Deciliter, centiliter, and milliliter are parts of a liter.
- A kiloliter is the most commonly used multiple of a liter.
- A gram is the basic unit of measure for mass or weight and a milligram is a commonly used part of a gram.
- Kilogram and metric ton are used to measure large weights.
- Change a capacity or weight metric unit to a smaller unit by moving the decimal point to the right one place for every unit of change.
- Change a capacity or weight metric unit to a larger unit by moving the decimal point to the left one place for every unit of change.
- Change customary and metric measurements by finding their equivalent value and multiplying.

21. Write the metric symbol for liter.

22. Write 1 milliliter in liter terms.

23. Write the metric symbol for gram.

24. Write 1 kilogram in gram terms.

25. $3 \text{ L} = ? \text{ mL}$

26. $50\ 000 \text{ mL} = ? \text{ kL}$

27. $670 \text{ cL} + 0.8 \text{ L} = ? \text{ L}$

28. $0.74 \text{ kg} = ? \text{ g}$

29. $0.25 \text{ t} = ? \text{ kg}$

30. $1\ 200 \text{ g} + 0.03 \text{ kg} = ? \text{ g}$

APPLICATIONS

5-1

After being made, cars are parked in a holding lot until they are shipped. Each car is exactly 5.2 m long. Cars are parked with their bumpers touching.

31. How many meters of space would be needed to park 4 cars?

32. If 80 meters of space was available to park cars, what total number of cars could be parked in this way?

33. If a meter is equal to 3.28 feet, what is the length of a car to the nearest tenth of a foot?

5-2

A homeowner plans to put in a new driveway measuring 25 m by 6 m. He also wants a new walkway measuring 40 m by 0.9 m. Both are to be paved with concrete.

34. How much area in square meters will be taken up by the driveway and walkway combined?

35. If 1 square meter is equal to 10.8 square feet, find the area to be paved, to the nearest square foot.

A can holds 9.5 L of gasoline. Mary Burke uses 7 full cans of gasoline in her lawn mower to cut her lawn each year. She uses 2 full cans to operate her snow blower.

36. How many liters of gasoline are used each year in the lawn mower and the snow blower?

37. If 1 liter is equal to 0.26 gallons, how many gallons of gasoline does Mary use each year for mowing her lawn and operating the snow blower?

Don Greer unloaded 80 bags of potatoes from a delivery truck. Each bag weighed 22.5 kg.

38. Find the total weight in kg of the potatoes unloaded.

39. At 2.2 pounds to a kilogram, what was the total weight in pounds of the unloaded potatoes?

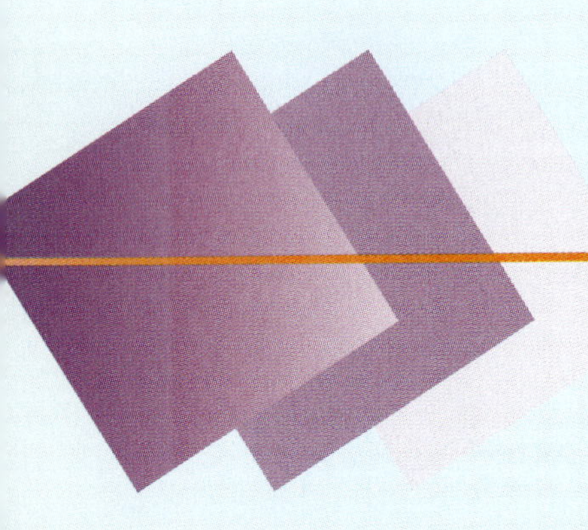

WORKPLACE KNOWHOW: **SEEING THINGS IN THE MIND'S EYE**

Seeing Things In The Mind's Eye To see things in your mind's eye is to be able to understand symbols, visuals, and graphs. For example, a woodworker can see a finished cabinet in wood; a chef can imagine how a mix of spices will flavor food; an employee may list the steps to finish a job from a written description of a project; and a seamstress can see a dress in cloth.

Assignment Have a class discussion about seeing in the mind's eye. Does everyone have this ability? Is it easier for some students than others? Why is this a valuable skill? If you can, think of one example where you can see something in your mind's eye and share it with the class.

Work Skills Portfolio Include an assessment of your ability to visualize and use metric measurements as well as your skill at converting between the metric and customary systems. If you completed the Integrated Project in the Student Workbook, include your worksheets as additional evidence of your skill.

CHAPTER 5 TEST

Math Skills

1. $3\ m^2 = ?\ cm^2$
2. $400\ mm^2 = ?\ cm^2$
3. 2 m = ? cm
4. 182 mL = ? L
5. 9 kL = ? L
6. 6 mm = ? m
7. 18 mm = ? cm
8. 582 g = ? kg
9. 1 500 mg = ? g
10. 0.8 km = ? m

Applications

11. To make orange juice from frozen concentrate, use 1 part concentrate to 4.5 parts water. Each part is 354 mL. How much juice in liters can you make from 1 part concentrate?
12. The heights of 5 members of a basketball team are: 1.97 m, 1.85 m, 2.08 m, 1.81 m, and 1.94 m. What is the average height, in meters, of these basketball players?
13. A bottle of rubber cement weighs 210 grams. One gross (12 dozen) of the bottles is being shipped to a store in 2 boxes that weigh 1.1 kg each when empty. What is the total weight in kilograms of the shipment?

You need to carpet a room that measures 4 m by 5.4 m.

14. How many square meters of carpeting will you need if carpeting is sold only in whole square meters?
15. Find the carpeting cost if it is $14 per square meter.
16. Danuta Afton is using a stencil to make a sign. She figures that each character will need an average of 20 mm of space. The longest line will contain 18 characters. Danuta also wants to leave 40 mm of space at each end of the longest line. What exact width must her sign be in centimeters so that the sign will be done as planned?
17. Les Wilkins delivers drinking water in his truck which holds 9 kL of water. When he began his deliveries one day, the water tank was 90% full. On that day, he delivered these amounts of water: 2 kL, 0.8 kL, 1.3 kL, 1.4 kL, 0.9 kL. How much water was left in the truck after Les's last delivery?

A homeless shelter plans to serve dinner to 80 people. Each person is to be served 160 grams of meat. What total amount of meat is needed

18. in kilograms?
19. in pounds?
20. A family built a new home on a lot that measures 30 meters wide by 50 meters long. The house and garage take up 210 square meters of the lot. The driveway and sidewalk take up 20% of the lot. If the family decides to plant grass on the remainder of the lot, how much space in square meters will they have to use?

CHAPTER 6

Building a Five-Star Career in *food* SERVICE

☆ ☆ ☆ ☆ ☆

Luxury and excellence describe the world-renowned five-star French restaurant where Manuel is a waiter. The five-star rating means the restaurant has been recognized for outstanding food and service.

Manuel and the other waiters here speak well and are polite, attentive, and enthusiastic. They are well-groomed and well-dressed. They know how to "read" their customers. (He's in a hurry. They want to be left alone.) They are also quick-thinking and quick-moving and work well under pressure.

There is a lot more to providing five-star service than putting good food on the table. The trick is to provide excellent service and make it look easy. Good planning is the key. Manuel starts the day early, reviewing his station and duties with his service team. Each team consists of a captain, two waiters, and a waiter's assistant. Team members double-polish the glasses, china, and silverware on the tables. Side duties, such as making coffee and bringing items from the pantries, are coordinated so that everything is available when it is needed.

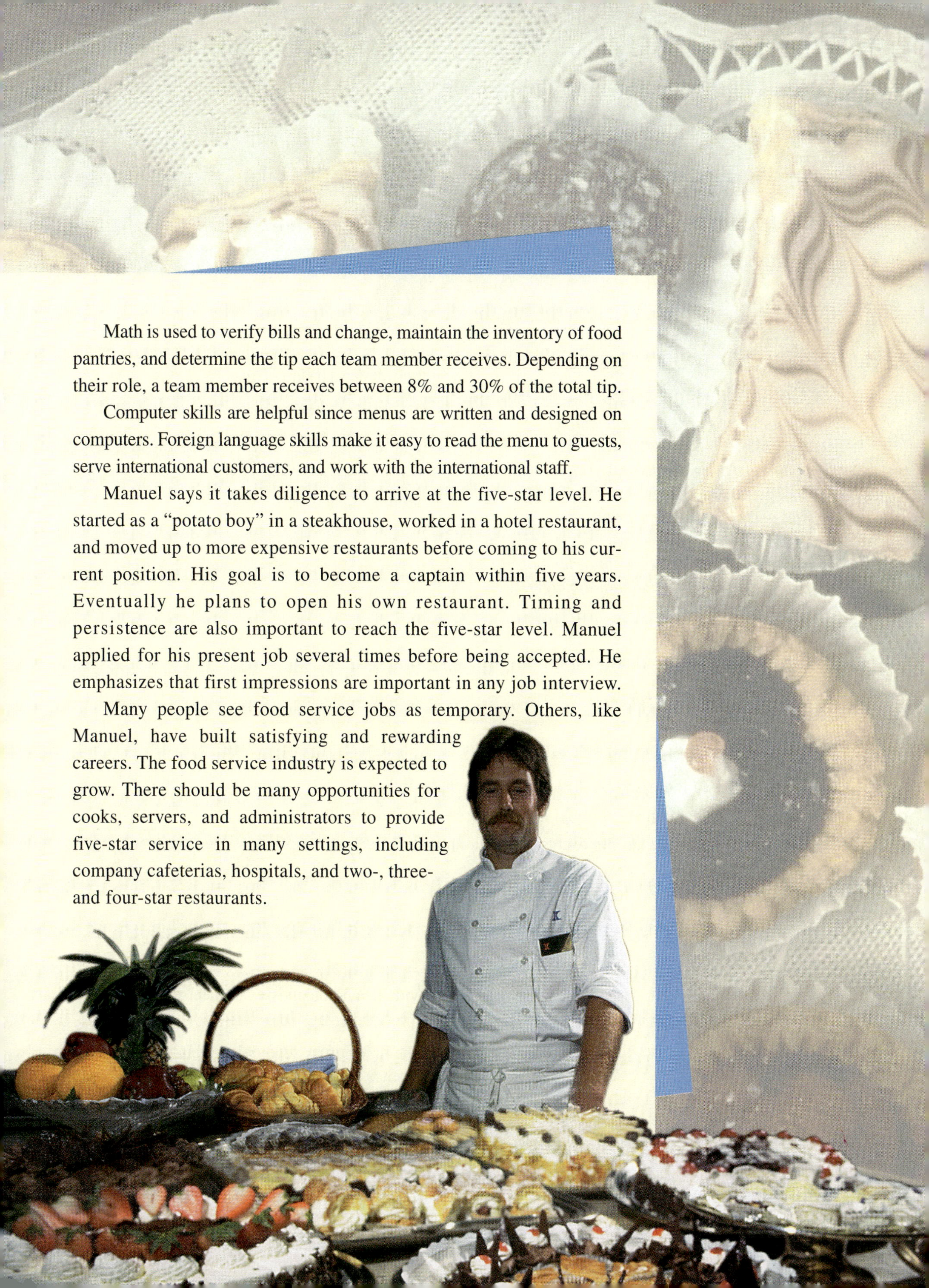

Math is used to verify bills and change, maintain the inventory of food pantries, and determine the tip each team member receives. Depending on their role, a team member receives between 8% and 30% of the total tip.

Computer skills are helpful since menus are written and designed on computers. Foreign language skills make it easy to read the menu to guests, serve international customers, and work with the international staff.

Manuel says it takes diligence to arrive at the five-star level. He started as a "potato boy" in a steakhouse, worked in a hotel restaurant, and moved up to more expensive restaurants before coming to his current position. His goal is to become a captain within five years. Eventually he plans to open his own restaurant. Timing and persistence are also important to reach the five-star level. Manuel applied for his present job several times before being accepted. He emphasizes that first impressions are important in any job interview.

Many people see food service jobs as temporary. Others, like Manuel, have built satisfying and rewarding careers. The food service industry is expected to grow. There should be many opportunities for cooks, servers, and administrators to provide five-star service in many settings, including company cafeterias, hospitals, and two-, three- and four-star restaurants.

6-1 PLANNING A BUDGET

Budgets are spending plans that help you allocate your income to meet financial obligations and save for things you want. Making a budget is easy.

◆ Look back at the income, expenses, and savings you have had recently. Think about what you would like to have or do. Then use the cash payments records you developed in Chapter 1 and your own experience to make a list of items you think should be included in your budget.

OBJECTIVES

In this lesson, you will learn to

- *use money records to define budget categories;*
- *distinguish between fixed, variable, and nonessential budget items;*
- *find percents of income spent on family expenses; and*
- *relate income to expenses in order to prepare a family or individual budget.*

WARM UP

a. Multiply $3\frac{1}{3}$ by $2\frac{1}{6}$

b. Subtract $3\frac{1}{2}$ from $7\frac{1}{8}$

c. 344 increased by 20% of itself is what number?

d. What number is equal to 342 decreased by $\frac{1}{3}$ of itself?

e. Write the answer to the subtraction in meters: 3.5 m − 250 cm

FIXED AND VARIABLE BUDGET EXPENSES

Budgets

Budgets are used by individuals and businesses to meet financial obligations and goals.

Developing a Budget The first step in developing a budget is to identify all your expenses. Include expenses that you have on a regular basis, and those that may happen periodically, such as semiannual insurance payments. You should also try to plan for expenses that cannot be scheduled or accurately predicted, such as car repairs.

Cash receipts records like the one you prepared in Chapter 1 can be used to identify expenses. The Lewis family used their cash receipts records to develop the monthly cash record shown in Illustration 6-1.1. After writing in the December amounts, the Lewis family totaled and ruled the summary to show their total receipts and cash payments for the year. They can use this information to plan their budget for next year. Cash summaries should be reviewed on a regular basis to be sure the budget is realistic.

Critical Thinking Why should items not usually paid monthly, such as insurance and taxes, be included in a monthly budget?

Budgets include **fixed expenses** and **variable expenses.** With fixed expenses, a set amount of money is allocated each month for that expense. Fixed expenses usually include money set aside for insurance and taxes; regular monthly payments such as savings and mortgage or rent payments. Other expenses, such as telephone bills, transportation expenses, and personal expenses, while predictable, vary. Those expenses are called **variable expenses.**

Some people track expenses with very specific categories, such as house insurance, home maintenance, housing expenses, and taxes. Others use broader categories. The Lewis' broad category of housing expenses would include mortgage payments, property taxes, insurance, and a maintenance budget.

Critical Thinking Why is savings included as a fixed expense?

Cash Record Summary								
			Types of Payment					
Month	Receipts	Payments	Food	Clothing	Housing	Transportation	Other	Savings
Jan.	$5,160.00	$4,953.60	$825.60	$365.75	$1,345.20	$616.46	$1,040.59	$760
Feb.	5,989.15	6,089.15	793.55	867.79	1,458.47	857.22	1,356.12	756
Nov.	6,123.64	5,823.64	785.62	275.79	1,985.86	563.10	1,438.27	775
Dec.	6,084.21	6,484.21	869.34	920.50	1,619.48	777.99	1,546.90	750
Total	$66,800.00	$66,800.00	$9,700.00	$4,680.00	$18,200.00	$8,200.00	$16,900.00	$9,120

Illustration 6-1.1. Cash Record Summary of the Lewis Family

In budgeting, the amounts allowed for each type of payment are often shown as a percent of income or of total payments. You can use those percents to find future allowances and to compare your payments with other people's payments.

EXAMPLE Find the percent of income that Laverne and Ramon Lewis (Illustration 6-1.1) spent for food.

SOLUTION Divide food expense for the year, $9,700, by income, $66,800. Round the result to the nearest whole percent.

$$\frac{\$9{,}700}{\$66{,}800} = 0.14521 \text{ or } 0.15 = 15\%$$

Percents for other types of payments are found in the same way.

Critical Thinking Joan's net annual income is $18,000 a year. On her 25th birthday, she spent $1,500 to celebrate with friends and family. Roberta makes $200,000 a year. When she celebrated her 50th birthday, she spent $5,000 for her party. Which one of them do you think overspent for the party? Explain your answer.

One way to begin a budget is to project next year's expenses based on the percent of income spent in the current year. To do this, you simply multiply the expected income by the percent allowed for each type of expense.

EXAMPLE Last year the Lewis' spent 15% on food. If they estimate the same percent of income allocated to food next year, how much will they spend on food if their income is expected to be $68,800?

SOLUTION 15% of $68,800 =
$0.15 \times \$68,800 =$
$10,320 food budget for the year.

Critical Thinking What other factors, besides the percent of money spent previously, should be considered in planning a new budget?

EXERCISE YOUR SKILLS

Use the data from the Lewis family's cash record summary shown in Illustration 6-1.1 to answer Exercises 1–11.

What fractional part of their January receipts were the Lewis'

1. January payments?
2. January payments for food?

To the nearest whole percent, what percent of the Lewis' November payments were their payments for

3. Housing?
4. Transportation?
5. Savings?

The Lewis family expects their income next year to be $69,500. Assume they budget equal percents for each month, based on last year's percents. Show the percent, rounded to the nearest whole percent, and amount of money they will spend during the year for

6. Food

7. Clothing

8. Housing

9. Transportation

10. Other

11. Savings

12. Last year Allon Company's expenses totaled $475,800. If $338,723 was spent for salaries, what percent of the total expenses was salary expense, to the nearest whole percent?

13. Myron Security, Inc., had total sales for one year of $945,860. Their advertising expenses were $57,370. Find the percent of total sales, that advertising expenses were to the nearest tenth of a percent.

Tri-County Medical Center budgets 2% of its total receipts for advertising and 7.75% for power. For receipts of $5,500,000, what amount would the Center budget for

14. Advertising?

15. Power?

16. ◆ Suppose Ramon and Laverne decide they want to adjust their budget so they can save for a special summer vacation. How would you suggest they go about revising their budget to meet their goal?

MIXED REVIEW

1. $2\frac{3}{4} + 6\frac{1}{2} + 3\frac{2}{3}$

2. $3\frac{3}{5} \div \frac{2}{3}$

3. $648 increased by $\frac{2}{3}$ of itself equals what amount?

4. What amount equals $108 decreased by $\frac{3}{8}$ of itself?

5. Dylor Metals, Inc., had total sales of $975,000 last year and spent $650,000 on salaries and wages. What percent, to the nearest percent, of total sales was the amount they spent on salaries and wages?

6. An office worker is paid $242 for a $37\frac{1}{2}$-hour week. Find the office worker's hourly rate to the nearest cent?

7. Find the FICA tax, at 7.65%, on wages of $436.

6-2 CHECKING SALES RECEIPTS

OBJECTIVES

In this lesson, you will learn to

- *find extensions;*
- *find sales tax amounts;*
- *find sales slip totals; and*
- *count change.*

When John got home from the store he found that he had been charged $9.25 for $8.75 worth of purchases. To add to his frustration, he realized he had paid for his purchases with a $10 bill and received only $0.25 in change.

◆ Think about times when you have been overcharged or undercharged at the cash register. Analyze why those errors happened and how you can avoid them in the future. Based on your analysis, organize a list of steps or procedures to be sure sales transactions are accurate.

WARM UP

a. 16 L + 12.5 L = ? L

b. 24 kg − 1 400 g = ? kg

c. 2 500 mg + 12 g = ? g

d. 22 cm + 16.5 cm + 126 mm = ? cm

e. 5 L − 115 mL = ? L

f. 34.1 m − 127 mm = ? m

THE SALES RECEIPT

Most store employees give customers **cash register receipts,** or **sales slips,** after a transaction is completed. Customers can use these receipts, or slips, to check that all the goods bought were received and that the prices and calculations were correct. Receipts, or sales slips, can also be used to keep expense records or as proofs of purchase if the goods are returned.

Like the sales slip in Illustration 6-2.1, most sales slips show the

- number of units, or **quantity,** and a description of each item.
- **unit price,** or the price of one item or group of items treated as one. For example, a dozen eggs is one unit.
- **extension,** or the product of the unit price and the quantity of each item. The amount, $159.80, is the extension of 10 boxes at $15.98.
- **subtotal,** or the sum of the extensions.
- **sales tax.**
- **total sale,** or the sum of the subtotal and the sales tax.

STATE ELECTRONICS
16 State St., Bellingham, WA 98225-7981
206-555-0100

SOLD TO: *Jessie Podanski*
STREET: *245 South Brevard St.*
DATE: *March 22* 19- - CITY, STATE, ZIP: *Bellingham, WA 98225-9080*

SOLD BY *RG*	CASH	CHARGE √	C.O.D.	DELIVERY BY *Taken*

QUANTITY	DESCRIPTION	UNIT PRICE		AMOUNT	
10 boxes	*Computer diskettes, 3 1/2"*	*15*	*98*	*159*	*80*
1	*Printer cable, parallel*	*14*	*95*	*14*	*95*
12	*Diskette caddies*	*8*	*95*	*107*	*40*
2 boxes	*Printer paper, white*	*49*	*95*	*99*	*90*
	SUBTOTAL	---	--	*382*	*05*
	SALES TAX	5%		*19*	*10*
	TOTAL	---	--	*401*	*15*

Everything You Ever Wanted in Electronics!

Illustration 6-2.1. Sales Receipt

Finding Extensions

To find the extension of an item on a sales slip, multiply the quantity by the unit price.

Quantity × Unit Price = Extension

For example, in the sales slip in Illustration 6-1.1, the extension of 2 boxes of printer paper at $49.95 is $99.90 (2 × $49.95 = $99.90).

"At" or "@" means the price of a single unit. For example, "1 dozen rolls @ $1.25" means "1 dozen at $1.25 *per dozen"*; and "3 lb of turkey at 79 cents" means "3 lb at 79 cents *per pound.*" But a price of "3 lb of ground beef, $5.99" means that the price is $5.99 for 3 pounds.

If the unit price is stated for a group of items, such as "2 rolls of foil, $2.29," you may need to find the extension for 12 rolls. To do this, first find the number of 2-roll units; then multiply that number by the unit price.

12 divided by 2 = 6 two-roll units
6 × $2.29 = $13.74 total price

When working with extensions, it is wise to make a rough estimate before calculating. Then compare your final result with the estimate to see if your answer is reasonable.

Sales slips can be easily checked using your calculator's memory keys. However, a calculator might not give you accurate results if one or more of the extensions is not rounded to the nearest cent. For example, if the extensions for the first two items on the sales slip in Illustration 6-2.1 were $159.803 and $14.954, added on a calculator their subtotal would be $174.757, which rounds to $174.76. On a sales slip, they are added, as shown, as $159.80 and $14.95. Their subtotal is $174.75. This kind of error usually occurs when the unit price includes a fraction of a cent. For example, 7 bunches of celery @ $37\frac{1}{2}$ cents = $2.625. To avoid a rounding error when using a calculator, key in the rounded amounts for each extension and then add those amounts to memory.

EXERCISE YOUR SKILLS

Find the estimated and exact extensions of each of these items.

1. 5 boxes cards at $4.89
2. 4 qt milk @ $2.19
3. 7 kg flour @ 32 cents
4. 1.95 m fabric at $6.89

Find the extensions of each of the items.

5. 5 videocassettes @ $7.99
6. 10 patio blocks at 69 cents
7. 12 qt of oil, $9.89
8. 12 cans of apple juice @ $3.49
9. 3.5 kg meat, $17.99
10. 3 L gas at $0.24

Create a sales slip for each of the following problems. Calculate the extensions for each item and the subtotal for each slip.

11. 4 m fabric @ 2 m for $6
    5.5 m lining @ $3.12
    7 spools thread @ 83 cents
    9 pkg. tape @ 69 cents
    3 pkg. buttons @ 1.28 cents
    1 zipper, $2.99
12. 3 bags charcoal @ $2.88
    1 bottle shampoo, $3.15
    8 jars of pickles @ 2 for $3.69
    8 cards at $1.25
    3 cans sweet potatoes, $1.69
    4 oranges @ 49 cents
    12 apples at 3 for 89 cents
    3 cans plant food, $3.99

Finding Sales Taxes

Many states, cities, and counties charge a **sales tax** on certain items. Sales tax is a percentage of the base price of an item or a percentage of the total of the base prices (subtotal) of all taxable items. Sellers collect sales tax from consumers for the government. Sales tax rates differ from community to community and state to state.

Sales Tax Rate × Subtotal = Sales Tax

Sales tax is always rounded to the nearest cent. If more than one tax is charged, the tax rates are combined into one rate. For example, if there is a state tax of 5% and a city tax of 3%, the tax shown on the sales slip is 8%.

The buyer pays the sum of the subtotal and the sales tax.

Subtotal + Sales Tax = Total

The sales slip in Illustration 6-2.1 shows that the sales tax rate is 5% (0.05) of the subtotal, $382.05. The amount of the sales tax, rounded to the nearest cent, is $0.05 \times \$382.05 = \$19.1025$, or $19.10. The total paid by the buyer is the subtotal, $382.05, plus the tax of $19.10, or $401.15.

If you want to quickly calculate the total cost of an item, including tax, add 100% to the tax rate and multiply that amount by the base price. For example, using the amounts in the previous paragraph, the base price times 105% is $\$382.05 \times 1.05 = \$401.1525$, or $401.15.

EXERCISE YOUR SKILLS

Mental Math Find the sales tax.

13. Subtotal, $40; sales tax rate, 5%

14. Subtotal, $15; sales tax rate, 8%

The subtotal of a sales slip is $49.39. The sales tax rate is 8%.

15. Estimate the amount of the sales tax.

16. Find the amount of the sales tax.

17. Find the total on the sales slip.

18. Where Tony lives, state sales tax is 5.5% and city tax is 2%. What amount of sales tax must he pay on a $78 purchase?

19. Mona wants to buy a bed that costs $795. The city sales tax rate is 8%. In a nearby city the sales tax rate is 5%. How much less would the bed cost if Mona bought it in the neighboring city?

Counting Change

Many cash registers show the amount of the sale, the amount paid by the customer, and the amount of change to be given to the customer. Other registers show only the amount of the sale. In either case, both the clerk and the customer should make sure that the amount of change is correct.

Change is usually given in the fewest possible bills and coins. If a $10 bill is given for a $7.78 purchase, the change is $2.22 and given as two pennies, two dimes, and two $1 bills.

A good way to give change is to state the amount of the purchase, then add the change to that amount as the change is given. For the purchase described in the previous paragraph, the clerk could say, "$7.78" while giving the customer two pennies and saying "$7.80," then say "$7.90, $8" while giving the two dimes, then say "$9, $10" while giving the two $1 bills.

EXERCISE YOUR SKILLS

Copy the form below. For each problem, show the number of each type of coin and bill the clerk should give the customer.

	Amount Received	Amount of Sale	Change					
			1¢	5¢	10¢	25¢	$1	$5
20.	$5.00	$3.26						
21.	$1.00	$0.56						
22.	$10.00	$3.49						
23.	$10.00	$6.15						

24. ♦ Use what you have learned in this lesson to make up a word problem that involves purchasing at least four items and requires that change be given. Be sure to include the sales tax rate and the amount given the clerk.

MIXED REVIEW

1. $12\frac{2}{3} \times 6\frac{1}{2}$

2. $7\frac{1}{5} \div 2\frac{1}{4}$

3. You buy these items: 6 cans of peas @ 3 for $0.99; 2 packages of frozen vegetables, $4.99; 6 bunches of celery @ 3 for $1.19. What is your subtotal?

4. Altors, Inc., had sales of $3,250,000 last year. The company spent 75% of its sales on wages and salaries and 5% on advertising. How much was spent by the company on wages and salaries? on advertising?

5. Sandra Reade wants to carpet the floor of her room. The room is 2.4 m wide, 2.2 m high, and 3.5 m long. If 10% more carpet is needed to allow for waste, how many square meters of carpet must Sandra buy?

6. The subtotal of a sale is $28.79. A 5% state and 2% city sales tax are charged. What is the correct total on the sales slip?

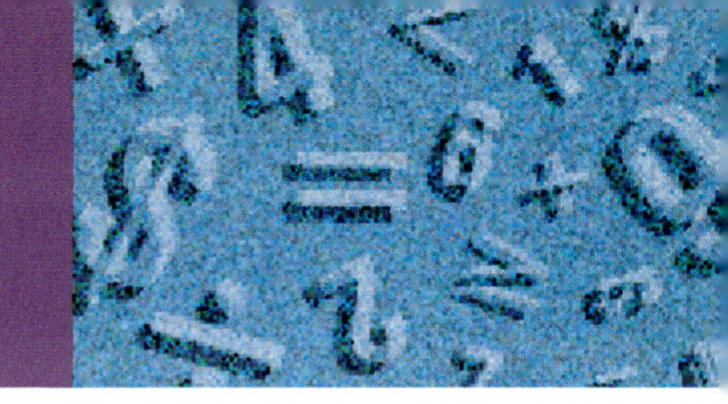

6-3 MORE SALES EXTENSIONS

Many items are sold in units that are not singular. Eggs are sold by the dozen; soft drinks are sold in 2-liter bottles; nails may be sold by the hundredweight.

◆ Make a list of the different units that are used in selling items. List the unit and then give examples.

Unit	Sample items
Gross	pencils, pens
Ream	paper

OBJECTIVES

In this lesson, you will learn to

- *find extensions when quantities or prices are mixed numbers;*
- *find extensions when large quantities are involved; and*
- *use shortcuts when finding extensions.*

WARM *UP*

a. $7\frac{1}{2} \times 3\frac{1}{5}$

b. $24\frac{1}{3} \div 2\frac{2}{3}$

c. 88 is what percent more than 76, to the nearest tenth of a percent?

d. 34 m − 2 400 mm = ? m

e. 20 is what percent less than 24?

f. $16,000 × 0.01

FINDING EXTENSIONS

Sales clerks must know how to find extensions on sales slips. Buyers should also know how to find extensions so they can check the accuracy of their sales slips. Sometimes sales slips contain:

- Fractional quantities, such as $\frac{1}{2}$ dozen.
- Mixed number quantities, such as $3\frac{1}{2}$ yards.
- Large quantities, such as a hundred or a thousand.

Finding the Cost of a Fraction of a Unit

To find the cost of a fraction of a unit, multiply the unit price by the fraction. For example, suppose computer cable is priced at 63 cents a yard and you want to buy two feet, or $\frac{2}{3}$ of a yard. Multiply the cost per yard by the quantity needed.

$$\overset{21}{\cancel{63}} \text{ cents} \times \frac{2}{\underset{1}{\cancel{3}}} = 42 \text{ cents} = \text{the cost for } \frac{2}{3} \text{ of a yard.}$$

Hint: Since 63 cents is exactly divisible by the denominator, 3, you could divide 63 cents by 3 to get 21 cents, the cost of $\frac{1}{3}$ yard, then multiply by 2 to get the cost of $\frac{2}{3}$ of a yard. If the price were 64 cents rather than 63 cents, which is not exactly divisible by 3, it would be easier to multiply the 64 cents by 2.

2 × 64 cents = $1.28

$1.28 divided by 3 = $42\frac{2}{3}$ or 43 cents, cost of $\frac{2}{3}$ yard

Some stores treat any fraction of a cent as an added cent. Other stores round fractions to the nearest cent and drop fractions of less than $\frac{1}{2}$ cent and raise fractions of $\frac{1}{2}$ cent or more to the next full cent. For example, $42\frac{2}{3}$ cents would be treated as 43 cents and $42\frac{1}{3}$ cents would be treated as 42 cents.

Finding the Cost When Quantities or Prices Are Mixed Numbers

The quantity or the unit price on a sales slip may be a mixed number. For example, $1\frac{1}{2}$ lb of potatoes or $2\frac{1}{2}$ dozen biscuits. Rope may be priced at $21\frac{1}{2}$¢ per ft, floor tiles at $51\frac{1}{4}$ cents each, or screws at $3\frac{1}{4}$ cents each. As you learned in previous lessons, you can deal with mixed numbers in several ways.

EXAMPLE You want to buy $3\frac{1}{4}$ tons of sand at $8.89 per ton. What is the cost?

SOLUTION 1
Step 1 $\$8.89 \times \frac{1}{4} \text{ ton} = \frac{8.89}{4} = 2.2225 = 2.22$ rounded

Step 2 $\$8.89 \times 3 \text{ tons} = \$26.67$

Step 3 Cost of sand $= 2.22 + 26.67 = \$28.89$

SOLUTION 2
Step 1 $3\frac{1}{4} = \frac{13}{4}$

Step 2 $\$8.89 \times \frac{13}{4} = \frac{(\$8.89 \times 13)}{4}$

Step 3 $= \frac{\$115.57}{4}$ or $\$28.89\frac{1}{4}$ or $28.89 rounded

SOLUTION 3
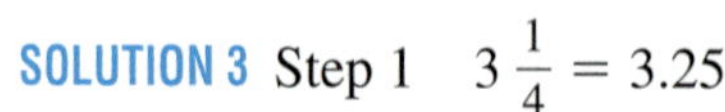
Step 1 $3\frac{1}{4} = 3.25$

Step 2 Use a calculator.
$\$8.89 \times 3.25 = \$28.8925$ or $28.89 rounded

EXERCISE YOUR SKILLS

Round answers to the nearest cent unless you are told not to.

1. Ana Diaz buys the items below. If she pays with two $20 bills, how much change should she receive?

   $\frac{1}{3}$ roll of baling wire, $4.89

   $\frac{3}{4}$ roll fencing @ $24.96

   $\frac{1}{5}$ length, plastic culvert @ $65.90

2. Find the total cost of $\frac{1}{2}$ dozen note pads @ $8.49 dz, $\frac{1}{3}$ box of folders @ $7.99 box, and $\frac{5}{8}$ box of ribbons @ $10.82 box.

Fujio Komuro bought 27 inches of cloth at $6.16 a yard.

3. What fraction of a yard did he buy?
4. Find the total cost of the cloth, including 5% sales tax.
5. Rick Lenoux bought these items at a grocery store: 14 oz salami @ $3.99 per lb, 8 oz bacon @ $2.49 a lb, 6 oz cauliflower @ $1.49 a lb, $\frac{2}{3}$ dozen bagels @ $3.99 a dozen. Sales tax is 3%. Find Rick's total cost.
6. Find the total price of these items, including an 8% sales tax: 75 ft of field tile @ $73.72 per 100 ft., 6 hay sleeve covers at $29.88 per dozen, 4 fasteners at $1.69 per dozen, and 12 oz seed at $8.95 per lb.
7. Philip Gotlieb bought the fabrics below. What was the subtotal on Philip's sales slip?

   $2\frac{2}{3}$ yd. cotton @ $1.99

   $3\frac{1}{2}$ yd. muslin @ $2.88

   $3\frac{1}{4}$ yd. wool @ $5.89

   $4\frac{5}{8}$ yd. lace @ $0.75

8. Find the total cost of $12\frac{1}{3}$ ft of wire at $0.28 per ft and $\frac{2}{3}$ yard of chain at $0.98 per ft.
9. Cassie Werner bought the items below at a market. How much change did Cassie get from two $20 bills?

   $3\frac{3}{4}$ lb potatoes @ $1.29

   $1\frac{7}{8}$ lb grapes @ 99¢

   $5\frac{1}{8}$ lb apples @ $0.59

   $2\frac{1}{3}$ lb of pecans @ $6.98

Finding the Cost for Large Quantities

Many items bought in large quantities are priced by the **hundred (C),** the **thousand (M),** the **hundredweight** or **hundred pounds (cwt),** the **ton (T),** or the **metric ton (t).** The symbols for hundred and thousand come from Roman numerals.

Office supplies such as envelopes and paper, building supplies such as cement and lumber, and farm products such as grain and fertilizer are often bought in large quantities.

To find the cost of goods priced by the C, M, or cwt, divide the quantity by 100 or 1,000 to find the number of hundreds or thousands you are buying. Then multiply by the unit price.

EXAMPLE Find the cost of 2,560 computer diskettes at $89 per hundred.

SOLUTION 2,560 ÷ 100 = 25.6
25.6 × $89 = $2,278.40

To find the cost of goods priced by the ton (T) when the weight is given in pounds, divide the weight by 2,000 (the number of pounds in a ton). The answer is the number of tons. Then multiply the number of tons by the price per ton.

EXAMPLE Find the cost of 9,000 lb of rock at $88 per T.

SOLUTION 9,000 ÷ 2,000 = 4.5 tons
4.5 × $88 = $396

To find the cost of goods priced by the metric ton (t) when the weight is given in kilograms, divide the weight by 1,000 (the number of kilograms in one metric ton). Then multiply that answer by the price per ton.

EXAMPLE Find the cost of 2 500 kg of sand at $8 per t.

SOLUTION 2 500 ÷ 1,000 = 2.5 metric tons
2.5 × $8 = $20

EXERCISE YOUR SKILLS

Find the estimated and exact costs

10. 450 ties @ $89.78 per 100

11. 219 lb seed @ $48 per cwt

12. 189 panels @ $83.50 per C

13. 8,207 tiles @ $571.99 per 1,000

14. 1,298 lb @ $64 per T

15. 12 450 kg @ $9.88 per t

Using Fractional Parts of $1

When the unit price is a fractional part of $1 you can calculate many extensions quickly and easily. Simply change the unit price to its fractional equivalent based on $1. Then multiply that fraction and the quantity.

EXAMPLE What is the cost of 24 clamps @ $12\frac{1}{2}$ cents each?

SOLUTION

$\frac{(\$0.12\frac{1}{2})}{\$1} = \frac{1}{8}$	Change the unit price to its fractional equivalent based on $1.
$\frac{1}{8} \times 24 = 3$	Multiply the quantity by the fraction.
$3 \times \$1 = \$3$	Multiply that product by $1.

This shortcut works best if you memorize the following equivalencies.

$25¢ = \frac{1}{4}$	$20¢ = \frac{1}{5}$	$12\frac{1}{2}¢ = \frac{1}{8}$	$33\frac{1}{3}¢ = \frac{1}{3}$
$50¢ = \frac{1}{2}$	$40¢ = \frac{2}{5}$	$37\frac{1}{2}¢ = \frac{3}{8}$	$66\frac{2}{3}¢ = \frac{2}{3}$
$75¢ = \frac{3}{4}$	$60¢ = \frac{3}{5}$	$62\frac{1}{2}¢ = \frac{5}{8}$	
	$80¢ = \frac{4}{5}$	$87\frac{1}{2}¢ = \frac{7}{8}$	

EXERCISE YOUR SKILLS

Mental Math Find each extension.

Hint: Use $\frac{1}{4}, \frac{1}{2}, \frac{3}{4}$

16. 12 @ 25¢ **17.** 32 @ 25¢ **18.** 26 @ 50¢

19. 20 @ $0.75 **20.** 48 @ $0.75

Hint: Use eighths.

21. 80 @ $12\frac{1}{2}¢$ **22.** 32 @ $\$0.12\frac{1}{2}$

23. 40 @ $37\frac{1}{2}¢$ **24.** 16 @ $\$0.62\frac{1}{2}$

25. 72 @ $62\frac{1}{2}¢$ **26.** 24 @ $\$0.87\frac{1}{2}$

27. 48 @ $37\frac{1}{2}¢$ **28.** 16 @ $\$0.87\frac{1}{2}$

Hint: Use thirds.

29. 18 @ $33\frac{1}{3}¢$ **30.** 24 @ $66\frac{2}{3}¢$ **31.** 33 @ $\$0.33\frac{1}{3}$

32. 30 @ $\$0.66\frac{2}{3}$ **33.** 96 @ $66\frac{2}{3}¢$

Hint: Use fifths.

34. 15 @ 20¢ **35.** 35 @ 40¢ **36.** 25 @ $0.60

37. 35 @ 80¢ **38.** 75 @ $0.60

39. ♦ Jay Muir, an agent for United Fruit Distributors, purchased the items shown in the following invoice. Look at the invoice and make a list that identifies and explains any errors. Then prepare a corrected invoice.

Invoice

NAME *United Fruit Distributors* ACCOUNT NO. *JMUFD*

ADDRESS

QUANTITY	DESCRIPTION	UNIT PRICE	AMOUNT
4,880 lb	*vegetable oil*	*12½¢*	*$610 00*
2,760 lb	*bananas*	*33⅓¢*	*920 00*
600 lb	*grapes*	*62½¢*	*375 00*
1,600 lb	*tomatoes*	*87½¢*	*1400 00*
	SUBTOTAL		*$3,305 00*
	TAX	2%	*66 10*
	TOTAL		*$3,371 10*

MIXED REVIEW

1. 19.6 m − 925 mm

2. 3% of $2,078 = ?

3. $24 is what percent of $75?

4. Write $\frac{6}{20}$ as a decimal.

5. Eric Sutton started the day with $32.77 in cash. During the day, he received $10.00 from a friend to repay a loan. Eric spent these amounts on the same day: $2.89, $5.15, $0.78, $12.99. How much cash should he have at the end of the day?

6. Beth bought these items at an office supply store: $1\frac{1}{4}$ dozen pencils @ $1.89 per dozen, $\frac{2}{3}$ dozen boxes of correction tape @ $6.10 per box, $1\frac{1}{3}$ cartons of envelopes @ $12.73 per carton. What was Beth's subtotal?

7. Kerry Durr earns a salary plus commission. Her total earnings last year were $23,800, of which $8,400 was salary. If her total sales last year were $308,000, what rate of commission did Kerry receive?

8. On June 30, Jerry Devoe's check register balance was $742.16. When he balanced his account, he had not recorded a service charge of $4.50, a deposit of $37.50, and a check for $9.15. Find his correct check register balance.

6-4 FINDING UNIT PRICES

Many items are packaged and priced in ways that make it difficult to compare prices. For example, crackers may be packaged in $7\frac{1}{2}$-oz, 8-oz, 12-oz, 15-oz, 1-lb, or 18-oz packages. Soap powder may be sold in 12-oz, 1-lb 2-oz, 24-oz, $2\frac{1}{2}$-lb, or 10-lb boxes. Baked goods may be packaged in quantities of 2, 6, 8, 9, 12, or more items to the box.

◆ Write a paragraph describing two ways you could compare prices to see which size and brand of a product sold for the lowest price.

OBJECTIVES

In this lesson, you will learn to

- *use unit prices to compare the cost of goods;*
- *find unit price from a group price; and*
- *find unit price when an item price is for a fraction of a unit.*

WARM UP

a. 320 @ $\$0.12\frac{1}{2}$ = **b.** 160 @ $\$0.87\frac{1}{2}$ = **c.** 24 @ $\$0.16\frac{2}{3}$ =

d. 40 @ $\$0.62\frac{1}{2}$ = **e.** 125 @ \$0.40 = **f.** 1,250 @ \$25 per M =

g. 1,600 lb @ \$3 per cwt = **h.** 8% of \$1,720 =

UNIT PRICES

To help purchasers compare the real costs of goods, many storeowners post unit prices on their shelves. The **unit price** is the price of one item or one measure of the item. It may be an ounce, a pound, a quart, a dozen, a hundred feet, or some other measure. If unit prices are not posted, the buyer who wants to compare must calculate the unit prices.

Finding a Unit Price from a Group Price

Stores may have one price for a single unit and another price for a group of two or more units. For example, soup may be sold for 89 cents a can or 3 cans for

$2.60. If you buy 3 cans, you pay only 87 cents a can ($\$2.60 \div 3 = 86\frac{2}{3}¢$ or 87¢). In this way, vendors encourage consumers to buy more of their products.

To compare prices of items priced by the group, find their unit price by dividing the group price by the number of units in the group. **If there is a fraction left over, count it as a whole cent.** For example, if beans are selling at 3 cans for $1, the price per can is $1 divided by 3, or $\$0.33\frac{1}{3}$. The unit price is **rounded up** to 34 cents.

Finding the Unit Price When the Item Price Is for a Fraction of a Unit

Rolls may be packaged in fractions of a dozen, such as 6 ($\frac{1}{2}$ dozen) or 8 ($\frac{2}{3}$ dozen). Toothpaste may be sold in 3 oz ($\frac{3}{16}$ lb) or 5 oz ($\frac{5}{16}$ lb) tubes. To compare prices in these cases, you must find the price of the appropriate unit. To find the unit price, divide the price by the quantity. You may first have to show the quantity as a fraction of a unit.

EXAMPLE A 5-oz tube of toothpaste costs $1.50. Find the price per pound. Remember: 16 oz = 1 lb.

SOLUTION $5 \text{ oz} = \frac{5}{16} \text{ lb}$ $\quad$ \$1.50 divided by $\frac{5}{16} = \overset{0.30}{\cancel{1.5}} \times \frac{16}{\underset{1}{\cancel{5}}} = 4.80$

The price per pound is $4.80.

CHECK $\overset{0.30}{\cancel{\$4.80}} \times \frac{5}{\underset{1}{\cancel{16}}} = \$1.50$

Critical Thinking What other unit could you use to compare the price of toothpaste?

EXERCISE YOUR SKILLS

Find the unit price. *Round up* to the next higher cent.

1. 4 batteries for $5.59
2. 3 packages of crackers for $5.89
3. 4 bottles of vitamins for $10.49
4. 6 boxes of candles for $9.47

Write each of these as a fraction of the unit shown.

5. 6 inches, as feet
6. 2 feet, as yards
7. 6 ounces, as pounds
8. 500 pounds, as tons
9. 16 fluid ounces, as quarts
10. 1 pint, as quarts
11. 3 quarts, as gallons
12. 4 items, as dozen
13. 8 items, as dozen
14. 5 square feet, as square yards

Find the price per pound. Round up to the *nearest cent.*

15. Peanuts at 14 oz for \$3.19

16. Birdseed at 12 oz for 36¢

17. Yogurt at 11 oz for \$1.56

18. Crabmeat at 6 oz for \$2.43

19. A package of 36 paper plates is marked to sell for \$0.69. What is the price, rounded to the *nearest cent,* of the plates per 100?

20. A 7-oz jar of cashews costs \$2.49. A 14-oz jar of the same cashews costs \$5.49. Which jar is less expensive to buy, and how much less per ounce is it?

Critical Thinking When comparing unit prices, is it better to compare a price rounded to the nearest cent or the nearest tenth of a cent?

Finding the Unit Price for Metric Quantities

To find the unit price of items sold in metric quantities, first show the quantity as a decimal value of the unit you want. Then divide the price by the decimal.

EXAMPLE A 1650 gram can of juice sells for \$5.99. What is the cost per kilogram?

SOLUTION $1650 \text{ g} = 1.65 \text{ kg}$ $\qquad \$5.99 \div 1.65 = \$3.6303$

CHECK $\$3.6303 \times 1.65 = \$5.9895$ or \$5.99

EXERCISE YOUR SKILLS

Find the unit price to the nearest cent.

21. A bag of pecans weighing 250 g is sold for \$1.89. What is the price of the pecans per kilogram?

22. A 750 mL carton of orange juice that usually sells for \$1.78 is on sale for \$1.58. What is the price per liter of the sale juice?

23. Two liters of cola costs \$0.89 on sale and \$1.29 when not on sale. How much per liter do you save by buying this cola when it is on sale?

Finding the Unit Price When the Item Price Is For a Mixed Number Quantity

When the price of an item is for a mixed number quantity, such as $3\frac{1}{4}$ lb, find the unit price by dividing the unit price by the quantity.

EXAMPLE A $1\frac{1}{4}$-qt bottle of bleach sells for \$2.29. What is the price per quart?

SOLUTION $1\frac{1}{4} \text{ qt} = \frac{5}{4} \text{ qt}$ $\qquad \$2.29 \div \frac{5}{4} = \$2.29 \times \frac{4}{5} = \$1.832$

CHECK $\$1.832 \times 1\frac{1}{4} = \$2.29$

EXERCISE YOUR SKILLS

Find the answers correct to the nearest cent.

24. A carpet that is $18\frac{1}{2}$ square yards in size is priced to sell for \$296. What is the price of the carpet per square yard?

25. A $2\frac{5}{8}$-lb box of laundry soap is priced at \$2.19. What is the equivalent price per pound?

26. A $1\frac{1}{4}$-lb box of oats sells for \$2.19. What is the price of the oats per pound?

27. A 28-oz can of sweet corn sells for \$0.59. What is the price of the corn per pound?

28. A 24-oz can of cranberries costs \$1.29. What is the price of the cranberries per pound?

29. An 18-oz can of peaches sells for \$1.09. What is the price of the peaches per pound?

30. ◆ Make a list of 10 items and find the cost for each item in at least two different stores. Is one store cheaper for all items? Which store would you prefer to shop in? Why?

MIXED REVIEW

1. 45 is what percent of 225? **2.** $32 \div \frac{3}{4}$

3. What percent of a kilogram is 400 grams?

4. Find the cost of $1\frac{1}{2}$ ft of cloth priced at \$1.89 per yard.

5. Find the unit price of 1 can of antifreeze @ 3 for \$8.59.

6. Ivy Mitchem works Monday through Friday from 8:30 A.M. to 5:00 P.M. with an hour off for lunch. Last week she worked full time at \$5.67 an hour. Find her earnings for the week.

7. Bart's check register balance on April 1 was \$405.58. He found that a check for \$8.50 was recorded in his check register as \$5.80 and that he had not recorded a service charge of \$3.25 or a deposit of \$16.37. Find Bart's correct check register balance.

6-5 BUYING WISELY

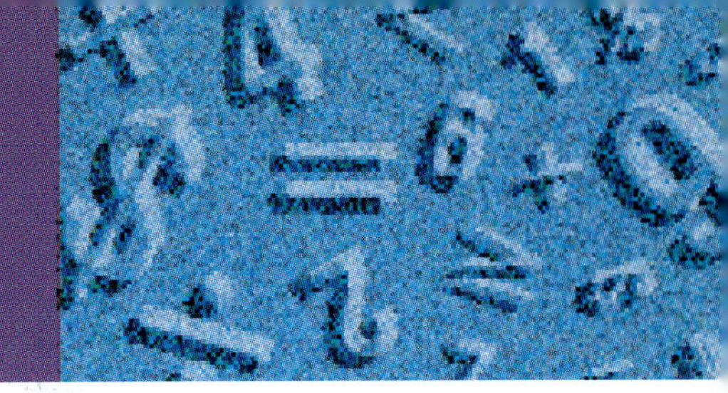

Buying wisely can save you money. Often a product costs less when it is purchased in large amounts. You can save money by buying products on sale or by shopping at stores that sell items at a discount. In addition, you may save money by renting items you do not use often.

◆ Do you shop wisely? Explain your evaluation. Be specific about what makes you a wise or unwise shopper.

OBJECTIVES

In this lesson, you will learn to compare

- *prices;*
- *rental options; and*
- *the cost of renting versus buying.*

WARM UP

a. 180 is what percent of 120?

b. Divide 24 by $\frac{3}{8}$

c. Multiply 24 by $\frac{3}{8}$

d. Write 2.6 as a percent.

e. Write $\frac{3}{17}$ as a percent, to the nearest tenth of a percent.

f. What percent of a meter is 500 mm?

COMPARING PRICES AND BUYING AT SPECIAL PRICES

To save money, you should know how to calculate the amount you can save by buying in large amounts, at sales, or at stores that sell at discounts. Remember, many small savings can add up to a lot of money.

Many of the items you buy come in different sizes and at different prices. Unit prices allow you to compare prices and identify the lowest priced item.

EXAMPLE 1 How much will you save by buying a case of motor oil at Dan's Discount Store rather than buying it at Nashville Auto Supply?

SOLUTION

\$29.99	← cost at Nashville Auto Supply
− 19.95	← cost at Dan's Discount Store
\$10.04	← amount saved

DAN'S DISCOUNT STORE

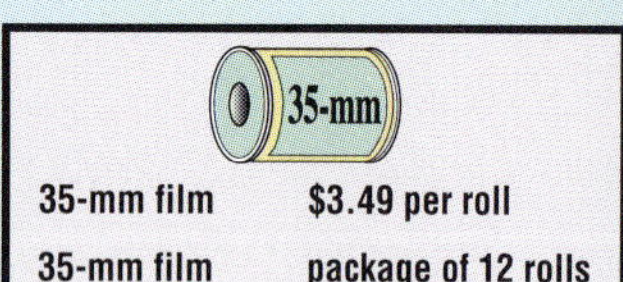

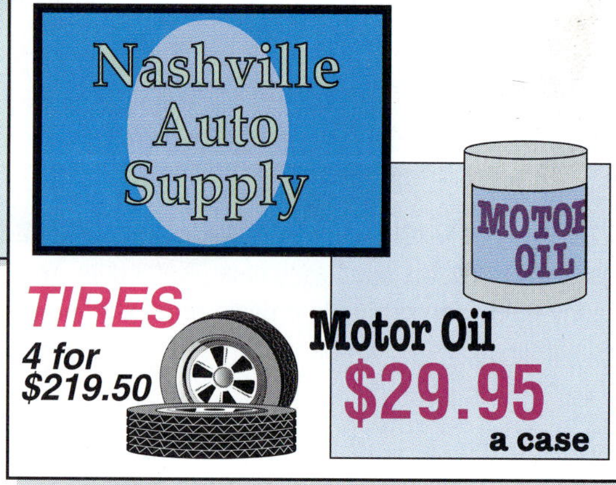

EXAMPLE 2 How much will you save by buying 12 rolls of film now rather than buying 1 or 2 rolls at a time?

SOLUTION

$12 \times \$3.49 =$	$41.88	← cost of 12 rolls @ $3.49 each
	− 34.95	← cost of 12 roll package
	$ 6.93	← amount saved

Critical Thinking Why might you buy the motor oil at Nashville Auto Supply anyway? List two possible disadvantages of buying the 12 rolls of film to save money.

EXERCISE YOUR SKILLS

1. Kevin can buy digital cassettes at $16.95 each, or 5 for $76.95. How much would he save by buying five disks now instead of one at a time?
2. Eva found an automatic washer priced at $429.97 and a dryer for $299.97. They are on sale for only $649.99 for the set. How much can Eva save if she buys the items as a set?
3. Miguel Romano can buy one office floor mat for $219.99 or a set of 3 for $629.85. If Miguel buys the set, how much will he save per mat?
4. At an end of season sale, Vi Carr bought a room air conditioner that was reduced from $298.99 to $259.75 and a fan that was reduced from $49.95 to $37.99. What total amount did she save?
5. Boxes of ten 3.5-inch computer diskettes regularly priced at $14 are on sale for 3 boxes for $35.85. How much would you save if you bought 6 boxes?

A department store offers senior citizens a courtesy discount of 5% on all items. Bob Vallow, a senior citizen, buys a portable TV priced at $295.99.

6. What is the estimated amount of Bob's discount?
7. What exact amount does Bob pay for the TV?

At a preseason sale, Gil Abramson bought 35 L of kerosene at 15% off the regular price of $0.37 per L.

8. How much did Gil save by buying the kerosene on sale?
9. How much did Gil pay for the kerosene?

A $1\frac{1}{4}$-lb box of Brand A cereal sells for $4.49.

A $1\frac{1}{2}$-lb box of Brand B cereal sells for $4.79.

10. What is the difference in the price per pound?
11. Which brand costs less per pound?

PROBLEM SOLVING

One strategy to use when trying to solve a problem is to act out the situation.

Use Exercises 8–9 to apply this strategy. Work with a partner. One person should be a customer; the other, a clerk. Imagine you are in a store trying to figure out if you should buy the kerosene on sale. The clerk comes along to help you. The customer should estimate what the total sale will be. The clerk should use a calculator to determine the exact cost of the items. Switch roles for Exercises 10–11.

How did this strategy help you solve the problems?

A 1-lb 10-oz box of Deli Fish Sticks costs $4.99. Seaview Fish Sticks in a 2-lb 4-oz box are priced at $6.49. A 2-lb 12-oz box of Star Fish Sticks sells for $6.59.

12. To the nearest cent, what is the price per pound of the least expensive brand of fish sticks?
13. To the nearest cent, find the difference in price per pound of the least expensive and most expensive brands.

You can buy a set of 2 computer diskette storage cases for $25.90 from Store A or a set of 6 for $71.70 from Store B.

14. What is the price per case from Store A? Store B?
15. How much would you save by buying 12 cases from the lower-priced store?

Comparing Rental Options

Many items can be rented. You may want to rent items you do not not use often or only need for a limited amount of time. Rental companies usually offer different rental rates for different time periods, such as hourly, daily, weekly, and monthly.

To determine the best rental option, you should know how long you need the item and then find the most economical rate.

EXAMPLE A rug cleaner can be rented for $5.50 an hour or $25 per day. If it takes you 6 hours to clean your rugs, would you be better off choosing the hourly rate or the day rate?

SOLUTION Step 1 Calculate the cost of renting the item by the hourly rate: 6 hours × $5.50 an hour = $33.

Step 2 Compare the cost of renting by the hour to the cost of the day rate: $33 − $25 = $8

SOLUTION In this case, it is more economical to choose the day rate.

EXERCISE YOUR SKILLS

16. Yi Sun wanted to refinish the floors in his home. He can rent the sander and polisher by the hour, the day, or the week. The costs are as follows:

Item	Hour	Day	Week
Sander	$12	$40	$150
Polisher	$10	$25	$100

If Yi Sun needs the sander for four 7-hour days and the polisher for two 7-hour days, calculate the most economical rental option for each item.

Comparing Renting to Buying

If you rent something on a regular basis, you may wonder if it would be more economical to buy than rent. To make the decision, you would need to:

- Calculate the annual cost of renting the equipment.
- Calculate how many years it would take for the rental price to equal or exceed the purchase price.
- Make a purchase decision based on price and other relevant factors.

EXAMPLE You rent a rug cleaning machine four days a year at a cost of $25 a day. You see a rug cleaner on sale for $225. Do you think you should buy the rug cleaner?

SOLUTION Step 1 Calculate the cost of renting the rug cleaner each year. 4 days × $25 = $100 year.

Step 2 Calculate the number of years it would take for the cost of the rental to equal or exceed the cost of the purchase. $225 ÷ 100 = 2.25 years.

Step 3 Make a purchase decision. There is no correct answer. In this case, you may want to consider purchasing the rug cleaner because in two years you would "break even" on the purchase. However, if you don't have any room to store the cleaner or if the machine is likely to break down after $1\frac{1}{2}$ or 2 years, you may decide to continue renting.

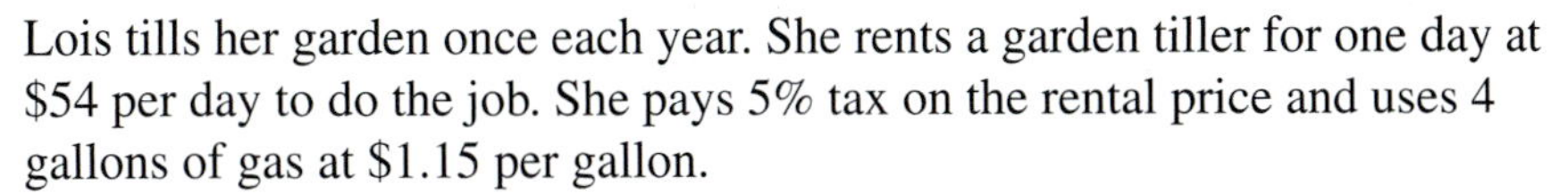

EXERCISE YOUR SKILLS

17. A home owner can rent a chain saw from a rental agency at $36 a day. The same saw can be bought new for $252. For how many days could the homeowner rent the saw before renting would cost more than buying?

Lois tills her garden once each year. She rents a garden tiller for one day at $54 per day to do the job. She pays 5% tax on the rental price and uses 4 gallons of gas at $1.15 per gallon.

18. What is her total cost for using the tiller?
19. If a similar tiller costs $264.60 including tax, for how many years could Lois rent it before renting would cost more than buying?

A computer can be rented for $174.50 a week or $34.75 a day. You need the computer for 6 days.

20. At which rate, daily or weekly, would it be cheaper to rent?
21. How much would you save by renting it at that rate?
22. If a similar computer costs $1,657.75 to buy, for how many weeks could you rent the computer at the weekly rate before renting would cost more than buying?

PUT IT ALL TOGETHER

23. ♦ A friend sees an ad that would allow him to rent a stereo for $20 a week, for 152 weeks. At the end of the rental period, he would own the stereo. He could buy the same stereo at the store for $2,000. He thinks the rent-to-own price is a good buy. You don't.

    Make a list of the advantages and disadvantages of the rent-to-own option. Include 3 mathematical comparisons that support your position that the rent-to-own option is not a good buy.

MIXED REVIEW

1. Write $3\frac{3}{5}$ as a decimal.
2. $42 is what part greater than $24?
3. What amount is $\frac{1}{3}$% of $2.28?
4. On June 1, Linda's bank statement balance was $372.57, and her check register balance was $307.65. While comparing the statement and her check register, she found a service charge of $4.23; earned interest of $1.79; and outstanding checks for $17.87, $2.97, $45.33, and $1.19. Prepare a reconciliation statement for Linda.
5. Gus Pinsker worked a total of 50 hours in one week. Of that time, 40 hours was at the regular rate of $10.25 an hour and 10 hours was at time-and-a-half for overtime. What was Gus's gross pay for the week?
6. Jake earns an annual wage of $32,750. He estimates his benefits at 28% of his wages. He also estimates that his job expenses are insurance, 6% of wages; commuting, $385; dues, $275; other, $125. Find his annual net job benefits.
7. A real estate agent sold a house for $213,400, and the agency earned a commission of $12,804. Find the rate of commission.
8. Delia wants to buy a car that costs $12,788 in a state with a 6% sales tax. How much sales tax will she pay?
9. In six weeks, Lois drove 1,920 miles in her new car. At the same rate, how many miles will Lois drive in a year?

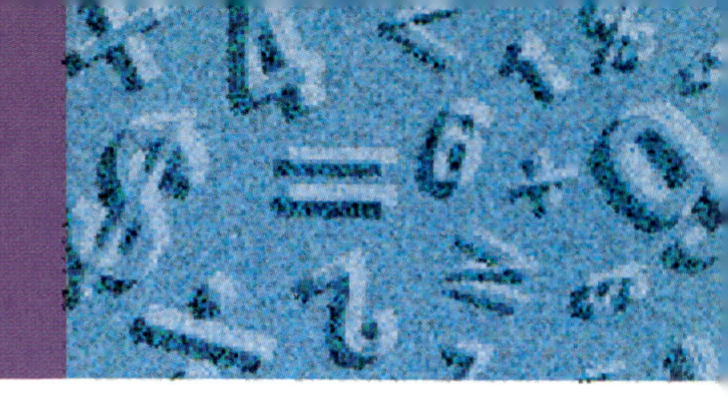

6-6 CHECKING ENERGY COSTS

Heating, cooling, cooking, lighting, and running appliances all take energy. Some landlords include the cost of energy (such as electricity and gas) in the rent they charge to tenants. Other owners expect tenants to establish an account with local power companies who bill the tenant for their power usage.

◆ If you were renting an apartment, which way would you prefer to be billed? Make a list of the advantages and disadvantages of each type of billing system.

OBJECTIVES

In this lesson, you will learn to

- *read an electric meter;*
- *check an electric bill;*
- *read a gas meter; and*
- *check a gas bill.*

WARM UP

a. \$400 is what part smaller than \$640?

b. \$196.50 less $66\frac{2}{3}\%$ of itself is what amount?

c. Write $1\frac{4}{5}$ as a decimal.

d. Find $33\frac{1}{3}\%$ of 183 L.

e. 8 kg is ?% of 5 kg?

f. Write $\frac{7}{8}\%$ as a decimal.

ENERGY SOURCES

People use many types of fuels for their home energy needs, including kerosene, wood, the sun, and wind. Electrical power and gas are among the most common energy sources.

In some communities, one power company provides both gas and electric service to their customers. In others there are separate utility companies. Illustration 6-6.2 on page 225 shows a utility bill that comes from a company that provides both gas and electric service.

Meter Gauges Power companies measure the amount of a resource used with meter gauges. They use the meter readings to bill customers. Customers are usually billed on a monthly basis though utility companies may read the meters on a bimonthly, quarterly, or annual basis. When meters are read periodically, monthly usage is estimated by the power company based on past meter readings.

Billing Options Many companies will let you average your power bill over one year. That means they estimate your annual bill based on the previous year's usage. The estimated annual bill is divided by 12, and you are billed each month for $\frac{1}{12}$ of the estimated bill. At the end of the year, after the meter has been read, you will receive a bill that includes an adjustment for over-or underpayment.

Critical Thinking What are the advantages and disadvantages of paying an estimated annual utility bill in 12 equal payments? How can you minimize the disadvantages?

ELECTRICITY

The amount of electricity you use is measured by an electric meter in kilowatt-hours, or **KWH.** A **kilowatt** is 1,000 watts of electric current. A **kilowatt-hour** is the flow of 1,000 watts of electricity for one hour.

Reading the Electric Meter To find out the amount of energy your household uses, you can read your electric meter. Digital meter displays are easy to read. Dial meters display energy use on four dials that are read from left to right. The figures read from each dial are the last ones the pointer has passed. For example, in Illustration 6-6.1, the last figures passed are

8 2 5 1

The reading 8,251 means 8,251 kilowatt-hours.

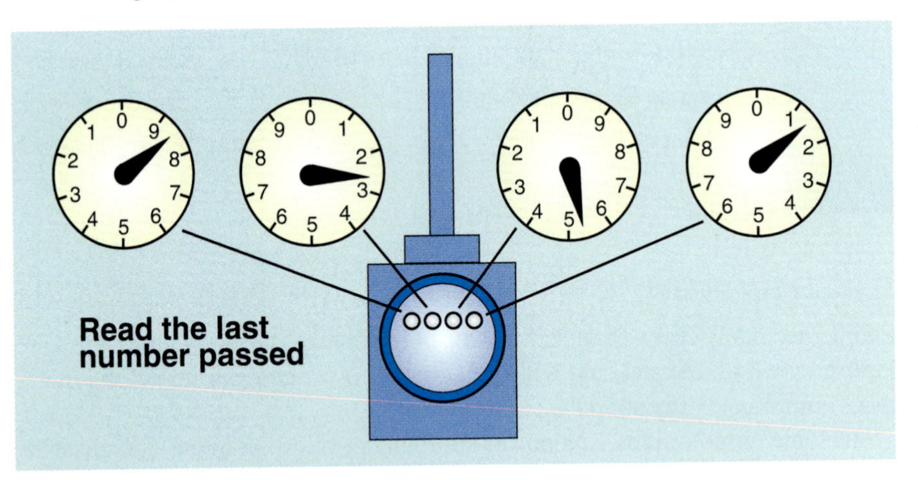

Illustration 6-6.1. Electric Meter Dials

Critical Thinking Look at each dial. Does the dial at the far right turn clockwise or counterclockwise? What about the other dials? What kinds of errors do you think are easy to make when reading one of these meters?

Reading An Electric Bill

The electric portion of the bill in Illustration 6-6.2 shows a May 14 meter reading of 4,892 and a June 14 reading of 5,717. The difference, 825, is the number of kilowatt-hours used during the billing month.

Velma Wense
1756 Bridge St.
Lima, OH 45806-8977

ACCOUNT NO.	DUE DATE	AMOUNT TO BE PAID	150.29
798-334-890	JUL 14		

BILLING SUMMARY

Amount due as of JUN 14	**$ 165.06**	
Payment Received by JUN 13 – Thank You	**165.06 CR**	
Balance (Payments after JUN 13 not included)		**0.00**
Current Gas Charge		**51.77**
Current Electric Charge		**99.02**
AMOUNT TO PAY DUE DATE – JUL 14		**$ 150.79**
TOTAL AMOUNT OWED IS $150.79 — **AMOUNT DUE IF PAYMENT RECEIVED AFTER JUL 14**		**$ 152.35**

EXPLANATION OF BILLING CHARGES

ITEM	METER NUMBER	BILLING PERIOD From	To	Days	METER READINGS Previous	Present	ENERGY USAGE
Gas	00000	MAY 14	JUN 14	30	802	906	104
Elec	00000	MAY 14	JUN 14	30	4892	5717	825

THERM FACTOR: 1.582 THERMS USED: 165

GAS RATE 202 – Residential Service		Facilities Charge	$ 19.25
Gas Usage Charge:	90 THMs @ $0.2107 =		18.96
	75 THMs @ $0.1478 =		11.09
		Current Gas Charge	**$ 49.30**
		Taxes @ 0.05	**2.47**
		Current Gas Charge	**$ 51.77**
ELEC RATE KR – Residential Srv – Summer			
Elec Usage Charge:	500 kWh @ $0.1011		$ 50.55
Elec Fuel Component:	300 kWh @ $0.1216		36.48
	Over 800 kWh @ $0.1471		3.68
Fuel Adjustment @ 0.00435			3.59
		Total Energy Charge	**$ 94.30**
		Taxes @ 0.05	**4.72**
		Current Electric Charge	**$ 99.02**

Illustration 6-6.2. Utility Bill for Electric and Gas Service

Notice that there are two components to the electrical charges. The first one is a usage charge. To find the usage charge, the power company multiplied the number of kilowatt-hours used in the month by the rate for each amount. The company charges $0.1011 each for the first 500 KWH, $0.1216 each for the next 300 KWH used, and $0.1471 on all KWH used over 800.

The second charge for usage is the **fuel adjustment rate.** The fuel adjustment rate, also known as the Electric Fuel Component (EFC), is the cost per kilowatt hour of the fuel used to generate electricity. This rate may vary based on the price of coal and oil. When the cost of the fuel used increases, the company multiplies the KWH by the fuel adjustment rate and *adds* that amount to your bill. When the cost of the fuel used decreases, they *subtract* that amount from your bill.

The fuel adjustment charge is calculated by multiplying the total kilowatt hours used by the adjustment rate. For example, the total KWH used on the sample bill is 825 and the fuel adjustment rate is \$0.00435. So the fuel adjustment amount was:

825 × \$0.00435 = \$3.58875 or \$3.59

Notice that Velma Wense had to *pay* the fuel adjustment on her bill. So the fuel adjustment was *added* to her bill. If she had *received* the fuel adjustment, that amount would have been *subtracted* from her bill.

Some state and local governments tax utility bills. Taxes vary from location to location and might include sales taxes, regulatory taxes, public utility taxes, and user taxes. The tax on Velma's bill is 5%. Velma's total electrical charges, including the fuel adjustment, are \$94.30. This amount is multiplied by 5% state tax to find the tax of \$4.72. The tax is added to the total energy charges to find the total amount of the electric bill, \$99.02.

EXERCISE YOUR SKILLS

Find the KWH used, the cost of the KWH, the fuel adjustment amount, the tax, and the total bill for each exercise. Use the rates from the Buckeye Power Company bill in Illustration 6-6.2. Assume that each customer must pay the fuel adjustment amount.

	Meter Readings		KWH	Cost of	Fuel		Total
User	**May 10**	**June 10**	**Used**	**KWH**	**Adj.**	**Tax**	**Bill**
A.	278	925	**1.**	**2.**	**3.**	**4.**	**5.**
B.	730	1,940	**6.**	**7.**	**8.**	**9.**	**10.**
C.	624	1,175	**11.**	**12.**	**13.**	**14.**	**15.**
D.	5,308	6,845	**16.**	**17.**	**18.**	**19.**	**20.**

21. Eda's electric reading was 8,124 for June and 7,280 for May. She pays 12.82¢ per KWH for the first 500 KWH; 14.17¢ for KWH over 500. A fuel cost adjustment of $0.0029 per KWH is *subtracted*. Sales tax is 6%. Find the current bill.

22. Karl Schmidt used 1,230 KWH last month. Because he is over 65, he pays a special senior citizen rate of 7.05¢ per KWH for the first 1,000 KWH and 6.12¢ for each KWH over 1,000 used in one month. He also *pays* a fuel adjustment rate of $0.0143 on the total KWH and a 3% state sales tax on his total energy charges. What amount must Karl pay for his current bill?

NATURAL GAS

Reading the Gas Meter Many homes use gas for cooking and heating. Gas usage is usually measured in cubic feet by a gas meter. As with electrical meters, there are both digital and dial display gas meters.

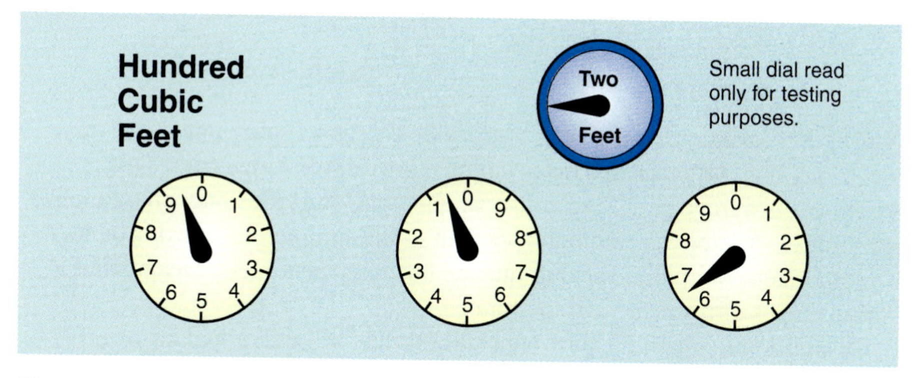

Illustration 6-6.3. Gas Meter

Checking the Gas Bill

The gas service portion of Velma's bill shows the present meter reading is 906. Her reading last month was 802. So she used 104 hundred cubic feet of gas this month.

THMs The amount of heat you get from gas varies. So many companies charge you for the heat you use instead of the cubic feet of gas you use. The amount of heat in gas is measured in British Thermal Units (BTUs), or **therms (THMs).**

The gas company finds the amount of heat in the gas by testing it. Using these tests, the company finds a therm factor, which shows the average amount of heat in the gas you bought. To find the cost of gas, multiply the number of cubic feet of gas used by the therm factor. The product is the number of therms used, rounded to the nearest whole therm. Once you have the product, multiply by the appropriate rate shown in the company's rate schedule.

For example, Velma's bill shows that she used 104 hundred cubic feet of gas with a therm factor of 1.582. The number of whole therms she used was:

104 × 1.582 = 164.528, or 165 whole therms

A gas bill may also include an adjustment rate, called a **Gas Cost Recovery Rate.** This is the cost per 100 cubic feet of gas purchased from suppliers. The rate varies based on current gas costs. Velma's company does not charge a gas cost recovery rate.

EXAMPLE **Rates:**

Rate schedule is:

Facilities charge	$19.25
First 90 whole therms	$0.2107
All therms over 90	$0.1478

Velma's gas bill is calculated this way.

Computation of Gas Bill:

Facilities charge	$19.25
First 90 whole therms (90 × $0.2107)	$18.96
Next 75 whole therms (75 × $0.1478)	$11.09
Subtotal	$49.30
5% Sales tax	2.47
Total bill	$51.77

Billing Practices Some utility companies charge a higher rate as more electricity and gas is used. This is called a **graduated rate scale.** This is meant to encourage people to conserve energy. Companies may charge lower rates for electric power in the winter than in the summer or give discounts to senior citizens or charge varying rates to business customers. Discounts for various types of users are also common.

In addition to charging for usage and service, utility companies often collect other fees. For example, some companies charge a monthly facilities fee, which is a rental fee for the use of the company's equipment and lines. Other common fees are gas pipeline fees and fees for environmental protection practices, such as cleaning power plant emissions. These fees are itemized on the utility bill and included in the total amount due.

EXERCISE YOUR SKILLS

23. On Brandon's May gas bill, the present reading is 896 and the last is 784. The therm factor is 1.377. The per therm rates are first 50 therms, $0.5772; next 50 therms, $0.3798; over 100 therms, $0.1365. Find the amount of his gas bill, including a utility tax of 5%.

24. Bea's gas meter read 357 hundred cubic feet on May 10 and 496 hundred cubic feet a month later. The therm factor was 1.023. The per therm rates were first 70 therms, $0.6011; over 70 therms, $0.7831. There was a facilities charge of $21.97 and a sales tax of 3%. What was Bea's total bill?

For each user, find cubic feet of gas used, the therms used, the charge for gas, a 5% sales tax, and the total bill. The per therm rates are first 90 therms, $0.4208; over 90 therms, $0.3568.

User	Meter Readings Present	Last	Hundreds of Cubic Feet of Gas Used	Therm Factor	Therms Used	Charge for Gas Used	Sales Tax	Total Bill
A.	380	260	**25.**	1.486	**26.**	**27.**	**28.**	**29.**
B.	784	697	**30.**	1.578	**31.**	**32.**	**33.**	**34.**
C.	189	78	**35.**	1.108	**36.**	**37.**	**38.**	**39.**
D.	573	489	**40.**	1.499	**41.**	**42.**	**43.**	**44.**
E.	205	132	**45.**	1.228	**46.**	**47.**	**48.**	**49.**

50. ♦ Make a list of questions you would ask a landlord before agreeing to rent an apartment that included gas and electric costs in the rent?

MIXED REVIEW

1. Write $1\frac{1}{8}$ as a decimal.
2. Write $\frac{7}{8}\%$ as a decimal.
3. Find 9,256 @ $0.23 per thousand.
4. $200 is what part smaller than $320?
5. $98.25 less $66\frac{2}{3}\%$ of itself is what amount?
6. Rose's electric meter read 6,361 KWH on May 1 and 6,987 KWH on June 1. She pays 12.7¢ each for the first 100 KWH, 10.5¢ each for the next 100 KWH, and 8.9¢ each for all KWH used over 200. She pays a fuel adjustment rate of $0.00132 per KWH and sales tax of 4%. Find Rose's total electric bill.
7. Bell's Sports Center sells 2 racquetball rackets with covers at a sale price of $19.99 for both. At Vito's Outdoor Market, similar rackets are priced at $12.99 each with a manufacturer's rebate of $5 each. The covers cost an additional $2.89 each. Which store offers the best price for 2 rackets and covers? How much lower is the price?
8. Bud Cross bought these items at the grocery store: $1\frac{1}{4}$ lb apples @ $0.45, $\frac{1}{2}$ lb cream cheese @ $1.19, 2 cans of carrots @ 3 for 1.09. He paid a 5% sales tax. What was the total amount of Bud's bill?

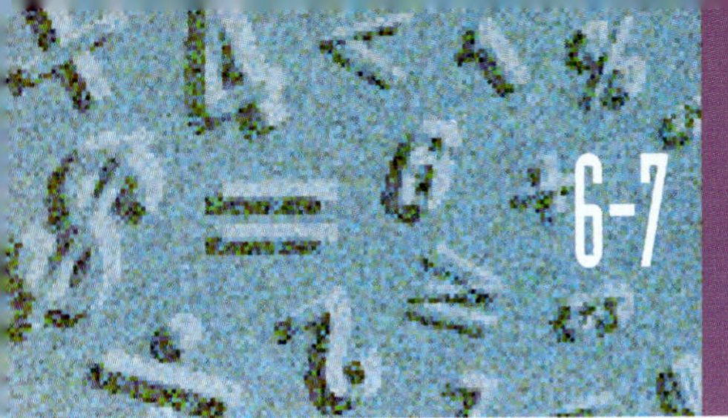

6-7 REDUCING ENERGY COSTS

OBJECTIVES

In this lesson, you will learn to

- *conserve home fuels;*
- *find net savings from conservation; and*
- *calculate payback periods.*

The cost of heating, lighting, cooling, and running appliances is a significant expense in operating a home today.

◆ Work in small groups to identify 5–10 common household appliances. Make a table and estimate the cost of running the appliance for one hour, one week, and one year. Assign each person in the group the responsibility of getting an actual estimate on the cost of running one of the appliances on your list from a library, power company, manufacturer, store, or the Internet. Plan to meet in a week to compare your estimates to the information you find in your research.

WARM UP

a. Write $2\frac{2}{5}$ as a decimal.

b. Write $\frac{1}{8}\%$ as a decimal.

c. 398 lb @ $12 per cwt

d. 1 350 mm is ?% of 2 m?

e. Divide 817 by 95,000

f. Write 0.75 as a common fraction in lowest terms.

SAVING MONEY BY CONSERVING HOME FUEL COSTS

Reducing energy costs and conserving energy are important ways of saving money. There are many ways to save energy in your home. In the winter, you can save energy by lowering the thermostat at night or when no one is at home. You can also close off unused rooms, close the fireplace damper, or lower the temperature setting on the hot water heater.

In the summer you can reduce cooling costs by keeping the shades drawn, raising the thermostat on the air conditioner, and opening the windows and doors when there is a breeze at night. You can also serve more cold meals to avoid using the oven, use a fan instead of the air conditioner when possible, dry clothes on a line instead of in a dryer, and use cold water to wash and rinse clothes.

Measuring the Value of Energy-Saving Appliances

You can save energy by buying and installing energy-saving products. Before you buy these kinds of products, however, you should estimate how much money they can save you. That way you can see if what you want to buy will save you enough energy to make it worth buying.

Finding Net Savings

One way to measure the value of energy-saving products is to figure the net savings over a period of time. You find the net savings by subtracting the cost of the product from the gross amount saved for the period of time.

Gross Savings − Cost of Product = Net Savings

For example, suppose the cost of buying and installing storm windows in your home is $750. If the windows are guaranteed to last 20 years and save you $125 a year in heating and cooling costs. Your net savings would be:

20 × $125 = $2,500 Total Savings

$2,500 − $750 = $1,750 Net Savings

Finding the Payback Period

Another way to measure the value of energy-saving products is to find the payback period. The **payback period** is the number of years it will take for the products to save enough money to pay you back for their cost.

To find the payback period, divide the cost of the product by the amount of energy costs saved in one year:

Cost ÷ Estimated Savings Per Year = Payback Period

For example, a new heating system was installed at a cost of $1,650. It was estimated that the new system would save $264 a year in heating costs. The payback period would be:

$1,650 divided by $264 = $6\frac{1}{4}$ **years**

EXERCISE YOUR SKILLS

1. During the spring, summer, and fall, you dried your clothes outdoors instead of using the dryer. This reduced your electric power usage by 2,700 KWH. If you pay $0.129 per KWH, how much did you save?
2. By lowering the temperature setting on your hot water heater from 180 degrees to 120 degrees, you estimate that your April fuel costs will drop from $175 to $140. What percent of your April fuel costs do you estimate you will save?

You usually set your thermostat at 72 degrees. Raising it in the summer may reduce your cooling costs by as much as 3% for each degree it is raised. If you raised your thermostat to 77 degrees

3. What percent of your fuel costs could you save?
4. How much could you save if your cooling costs are usually $212 in July?
5. How many KWH of electricity could you save if it usually takes 2,100 KWH of energy to cool your house in July?

6. Alice installs a clock thermostat for $59 to adjust the heating and cooling automatically at night and during the day when no one is home. She estimates that this will save her $216 a year. What will be Alice's net savings over ten years?
7. Haru installs insulation around her hot water pipes and water heater for $18. She estimates that this will save her 7% of her heating oil costs each year. If she spends $972 on heating oil a year, find her net savings after 15 years.
8. Kip pays $320 to have a whole-house fan installed in his attic to keep his house cool. He estimates that the fan will reduce his air conditioning costs by 15% a year. If his annual air conditioning costs are $1,215, what will be his net savings over 10 years?
9. Vicente installs insulation in his attic at a cost of $162. He estimates that this will reduce his heating costs by 8%. If his heating costs average $1,090 a year, what are his estimated net savings after 15 years?
10. John Posek has a solar hot water heating system installed for a total cost of $3,100. He estimates that this will save him $375 each year in fuel costs. What is the payback period for the system, to the nearest whole year?
11. Rosa Velez has storm doors installed at a cost of $820. She estimates that the storm doors will save her 3% of her winter fuel bill. If her winter fuel bill averages $1,030, what is the payback period for the doors to the nearest whole year?

Ruth Lucas weather-strips her windows and doors for $41.25. She estimates that this will reduce her annual heating costs by 2.5%. Her annual heating costs are $1,162.

12. How much will she save each year in heating costs?

13. What is the payback period for the weather-stripping, to the nearest tenth of a year?

THE COST OF USING ELECTRIC APPLIANCES

Many electric appliances are marked to show the number of watt-hours of electricity they use in one hour. An electric light bulb, for example, may be marked "100 WATT." This is the *rating* of the bulb. It means that the bulb uses 100 watts of electricity in one hour.

If a 75-watt bulb burns for 10 hours, it will use 10 × 75 or 750 watt-hours of electricity. This is equal to 0.75 KWH (750 ÷ 1,000 = 0.75). If the rate is 11 cents per KWH, the cost of the electricity used is 0.75 × $0.11, or $0.0825, or $8\frac{1}{4}$ cents.

Appliances using electric motors are rated in terms of horsepower (HP). One horsepower is equal to 746 watts. So a 1 HP motor uses 746 watt-hours of electricity in one hour.

Some heating and cooling units are marked with energy guide labels, such as the one shown in Illustration 6-7.1. These labels show the average annual operating cost. The labels also show the unit's EER, or **Energy Efficiency Ratio.** The higher the EER, the more efficient the unit, and the lower the energy cost.

ENERGYGUIDE

ESTIMATES ON THE SCALE ARE BASED ON A NATIONAL AVERAGE NATURAL GAS RATE OF 3.627 PER THERM

ONLY MODELS WITH FIRST HOUR RATING OF 65 TO 74 GALLONS ARE USED IN THE SCALE

Model with lowest energy cost $185 #

$222 # THIS MODEL

Model with highest energy cost $273 #

Estimated yearly energy cost

Your cost will vary depending on your local energy rate and how you use the product. The energy cost is based on U.S. Government standard tests.

How much will this model cost you to run yearly?

		Yearly Cost
COST PER THERM	10	$035
	20	$071
	30	$106
	40	$142
	50	$177
	60	$212

Ask your salesperson or local utility for the energy rate in your area.

IMPORTANT Removal of this label

Illustration 6-7.1.

EXERCISE YOUR SKILLS

Jane uses ten 100-watt bulbs in her house. She burns these bulbs an average of 5 hours each day.

14. How many KWH do these bulbs use each day?

15. At 11¢ per KWH, what do these bulbs cost to use each day?

16. If she replaces the 100-watt bulbs with 60-watt bulbs, how many KWH would she save in one day?

17. If she replaces the 100-watt bulbs with 60-watt bulbs, how much money would this save each day? each year?

At 10.5 cents per KWH, find the hourly cost of running these appliances to the nearest cent:

18. Toaster: 1,100 watts

19. Clothes dryer: 5,000 watts

20. Microwave: 1,450 watts

21. Hair dryer: 1,000 watts

22. ♦ Make a survey of your home. How could you conserve energy? Make a tentative list and calculate how much money could be saved using your energy conservation measures.

MIXED REVIEW

1. Write $8\frac{5}{8}$ as a decimal.

2. Write 0.3% as a fraction.

3. 45 @ $0.10

4. $3\frac{1}{2} \times 6\frac{2}{3}$

5. $2.28 is $\frac{3}{4}$ of ?

6. $382.50 is 85% of ?

An electric range uses 12,200 watts per hour and is run an average of 60 hours a year.

7. On average, how many KWH per year does the range use?

8. At 11 cents per KWH, what is the cost of running the range for one year?

9. Maura is paid $7.50 an hour and time and a half for hours worked beyond 8 in any day. Last week she worked: Monday, 8; Tuesday, $7\frac{1}{2}$; Wednesday, $9\frac{1}{4}$; Thursday, 10; Friday, $8\frac{3}{4}$. What was Maura's gross pay for the week?

10. Michi is paid $437.50 a week. How much is that a month? a year?

11. A real estate agent sold a 10-room house for Donna for $135,600. The listing price of the house was $150,000. The agent deducted a 6% commission. What net proceeds did Donna receive?

12. Betty had an automatic thermostat installed for $110. On it, she lowered the temperature at night and when the home was empty. She estimates that it saved 18% of her heating costs last year. If her yearly heating costs before the thermostat was installed were $800, what were Betty's net savings for the year?

6-8 CHECKING AND REDUCING WATER COSTS

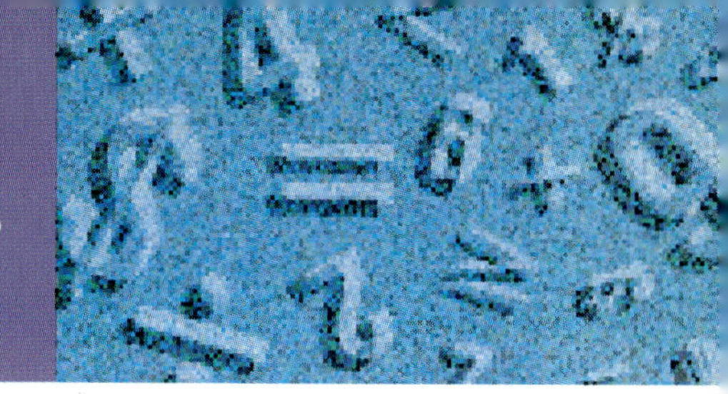

Water is another expense of running a home. It takes about 20 gallons of water to fill a bath; 12 gallons of water to do laundry; 4 gallons of water to wash your hair, and 2 gallons of water to wash your hands.

◆ Make a list of how water is used in your household. Categorize uses in some way—such as by type of use (cooking/drinking; household cleaning) or room. Which category uses the most water? the least? similar amounts? Make a bar chart that compares water usage in each category.

OBJECTIVES

In this lesson, you will learn to

- *read a water meter;*
- *check a water bill; and*
- *identify ways to conserve water.*

WARM UP

a. Write $4\frac{7}{8}$ as a decimal.

b. \$640 is what part greater than \$400?

c. $\$3,500 \times 0.0312$

d. \$105 is what percent less than \$120?

e. 30 m^2 is what percent greater than 20 m^2?

WATER COSTS

Water companies measure the water a household uses by the gallon or cubic foot. The amount used is recorded by a water meter. Water companies usually read meters and bill households every 3 or 4 months.

Reading the Water Meter Like electric and gas meters, water meters come in digital or dial models. An example is shown in Illustration 6-8.1. If you have a dial model, start reading the meter at the 100,000 dial. The last digit read is the last one the pointer has passed. The meter shows a reading of 17,780 cubic feet. The small dial measures parts of a cubic foot and is used only for testing.

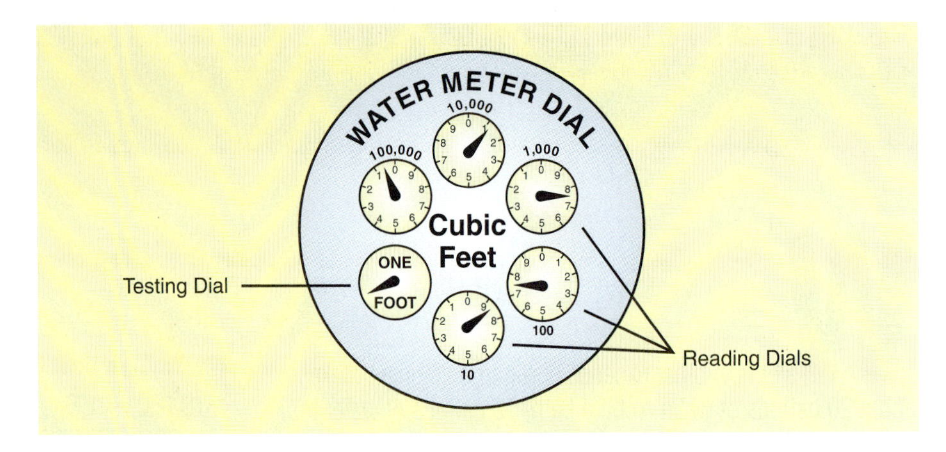

Illustration 6-8.1. Water Meter

Reading the Water Bill The bill in Illustration 6-8.2 is based on the meter in Illustration 6-8.1. The bill shows present reading: 17,780 cubic feet; last reading: 9,280; and water used: 8,500 cubic feet. The amount of water used during the period was calculated by subtracting the current reading from the last reading.

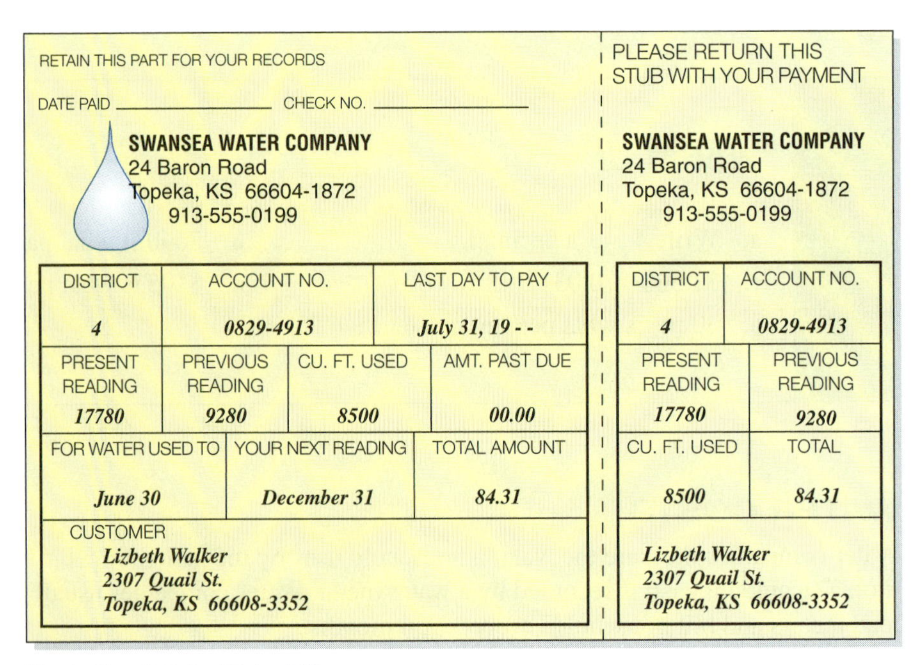

RETAIN THIS PART FOR YOUR RECORDS

DATE PAID ______________ CHECK NO. ______________

SWANSEA WATER COMPANY
24 Baron Road
Topeka, KS 66604-1872
913-555-0199

DISTRICT	ACCOUNT NO.	LAST DAY TO PAY	
4	*0829-4913*	*July 31, 19 - -*	
PRESENT READING	PREVIOUS READING	CU. FT. USED	AMT. PAST DUE
17780	*9280*	*8500*	*00.00*
FOR WATER USED TO	YOUR NEXT READING	TOTAL AMOUNT	
June 30	*December 31*	*84.31*	

CUSTOMER
Lizbeth Walker
2307 Quail St.
Topeka, KS 66608-3352

PLEASE RETURN THIS STUB WITH YOUR PAYMENT

SWANSEA WATER COMPANY
24 Baron Road
Topeka, KS 66604-1872
913-555-0199

DISTRICT	ACCOUNT NO.
4	*0829-4913*
PRESENT READING	PREVIOUS READING
17780	*9280*
CU. FT. USED	TOTAL
8500	*84.31*

Lizbeth Walker
2307 Quail St.
Topeka, KS 66608-3352

Illustration 6-8.2. Water Bill

The charges on the bill are based on the schedule of rates for the Swansea Water Company shown in Illustration 6-8.3. Note that the rates for the Swansea Water Company are for three months, or one quarter of a year.

Checking the Water Bill The water bill lists the basic quarterly charge, which is the same regardless of how much or how little water is used, and the amount and cost for water used in each rate category.

Basic quarterly charge	$ 4.75
Charge for first 5,000 cubic feet	
5,000 ÷ 100 × $0.987	49.35
Charge for next 3,500 cubic feet	
3,500 ÷ 100 × $0.863	30.21
Total water bill	**$84.31**

Swansea Water Company Quarterly Meter Rates for Water	
Type of Charge	**Rate**
Basic quarterly charge	$4.75
First 5,000 cubic feet	0.987 per 100 cubic feet
Next 10,000 cubic feet	0.863 per 100 cubic feet
Next 15,000 cubic feet	0.724 per 100 cubic feet
All over 30,000 cubic feet	0.563 per 100 cubic feet

Illustration 6-8.3.

Conserving Water and Reducing Water Costs Water is a finite resource. In some parts of the country, water is scarce and very expensive. Wherever you live, there are many ways you can waste or conserve water.

For example, letting the faucet run when you brush your teeth may use 10 gallons. Running the water only to wet the brush and rinse your mouth may save $9\frac{1}{2}$ gallons. A shower may use 15 gallons. Using the shower to wet down and rinse off only may cut that use to 4 gallons. Fixing a dripping faucet may save as much as 20 gallons per day.

EXERCISE YOUR SKILLS

Use the Swansea Water Company's schedule of rates shown in Illustration 6-8.3 for Exercises 1–9.

Find the total water bill for each amount of water used.

1. 2,170 cubic feet
2. 5,930 cubic feet
3. 17,640 cubic feet
4. 32,590 cubic feet

Find the total water bill for each problem.

5. Last reading, 8,920; present reading, 12,220.
6. Last reading, 16,000; present reading, 27,740.
7. Last reading, 81,090; present reading, 98,370.

8. Scott's water meter reads 48,170 cubic feet now. Three months ago it read 35,180 cubic feet. Find Scott's total water bill for the quarter.
9. Ria's water meter read 72,110 cubic feet on April 1 and 96,700 cubic feet on July 1. Find her water bill for the quarter.

Frank uses the full cycle of his washing machine to clean two loads of laundry each week. Each load uses 60 gallons of water @ \$0.87 per thousand gallons.

10. How many gallons of water does Frank use a year washing clothes?
11. How much does the water used each year cost him?
12. If Frank uses the short cycle of his washing machine, he uses 27 gallons of water per load. How many gallons per year would Frank save using the short cycle? How much money per year would Frank save this way?

Sheila takes a 10-minute shower each day. Her shower uses 4 gallons of 140° water per minute. Water costs \$1.182 per thousand gallons.

13. How many gallons per day does Sheila use for her shower?
14. What is the yearly cost of the shower water used, to the nearest cent?
15. If Sheila cuts her shower to 5 minutes per day, how many gallons would she save per year?
16. How much money would she save in water costs per year?

Vince had a leaking faucet that wasted 18 gallons of water a day in August.

17. Estimate the number of gallons of water Vince wasted in August.
18. What is the exact number of gallons Vince wasted in August?
19. At $0.98 per thousand gallons, what was the cost of the wasted water?
20. ♦ Make a list of the ways water is used in your household. Identify ways you can conserve water at home. Find your local water rates, and calculate the amount of water and money that could be saved in one year by following the practices you identifed.

MIXED REVIEW

1. $\$1{,}200 \times 0.0415$
2. $\$580 \div 0.1$
3. Write $7\frac{1}{8}$ as a decimal.
4. 248 @ $\$0.37\frac{1}{2}$
5. 23,450 lb @ $16 per ton
6. $48 is $\frac{1}{6}$ of ?
7. Alberto runs his automatic dishwasher each day using the full washing cycle. The full cycle takes 16 gallons. If he used the short cycle, it would only take 7 gallons. At $0.835 per hundred gallons, how much could Alberto save in June by using the short cycle?
8. Velma used 231 therms of gas last month. Her gas company charged her $0.86 per therm. The company also charged a fuel adjustment rate of $0.0041 per therm and a state sales tax of 5%. What was Velma's gas bill last month?
9. A salesperson was paid $4,432 commission on sales of $55,400. What was the salesperson's rate of commission?
10. Edgewood Industries' gross sales last year were $514,678,120. They estimate that gross sales this year will be 115% of last year's sales. What are the estimated gross sales for this year?
11. Tim worked $36\frac{1}{4}$ hours @ $8.50 an hour. His deductions were federal withholding taxes, $37; state withholding taxes, $6.16; FICA, 7.65%; union dues, $12.50. What was Tim's net pay?
12. A holiday light string has only blue and white lights. The lights are spaced 1 foot apart. For each white light on the string there are 2 blue lights. How many white and blue lights should there be on a 5-yard light string?

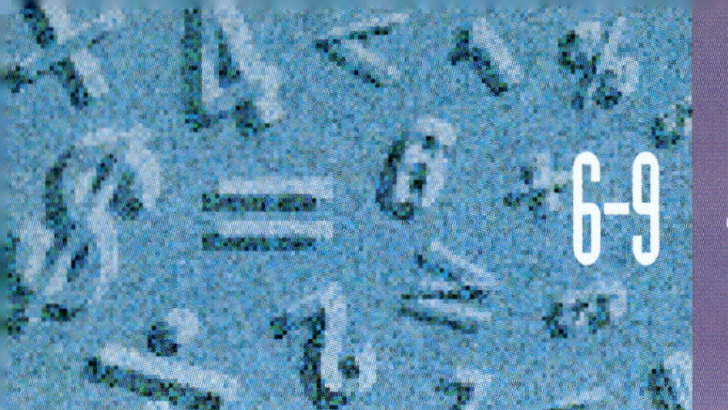

6-9 CHECKING AND REDUCING TELEPHONE COSTS

OBJECTIVES

In this lesson, you will learn to

- *read telephone bills;*
- *check the accuracy of telephone bills; and*
- *identify ways to reduce telephone expenses.*

Technology has made it easy and convenient to pick up the phone and call someone across town, across the country, or across an ocean. Many companies offer services that make it even easier to make and receive calls. Do you know what costs are incurred for different types of calls and services?

◆ Make a table with three columns. In the first column, record factors you are aware of that may affect the cost of a phone call. In the second column use a code to distinguish controllable costs from required charges. In the third column, identify ways you can control costs.

WARM UP

a. 24,560 ÷ 250,000

b. Write $7\frac{3}{8}$ as a decimal.

c. \$11.52 less $37\frac{1}{2}\%$ of itself

d. 2000 × \$0.1387

e. 170 ÷ \$0.0085

f. \$270 is what part smaller than \$315?

TELEPHONE SERVICE

Local and long-distance telephone services are available to most consumers. Local phone service is usually purchased from the local phone company. Consumers can buy long-distance service from their local company or choose a plan offered by a long-distance service company. For example, Angela Wilson uses the Comtel Telephone Company for local phone service and SuperNet Communications Company for long-distance service. The bill she received from the local phone company is shown in Illustration 6-9.1.

Typical items on a telephone bill include:

Local Telephone Service Charges for local phone service depend on the services you rent from your phone company. *Local service charges* may include fees for phone lines, phone equipment, and special phone services. Angela's local service charges are for the one phone line she rents. The total, \$3.50, is a flat fee charged to all phone users for one line.

COMTEL TELEPHONE CO.
Alton, Illinois 62002

ANGELA WILSON
1250 RODIL STREET
ALTON, IL 62002

DATE OF BILL	PAGE	BILLING NUMBER
5/1/--	1	618-555-0102
LAST AMOUNT DUE	PAYMENTS	BALANCE
$21.57	$21.57	0.00

LOCAL SERVICE	3.50
ACCESS CHARGE	9.15
LOCAL CALLS	1.80
U.S. TAX (3%)	.43
STATE TAX (5%)	.73
	AMOUNT DUE
DUE ON OR BEFORE 6/1/- -	**15.61**

COMTEL TELEPHONE CO. ITEMIZED LOCAL CALLS **PAGE 2**

DATE	TIME	CALLED PLACE	AREA–NUMBER	RATE	MINS	
4-3	915A	EDWARDSVILLE	618-555-0178	AD	3	.17
4-5	217P	BELLEVILLE, IL	618-555-0165	AD	6	.27
4-19	857P	HIGHLAND, IL	618-555-0144	AE	38	1.36
				TOTAL ITEMIZED CALLS		1.80

RATE: A – DIAL, O – OPERATOR ASSISTED, P – PERSON TO PERSON, D – DAY, E – EVENING, N – NIGHT

Illustration 6-9.1. A Telephone Bill for Local Service

Telephone Access Charges These are charges to access, or use, your company's phone switches. Comtel customers pay a $9.15 *access charge* to use Comtel's telephone switches.

Local Telephone Calls Some companies have a flat fee for all local service. Others charge for calls made locally. The costs of these calls depend on where in the calling region a call is received, the length of a call, and when (day and time) a call is made. Angela was charged $1.80 for local calls.

Telephone Taxes Federal, state, and local taxes are added to many phone bills. Each tax is figured on each part of the bill separately. For example, Angela's taxes were computed as:

	Charges	U.S. Tax at 3%	State Tax at 5%
Local service	$ 3.50	$0.11	$0.18
Access charge	9.15	0.27	0.46
Local calls	1.80	0.05	0.09
Totals	$14.45	$0.43	$0.73

Special Telephone Services

In addition to basic services, most companies offer a variety of special services such as:

Automatic callback: Lets you return the last call received —whether you answered it or not.

Call waiting: Lets you put the caller you are talking to on hold when you receive another call.

Call forwarding: Automatically transfers calls made to your phone number to another phone number.

Call screening: Lets you block calls. Screened callers hear a tone and a message that you are not taking calls.

Three-way calling: Lets you talk to two or three callers at the same time.

Speed dialing: Lets you code frequently used numbers so they can be dialed using only one or two keys.

Caller ID: Displays the number of the person who is calling.

Repeat dialing: Redials the last number dialed.

Of course, these services involve additional fees. Angela's phone company charges the following monthly fees for:

Automatic callback	$3.50	Three-way calling	$2.50
Call waiting	$2.50	Speed dialing	$2.50
Call forwarding	$2.50	Caller ID	$6.50
Call screening	$5.50	Repeat dialing	$3.50

Charges for special services are added to the basic service charges. For example, if Angela had taken call waiting, her local service charges would have been $6 ($3.50 + $2.50). If Angela ordered all the additional services, her local monthly service charges would total $32.50 ($3.50 local charge plus $29 for all the additional services).

EXERCISE YOUR SKILLS

Use the service and tax rates for the Comtel Telephone Company for Exercises 1–3.

1. Jim paid $3.50 for local phone service, $3.50 for automatic callback, and $2.50 for call forwarding in May. He also paid $9.15 for access charges and $22.45 for local calls that month. What was his total phone bill from Comtel Telephone Company for May including taxes?
2. Mario pays for one telephone line, access charges, call forwarding, call waiting, and call screening. He makes $17.49 in local calls in March. What is his March phone bill from Comtel including taxes?
3. Darlene pays for one line, access charges, call waiting, call forwarding, automatic callback, and caller ID. She makes local calls costing $37.55 in October. What is her October phone bill from Comtel including taxes?

4. Lavonda uses Via-Phone, Inc., as her phone company. She pays $3.90 for one line, $8.75 for access charges, $2.25 for call forwarding, $2.25 for call waiting, $3.75 for speed dialing, $3.75 for automatic callback, and $4.35 for three-way calling. If she makes $32.44 in local calls in June, what is her phone bill for the month including 3% federal and 8% state taxes?

LONG-DISTANCE PHONE BILLS

Long-distance calls are calls made outside a local service area. The cost of these calls depends on the rates of your long-distance service provider, the location of the receiving phone call, the length of the call, when the call is made, and if an operator assisted the caller.

It pays to compare the rates and services charged by all the long-distance companies that serve your area. Usually you can get the best price if you select the fewest extra charges and purchase a package that meets your phone pattern needs economically. For example, some companies offer discount rates to customers whose long-distance calls are to one or a few locations.

The long-distance rates charged by SuperNet Communications Co. are shown in Illustration 6-9.2. The charges shown are for direct-dial service; that is, when you dial the number yourself instead of asking for operator assistance.

	DAY RATES (D)		EVENING RATES (E)		NIGHT RATES (N)	
MILEAGE BAND	FIRST MIN.	ADDED MIN.	FIRST MIN.	ADDED MIN.	FIRST MIN.	ADDED MIN.
1–124	0.2760	0.2430	0.1560	0.1471	0.1199	0.1099
125–430	0.2810	0.2540	0.1670	0.1531	0.1380	0.1299
431–3000	0.2950	0.2658	0.1780	0.1551	0.1480	0.1389
3001 & over	0.3160	0.2680	0.1830	0.1701	0.1630	0.1420

Illustration 6-9.2. Long-Distance Rates Charged by SuperNet Communications Co.

Day rates are charged for calls made between 8 A.M. and 5 P.M. weekdays. Evening rates apply from 5 P.M. to 11 P.M. weekdays and Sundays. Night rates apply from 11 P.M. to 8 A.M. weekdays, all day Saturdays and holidays, and up to 5 P.M. on Sundays.

Angela's long-distance charges from SuperNet are shown in her bill in Illustration 6-9.3, shown on the next page. Her charges for the Peoria call were found by looking at the evening (E) rates for a call to a person between 125–430 miles away. She was charged $0.1670 for the first minute and $0.1531 for each of the other five minutes, or $\$0.1670 + (5 \times \$0.1531) = \$0.9325$ or $0.93. When calculating long-distance rates, round the final answer. The other long-distance calls were computed in a similar manner.

Reducing Long-Distance Telephone Costs

Long-distance costs may be reduced if you call in the evening or on weekends, keep your calls as short as possible, and dial all numbers direct. Operator-assisted calls are usually billed at a higher rate.

SUPERNET COMMUNICATIONS CO., INC.

ACCOUNT NO. 22-0000397587

APRIL 30, 19 --

PAGE 1

PREVIOUS BALANCE	.00
04/30/- - TOTAL LONG-DISTANCE SERVICE	10.65
04/30/- - FEDERAL TAXES (3%)	.32
04/30/- - STATE TAXES (5%)	.53
BALANCE DUE UPON RECEIPT	**11.50**

PLEASE RETURN THIS PORTION

SUPERNET COMMUNICATIONS CO., INC.

PAGE 2

DATE	TIME	# CALLED	DESTINATION		MIN	AMOUNT
03/03	0704P	218-555-0120	PEORIA	IL	6E	.93
03/12	0148P	309-555-0180	SPRINGFIELD	IL	30N	3.31
03/28	0413P	203-555-0175	HARTFORD	CT	24D	6.41
			TOTAL LONG DISTANCE SERVICE			10.65

D – DAY, E – EVENING, N – NIGHT

KEEP THIS PORTION FOR YOUR RECORDS

Illustration 6-9.3. A Long-Distance Telephone Bill

You may need operator assistance for certain kinds of calls, such as long-distance calls from a pay phone, collect calls, calls that are billed to another number, and time and charge calls (calls in which the operator tells you how long you have been calling and how much it has cost). Since operator-assisted calls are more expensive, don't assume you have to use an operator. Many kinds of calls that once required operator assistance no longer do.

EXERCISE YOUR SKILLS

Use the rates in Illustration 6-9.2 to solve the following problems. Round answers to the nearest cent.

Edna makes a 6-minute, direct-dial call to a friend 100 miles away at 2 P.M. on Monday.

5. What is the cost of the call?

6. How much could Edna have saved if she had called at 5:30 P.M.?

7. How much could she have saved by calling on Saturday?

Hector made a 27-minute call to his mother, who lives 300 miles away, at 3 P.M. on Friday. He dialed the number himself.

8. What did Hector's call cost?

9. What would have been the cost if Hector had called at 11:01 P.M.?

10. What would have been the cost if he had called at 4 P.M. on Sunday?

Find the cost of a 15-minute, direct-dial call to a person 50 miles away

11. On Wednesday at 9 A.M.
12. On Wednesday at 6 P.M.
13. On Sunday at 7 P.M.
14. On Saturday at noon.

Find the percent saved, to the nearest whole percent, if you make a 5-minute, direct-dial call to a person 400 miles away at the

15. Evening rate rather than the day rate.
16. Night or weekend rate rather than the day rate.
17. ♦ Assume you want to find the best rates for your long-distance phone service. Identify your long-distance phone service requirements. Then make a checklist of questions you would ask service providers to be able to compare prices and find the lowest cost service available for your needs.

PUT IT ALL TOGETHER

18. Based on what you have learned about reading and checking utility and phone bills, write procedures to follow to ensure the accuracy of these bills.

MIXED REVIEW

1. 13,240 ÷ 150,000 to the nearest ten thousandth
2. \$540 is what part smaller than \$630?
3. Write $9\frac{2}{5}$ as a decimal.
4. $37\frac{1}{2}\%$ less than \$3.60 is what amount?
5. In June, Hayato's local service charges were \$12.50, his access charges were \$2, he made 36 local calls totaling \$44.23, and he paid 3% federal and 4% state taxes. What was Hayato's total June bill?
6. Tom's water meter read 12,880 cu. ft. on April 1 and 18,450 cu. ft. on July 1. He pays \$6.34 for the first 500 cu. ft. and \$0.79 for each 100 cu. ft. over 500 cu. ft. What is Tom's current quarterly water bill?

Val worked eight hours daily, Monday through Sunday, producing 698 valves that passed inspection. Her employer paid her a regular-time rate on weekdays, a time-and-a-half rate on Saturday, and a double-time rate on Sunday. Val's regular pay rate was \$9 an hour.

7. What were Val's gross earnings?
8. At 7.65%, find the estimated and exact deduction for FICA taxes.

EXPLORE TECHNOLOGY

TEMPLATE: Spreadsheet to Compare Renting to Buying

Evelyn Franke is considering renting a snow blower for $70 a week, $25 a day, or $10 an hour. If she purchased the same snow blower it would cost her $350. Use the spreadsheet labeled ET6 on the applications diskette to help Evelyn decide how to get the most for her money.

After you load the spreadsheet, your computer screen should look like the one shown below.

	A	B	C
1	**Renting vs. Buying Worksheet**		
2			
3	Description	Amount	
4			
5	Rent per week		
6	Rent per day		
7	Rent per hour		
8	Cost of item		
9			
10	Rent per week equals or exceeds cost in:		weeks
11	Rent per day equals or exceeds cost in:		days
12	Rent per hour equals or exceeds cost in:		hours

Formulas within the spreadsheet will calculate the number of weeks, days, and hours it will take for the rent to equal the original cost of the snow blower.

Enter the bold-faced data into the specified cell of your template.

B5 **70**　　B6 **25**　　B7 **10**　　B8 **350**

Now answer these questions about the completed spreadsheet.

1. How many weeks will it take the cost of the weekly rental fees to equal or exceed the cost of the snow blower? how many days? how many hours?
2. If Evelyn needs to rent the snow blower for 3 consecutive days, what rental plan do you recommend that she choose, and why?
3. If she expects to rent the snow blower for 4 non-consecutive days, (spread out during the winter season), what rental plan would you recommend, and why?
4. Suppose the snow blower goes on sale at the end of the winter season for $295 and the rental fees will be raised to $73.50 weekly, $26.25 daily, and $10.50 hourly next year. Given this new pricing, answer the three questions posed in Question 1 above.
5. What is the spreadsheet doing arithmetically to get the value in cell B10?

Design Your Own

There are many useful software solutions you can design to help you manage your spending and make good financial choices. For example, you could set up and monitor a budget on a spreadsheet or in a database. You could also set up spreadsheet templates to check the accuracy of your utility bills. You may wish to explore some of these ideas on your own. For this activity, you will design a spreadsheet to help you calculate and interpret payback periods on energy saving appliances.

Anne Martin is considering replacing her old central air conditioning unit for a new energy efficient one. She estimates that the old unit costs her $648 to run in the summer. Before making a decision, she wants to compare payback periods for the following three units:

	Landair	**Belair**	**Sunray**
Price	$899	$1,189	$1,059
Estimated % Saved	15%	22%	19%

Based on what you learned in Lesson 6-7, how would you set up a spreadsheet to compare the cost, payback periods, and savings for each unit?

6. What information do you need to know to calculate payback period?
7. What columns and rows should your spreadsheet contain?

Create your spreadsheet and answer the following questions:

8. What is the nearest payback period, to the nearest hundredth, for the Landair unit? the Belair unit? the Sunray unit?
9. How much would Anne save in one summer by buying the Landiar unit? the Belair unit? the Sunray unit?
10. If the cost of energy rose and the unit Anne now has cost $712 to run in the summer, how much would Anne save by buying the unit that saves the most?
11. What mathematical process do you need to do to find the payback period for each unit? How did you write the formula for this process for the Belair unit?

CHAPTER 6 SUMMARY AND REVIEW

VOCABULARY REVIEW

budget *6-1*
extension *6-2*
fixed expense *6-1*
fuel adjustment rate *6-6*
hundredweight *6-3*
income *6-1*
kilowatt-hour *6-6*
payback period *6-7*
sales slip *6-2*
subtotal *6-2*
therms *6-6*
unit price *6-2*
variable expense *6-1*

MATH SKILLS

6-1 ♦ Find the part of a number by multiplying the percentage by the whole number.

1. Find 18% of $43,000
2. Find 45% of $50,000

6-2 ♦ Multiply quantity by unit price to find the extension.

Find each extension.

3. 6 cans @ $0.89
4. 3 kg @ $5 per kg

6-3 ♦ Find the cost of a fraction of a unit by multiplying the unit price by the fraction.
♦ Find the cost for large quantities by dividing the quantities by 100, 1,000, or 2,000 as appropriate and then multiplying by the unit price.
♦ Use fractional parts of $1 to find extensions quickly.

5. Find the cost of $\frac{1}{2}$ dozen pens @ $1.88 per dozen.
6. Find the cost of 11,500 lb of gravel @ $32.50 per ton.
7. Find the cost of 320 lb @ 0.37\frac{1}{2}$.

6-4 ♦ Find unit prices from group prices by dividing the group prices by the number in the group and rounding up.
♦ Find the unit price when the item price is a fraction of a unit by dividing the item price by the quantity.

Find the cost of one unit for each item:

8. A 400-cm length of tubing costs $0.36. What is the cost of the tubing per meter?
9. A 2$\frac{3}{4}$ ton load of land fill costs $28. What is the cost per ton?

APPLICATIONS

6-1 ♦ Find percent spent on a budget category by divding the amount spent by the total income.
♦ Find total amount to be be budgeted for a category by multiplying the percent allotted by the total income.

Kay's total income last year was $22,000. She expects to earn 10% more this year and wants to budget that income using the percentages below. What yearly amount should Kay budget for each type of payment?

10. Food, 24%

11. Transportation, 17%

12. Personal, 13%

6-2 ♦ Add extensions to find the subtotal.

♦ Multiply the subtotal by the tax rate to find sales tax.

♦ Add subtotal and sales tax to find total of sales slip.

13. Tara bought 6 cans of soup @ $0.94, 10 bulbs @ $0.34, and 2 kg of meat @ $3.20. Find each extension and the subtotal.

6-5 ♦ Compare prices of similar items sold in different quantities and prices by finding unit prices of each item.

♦ Compare the cost of renting or buying by finding the cost of renting and the cost of buying for the same periods of time.

14. A set of four tires costs $326.80. One tire costs $99.95. How much is saved by buying the set of tires?

6-6 ♦ Check electric and gas bills by multiplying the amounts of fuel used by the rates, computing all service charges, finding a subtotal, computing taxes and other charges, and adding the results.

15. Hilda Navarro's electric bill for May 1 shows a reading of 4,560 for April 30 and 3,140 for March 31. She pays 12.5¢ per KWH for the first 500 KWH and 14.2¢ per KWH over 500. She pays a fuel cost adjustment of $0.003 on the total KWH used and pays a 5% regulatory tax on the total energy charges. What is the total of her bill?

16. On Reggie's July gas bill, the current reading is 945 and the last is 822. The therm factor is 1.022. The per therm rates are first 50 therms, $0.312; next 50 therms, $0.267; over 100 therms, $0.223. Find the total bill, including a sales tax of 4%.

6-7 ♦ Find savings from energy-savings steps by calculating the percent saved, the dollars saved, or the fuel saved.

♦ Find net savings from energy-saving products by subtracting the cost of the products from the gross amount saved in dollars.

♦ Find the payback period for energy-saving products by dividing cost of product by estimated savings per year.

♦ Find the cost of electric appliances by calculating the amount of the energy used and multiplying the amount by the rate of cost.

17. Raising your thermostat in the summer can save an estimated 3% for each degree over 72°. How much is saved by raising your thermostat to 78° if your usual August cooling costs are about $180 when the thermostat is set at 72°?

18. An energy-saving thermostat costing $85 is estimated to save a homeowner $174 a year. What are the estimated net savings over ten years?

6-8 ◆ Check water bills by finding the basic charges, multiplying amounts of water used by water rates, and adding results.

◆ Find savings from water-saving steps by calculating the amounts of water saved and multiplying by the water rates.

19. A water meter read 14,280 cu. ft. on July 1 and 18,230 cu. ft. on October 1. The water company charges $7.35 for the first 500 cu. ft. and $0.89 for each 100 cu. ft. over 500 cu. ft. What is the water bill for the quarter?

6-9 ◆ Find the cost of local telephone service by finding the access charges, local service charges, and charges for local calls; by calculating taxes; and by adding the results.

◆ Find the cost of long-distance service by finding the cost of each call using a rate chart, adding the cost of the calls, multiplying the sum by the tax rates, and adding all the costs and taxes.

◆ Find the amount saved from steps taken to reduce long-distance costs by comparing the cost of a call made at different times of the day.

20. James pays $12.55 for access charges, $4.50 for a line, and $3.50 for call waiting. He also pays $23.16 for local calls this month. What is his telephone bill, including 3% federal and 5% state taxes?

21. Victoria called a friend during the day. The call cost $4.55. If she had waited until after 5 P.M., she could have saved 35%. What would the call have cost if she had waited?

WORKPLACE KNOWHOW: **PROBLEM SOLVING**

Problem-solving abilities include knowing when a problem exists, determining the cause of a problem, coming up with solutions, and monitoring solutions to ensure the problem is solved.

Assignment This chapter involves several problem-solving activities. For example, you were asked to identify steps you could take to prevent overpaying for items or services. Look back at the problem-solving strategies that have appeared in each chapter. Did you use any of these strategies? Did you come up with your own? Have a class discussion about techniques used. Create your own class list of problem-solving strategies.

Work Skills Portfolio The ability to look for and correct errors in financial and other kinds of documents is a critical one. Include an assessment of your ability in this area in your portfolio. You may want to include your guidelines for insuring the accuracy of bills as an example of your writing and problem-solving skills. If you completed the integrated project from the workbook, you may also wish to add the corrected sales slip from Vicker's Department store.

CHAPTER 6 TEST

Math Skills

Find each cost. Use mental math when you can.

1. 6 cans of soup @ 2 for $0.72
2. 24 yd of cloth @ $0.375 per yd
3. 34,500 lb of cement at $27.98 per ton
4. 15 ft of rope @ $1.80 per yard
5. 3,600 lb of potatoes at $32 per hundred pounds

Applications

6. Last summer Arnie spent 40% of his income on food and entertainment and 20% on clothes; he also saved 40%. Arnie expects to earn $250 per week for 10 weeks over the summer. Make a budget for Arnie based on last year's information.
7. Your budget says that savings is 12% of income. If you earn $1,400 in a month, how much should you save?
8. High-density computer diskettes, regularly priced at $15 a box, are on sale for 3 boxes for $39.99. How much would you save by buying 12 boxes at the sale price?
9. Find the total cost of a $289.95 VCR plus $5\frac{1}{2}$% sales tax.

A camcorder can be rented for $25 a day, or $100 a week. You need it for 6 days.

10. How much cheaper is it to rent it for a week instead of by the day?
11. If you could buy the camcorder for $989, what is the payback period, to the nearest tenth of a week, for buying rather than renting it by the week?
12. Bregnev's electric meter read 42,789 on October 1 and 44,885 on January 1. Bregnev's electric company charges $0.1052 for the first 500 KWH and $0.1247 for all KWH over 500? What was Bregnev's electric bill for the quarter if the tax is 5%?
13. Jarret's call to a friend cost $12.76 on Wednesday morning. By waiting until Saturday, he could have saved 55%. What would the cost of the call have been on Saturday?
14. Tina buys a space heater for $246. She plans to lower her thermostat and save 20% of her winter heating bills. If her winter heating bills average $820, what is the payback period for the heater?
15. Rod's business used 5,600 cubic feet of gas last month. The therm factor is 1.115. The per therm rates were first 50 therms, $0.2011; over 50 therms, $0.1892. There was a facilities charge of $15 and a tax of 3%. Find the total bill.
16. Rose used 1,200 gallons of water last quarter. Find her total water bill if she pays $0.895 per 100 gallons.

CHAPTER 7

CUSTOMER

Get Your CAREER Off to a FAST START

Customer service is critical to most companies. Customer service staff provide product information, fill orders, and solve problems. A good customer service person can make the difference between a company keeping or losing a customer.

Customer service can be a satisfying career or a good entry-level position. Osami, manager of the customer service department of a magazine subscription service, says most of her staff is promoted (usually to marketing or sales) within six months to a year.

Customer service representatives (reps) earn their promotions. Each member of Osami's staff handles a minimum of 200 subscription-related phone calls and letters a day. Each rep has a computer on his or her desk. They use the computer to look up information on a subscription and to enter new orders. Reps must be at their desks most of the day. Lunches, breaks, and vacations are carefully scheduled to make sure that there are always enough people to service customers who call in. Reps must be familiar with all the magazine titles the company services. They must have good telephone skills and enjoy working with people. They must also key at least 35 words a minute and proofread accurately.

SERVICE

Reps are evaluated on the quality of the service they give customers and on their contribution to the department's goals, which include answering a minimum of 96% of all incoming calls. (A computer system tracks the number of calls received and the number of hang-ups by customers on hold.) In addition, all orders received by mail must be entered into the computer within 72 hours of receipt.

Math skills are needed to add invoices and calculate percentages for discounts and shipping charges. Problem-solving skills are also important. Osami says good reps detect patterns and track down the source of problems. For example, one rep noticed a high number of calls from customers who did not receive an issue of their magazine. He discovered that address labels were falling off the magazine because the amount of glue used on each label had been reduced. Thanks to the rep's efforts, more glue was used when the next issue was mailed. That saved the company the cost of mailing replacement issues to unhappy subscribers.

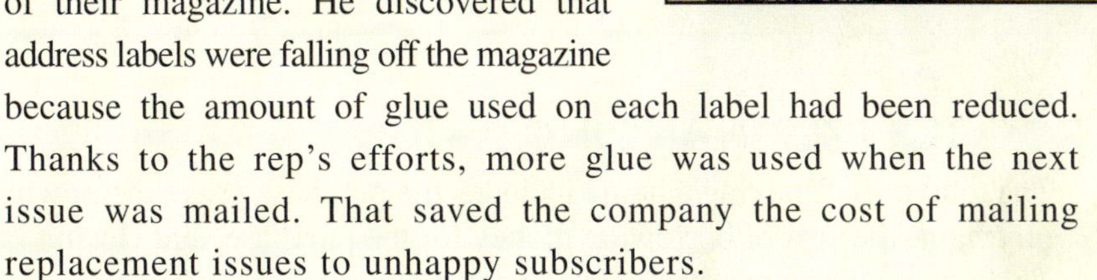

In addition to pay, the rewards to the customer service representatives are personal satisfaction and the possibility of advancement. Osami tells her staff that "a positive attitude and motivation will take you far in any position."

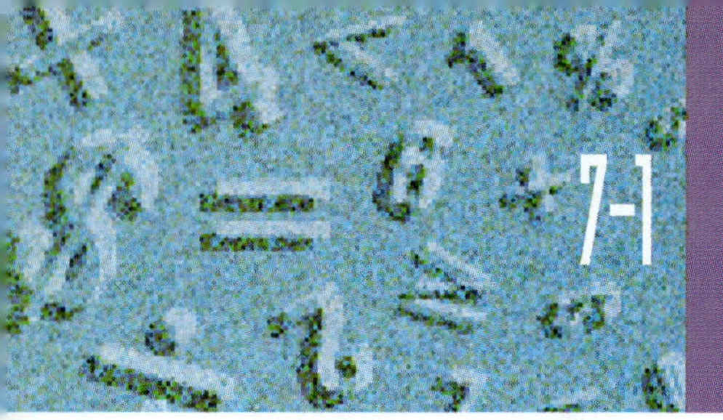

7-1 BUYING A HOME

OBJECTIVES

In this lesson, you will learn to

- *calculate down payments;*
- *calculate closing costs;*
- *use a table to find monthly payments;*
- *find total interest cost for a loan; and*
- *compare costs when refinancing a mortgage.*

The decision to buy a home is a major personal and financial decision.

◆ In a small group, discuss the personal and financial factors that you think are important in making a decision to buy a home.

WARM UP

a. 8 is what part less than 24?

b. Write $8\frac{1}{4}$ as a decimal.

c. $18 less 25% is what amount?

d. Write 150 km in meters.

e. $2,500 divided by $18,000 to the nearest hundredth

f. Find the cost of 18,000 bushels at $1.50 per C.

COSTS OF FINANCING A HOME

The total cost of buying a home includes the purchase price, the **down payment,** the cost of borrowing money for the purchase, and **closing costs.** There may also be costs associated with looking for a house, such as travel, which are not discussed in detail in this lesson.

Down Payment Most people pay a cash down payment, a percentage of the total cost of the house, to their lender. Some lenders require no down payment; others ask for as much as 30%. You should know how much you can afford as a down payment before you start to look for a home. The more money you can put as a down payment, the lower the amount of money that needs to be borrowed.

Financing The balance of the purchase price (after the down payment) is usually paid from a mortgage loan taken with a bank or other lender. The amount of money borrowed is called the **principal.** In addition to the principal, **interest** on the loan is charged. Interest is the cost of using the lender's money. A mortgage gives the lender the right to take the property if the loan is not repaid as agreed. The length and terms of mortgages varies; 15-, 20-, and 30-year mortgages are common.

Closing Costs are fees and expenses paid by the buyer to complete the transfer of ownership of a home. Closing costs may be 1% to 4% of the purchase price of the home. They are usually paid to the person or institution financing a mortgage loan. Typical closing costs include legal fees, title insurance, loan origination fees, appraisal and inspection fees, land surveys, prepaid taxes, and prepaid interest charges known as *points*. Interest rates and closing costs may vary among lenders, so it pays to compare when you are looking for a lender.

Critical Thinking What do you think are the advantages and disadvantages of low down payments and long-term loans?

There are many different types of mortgages. Two of the most common types are fixed rate mortgages and variable rate mortgages. With a **fixed rate mortgage,** the same rate of interest is paid for the life of the loan. With a **variable rate, or adjustable rate, mortgage,** the rate of interest is not guaranteed and can be increased.

Fixed Rate Mortgages

Most mortgages are repaid gradually, or *amortized*, over the life of the mortgage. The mortgage payments are usually equal monthly payments. Each payment pays off part of the principal plus the interest due each month.

At first, most of the monthly payment is to pay interest. Later, the amount that goes to repay the principal increases. Most lenders allow customers to make additional payments toward principal so the mortgage can be paid off earlier. This reduces the total interest due.

Illustration 7-1.1, at the top of the next page, shows how much of a $442.48 equal monthly payment goes for interest during the term of the mortgage. The monthly payment is figured on a $40,000 mortgage taken for 30 years at 13%.

Using the Amortization Table Illustration 7-1.2, at the bottom of the next page, shows the monthly payments needed to amortize mortgage loans over different periods of time using interest rates of 7%, 9%, and 11%.

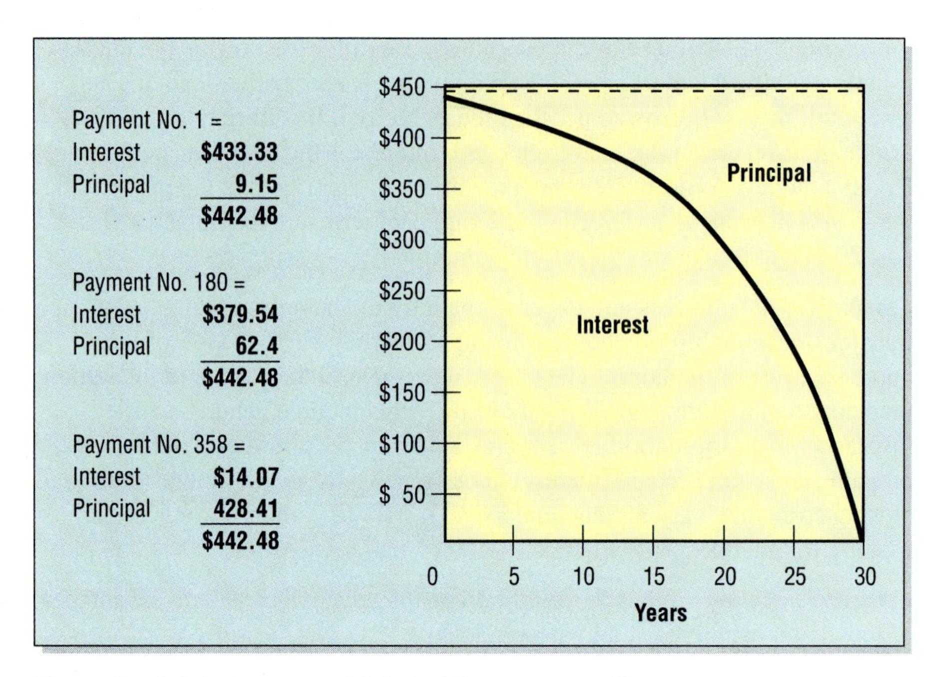

Illustration 7-1.1. Interest and Principal Payments on a Mortgage

MONTHLY PAYMENTS NEEDED TO PAY A LOAN									
Amount of Loan	Interest Rate								
	7%			9%			11%		
	Time of Loan								
	20 yrs	25 yrs	30 yrs	20 yrs	25 yrs	30 yrs	20 yrs	25 yrs	30 yrs
$30,000	$232.59	$212.03	$199.59	$269.92	$251.76	$241.39	$309.66	$294.04	$285.70
35,000	271.35	247.37	232.86	314.91	293.72	281.62	361.27	343.04	333.32
40,000	310.12	282.71	266.12	359.90	335.68	321.85	412.88	392.05	380.93
45,000	348.88	318.05	299.39	404.88	377.64	362.09	464.49	441.06	428.55
50,000	387.65	353.39	332.65	449.87	419.60	402.32	516.10	490.06	476.17
55,000	426.41	388.73	365.92	494.85	461.56	442.55	567.71	539.07	523.78
60,000	465.18	424.07	399.18	539.84	503.52	482.78	619.32	588.07	571.40
65,000	503.94	459.41	432.45	584.83	545.48	523.01	670.93	637.08	619.02
70,000	542.71	494.75	465.71	629.81	587.44	563.24	722.54	686.08	666.63
75,000	581.47	530.08	498.98	674.80	629.40	603.47	774.15	735.09	714.25

Illustration 7-1.2. Sample Amortization Table

EXAMPLE Tanisha wants to buy a home that costs $72,000. She has $17,000 for a down payment, and her bank will lend her $55,000 on a 25-year, 11% mortgage. Find Tanisha's monthly payments and the total amount of interest she would pay over the length of the mortgage.

SOLUTION On the $55,000 line of the table in the 25-year column under 11% is the amount, $539.07. This is the monthly payment Tanisha must make to pay off the loan in 25 years.

In 25 years, there are 25 × 12, or 300, months.
300 × $539.07 = $161,721 ←Total amount paid over the 25 years

$161,721	←Total amount paid
− 55,000	←Amount of mortgage
$106,721	←Interest paid over the 25-year period

EXERCISE YOUR SKILLS

The Greenwalds want to buy a condominium priced at $90,000. They will need to make a down payment of 20% and pay closing costs of 3% of the purchase price.

1. How much cash will they need for the down payment?
2. How much of the purchase price must they borrow?
3. How much cash will they need for the closing costs?

Ed and Roberta Rogers bought a house at its market value of $70,000. They made a 25% down payment and paid these closing costs: legal fees, $280; property survey, $150; title insurance, $220; inspection fees, $350; loan origination fees, $400.

4. What was the amount of the down payment?
5. What was the total of the closing costs?
6. What percent were the closing costs of the purchase price, to the nearest whole percent?

Ann Doyle will purchase a house for $53,500 on August 1. The house was appraised at $58,000. On that date, she must make a down payment of 15% of the purchase price and pay these closing costs: loan origination fees, $340; termite inspection fee, $80; legal fees, $420; land survey costs, $115; title insurance, $210; prepaid taxes, $600; points, $535.

7. What total amount of cash must Ann have on August 1 to make the purchase?
8. What percent are the closing costs of the purchase price, to the nearest tenth of a percent?

Use the table in Illustration 7-1.2 to find (a) the monthly payment needed to amortize the loan and (b) the total interest cost.

9. Marilee Lenfield borrows $60,000 from a savings and loan association on a 20-year, 9% mortgage.
10. Armondo borrows $40,000 from a credit union on a 25-year, 7% mortgage.
11. Aurelia Beane borrows $75,000 from a mortgage company on a 30-year, 11% mortgage.
12. Although he wanted a 20-year, 7% mortgage, the best Howard could negotiate was a 25-year, 9% mortgage for the $45,000 he needed to buy a home. How much more interest will the loan he received cost him than the loan he wanted?

Variable Rate Mortgages

As their name implies, variable, or adjustable, rate mortgages do not guarantee an interest rate for the life of the loan. Usually the initial interest rate is lower for a variable rate than for a fixed rate mortgage. However, the terms of the loan agreement allow the lender to increase or decrease the interest rate at specified intervals. Changes in the rate should be tied to changes in the prime interest rate or other public economic indicators.

It pays to shop around for the best variable rate mortgage. Some lenders offer loans on which the interest rates can be changed annually; others, every three or five years. Some lenders may have a cap on the minimum and maximum interests they can charge you. Some lenders allow you to convert a variable rate mortgage to a fixed rate; others do not.

Other Mortgage Options Another mortgage option is a **graduated payment mortgage** (GPM), which starts out with low initial monthly payments that rise gradually over the life of the loan. **Balloon payment mortgages** are another option. With these mortgages, the entire monthly payment for the first 20–25 years goes to pay off the interest on a home; the payments for the last years pay off the principal.

Critical Thinking Why do lenders offer a variety of mortgage options?

Refinancing a Mortgage

When interest rates go down, business firms and property owners often refinance, or replace, their higher interest rate mortgages with lower interest rate mortgages. When you **refinance** a mortgage, you take out a new mortgage and use that money to pay off the old mortgage. When you refinance your mortgage, you also pay closing costs on the new loan. There may also be other fees, such as the application costs for a loan or a *prepayment* penalty charged for paying off the first mortgage before it is due.

EXAMPLE The Rowes had a fixed rate mortgage at 16.25% with an unpaid balance of $50,000. The monthly payment on the old mortgage was $845.86. They got a new mortgage at 13% for the amount of the unpaid balance from another lender. Their new monthly payment is $585.79. To get the new mortgage, they had to pay closing costs of $935. To pay off the old mortgage before it was due, they had to pay a prepayment penalty of $500. How much did they save during the first year by getting the new mortgage?

SOLUTION

12 × $845.86 =	$10,150.32	One year's payment under old mortgage
12 × $585.79 =	7,029.48	One year's payment under new mortgage
	$ 3,120.84	Difference in yearly payments
	− 1,435.00	Closing costs and prepayment penalty ($935 + $500 = $1,435)
	$ 1,685.84	Net amount payments saved first year

EXERCISE YOUR SKILLS

13. After being in effect for 4 years, the rate of interest on Sol Retal's variable rate mortgage has changed from 12.5% to 10%. Sol's old monthly payment was $763.25. Sol's new monthly payment is $655.01. How much less does Sol pay in one year at the new mortgage rate?

14. Rita Metarel's monthly payment on her original 11.75% variable rate mortgage is $656.12. Eight years later, the interest rate on her mortgage has been raised to 13.25%, and her new monthly payment is $731.74. How much more will Rita pay in one year at the new mortgage rate?

Your old $45,000, 30-year, 13.6% mortgage has a monthly payment of $518.98. Over the 6 years since you took out the loan, mortgage rates have dropped. You can now get a mortgage at 9.15%, which will result in a new monthly payment of $377.37. To refinance, you must pay $990 in closing costs and a $510 prepayment penalty. By getting a new mortgage, find the net amount you will save in

15. the first year.

16. the second year.

The Greens have a fixed rate mortgage at 11.25% with a monthly payment of $529.07. They can refinance their loan with a variable rate mortgage at 8.9% from another lender with a monthly payment of $447.79. The rate of interest on the new mortgage will stay fixed for 3 years. If they refinance the mortgage, they must pay closing costs of $1,630. Find the

17. total monthly payments under the old mortgage for 3 years.

18. total monthly payments under the new mortgage for 3 years.

19. net amount saved in 3 years under the new mortgage.

Critical Thinking When do you think refinancing a mortgage would not be a good idea?

20. ♦ Look at newspaper ads or contact local banks to find current mortgage loan rates and closing costs. Pick one bank. Using that bank's rates, redo the exercises pertaining to the Greens. Would they have saved money at your bank? If not, how much would they spend at your bank? If so, find the net amount saved in 3 years under the new mortgage.

MIXED REVIEW

1. What percent of $24 is $84?

2. What is 140% of $85?

3. What is $\frac{1}{8}$ of $24.56?

4. Find the cost of 450 frames at $580 per C.

5. A 30-ounce jar of spaghetti sauce sells for $2.09. What is the price of the sauce per pound?

6. A space heater uses 4,000 watts per hour. During a cold month, Tom used the space heater 2 hours each day for 26 days. How much did it cost to use the space heater for one month if electricity costs 15 cents per KWH?

Shirley buys a house for $85,000. She pays 25% as a down payment and these closing costs: prepaid interest, $799; insurance, $340; legal fees, $280; land survey, $150; city inspection fee, $80.

7. What down payment did Shirley make?

8. What were the total closing costs?

9. What percent were the closing costs of the purchase price, to the nearest tenth of a percent?

7-2 THE COST OF OWNING A HOME

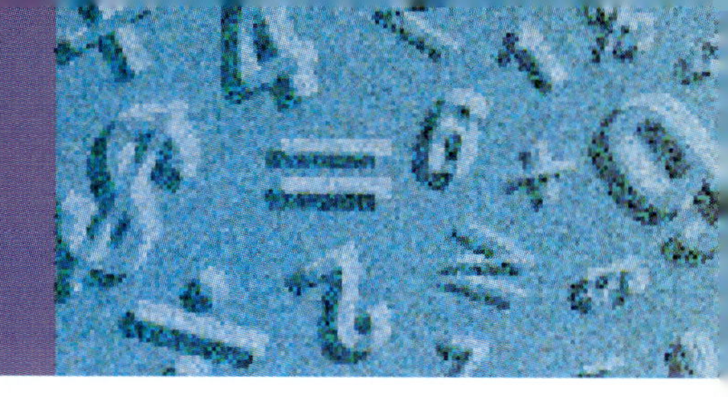

Many people want to own their home. Other people prefer to rent a residence. There are economic and personal advantages associated with both choices.

♦ Assuming your monthly rent and mortgage payments would be the same, would you prefer to rent or buy a home? Make a list of reasons for your choice. Rate the list in order of importance. Use 1 to indicate your most important reason.

OBJECTIVES

In this lesson, you will learn to

- *calculate depreciation of a home; and*
- *compare renting and owning a home.*

WARM UP

a. Find 25% of $42,000.

b. What percent of 800 is 320?

c. 52 + 68 + 110 + 37 + 63

d. What is $\frac{1}{4}$ of 12,500?

e. Estimate the product of $189.76 and 1.95.

f. Estimate the quotient of $4,540 ÷ 31.2.

THE COSTS OF HOME OWNERSHIP

In the last lesson, you learned how to calculate the cost of buying a home. A homeowner has many ongoing expenses. Cash has to be paid out for property taxes, repairs, maintenance, insurance, and mortgage interest. Two other less obvious expenses are **depreciation** and the loss of income on the money invested in the home. When making the decision to invest in a home, you should compare the financial advantages and disadvantages of both buying and renting.

Financial Costs of Home Ownership

Home Depreciation is a loss in the value of property caused by aging and use. The loss in value may be caused by the wearing out of parts of the home, such as the roof. It may also occur as home styles change or if the home becomes too expensive to heat and cool as energy costs rise. Most housing and business property depreciates slowly at about 2% to 4% of its original value per year.

Actual loss in value cannot be calculated until a house is sold. Until that time, depreciation must be estimated. Estimates of depreciation are often shown as a percent of the original purchase price.

EXAMPLE Elvita's home cost her $70,000. She estimates that depreciation each year reduces the value of the home by 3%. Find the depreciation for one year.

SOLUTION $0.03 \times \$70{,}000 = \$2{,}100$
The estimated depreciation is $2,100 per year.

Critical Thinking When may a home value appreciate or increase in value? When making the decision to buy a home, should you assume that it will appreciate in value? Explain your answer.

Loss of Income on Money Invested in a Home If you take money out of an interest-earning investment to make a down payment, you lose interest on that money. The interest lost each year is another cost of home ownership.

Financial Benefits of Home Ownership

Homeowners may deduct from their income the interest they pay on their home mortgage and the property taxes they pay on their property. This reduces the income tax they pay. Homeowners also build equity on their homes. **Equity** is the difference between what you owe on a home and its value.

Estimating the Net Cost of Home Ownership

Before you buy a home, you can estimate the net cost of owning it. First, add the expenses of mortgage interest, taxes, insurance, maintenance, repairs, depreciation, and the income you will lose by investing in the home. Then subtract the income tax savings you estimate you will get for owning the home.

EXAMPLE The Lingalis want to buy a home. The interest they will pay on their mortgage in the first year will be $4,810. The annual property taxes on the home are $1,350, and an insurance policy on the home will cost $320 a year. They estimate that the home will depreciate $1,240 the first year and that maintenance and repairs will cost $1,450. They lose $1,270 interest on their down payment. They estimate that they will save $1,428 in income taxes in the first year because they can subtract mortgage interest and property taxes from their income. What will be the net cost of owning the home in the first year for the Lingalis?

SOLUTION

$4,810	mortgage interest
1,350	property taxes
320	insurance
1,240	depreciation
1,450	maintenance, repairs
1,270	interest on down payment
$10,440	total expense

$10,440	total expenses
− 1,428	tax reductions
$9,012	net cost

The net cost of owning this home for the first year is $9,012.

EXERCISE YOUR SKILLS

1. Tom and Jane Yurko estimate that their home depreciates 2% each year. If the home originally cost $75,300, how much do they estimate for depreciation this year?
2. The home of the Collins family is estimated to depreciate at the rate of 3.2% each year. If the home originally cost $52,700, how much will this home depreciate in 1 year?
3. Artie Brouillard bought a condominium. The former owner originally wanted $85,000 for the home before selling it to Artie for $74,900. If the annual depreciation is estimated at 2.5%, how much will it depreciate this year?
4. Marie Santos bought a house for $55,200 ten years ago. If she estimates depreciation at 1.8% a year, how much has the house depreciated so far?
5. Ezra Wilkins bought a mobile home for $32,000 three years ago. He estimates that the home depreciates at a 5% rate each year. How much does Ezra estimate the mobile home has depreciated so far?
6. The Kerns want to buy a home. Their estimated first-year expenses are mortgage interest, $7,128; property taxes, $1,800; insurance, $515; depreciation, $1,680; maintenance and repairs, $1,237. They estimate lost interest income on savings as $1,472. They also expect to save $3,112 on income taxes. Find their net cost of owning their home the first year?
7. Kirk and Edna Rowe own a home. They estimate their expenses to be mortgage interest, $10,180; property taxes, $3,690; insurance, $833; depreciation, $3,800; maintenance and repairs, $900; lost interest income, $2,375. They expect to save $5,700 in income taxes from home expense. What is their net cost of home ownership?
8. The Sotos want to buy a home that costs $55,000. They must make a 20% down payment. They estimate their expenses for the first year to be mortgage interest, $4,914; property taxes, $1,208; insurance, $260; lost interest income, $1,320; depreciation, 2% of the purchase price; maintenance and repairs, $1,050. They expect to save $1,280 in income taxes because of home expenses. What will be the net cost of the home the first year?
9. Trudy Hertz wants to buy a home for $64,800. She estimates her first-year expenses to be mortgage interest, $6,005; lost interest of $1,377; property taxes, $850; insurance, $395; depreciation at 2.5% of the purchase price; maintenance and repairs, $1,400. She expects to save $1,847 in taxes. What will be the net cost of owning the home in the first year?

RENTING VERSUS OWNING

For some people, renting is something they do until they can afford to buy a home. For others, renting is their preference. Many people prefer not to worry about the expense and effort of maintaining a property. There are financial advantages and disadvantages to renting as well.

Financial Advantages of Renting Advantages include: not having to pay a significant down payment to rent an apartment; earning interest on money that would be used for a down payment; and more predictable housing costs. Renters do not have the maintenance costs associated with home ownership. Renters also do not have to worry about major unplanned housing expenses, such as a leaking roof or a cracked foundation.

Financial Disadvantages of Renting Renters receive no federal income tax benefits. (Some states let renters claim a portion of their rent as property tax.) In addition, renters do not earn equity on the housing they rent.

EXERCISE YOUR SKILLS

10. Art Emmons can rent an apartment for $475 a month, or he can buy a home. If he buys the home, he estimates his net annual cost of owning the home will be $7,250. How much will Art save in one year by renting?

11. Nina rents a home for $850 a month. If she buys the home, her estimated yearly expenses would be mortgage interest, $4,800; property taxes, $1,400; depreciation, $1,980; maintenance, $900; insurance, $410 and $1,100 interest income. She would save $1,736 in income taxes. Are net annual costs lower by buying or renting? How much is saved?

Amelia Burgos rents an apartment for $640 a month. She can buy a home with about the same space for $52,000. If she buys the home, she must withdraw $10,400 from her savings account and lose $728 interest. Her other home ownership expenses are estimated to be $9,300. She also estimates that she will save $1,428 in income taxes from home expenses.

12. What is her total rent for the year?

13. How much less is it for Amelia to rent than own for a year?

PUT IT ALL TOGETHER

♦ Jennifer Mills presently rents an apartment for $510 a month. She estimates that her rent will increase 4% each year and calculates total rent payments of $186,705.12 after 20 years. Ms. Mills just received a promotion at work. She could buy a home with the same space as her apartment for $44,000. If she takes a 20-year mortgage, she has estimated her average annual home ownership expenses to be depreciation, $570; maintenance and repairs, $1,400; insurance, $200; mortgage $368.03 per month; property taxes, \$1,120. She would save an average of \$1,120 in income taxes each year. She estimates she would have earned an average of \$495 interest each year on the money used as a down payment. Ms. Mills asks for your help in determining if it would be financially advantageous for her to buy a house.

14. Make a list of the steps you would recommend she take to compare the costs of renting versus buying. Then calculate the actual costs.

15. Is it financially advantageous for Ms. Mills to buy the house?

MIXED REVIEW

1. Add: $2\frac{3}{4}$, $5\frac{1}{2}$, $7\frac{5}{6}$

2. Subtract: 17.736 from 28.94

3. Multiply: $6\frac{1}{4}$ by $8\frac{1}{2}$

4. Divide: 24 by $\frac{2}{3}$

5. Multiply: 5.65 by 0.58

6. Find $\frac{1}{2}$% of \$56,000.

7. Julio is paid a monthly salary of \$1,350, plus 4% commission on all sales over \$38,000 made in a single month. For March, April, and May, his sales were \$35,600, \$39,400, \$42,800, respectively. Find his total earnings for the 3 months.

8. Daphne earns a salary of \$900 per month, a commission of 5% on all sales over \$4,000, and an additional commission of 2% on sales over \$7,000 a month. Her April sales were \$8,300. Find her total salary and commission for April.

9. Mark Champlain's regular hourly pay rate is \$8.80. He is paid time and a half for overtime work from Monday through Friday and double time for all weekend work. What is his gross pay for a week in which he worked 38 regular hours, 6 overtime hours, and 8 hours of work on the weekend?

7-3 DEPRECIATING A MOTOR VEHICLE

OBJECTIVES

In this lesson, you will learn to

- *calculate depreciation on a motor vehicle; and*
- *calculate average annual depreciation on a motor vehicle.*

When cars or trucks wear out from being used or getting old, they are usually not worth as much as when they were new. This loss of value is called depreciation.

◆ What do you think has more value: a 3-year-old car with 30,000 miles or a 1-year-old car of the same make and style with 80,000 miles? Identify the information you need to answer this question. List 3 places you can go to find the information you need.

WARM UP

a. $6\frac{3}{4} + 4\frac{1}{2} + 6\frac{1}{5}$

b. $238.333 - 13.875$

c. $2\frac{1}{3} \times 10\frac{1}{2}$

d. $36\frac{1}{4} \div \frac{1}{2}$

e. 0.78×0.657

f. Find $\frac{1}{4}\%$ of \$792.

g. Find 0.1% of \$876,000.

h. Find 160% of \$550.

DEPRECIATION

A car loses value as it grows older. This loss of value is called *depreciation*. The total depreciation on a car is the difference between its original cost and its resale, or trade-in, value. **Resale value** is the market value, or the amount you get when you sell the car to someone else. The **trade-in value** is the amount you get for your old car when you trade it in to buy a new car.

EXAMPLE Alfonzo buys a car for $16,500 and four years later trades it in for $7,500. Find the total depreciation.

SOLUTION

$16,500	original cost
− 7,500	trade-in value
$ 9,000	depreciation

Finding Average Annual Depreciation The depreciation on a car or truck is usually calculated for one year. A year's depreciation, or **average annual depreciation,** can only be estimated because the actual depreciation cannot be known until the car or truck is sold or traded in. To calculate the average estimated annual depreciation on a car or truck, follow these steps:

1. Estimate the number of years the car or truck will be kept.
2. Estimate the value of the vehicle when it is resold or traded in.
3. Subtract the resale or trade-in value from the original cost to find the estimated total depreciation.
4. Divide the total depreciation by the number of years the car or truck will be kept.

EXAMPLE A car that cost $14,800 has an estimated trade-in value of $5,900 at the end of 4 years. Find the average annual depreciation.

SOLUTION

$14,800	original cost
− 5,900	trade-in value
$ 8,900	total depreciation

$8,900 \div 4 = \$2,225$ average annual depreciation

This way of estimating the average annual depreciation is called the **straight-line method.** It spreads the total depreciation expense evenly over the time the car is to be kept.

Rate of Depreciation When the straight-line method of finding depreciation is used, the average annual depreciation is often shown as a percent of the original cost. When finding the **rate of depreciation,** the average annual depreciation is the part, and the original cost is the whole with which the part is compared.

$$\textbf{Rate of Depreciation} = \frac{\textbf{Average Annual Depreciation}}{\textbf{Original Cost}}$$

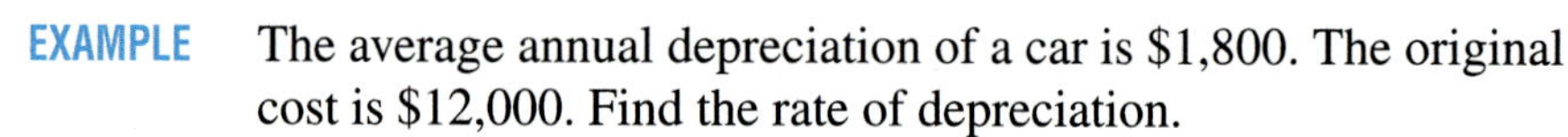

EXAMPLE The average annual depreciation of a car is $1,800. The original cost is $12,000. Find the rate of depreciation.

SOLUTION Rate of Depreciation $= \frac{\$1{,}800}{\$12{,}000} = 0.15$ or 15%

EXERCISE YOUR SKILLS

Find the total depreciation.

	Type of Vehicle	Original Cost	Resale Value or Trade-in Value	Total Depreciation
1.	Delivery Van	$24,500	$8,100	
2.	Compact Car	$10,050	$5,850	
3.	Luxury Car	$48,900	$12,600	
4.	Pickup Truck	$16,220	$3,850	

5. Manda bought a truck for $14,740. After using the truck for awhile, she sold it for $6,790. Find the total depreciation on the truck.

6. Stuart sold his car for $4,400. If he paid $10,100 for the car 4 years ago, what was the total depreciation on the car?

7. Yvette Tolbert paid $7,300 for a car 7 years ago. She bought a new car recently at a total cost of $11,100 after deducting the $700 she got as a trade-in for her old car. What was the total depreciation on the 7-year old car?

Find the average annual depreciation for each.

	Original Cost	Resale or Trade-in Value		Average Annual Depreciation
		At End of	Amount	
8.	$ 9,500	3 years	$6,200	
9.	$18,500	7 years	$3,800	
10.	$13,525	2 years	$8,775	
11.	$24,300	5 years	$6,075	

12. A used dump truck that sells for $20,250 today is estimated to have a resale value of $5,100 in 5 years. What is the average annual depreciation of the truck?

13. A paint supplier bought a truck for $18,420. After 4 years, a new truck was bought that cost $21,890. A trade-in value of $5,565 was given for the old truck. To the nearest dollar, find the average annual depreciation of the old truck.

Diedra Neff bought a truck for business use for $17,072. She used the truck for 7 years and then sold it. The average annual depreciation was 12%.

14. Find the dollar amount of average annual depreciation.

15. How much did she get when she sold the truck?

PROBLEM SOLVING

Work Backwards

You have learned to find the average annual depreciation as either a rate or dollar amount. How would you find the original cost of a vehicle if you know its trade-in value and the average annual depreciation rate?

EXAMPLE Nora sold her car for $3,100 after owning it for 5 years. She found the average annual depreciation was $1,860 and the rate was 15%. What did she pay for the car when it was new?

SOLUTION Think:

$$\text{rate of depreciation} = \frac{\text{average annual depreciation}}{\text{original cost}}$$

$$15\% = \frac{\$1{,}860}{?}, \text{ so } 15\% \text{ of what number is } \$1{,}860?$$

$$\$1{,}860 \div 0.15 = \$12{,}400$$

—OR—

$$\text{average annual depreciation} = \frac{\text{original cost} - \text{trade-in}}{\text{number of years}}$$

$$\$1{,}860 = \frac{(? - \$3{,}100)}{5}$$

$$5 \times \$1{,}860 = ? - \$3{,}100$$

$$\$9{,}300 + \$3{,}100 = ?$$

$$\$12{,}400 = \text{original cost}$$

Alphonse Carr bought a car for $8,850 three years ago. Alphonse has been offered $3,825 by a car dealer as the trade-in value, but he feels he can sell the car for $4,500. What will be the average annual depreciation

16. If he takes the trade-in offer?

17. If he is able to sell the car at the price he wants?

Find the rate of depreciation in each problem. Round answers to the nearest percent or nearest whole dollar.

	Original Cost	Resale or Trade-in Value		Rate of Depreciation
		At End of	Amount	
18.	$14,500	4 years	$ 5,600	
19.	$28,350	3 years	$14,700	
20.	$ 9,450	6 years	$ 1,800	
21.	$12,680	2 years	$ 7,700	

22. Rex Bell bought a new truck. Six years later he sold it for $6,160. He found the average annual depreciation was $2,640 at a rate of 12%. How much did Rex pay for the truck when it was new? Work backwards to solve the exercise.

23. Jorge Lanell bought a car for $15,300 and used it in his business for two years. His annual rate of depreciation was 24%. How much did he sell the car for?

24. A van that costs $17,460 is estimated to be worth $4,500 after four years. Find the rate of depreciation, to the nearest percent, on the van.

25. ♦ Write down the make and model year of a used car or truck you would be interested in buying. Visit an automobile dealership or check your library resources to find out the resale value of that car. Compare the resale value with prices you see listed in newspaper ads (classified or automobile). What are the differences? Why do you think they exist?

MIXED REVIEW

1. $63,000 × 10.5%

2. $18,480 ÷ 12

3. What percent less than 960 is 192?

4. $8.40 plus 8.5% of itself = ?

5. Marty Sill is paid $1.07 for each extension lamp he assembles that passes inspection. What is his pay for a week where he assembles 260 lamps, 13 of which were faulty when inspected?

6. Sheila McKinstry got $\frac{2}{5}$ of the 6,805 ballots cast in a primary election. How many votes did Sheila get?

7. Alice Broeker spent these amounts on a business trip: airfare, $347; lodging, $385.92; food, $137; rental car, $118.52; other expenses, $29.18. What total amount did Alice spend on the trip?

8. Lowell Krygol bought a motor home for $28,000. Six years later he sold the motor home for $12,880. What was the average annual depreciation of the motor home?

9. A van that cost $19,000 was traded in for $2,280 after eight years of use. What was the average annual rate of depreciation?

7-4 COST OF OPERATING A MOTOR VEHICLE

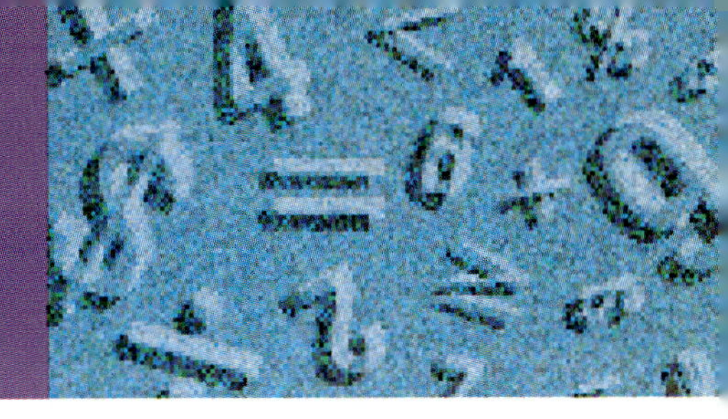

Roberto wants to buy a car to get to work. He will use it to drive a total of 50 miles each day to and from work. On his drive to work, he must pay a daily toll of $4, and he has a monthly parking expense of $65. He also must budget for the cost of insurance, gas, oil, registration, and repairs.

◆ Do you think it would be more economical for Roberto to buy an older, reliable car or a new car? To answer the question, create a chart that shows estimated expenses for the items mentioned for both the older and newer car for one year. What conclusions can you draw about car costs?

OBJECTIVES

In this lesson, you will learn to

- ▸ *calculate annual operating costs for a motor vehicle; and*
- ▸ *compare costs of owning and leasing a motor vehicle.*

WARM *UP*

a. $85,000 × 15%

b. Write $\frac{12}{100}$ as a decimal.

c. Estimate: 58 × 15,108

d. Estimate: 202 ÷ 0.49

e. $\frac{1}{2} + \frac{5}{6}$

f. 6 500 kg = ? t

g. 536 L = ? kL

h. Round $138.836 to the nearest cent.

COSTS OF OPERATING A MOTOR VEHICLE

Like home ownership, the cost of owning a car consists of the costs of the original purchase (purchase price plus interest on a loan, if one was used) and the cost of operating the vehicle.

The total operating cost for a car is the sum of all the annual expenses. These expenses may include insurance, gas, oil, license and inspection fees, tires, repairs, garage rent, parking fees, tolls, taxes, and general upkeep.

Critical Thinking Should you include in your cost calculations the interest lost on the money invested in the car and the costs of depreciation? Why or why not?

EXAMPLE Kay paid $14,200 for her car. Her annual payments for insurance, gas, oil, repairs, and other expenses total $1,900. The car depreciates 18% a year. Kay could have earned $568 interest on her investment in the car. What was her total annual cost of operating the car?

SOLUTION

Insurance, gas, oil, etc.	$1,900
Depreciation, 18% of $14,200	2,556
Loss of interest	568
Total annual operating cost	$5,024

EXERCISE YOUR SKILLS

Find the total annual operating costs.

	Original Cost	Interest Loss	Rate of Depreciation	Other Expenses	Total Operating Cost
1.	$18,600	$ 651	16%	$1,600	
2.	$15,400	$ 616	20%	$1,940	
3.	$12,800	$ 890	15%	$2,100	
4.	$31,300	$2,030	21%	$1,570	

5. Richard bought a used car for $13,500. His expenses for the first year were gas and oil, $755; repairs, $125; insurance, $562; license plates, $42; loss of interest on investment, $945; depreciation, 17%. Find the total operating cost for the year.

6. When buying a car for $16,800, Rita Roma estimates first year's car expenses at gas and oil, $670; license plates, $68; insurance, $780; loss of interest, $840; repairs, $30; depreciation, 17% of car cost. Find Rita's total cost of operating the car for the first year.

A new car is purchased for $14,000. The car's estimated resale value at the end of one year is 75% of its purchase price and 60% of the purchase price at the end of two years. Using these estimates,

7. What is the car's resale value at the end of one year?

8. What is the resale value of the car at the end of two years?

9. At the end of two years, what is the dollar amount of the average annual depreciation?

Critical Thinking What factors make it more or less expensive to own and operate a car. Consider both fixed expenses, such as insurance costs, and variable costs, such as the cost of gas, oil, and repairs. How would you compare the cost of operating different vehicles based on the number of miles you drive each year?

LEASING A MOTOR VEHICLE

When you buy a car or truck, you usually make a down payment and then make monthly payments for the balance of the purchase price. However, instead of buying, you may want to lease a car or truck for a monthly charge.

The monthly charge generally depends on the length of the lease and how much it would cost to buy the car or truck you are leasing. Leases usually run for 24, 36, or 48 months and limit the number of miles the car or truck can be driven. A per mile charge is made for excess mileage. Some leases may require a cash down payment in exchange for lower monthly payments. Others may require a security deposit to cover damage to the vehicle.

If you lease a car, you are responsible for insurance, gasoline, and maintenance. You also have to pay a charge for excess wear and tear.

Critical Thinking What do you think are some advantages and disadvantages of leasing a car rather than buying one?

Use these leasing charges to solve Exercises 10–15.

A-1 LEASING COMPANY MONTHLY LEASING CHARGES			
Type of Car or Truck	**Monthly Charge**	**Type of Car or Truck**	**Monthly Charge**
Compact car	$237	Sports utility vehicle	$349
Pickup truck	$256	Standard van	$315
Luxury car	$599		

Monthly charges are for a 36-month lease with no down payment.
For all miles over 15,000 miles annually, 12¢ per mi is charged.

10. What amount must be paid to the leasing company during a 12-month period for leasing a sports utility vehicle?

Tameko wants to lease a pickup truck from A-1 for three years. She estimates that insurance, maintenance, and fuel will cost $1,600 a year. She also estimates that she will drive an average of 14,000 miles each year. Including the leasing charges, find

11. The total cost of operating the truck for three years.

12. The operating cost per mile for three years, to the nearest cent.

Kirk Mandell is comparing the leasing charges for a standard van and a custom van that leases for $396.90 a month.

13. How much more per year will leasing the customized van cost?

14. What percent more per year will it cost to lease the custom van instead of the standard van?

15. Adam Voleck leased a compact car from A-1 Leasing for three years. The total cost of insurance, maintenance, repairs, and gas for three years was $4,434. He drove the car for 14,000 miles for the first year, 17,000 miles for the second, and 16,000 miles for the third. What was the total cost of operating the car for three years?

16. ◆ Add a third column to the chart you created at the start of the lesson. On that chart, itemize the costs associated with commuting to work by public transportation in your community. What costs would be eliminated? What would be added? Of the three items on the chart, what would be the most economical choice? The least economical? What are the advantages and disadvantages of the least and most economical choice?

PUT IT ALL TOGETHER

17. Make a worksheet or checklist that you could give to someone to help them determine if they could afford to buy and maintain a house or a car. If you chose a car purchase checklist, find a partner who has developed a house purchase checklist. Evaluate each other's list for clarity and completeness. Are any items missing? What do the checklists for each item have in common? What is different? What conclusions can you draw about making a major purchase?

MIXED REVIEW

1. 17.8 m × 8 m
2. 34.8 ÷ 0.04
3. $20,000 × 1.05
4. 2.5 m = ? cm
5. The trade-in value of a car that cost $11,900 seven years ago is $1,400. The same type of car sells for $15,200 today. Find the average annual depreciation of the seven-year old car.
6. The monthly payment on a 20-year, 12%, $45,000 mortgage is $495.49. How much interest will the buyer pay over the life of the loan?

Eli Fromme earns $280.50 per week. Find his equivalent pay for a

7. Year?
8. Month?
9. Polly Rhodes bought a car for $9,200. She drove the car 12,000 miles last year. Her car expenses for the year were insurance, $436; gas and oil, $670; repairs, $28; license plates, $42; loss of interest on original investment, $460; depreciation, 18%. Find last year's operating cost per mile, to the nearest cent.
10. You lease a car for $238 a month and 7¢ for every mile driven over 15,000 miles in one year. If you drive the car 16,500 miles each year, what amount will you pay for leasing charges in two years?

EXPLORE TECHNOLOGY

TEMPLATE: Spreadsheet to Compare Costs of Home Ownership and Renting

Louise Ruell has collected data about the expenses of owning and renting a home and wishes to compare the two. You can help her compare costs by loading ET7 from your applications diskette. Your spreadsheet should look like the one shown.

Figure her total cost of renting and owning a home by entering the bold-faced data into the spreadsheet cells specified.

B3 **420** B6 **40000** B10 **600** B13 **450**
B4 **12** B7 **0.0575** B11 **550** B14 **980**
B9 **1200** B12 **317**

Formulas built into the template will calculate the empty cells. Your completed spreadsheet will look like this.

	A	B
1	**Renting vs. Owning a Home**	
2		
3	Monthly Rent	$0.00
4	# of Mos. Rented	0
5		
6	Home Purchase Price	$0.00
7	Mortgage Int. Rate	.00
8	Mortgage Interest	$0.00
9	Property Taxes	$0.00
10	Depreciation	$0.00
11	Maint./Repairs	$0.00
12	Insurance	$0.00
13	Lost Interest	$0.00
14	Income Taxes Saved	$0.00
15		
16	Total Yearly Rental Cost	$0.00
17	Total Yearly Owning Cost	$0.00
18	Difference (Rent-Own)	$0.00

	A	B
1	**Renting vs. Owning a Home**	
2		
3	Monthly Rent	$420.00
4	# of Mos. Rented	12
5		
6	Home Purchase Price	$40,000.00
7	Mortgage Int. Rate	.0575
8	Mortgage Interest	$2,300.00
9	Property Taxes	$1,200.00
10	Depreciation	$600.00
11	Maint./Repairs	$550.00
12	Insurance	$317.00
13	Lost Interest	$450.00
14	Income Taxes Saved	$980.00
15		
16	Total Yearly Rental Cost	$5,040.00
17	Total Yearly Owning Cost	$4,437.00
18	Difference (Rent-Own)	$603.00

Refer to the spreadsheet above right to answer the following questions.

1. What is the total yearly cost of renting the home? What formula was used to calculate yearly rent?
2. What is the total yearly cost of owning the home? What formula was used to calculate yearly costs of owning the home?
3. Which costs less, renting or owning?
4. What formula was used to calculate cell B8?

Design Your Own

Suppose you are thinking about buying or renting a new car.

Leasing a car is similar to renting a home. For example you usually pay a monthly fee to lease a car. You also pay for miles you drive over the maximum number allowed in the lease.

Ownership costs include operating costs as well as the loss of the car's value over time and the loss of interest on the investment. You are responsible for all normal operating costs, such as gas, oil, and maintenance, whether you lease or own a car.

5. Set up a spreadsheet that will allow you to compare the total cost of leasing a car and the total cost of owning the same car over a 4-year period. The information you'll need is shown in the spreadsheet below. You may want to use a different layout than the one shown. You may also want to add additional rows or columns to calculate answers to the following questions.

	A	B
1	**Leasing vs. Owning**	
2		
3	Monthly Lease Charge	$215
4	Years Owned/Leased	4
5	Excess Mileage:	
6	# Excess Miles	2,000
7	Per Mile Charge	$0.11
8		
9	Total Operating Costs	$5,000
10		
11	Rate of Depreciation	.16
12	Total Depreciation	$7,424
13	Total Interest Lost	$2,320
14		
15	Total Cost of Leasing	
16	Total Cost of Owning	
17	Difference (Lease-Own)	

Write down the formulas you will use for each exercise. Then enter the formulas and calculate the answer.

6. Total cost of leasing.
7. Total cost of owning.
8. Which is more expensive, owning or leasing? By how much?
9. What formula can you use to find the purchase price of the car? Enter the formula into your spreadsheet and calculate the answer.

CHAPTER 7 SUMMARY AND REVIEW

VOCABULARY REVIEW

amortization table *7-1*
average annual depreciation *7-3*
closing costs *7-1*
depreciation *7-2*
down payment *7-1*
equity *7-2*
fixed rate mortgage *7-1*
interest *7-1*
lease *7-4*
mortgage *7-1*
operating costs *7-4*
principal *7-1*
resale value *7-3*
straight-line method of depreciation *7-3*
trade-in value *7-3*
variable rate mortgage *7-1*

MATH SKILLS REVIEW

7-1 ◆ Multiply whole numbers and decimals.

1. 12 × $1,024.35

2. 36 × $200.25

7-2 ◆ Find a percent of a number by changing the percent to a decimal and then multiplying.

3. 3% of $150,000 = ?

4. 6% of $9505 = ?

7-3 ◆ Find the average of a set of data by dividing the sum of the data by the number of data.

5. Find the average of $2,300, $3,400, $2,850, and $4,050.

APPLICATIONS

7-1 ◆ Purchase Price × Down Payment Expressed as a Percent = Amount of Down Payment.

◆ Total Closing Costs = Sum of All Costs of Buying a Home Excluding the Amount Borrowed and the Down Payment.

◆ Total Mortgage Loan Interest = (Number of Payments × Amount of Each Payment) − Amount Borrowed

The new monthly payment for a refinanced mortgage is $568.54. The old mortgage had a monthly payment of $681.28. To refinance, the buyer paid $675 in closing costs and a $560 prepayment penalty. What will be the net savings to the buyer

6. for the first year?

7. for the second year?

Use Illustration 7-1.2 to solve this problem. Myrna purchases a home priced at $81,250 and makes a 20% down payment. She borrows the rest of the money at 9% for 25 years.

8. How much must Myrna borrow to buy the house?

9. What monthly payment must she make?

10. Find the total amount of interest to be paid over 25 years.

7-2 ◆ Estimated Annual Depreciation = Value of Home × Depreciation Rate.

◆ Net Cost of Home Ownership = (Sum of All Expenses Including Lost Interest) − Income Tax Savings.

◆ Compare renting and owning of a home by finding the difference between the costs of ownership and of rental.

11. A home depreciates 2% a year. What will be the depreciation for one year if the home's original cost was $38,000?

Amelda rents an apartment for $470 a month. She can buy a home for $48,000 by using $7,200 in savings for the down payment. Interest lost on the down payment is $508. Other annual home ownership expenses are estimated to be $6,500. Amelda estimates that she will save $800 in income taxes by buying the home.

12. What is her total rent for the year?
13. What is her net cost of buying the home?
14. Which costs less, renting or buying? How much less?

7-3 ◆ Total Automobile Depreciation = Original Cost − Resale or Trade-in Value

◆ Average Annual Auto Depreciation = Total depreciation ÷ Number of Years of Automobile Ownership

◆ Rate of Depreciation = Average Annual Depreciation ÷ Original Cost

15. Terri Maynor bought an extended cab pickup truck 6 years ago at a cost of $13,400. She received a $3,200 trade-in allowance for the truck when she bought a new van. What was the total depreciation on the truck?
16. Charles Shein owns a van that cost $15,790 four years ago and has a resale value of $5,350 today. The van now has about 80,000 miles on the odometer. Find the average annual depreciation for Charles Shein's van and the rate of depreciation to the nearest tenth.

7-4 ◆ Total Annual Operating Cost of a Vehicle = Sum of All Annual Expenses (including depreciation and lost interest)

◆ Total Cost of Leasing = Monthly Lease Charge × Number of Months in the Lease + Excess Mileage Charges and Operating Costs.

◆ Calculate which is less expensive, leasing or buying, by comparing total purchase price to the costs for the full term of the leasing contract.

Al's Tile Shop had these truck expenses last year: depreciation, $4,100; lost interest, $345; gasoline and oil, $2,800; insurance, $1,560; license plates, $140; repairs, $675. The truck was driven 40,000 miles last year.

17. What was the average weekly cost of operating the truck last year?
18. What was the annual operating cost per mile, to the nearest cent?

The Instar van leases for \$356 a month with an 11¢ excess mile charge for all miles driven over 12,000 in a year. The Hi-Road van leases for \$418 a month with a 15¢ excess mile charge for miles driven over 15,000 per year. Compare the costs of leasing the vans over two years assuming that 14,000 miles are driven the first year and 18,500 the second year.

19. What will be the total excess mile charges for both vans over two years?
20. What are the total two-year lease costs for the vans?

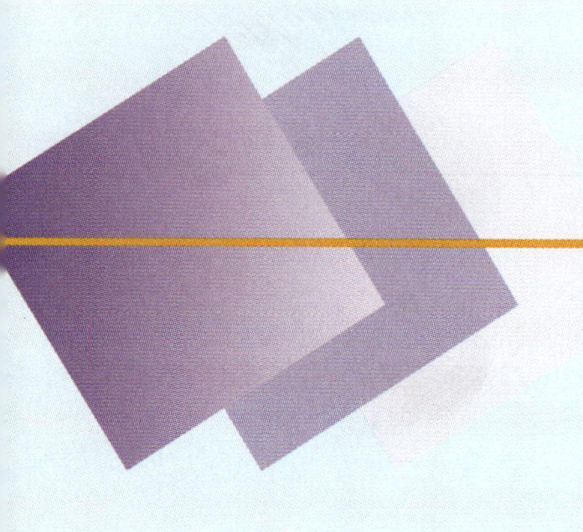

WORKPLACE KNOWHOW: **SERVING CUSTOMERS**

Serving Customers A customer is someone who must be satisfied with the product or service provided. Some businesses refer to external customers and internal customers. External customers are the people who buy a product or service. In many companies the job of a group of people within an organization is to provide a service to other groups within the same organization. For example, an advertising department creates advertising campaigns under the direction of the marketing department. The marketing people are the internal customers of the advertising department.

Satisfying customers requires the ability to listen, ask meaningful questions that help you understand what someone needs, communicate in a clear and positive manner, handle complaints, obtain information, and solve problems.

Assignment Work in groups. Review the career profiles in the first seven chapters of this book. Assume that you are doing the job of the person profiled. Answer these three questions for each job:

1. Who are my internal and external customers?
2. What do my customers need?
3. What kinds of customer service issues do I have?

As a group, select one profile and one customer service issue. Then prepare a scene where one person plays a customer and another person plays the person profiled. Be prepared to present your scene to the class. After several groups have acted out their scenes, have a class discussion about what you have learned about customer service.

CHAPTER 7 TEST

Math Skills

1. 15% of $63,000
2. $2,900 ÷ $14,500 = ?%
3. 240 × $343.04
4. $1.15 × 670
5. $693.72 − $580.17
6. 782.6 ÷ 36.4
7. 0.015 × $48,600
8. 11¢ × 1,877
9. $8,736 ÷ 4
10. $312.80 × 36

Applications

11. Matt Dixon bought a two-bedroom condo for $80,000. He paid 35% of the price in cash. He also paid $250 for a property survey, $472 for title insurance, and $320 for legal fees. What amount of cash did Matt need when he bought the condo?
12. On a 30-year mortgage of $71,400, Nicole Washburn makes monthly payments of $692.37. What total amount of interest will she pay on her mortgage over its 30-year term?
13. Sue and Ed Carter are buying a home. Estimated first-year expenses are: mortgage interest, $4,592; real estate taxes, $2,174; insurance, $589; depreciation, $890; maintenance, $780; lost income on the cash invested, $1,400. They estimate their income tax benefit at $2,300. Find the net cost of owning the home in the first year.

The Bowdell family rents a house for $735 monthly. They can buy that house for $70,000. If they buy the house, they estimate annual operating expenses of $9,300 and an income tax saving of $1,450. To make the down payment, the Bowdells would have to withdraw $10,600 from savings and lose $583 interest each year.

14. Is it cheaper to rent or buy the home? How much will be saved in one year by choosing the less expensive way?

Beverly bought a car for $10,370. After five years, she bought a new car for $12,790, trading in her old car and paying $9,590 in cash. What was the

15. Total depreciation on the old car?
16. Average annual depreciation of the old car?
17. Rate of depreciation, to the nearest tenth percent?
18. Ken leased a car for $236 a month and 12¢ per mile for any over 15,000 miles in a year. He drove 22,000 miles in a year. Find the total cost of leasing the car for the year.

CHAPTER 8

TEAMS at

Many businesses are organizing work groups into teams. This team approach assumes "The better we work together the more profitable our company becomes."

Chi-luan works as a manager of a distribution center for a major clothing manufacturer that has adopted the team concept in its warehouses. That means performance objectives and rewards are based on each team's efforts. Teams are evaluated and rewarded for a variety of performance factors. Accurate and on-time delivery is the most important performance standard. Teams are also evaluated on the number of boxes processed daily, the accuracy of their inventory, and safety records. All teams are expected to meet a specified performance standard. Teams that exceed that standard are rewarded with salary bonuses and other performance incentives.

The primary requirement of a successful team is workers who feel responsible for their team's performance. Chi-luan explains, "Attention to detail is critical. So is a good attendance record. You can't support your team if you constantly call in sick."

Successful teams practice good communication and listening skills. Chi-luan's team members help settle disputes so they can put their energy into processing over 5,000 boxes of clothing daily. Successful teams also help train new team members and identify ways to improve work processes.

Technology and math are used daily in the distribution center, Warehouse workers use

WORK

portable computers to keep track of inventory and operate a "crown picker," that lifts boxes and the driver up and down the levels of the warehouse's 40-foot shelves. Math skills are needed to record the number of boxes received and shipped, and to keep count of the number of garments added and removed from boxes to fill orders. As supervisor, Chi-luan uses percentages and equations for payroll duties and team efficiency ratings.

Chi-luan started at this company after high school. He enjoys working there because the company rewards good team workers. Chi-luan explains he was able to move from being a forklift operator to a department manager in just four years by earning good ratings, having an excellent attendance record, volunteering for new responsibilities, and having a will to learn. As a manager, he is evaluated on the performance of his teams. He reads books and takes courses in management, conflict resolution, and communication. He says "...if you have initiative and apply yourself, you can really go places."

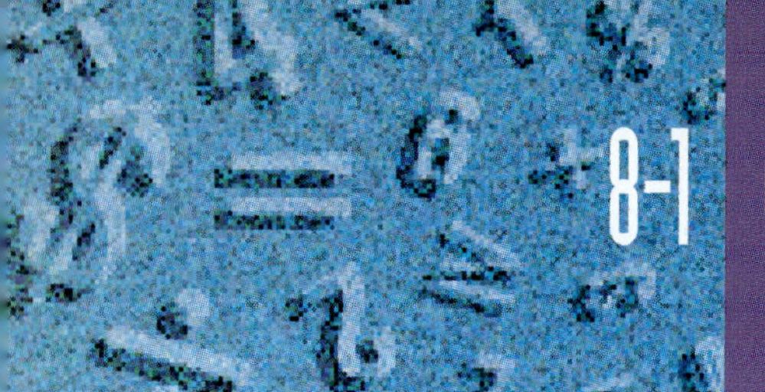

8-1 PROPERTY TAXES

OBJECTIVES

In this lesson, you will learn to

- *find assessed value of property;*
- *calculate property taxes due; and*
- *determine property tax rates.*

Property taxes are used to help pay for services to the community.

◆ Make a 2-column list. In the first column, list the services you believe your community's property taxes are used to support. In the second column, list the services you would like your property taxes to support.

WARM UP

a. \$210 ÷ \$240

b. 23.2 m × 12 m

c. 45.78 ÷ 0.07

d. $37\frac{1}{2}\%$ less than \$8.96 is what amount?

e. $66\frac{2}{3}\%$ of \$12,000 is what amount?

PROPERTY TAXES

Property taxes are taxes on real estate collected by local governments, such as cities and towns. These taxes are one of the expenses of owning farm land, business property, or a home. Property taxes are collected by the community in which the property is located. Services provided with the tax money vary by localities. Services that are often supported by taxes include schools, government operations, fire and police protection, and road maintenance.

Property owners are usually billed annually or semi-annually by the tax department within their community. Illustration 8-1.1 is an example of a property tax bill.

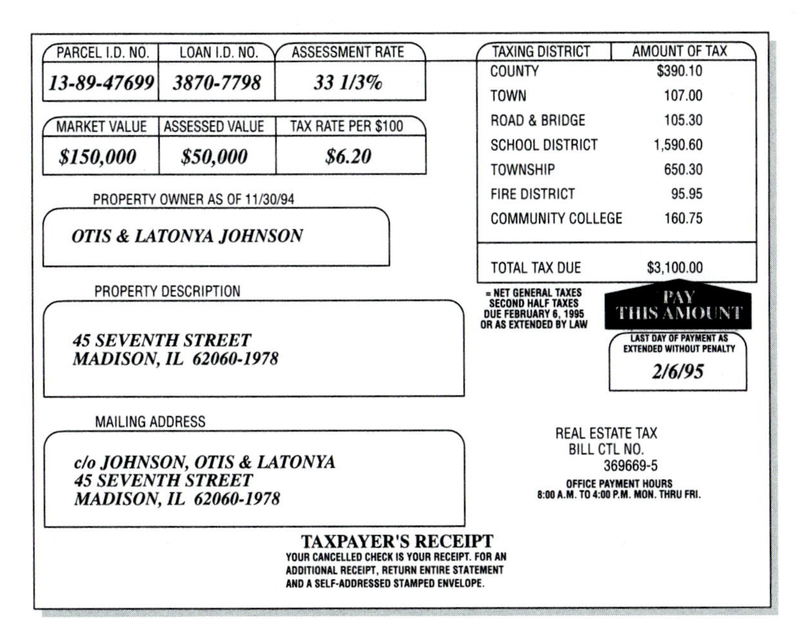

PARCEL I.D. NO.	LOAN I.D. NO.	ASSESSMENT RATE
13-89-47699	*3870-7798*	*33 1/3%*

MARKET VALUE	ASSESSED VALUE	TAX RATE PER $100
$150,000	*$50,000*	*$6.20*

PROPERTY OWNER AS OF 11/30/94

OTIS & LATONYA JOHNSON

PROPERTY DESCRIPTION

45 SEVENTH STREET
MADISON, IL 62060-1978

MAILING ADDRESS

c/o JOHNSON, OTIS & LATONYA
45 SEVENTH STREET
MADISON, IL 62060-1978

TAXING DISTRICT	AMOUNT OF TAX
COUNTY	$390.10
TOWN	107.00
ROAD & BRIDGE	105.30
SCHOOL DISTRICT	1,590.60
TOWNSHIP	650.30
FIRE DISTRICT	95.95
COMMUNITY COLLEGE	160.75
TOTAL TAX DUE	$3,100.00

= NET GENERAL TAXES SECOND HALF TAXES DUE FEBRUARY 6, 1995 OR AS EXTENDED BY LAW

PAY THIS AMOUNT

LAST DAY OF PAYMENT AS EXTENDED WITHOUT PENALTY

2/6/95

REAL ESTATE TAX
BILL CTL NO.
369669-5

OFFICE PAYMENT HOURS
8:00 A.M. TO 4:00 P.M. MON. THRU FRI.

TAXPAYER'S RECEIPT
YOUR CANCELLED CHECK IS YOUR RECEIPT. FOR AN ADDITIONAL RECEIPT, RETURN ENTIRE STATEMENT AND A SELF-ADDRESSED STAMPED ENVELOPE.

Illustration 8-1.1. Property Tax Bill

Assessed Value of Real Estate

The amount of property tax paid is based on the **assessed value** of a property. This value is calculated by local tax assessors. For example, the Johnsons' property in Illustration 8-1.1 has a fair market value of $150,000. It is assessed at $33\frac{1}{3}$% of its market value, or $50,000. Assessed values of properties are usually less than their market values. Similar properties in the same community should have similar assessed values. Homes may be reassessed after major improvements. Local governments may also periodically reassess all the homes in their area.

Critical Thinking Do people who do not own property pay property taxes?

The Tax Rate

The tax rate is calculated on the assessed value. In some communities, the tax rate may be shown as the number of *dollars* for each $100 or $1,000 of assessed value. In other communities, the rate is shown as either the number of ***mills*** or *cents* for each $1 of assessed value. A cent is one hundredth of a dollar ($0.01). A **mill** is one thousandth of a dollar ($0.001).

In other words, the Johnson's tax rate could be stated as

$62 per $1,000 of assessed value;
$6.20 per $100 of assessed value;
6.2 cents per $1 of assessed value; or as
62 mills per $1 of assessed value.

The tax due on the Johnson's bill in all cases is $3,100.

Finding the Tax for Rates Shown per $100 or $1,000 First, find the number of $100 units or $1,000 units in the assessed value. Then, multiply the numbers of units by the tax rate to find the tax due.

EXAMPLE 1 If the Johnson's tax rate is stated as $62 per $1,000, their tax would be calculated as:

SOLUTION

$50,000 ÷ $1,000 = 50	Find the number of 1,000 units in the assessed value
50 × $62 = $3,100 tax	Multiply the number of units times the tax rate for the property tax due

EXAMPLE 2 If the Johnsons' tax rate is stated as $6.20 per $100 of assessed value, their tax bill would be calculated by following these steps:

SOLUTION $50,000 ÷ $100 = 500 units of $100
500 × $6.20 = $3,100 of tax

EXERCISE YOUR SKILLS

	Assessed Value	Tax Rate	Tax Due
1.	$80,000	$5.20 per $100	
2.	48,500	$7.10 per $100	
3.	15,000	$34 per $1,000	
4.	20,800	$2.80 per $100	
5.	18,000	$3.085 per $100	

6. The tax rate for the town of Elton is $3.775 per $100. Find Norma's tax bill if her property in Elton is assessed at $60,500.

7. What tax must Ira Baum pay on his home, assessed for $25,900, if the tax rate is $38.20 per $1,000?

8. Find the tax on property assessed at $17,500 if the tax rate is $4.155 per $100.

Finding the Tax for Rates Shown per $1 To find the tax when the rate is in mills or cents per $1, change the rate to a rate in dollars. Then multiply that rate by the assessed value.

A mill is one tenth of a cent, one thousandth of a dollar, or $0.001. There are 10 mills in one cent and 1,000 mills in a dollar.

To change mills to dollars, divide the number of mills by 1,000.

56.8 mills ÷ 1,000 = $0.0568	← Move the decimal point 3 places to the left.

To change cents to dollars, divide the number of cents by 100.

5.68 cents ÷ 100 = $0.0568	← Move the decimal point 2 places to the left.

EXAMPLE If the Johnson's tax rate is expressed in mills per dollar or cents per dollar of assessed value, you would find the tax on their property this way:

SOLUTION

62 mills ÷ 1,000 = \$0.062	Change the tax rate from mills to dollars by dividing by 1,000.
6.2 cents ÷ 100 = \$0.062	Change the tax rate per cents to dollars by dividing by 100.
\$0.062 × \$50,000 = \$3,100 tax	Multiply the rate per \$1 by the assessed value of the property.

EXERCISE YOUR SKILLS

	Assessed Value	Tax rate Per \$1	Tax Due
9.	\$80,000	35 mills	
10.	48,500	82.1 mills	
11.	15,000	77.3 mills	
12.	20,800	9 cents	
13.	18,000	2.7 cents	
14.	40,500	4.54 cents	

Solve each problem. Be sure to notice how the tax rate is stated.

15. The tax rate in Daberg is 48 mills per \$1 of assessed value. Find the tax to be paid on property assessed at \$35,200.

16. What tax must Barby Tomlin pay on a home assessed at \$20,750 if her tax rate is 3.5 cents per \$1?

17. The town of Chester has a tax rate of 45.089 mills per \$1. Find the tax on property in Chester worth \$250,000, assessed at 60% of its value.

DETERMINING TAX RATES

Each tax district determines the tax rate needed to pay for the services they provide. Tax districts estimate their expenses for the coming year and prepare an expense budget. They also estimate income from sources other than the property tax, such as licenses, fees, fines, rents, state aid, and so on. The difference between the total budgeted expenses and the income from other sources is the amount that must be raised by the property tax. The next step is to determine the **decimal tax rate,** which is the tax rate at which property is to be taxed. To find the decimal tax rate, divide the amount to be raised by the property tax by the total assessed value of all property in the district. Districts may round decimal tax rates to 3, 4, or 5 places.

Critical Thinking Express the steps needed to find the decimal tax rate as a mathematical formula.

EXAMPLE The Peru School District's total budgeted expenses last year were \$6,000,000. Estimated income from other sources was \$1,800,000. The total assessed value of all taxable property in Peru last year was \$39,000,000. Find the tax rate needed to meet expenses.

SOLUTION

Total budgeted expenses	\$6,000,000
Income from other sources	− 1,800,000
Amount needed from property taxes	\$4,200,000

Divide amount needed by total assessed value of all taxable property: \$4,200,000 ÷ \$39,000,000 = 0.107692

Round to 5 places to find Peru's decimal tax rate. \$0.10769

Critical Thinking Could the tax rate for Peru be \$0.11? Explain.

EXERCISE YOUR SKILLS

For each problem, find the decimal tax rate correct to 5 places.

	Total Assessed Value	Amount to be Raised
18.	\$ 8,000,000	\$ 507,000
19.	6,000,000	483,200
20.	2,700,000	158,500
21	9,800,000	889,600
22.	1,750,000	100,500
23.	5,390,000	125,800
24.	15,320,000	1,279,000
25.	24,130,000	2,528,600

26. The West Allis Fire District must raise \$583,400 from property taxes. The assessed value of property in the district is \$24,780,000. What is the decimal tax rate needed, to 3 decimal places?

27. Property in the Boise School District has a total assessed value of \$89,500,000. The school's budget for next year shows expenses totaling \$4,000,000 to serve 3,500 students in 7 buildings. The district expects to earn \$2,900,000 from sources other than property tax. What decimal property tax rate, to 4 places, will the district use to raise enough money to meet the budgeted expenses?

Find the amount to be raised by property tax and the tax rate. Show the rate as a decimal, correct to 3 places.

	Assessed Value	Total Expenses	Other Income	Raised by Property Tax	Tax Rate
28.	\$36,000,000	\$878,000	\$97,500		
29.	22,750,000	382,700	68,400		
30.	7,900,000	396,300	45,600		

Central City plans to spend \$2,210,000 next year. Income from sources other than property taxes will be \$835,000. The taxable property in the city has a total assessed value of \$40,000,000.

31. Use mental math to estimate the amount of money the city must raise from property taxes. Then, find the exact amount.

32. Find the tax rate, correct to 4 decimal places.

Finding a Property Tax Using a Decimal Rate Multiply a property's assessed value by the decimal tax rate to find the tax due.

EXAMPLE The decimal tax rate in the Boise School District is 0.0123. Find the tax on a property in Boise, assessed at \$96,000.

SOLUTION 0.0123 × \$96,000 = \$1,180.80

Changing a Decimal Rate to a Rate Per \$1, \$100, or \$1,000 Multiply the decimal tax rate × (Base) = Rate in Other Terms.

For example:

0.04858 × \$100 = \$4.858 per \$100
0.04858 × \$1,000 = \$48.58 per \$1,000
0.04858 × 100 cents = 4.858 cents per \$1 (100 cents = \$1)
0.04858 × 1,000 mills = 48.58 mills per \$1 (1,000 mills = \$1)

EXERCISE YOUR SKILLS

33. Jose's property is assessed at \$82,000. The school tax rate in his district is 0.0145. What is Jose's school tax?

34. Cato Community College's tax rate is 0.04218. Find the tax on property in the college's district assessed at \$15,600.

Fox County's budget for a year is \$6,750,000. Of that, \$650,000 is from other income, and the rest from property taxes. The total assessed value of the county's property is \$80,000,000.

35. What amount is to be raised by property tax?

36. What is the decimal tax rate, rounded to 3 places?

37. What is the county tax on property assessed at $40,000?

38. Explain how these tax rates are the same: $40 per $1,000; $4 per $100; 4¢ per $1; and 40 mills per $1.

39. ◆ Property tax bills identify the many different agencies (taxing districts) that are supported by your taxes. Complete the property tax bill below. Find the assessed value, the dollar amount paid for each agency, and total tax due.

PARCEL ID NO. 09-19-23423		Taxing District	Rate	Amount of Tax
		County	0.0084	
Market Value	$160,000	Town	0.0025	
		Roads	0.0019	
Assessment Rate	25%	Schools	0.0355	
		Fire Dist	0.0035	
Assessed Value	$ _____	College	0.0052	
Tax Rate per $100	$5.70	TOTAL	0.0570	

MIXED REVIEW

1. 2,800 ÷ 175,000
2. What amount is 250% of $60?
3. $28,000 × 0.08076
4. $264 is what part less than $440?
5. The tax rate in a village is $82.19 per $1,000 of assessed value. Find the tax on a property assessed at $46,800.
6. Computer diskette caddies are reduced from $12.99 each to 2 for $20.00. How much would you save if you bought 12 diskette caddies at the sale price?
7. A lot measures 40 m by 50 m. The house, driveway, and parking area use 400 square meters. How many square meters of the lot is left for a lawn and gardens?
8. Mable MacIntyre earned total wages of $654 last week. At 7.65%, how much FICA tax was withheld from her gross pay?
9. A house with 2,800 square feet of living space is worth $225,000. It was assessed this year at 40% of its value. If the tax rate is 72.5 mills per $1, what is the tax bill?

8-2 SALES TAXES

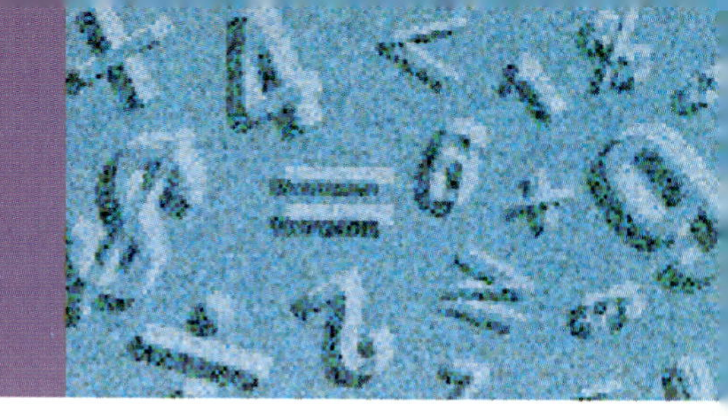

When you go to the store you may be charged sales tax only on certain items, such as a soft drink or a paperback book.

◆ Make a list of items that you know do and do not have sales tax added. Are some items taxed at a higher rate than others? Do the untaxed items have anything in common? Can you explain why some items are taxed and others are not?

OBJECTIVES

In this lesson, you will learn to

- *read and use a sales tax table; and*
- *find sales tax when not all items are taxable.*

WARM *UP*

a. $28\frac{1}{2} \div \frac{1}{4}$

b. $102.99 × 4.5%

c. 81.001 + 0.06 + 102.9 + 2.78

d. $200 − $25.62

e. Round: $12.08499 to the nearest cent.

SALES TAX TABLES

In chapter 6, you calculated sales tax by multiplying the subtotal by the sales tax rate. Sales tax tables make it easier and faster to look up sales tax. The tax table in Illustration 8-2.1 is used in a city where a 5% sales tax is charged.

Amount of Sale	Sales Tax
$0.01—$0.10	0
0.11—0.25	1¢
0.26—0.45	2¢
0.46—0.65	3¢
0.66—0.85	4¢
0.86—1.10	5¢
More than $1.10	5¢ on each dollar, plus amount in table for cents

Illustration 8-2.1. 5% Sales Tax Table

The table shows that the sales tax on a $0.60 sale is $0.03. To find tax on a sale of $8.60, follow the steps on the following page.

EXAMPLE Find the tax on a sale of \$8.60.

SOLUTION

$8 \times 0.05 = 40¢$	← $8 \times$ tax on \$1
Tax on \$0.60 = 3¢	← From table for \$0.46 to \$0.65
40¢ + 3¢ = 43¢	← Add both amounts to find total tax.

EXERCISE YOUR SKILLS

Using the table in Illustration 8-2.1, find the tax on each sale.

	Amount of Sale	Sales Tax		Amount of Sale	Sales Tax
1.	\$0.81		**2.**	\$3.47	
3.	\$0.11		**4.**	\$4.55	
5.	\$1.12		**6.**	\$3.01	

7. A video is priced at \$24.99. What is the sales tax?

8. An auto seat cover at \$99.95 and a floor mat at \$10.79 are purchased. What is the sales tax on the total sale?

A box of $3\frac{1}{2}$-inch diskettes sells for \$12.79. What is the

9. Sales tax?

10. Total price, including tax?

FINDING SALES TAXES WHEN SOME ITEMS ARE NOT TAXABLE

Sales taxes are imposed by local governments. Some cities and states do not tax certain items, such as food, prescription drugs, and services. To find the sales tax when some items are not taxable, first find the subtotal of all taxable items. Then calculate sales tax on that portion of the bill only. To find the total bill, add the subtotals of non-taxable and taxable items, and the sales tax.

For example, a service station mechanic took 4 hours to replace a car's fan belt and battery. The service charge was \$35 an hour. The belt cost \$12.98 and the battery, \$48.99. The state has a sales tax of 5% that is charged on goods, but not on services. Find the total bill following these steps.

• Find the subtotal of non-taxable items	$35 × 4 = $140
• Find the subtotal of taxable items	$12.98 + $48.99 = $61.97
• Find the tax on taxable items	$61.97 × .05 = $3.098 or $3.10
• Add the taxable and non-taxable subtotals and the tax to find the total bill.	$140 + $61.97 + $3.10 = $205.07

EXERCISE YOUR SKILLS

11. A carpenter built a bookcase for a customer. She estimated her time to be 3 hours @ $30, and the cost of materials to be $45. The state charges a 6% sales tax on goods but does not tax services. Find the total cost of the bookcase.

12. Sandra Brown bought these items at a grocery store: 3 cans of soup @ 3 for $1, 2 bunches of celery @ $0.87, 5 lbs. of potatoes for $2.88, 1 box of dishwasher soap, $2.79, and 2 potholders @ $1.29 each. The state has a 3% sales tax but does not tax food. What was Sandra's total bill?

Find the sales tax and total bill for each sale below. The state has a sales tax of 5% but does not tax all items sold.

	Total of Taxable Items	Total of Nontaxable Items	Sales Tax	Total Bill
13.	$145.70	$103.00		
14.	39.39	128.89		
15.	8.34	201.12		
16.	378.50	100.08		
17.	44.67	34.87		

PROBLEM SOLVING

Use a Table

Frequently, you may find it necessary to read a table in order to solve a problem. Some tables may provide too much information, or information that you do not need to solve the problem. When using a table, be sure to identify what information you need to find before reading the table. Then find the correct row and column to locate the needed information.

Look at the sales tax table shown in this lesson. How much more tax is there on $1.09 than on $0.47? To solve the problem, find the tax on $1.09 (5¢) and on $0.47 (3¢) and subtract (2¢).

18. Find another table in this text. Explain how you use that table to solve problems.
19. Using your local sales tax, write a word problem that requires calculating sales tax.
20. ♦ Write what you consider to be the advantages and disadvantages of sales tax. Consider the question from the perspective of individuals, businesses, and governments. Debate with a partner the advantages and disadvantages of sales tax. One partner should take a pro-tax position, the other an anti-tax position.

MIXED REVIEW

1. $450 × 0.2
2. $230 ÷ 0.01
3. $75.45 − $0.89
4. ? is $\frac{1}{2}$% of $380?
5. $150 is $\frac{2}{3}$ smaller than?
6. A bill for $18.56 for a television antenna was subject to sales tax of 6%. Find the total bill.
7. In May, Don started with 56 gallons of fuel oil. He bought 100 gallons @ $0.92 during May. At the end of May, he had 65 gallons left. How much fuel oil had he used during May?
8. An agent sold a house for $96,200 and charged a 6% commission. Find the agents commission for selling the house.

Jerry's school tax for a year was $1,419.12. His property was assessed at $21,600. What was the tax rate shown in

9. Dollars per $1,000?
10. Dollars per $100?

8-3 SOCIAL SECURITY TAXES AND BENEFITS

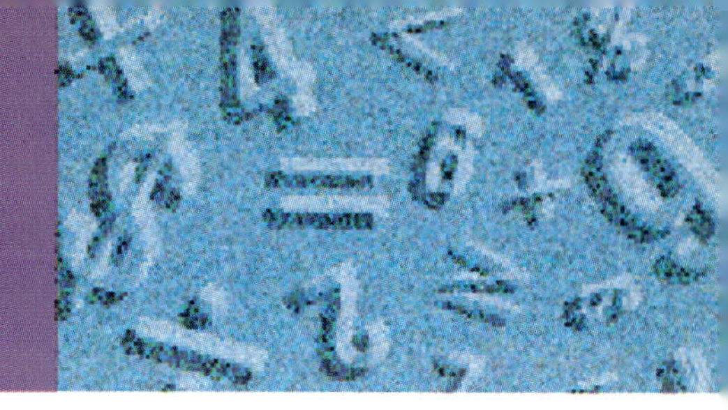

These programs are supported by the Federal Insurance Contributions Act (FICA tax): **Disability benefits** for workers who are disabled and unable to work; **Medicare,** which provides hospital insurance for some disabled people and for people over 65; and **retirement,** and **survivors' benefits.** Survivors' benefits are paid to spouses and dependent children when a retired or currently ensured worker dies.

◆ Write a paragraph explaining your understanding of the purpose of the Social Security system. Make a list of questions you have about the Social Security system.

OBJECTIVES

In this lesson, you will learn to

- *calculate FICA taxes on wages; and*
- *find estimated retirement benefits.*

WARM UP

a. 25 mL × 600 = ? L

b. $40 \div \frac{2}{5}$

c. $240 is what percent larger than $80?

d. Show a tax rate of $15 per $1,000 as a rate per $100.

e. Rounded to the nearest cent, $1,492 × 0.0054 = ?

SOCIAL SECURITY TAXES

Employees, employers, and self-employed persons are taxed under the **Federal Insurance Contributions Act.** This tax is called the ***FICA tax*** or **Social Security tax.**

Calculating FICA Taxes

The FICA tax rates and the maximum wages on which the taxes are charged are set by Congress and may change from time to time. When calculating taxes in this text, the overall tax rate of 7.65% is used. The Social Security tax rate used is 6.2% applied to a maximum wage of $70,000. The Medicare tax rate used, is 1.45%, applied to all wages.

When someone earns more than $70,000 a year from one job, Social Security taxes are collected each pay period at the 6.2% rate. When the maximum amount of Social Security tax for a year has been collected, no more Social Security tax should be deducted by the employer. If someone earns more than $70,000 a year from several jobs, each employer withholds 6.2% Social Security tax to the maximum $70,000 earning limit. The tax payer files for a return on Social Security tax overpayments when he or she files their federal income tax return.

For example:

Income	Social Security	Medicare	Total FICA
$82,000	$70,000 × 6.2% = $4,340	$82,000 × 1.45% = $1,189	$5,529.00
$70,000	$70,000 × 6.2% = $4,340	$70,000 × 1.45% = $1,015	$5,355.00
$35,000	$35,000 × 6.2% = $2,170	$35,000 × 1.45% = $507.50	$2,677.50

FICA Tax on Employees FICA taxes owed by workers are collected by their employers. Employers must deduct a certain percent of each employee's earnings for FICA tax. For example, in June the employees at Faber's Designs earned $3,000 in taxable wages. From these wages, FICA taxes of 7.65% were deducted. The owner of Faber's Designs deducted $229.50 (0.0765 × $3,000 = $229.50) from the total wages and paid it to the government for its employees.

FICA Tax on Employers Employers are also taxed on wages paid to employees. The rate is the same as the rate paid by employees. For example, the owner of Faber's Designs also had to pay $229.50 on the June wages. So, the total of the FICA taxes that Faber's Designs sent to the government for June was 2 × $229.50, or $459.

FICA Tax on Self-Employed Persons People who work for themselves must also pay a FICA tax on yearly net earnings. The FICA tax rate for the self employed is twice the rate paid by employees (7.65% × 2 = 15.3%). For example, Eli Faber, the owner of Faber's Designs, found that her net earnings from her company for the year were $35,000. She paid a **self-employment tax** of 15.3% on the earnings. The amount of the tax was $5,355 (0.153 × $35,000 = $5,355).

EXERCISE YOUR SKILLS

Find the FICA tax on each of the weekly wages. Each wage is fully taxable. Use a rate of 7.65%. Round the tax to the nearest cent.

1. Steven Dietz, $206.00
2. Wilma Drake, $399.70
3. Jose Mendoza, $516.20
4. Marge O'Brien, $422.50
5. Ying Pai, $380.80
6. Irving Rosen, $295.07

7. During June, Rayco Toy Co. paid a total of $38,746 in taxable wages to employees. At a rate of 7.65%, what amount of FICA tax did Rayco have to pay as an employer?
8. Sylvia is a self-employed accountant. Last year she paid 15.3% FICA tax on net income of $42,370. What was her self-employment tax?

For Exercises 9–19, assume FICA taxes include Social Security, which is taxed at a rate of 6.2% on maximum earnings of $70,000, and Medicare, which is taxed at a rate of 1.45% on all earnings.

Jim Moore is paid an annual salary of $88,800.

9. What amount of Jim's salary is subject to Medicare taxes?
10. What amount is subject to Social Security taxes?
11. Estimate Jim's Social Security tax.
12. Find Jim's Social Security tax.
13. What Medicare tax did he pay?
14. Find the total FICA taxes withheld annually from Jim's paycheck.

Eva Lavata earns an annual salary of $81,600, payable monthly. How much *Social Security tax* will be deducted from her salary

15. in January?
16. in November?
17. in December?

Find the total FICA taxes that will be deducted from Eva's pay

18. in January.
19. in December.

CALCULATING SOCIAL SECURITY BENEFITS

Social Security benefits for workers include retirement benefits, disability benefits, and **Medicare.** Medicare pays specified amounts for hospital and medical bills for qualified retired workers. Spouses and children of retired or disabled workers may also qualify for Social Security benefits.

Finding a Worker's Retirement Benefit Social Security retirement benefits are paid monthly. The amount of the monthly benefit depends on several factors, including age at which you retire, eligibility, number of years worked, the amount of salary you earned that was subject to Social Security tax, and the benefit plan in effect when you retire.

To determine exact benefits, an individual must contact the local office of the Social Security Administration. However, you can get an idea of how the benefit system works from the partial benefits table shown in Illustration 8-3.1. The table shows approximate monthly benefits for individuals or couples retiring in 1995 at age 65. The table assumes that both spouses are the same age and that payments are based on one spouse's benefits.

Average Annual Earnings	Approximate Monthly Benefits for You at 65	Approximate Monthly Benefits for You and Spouse at 65[1]
20,000	761	1,141
30,000	1,014	1,521
40,000	1,111	1,666
50,000	1,166	1,749
60,600 or more	1,199	1,798

[1]Assumes both spouses are the same age and benefits are based on one spouse's earnings record.

Illustration 8-3.1. Partial Table of Social Security Retirement Benefits

Average annual earnings are the amounts of yearly earnings covered by FICA. To find that amount, you must know how many years someone has worked and how much of that person's earnings were subject to FICA tax. (The Social Security Administration provides a free earnings history and benefit estimate to individuals who send in SSA Form 7004, Request for Personal Earnings and Benefit Statement.)

Impact of Retirement Age on Benefits Age 65 is now considered the full retirement age. That is, 65 is the age at which you receive no reductions in your retirement benefits. For those born in 1960 or later, full retirement age will be 67.

The earliest retirement age allowable for Social Security is 62. People retiring early receive reduced benefits. The Social Security Administration suggests that individuals retiring at 64 should estimate their benefits at 93.3% of the amounts shown in Table 8-3.1; individuals retiring at 63, should estimate benefits to be 14% less, and individuals retiring at 62, should estimate benefits at 20% less.

For example, if Paulo Generali's average covered annual earnings were $20,000 and he retired at age 65, his monthly benefit

would be $761. If he retired at 62, his monthly benefit would be 20% less than $761, or $608.80. The Social Security Administration always rounds down to the nearest whole dollar when calculating benefits. Therefore, Mr. Generali's monthly benefit would actually be $608.00.

People who delay their retirement past 65, receive increased monthly benefits. The increase depends on the year of birth. People born after 1943, will receive 8% more per year for each year retirement is delayed until age 70.

EXERCISE YOUR SKILLS

Use Illustration 8-3.1 to solve Exercises 20–32.

Ira Schaumberg estimates that his approximate average annual covered income will be $30,000 when he retires. Find his approximate monthly benefits, rounded down to the nearest whole dollar, if he retires at

20. 65. **21.** 62. **22.** 64.

Ramona and Mario Mendoza estimate that Mario's average annual covered income will be $40,000 when they retire. Assuming Ramona and Mario are the same age, what will be their approximate monthly benefits, rounded down to the nearest whole dollar, if they retire at

23. 65? **24.** 64? **25.** 66?

Craig Keller's average covered annual earnings are $50,000. What will be his monthly benefit, rounded down to the nearest whole dollar, if he

26. retires now at age 62? **27.** waits until age 65 to retire?

28. How much more per year will he receive if he waits until age 65 to retire?

Vera and Tom Esposito are both 63 years old. They estimate that their average annual covered earnings are $75,000. What will their monthly benefit be, rounded down to the nearest whole dollar, if they

29. retire now at age 63? **30.** wait until age 65 to retire?

31. How much more per month will they receive if they wait?

32. Jill Bonner worked for 5 employers over a 40-year period. She estimates that her monthly retirement benefits will be $1,500 when she retires at full retirement age. She can receive 8% more in monthly benefits for each year she delays retirement. If she waits two more years to retire, what will her approximate monthly benefit be?

33. ♦ Contact your local Social Security office and ask for the form to fill out to find out about your Social Security benefits. If you have had eligible earnings, send in this request. Also, find out why it is important to check on your earning records periodically during your lifetime.

MIXED REVIEW

1. Find $\frac{1}{4}\%$ of \$3,800.
2. \$700 is what part greater than \$420?
3. Show a tax rate of 67 mills on \$1 as a rate on \$1,000.
4. Divide 546 by 23.6, to the nearest tenth.
5. 15.3% of \$23,895 is what amount?
6. If Victor retires at 65, his monthly Social Security benefits will be \$633. How much would his monthly benefits be if he retired at 62 and his average monthly covered earnings were the same?

Veronica was paid \$82,000 for 1 year. Her employer deducted Social Security taxes at 6.2% on \$70,000, and 1.45% on all of her wages.

7. How much Social Security taxes did her employer pay for Veronica?
8. How much Medicare taxes did her employer deduct?
9. Hortencia is self-employed. Last year she earned \$56,000. She paid self-employment tax of 15.3% on a maximum of \$70,000. How much tax did she pay?
10. A town's total budget for last year was \$4,104,670. Income from other sources was \$376,800. The balance was raised by a property tax on a total assessed value of \$93,600,000. What was the decimal tax rate, correct to 3 places?
11. A car that sells for \$18,500 today is estimated to have a resale value of \$8,600 in 4 years. What is the average annual depreciation?
12. Sue bought a used car for \$10,200, cash. She drove the car 11,500 miles in one year. Her car expenses for the year were: insurance, \$512; gas and oil, \$720; repairs, \$145; license plates, \$54; loss of interest on original investment, \$550.80; depreciation, 16%. Find her year's operating cost per mile, to the nearest cent.
13. The tax rate in a city is \$121.23 per \$1,000 of assessed value. Find the tax to be paid on a property assessed for \$30,000.

8-4 FEDERAL INCOME TAX

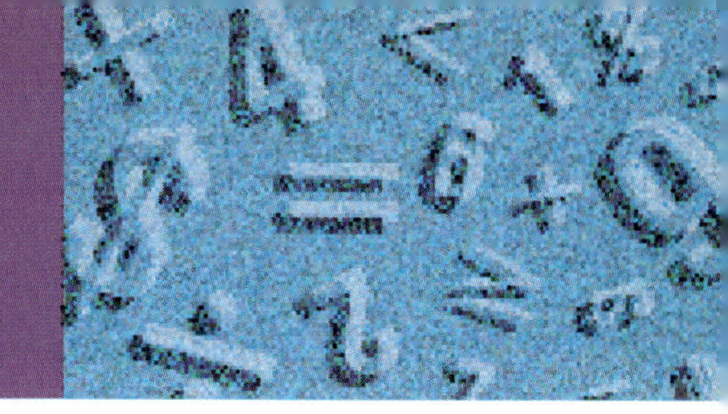

The incomes of United States citizens and others are taxed by the federal government. The tax is called the United States individual income tax, or **federal income tax.**

◆ Federal income tax is a progressive tax. That is, as people earn more money, they pay higher rates of taxes. Do you think that a progressive tax is fair? Explain your answer.

OBJECTIVES

In this lesson, you will learn to calculate

- *adjusted gross income;*
- *taxable income;*
- *refunds or amounts owed; and*
- *income tax refunds for dependents.*

WARM *UP*

a. \$120 is what percent smaller than \$150?

b. Show a tax rate of \$0.025 per \$1 as a rate per \$100.

c. $106 \text{ cm} \times 5 = ? \text{ m}$ **d.** $56 \times \frac{3}{8} = ?$ **e.** $92 \div 0.05 = ?$

CALCULATING FEDERAL INCOME TAXES

If you are an employee, your employer deducts money for the federal income tax from your pay. This is called a *withholding tax.* If you are self-employed, you must make **estimated tax** payments each quarter. In either case, the amounts withheld or paid are estimates of the tax you will owe when the year ends. You briefly studied federal income tax withholding in Chapter 4 when you found the amount withheld from a paycheck is based on earnings and withholding allowances claimed.

The tax year for individuals ends on December 31. You must calculate and pay any federal income tax due by April 15 of the next calendar year. Income earned and taxes due are reported and calculated on a **federal income tax return.** A completed return shows how much you actually owe in federal income taxes. If the amount withheld from wages during the year was larger than what was owed, you should claim a refund. If the withholding taxes paid were less than what you owe, you must pay the difference.

Calculating Taxable Income

The total of all your income in a year is called your **gross income. Taxable income** is the income that you actually pay tax on after making adjustments and taking all the **deductions** and **exemptions** you can claim.

Gross income includes income from all sources including wages, salaries, commissions, bonuses, tips, interest, prizes, pensions, and profit from a business or profession. Gifts and inheritances usually are not considered income and are not usually taxed by the federal government.

From your gross income, you may be eligible to subtract certain amounts, called **adjustments to income.** These include business losses, payments to approved retirement plans, alimony, and penalties on early withdrawal of savings. The amount left after subtracting adjustments to income from gross income is **adjusted gross income.** From adjusted gross income you subtract the deductions and exemptions you qualify for. The result is your taxable income.

Deductions are expenses that reduce the amount of your taxable income. For example, income tax rules allow you to deduct interest paid on a home mortgage for a primary residence, property taxes, state and local income taxes, part of medical and dental expenses, some casualty and theft losses, and contributions to charities. You may claim a fixed amount as a deduction, called a **standard deduction.** Or, if your actual deductions are more than the standard deduction, you may list, or *itemize* your deductions on your return.

An **exemption** is an amount of income per person that is free from tax. You may claim one exemption for yourself *unless* you are claimed as a dependent on another person's return. You can also claim one exemption for a spouse and one exemption for each dependent. For example, a couple with two dependent children can claim four exemptions. A single person with a dependent parent can claim two exemptions. A child listed as an exemption on their parent's return can not claim an exemption.

The amounts allowed for the standard deduction and exemptions change often. In this text, the amount used for the standard deduction is $4,000 and for an exemption is $2,500.

Critical Thinking Write a formula you can use to find taxable income.

EXAMPLES Nora Boone is single and has an adjusted gross income of $30,000. Nora has deductions of $6,400. She has two exemptions. What is her taxable income?

Alan Roche also has an adjusted gross income of $30,000. Alan takes the standard deduction of $4,000. He has two exemptions. What is his taxable income?

SOLUTIONS

	Nora Boone	Alan Roche
Adjusted Gross Income	$30,000	$30,000
Deductions	− 6,400	− 4,000
	$23,600	$26,000
Exemptions (2 × $2,500)	− 5,000	− 5,000
Taxable Income	$18,600	$21,000

EXERCISE YOUR SKILLS

1. Tina Barr's tax return last year showed gross income of $18,648 and adjustments to income of $1,380. What was Tina's adjusted gross income last year?
2. Ed and Ginny Threll's income last year was from these items: net income from business, $45,635.90; dividends, $1,926.12; interest, $4,101.73; rental income, $2,180. Adjustments to income totaled $5,108.33. Find their adjusted gross income.
3. In one year, Felix Gomez's wages totaled $27,850. His wife, Marta Gomez had a salary of $26,378 and a bonus of $750. The Gomezes also received $1,610 in interest, and $787.35 in dividends. They paid $4,330 into a retirement fund and were penalized $78 for removing money from a savings plan early. What was their adjusted gross income that year?
4. Jan Teel's adjusted gross income last year was $15,354. Jan's itemized deductions were $5,650. She claimed one exemption of $2,500. What was her estimated and actual taxable income?
5. On Jamal Jackson's return, the adjusted gross income was $32,007. His itemized deductions totaled $3,700, and he claimed 3 exemptions at $2,500 each. Find his taxable income.

In preparing their tax return, the Delgados claimed 4 exemptions at $2,500 each, and itemized deductions of $7,107. Their adjusted gross income was $45,208.

6. Estimate their taxable income.
7. What was their exact taxable income?

Calculating Tax Due Using a Tax Table

If taxable income is less than $100,000, a tax table can be used to find the tax. Parts of a tax table are shown in Illustration 8-4.1.

To use a tax table, find your taxable income in the "At least . . . but less than" columns. Then read across that line to the column that shows your filing status. The amount on that line and in that column is your tax. For example, if your taxable income is $22,950 and you are single, your tax is $3,476. If you are married and filing a joint return, your tax for the same taxable income is $3,446.

A "head of household" is an unmarried or legally separated person who pays more than half the cost of keeping a home for a dependent father, mother, or child.

Calculating the Refund or Amount Owed If you worked during the year, either your employer withheld money for income taxes or you paid self-employment tax. In either case, the amount of tax you paid was an estimate and is probably more or less than the tax you actually owe.

To find the tax due, you must complete a tax return. If too much withholding or self-employment tax has been paid, the government will pay back (refund) the difference. If too little tax has been paid, you must pay the difference to the government. For example, if your tax due was $3,304 and you paid $3,300 you would owe $4 ($3,304 − 3,300). If you had paid $3,546, you would be due a refund of $242 ($3,546 − 3,304).

If your taxable income is —		And your filing status is —			
At least	But less than	Single	Married filing jointly	Married filing separately	Head of a household
		Your tax is —			
0	5	0	0	0	0
5	15	2	2	2	2
15	25	3	3	3	3
25	50	6	6	6	6
50	75	9	9	9	9
75	100	13	13	13	13
100	125	17	17	17	17
125	150	21	21	21	21
150	175	24	24	24	24
175	200	28	28	28	28
200	225	32	32	32	32
225	250	36	36	36	36
250	275	39	39	39	39
275	300	43	43	43	43
300	325	47	47	47	47
325	350	51	51	51	51
350	375	54	54	54	54
375	400	58	58	58	58
400	425	62	62	62	62
425	450	66	66	66	66
450	475	69	69	69	69
475	500	73	73	73	73
500	525	77	77	77	77
525	550	81	81	81	81
550	575	84	84	84	84
575	600	88	88	88	88

If your taxable income is —		And your filing status is —			
At least	But less than	Single	Married filing jointly	Married filing separately	Head of a household
		Your tax is —			
13,000	13,050	1,954	1,954	1,954	1,954
13,050	13,100	1,961	1,961	1,961	1,961
13,100	13,150	1,969	1,969	1,969	1,969
13,150	13,200	1,976	1,976	1,976	1,976
13,200	13,250	1,984	1,984	1,984	1,984
13,250	13,300	1,991	1,991	1,991	1,991
13,300	13,350	1,999	1,999	1,999	1,999
13,350	13,400	2,006	2,006	2,006	2,006
13,400	13,450	2,014	2,014	2,014	2,014
13,450	13,500	2,021	2,021	2,021	2,021
13,500	13,550	2,029	2,029	2,029	2,029
13,550	13,600	2,036	2,036	2,036	2,036
13,600	13,650	2,004	2,004	2,004	2,004
13,650	13,700	2,051	2,051	2,051	2,051
13,700	13,750	2,059	2,059	2,059	2,059
13,750	13,800	2,066	2,066	2,066	2,066
13,800	13,850	2,074	2,074	2,074	2,074
13,850	13,900	2,081	2,081	2,081	2,081
13,900	13,950	2,089	2,089	2,089	2,089
13,950	14,000	2,096	2,096	2,096	2,096

If your taxable income is —		And your filing status is —			
At least	But less than	Single	Married filing jointly	Married filing separately	Head of a household
		Your tax is —			
22,000	22,050	3,304	3,304	3,697	3,304
22,050	22,100	3,311	3,311	3,711	3,311
22,100	22,150	3,319	3,319	3,725	3,319
22,150	22,200	3,326	3,326	3,739	3,326
22,200	22,250	3,334	3,334	3,753	3,334
22,250	22,300	3,341	3,341	3,767	3,341
22,300	22,350	3,349	3,349	3,781	3,349
22,350	22,400	3,356	3,356	3,795	3,356
22,400	22,450	3,364	3,364	3,809	3,364
22,450	22,500	3,371	3,371	3,823	3,371
22,500	22,550	3,379	3,379	3,837	3,379
22,550	22,600	3,386	3,386	3,851	3,386
22,600	22,650	3,394	3,394	3,865	3,394
22,650	22,700	3,401	3,401	3,879	3,401
22,700	22,750	3,409	3,409	3,893	3,409
22,750	22,800	3,420	3,416	3,907	3,416
22,800	22,850	3,434	3,424	3,921	3,434
22,850	22,900	3,448	3,431	3,935	3,431
22,900	22,950	3,462	3,439	3,949	3,439
22,950	23,000	3,476	3,446	3,963	3,446

Illustration 8-4.1. Partial Tax Table

EXERCISE YOUR SKILLS

Use the table in Illustration 8-4.1 to find the tax.

8. Carmen and Angel Diaz are married and file a joint tax return. Their taxable income is $22,378. What is their tax?

9. Roberta O'Toole is 25 years old and is single. Her gross income last year was $21,455 and her taxable income was $13,926. Find her tax.

10. How much tax does a head of household with taxable income of $22,798 owe?

11. Rob Duval is married but is filing a separate tax return. His taxable income is $22,624. Find the amount of his tax.

12. Carl Breese's tax return for last year shows a total tax of $8,278. Carl's employer had withheld $8,450 from his wages during the year. What refund should Carl receive?

13. Sara Wickam is a self-employed farmer. Last year she paid $16,118 in self-employment tax. Sara's total tax for the year, as shown on her tax return, was $17,448. What amount of tax did she owe?

14. On their joint tax return, the Pavliks had taxable income of $22,298. Their employers withheld $3,205.46. How much more will the Pavliks owe with their tax return?

Paul Dimitry is single. On last year's tax return, his taxable income was $13,832. His employer had withheld $2,040.75.

15. What was his actual tax for the year?

16. Does he owe money or have a refund due him? How much?

On his federal income tax return, Cesar Castillo, a single taxpayer, reported these items of income: wages, $23,820; tips, $5,495; and interest earned, $1,466. Cesar had these adjustments to income: payments to a retirement plan, $2,000; penalty for early withdrawal of savings, $19. His employer withheld $3,985 from his pay for federal income tax. Cesar claims the standard deduction of $4,000 and an exemption of $2,500.

17. What was Cesar's gross income?

18. What was the total of his adjustments to income?

19. What was Cesar's adjusted gross income?

20. What was his taxable income?

21. What is the amount of tax due from Cesar?

22. Does he owe money or have a refund coming?

23. How much?

FEDERAL INCOME TAXES FOR DEPENDENTS

Many people listed as dependents on someone else's income tax are employed. They are required to pay income taxes on their earnings, even though the tax they actually owe is usually very low. To claim a refund on taxes paid, you must file an income tax return. The rules for dependents filing returns are different than for people who are not dependents.

Finding the Standard Deduction for a Dependent A dependent's income is grouped into two categories: **earned income** and **unearned income.** Earned income is from the dependent's own labor, such as wages, salaries, and tips. Everything else is unearned income, including interest and dividends.

A dependent can claim as a standard deduction the higher of these two amounts: $600, or the amount of earned income up to $4,000, the standard deduction. For example, Jane Vistal is a senior at Allorton High School. Jane worked five weeks last summer and earned $1,785. Jane also earned $22 in interest on her savings account. Because she had no adjustments to income, the total ($1,785 + $22 = $1,807) was her adjusted gross income.

Jane wants to claim the standard deduction instead of itemizing her deductions. However, since Jane's parents claimed her as a dependent on their tax return, the maximum standard deduction she can claim is limited.

Jane's earned income of $1,785 in wages was more than $600 and not more than $4,000. So, $1,785 is the maximum deduction she can claim.

Finding the Tax Refund Since Jane's parents claim her as an exemption, she could not claim a $2,500 exemption for herself. Her taxable income was found this way:

Wages	**$1,785**
Interest	**+ 22**
Adjusted Gross Income	**$1,807**
Less deductions	**– 1,785**
Taxable Income	**$ 22**

To find her tax, Jane used the table in Illustration 8-4.1. Her tax was only $3, but her employer withheld $215 from her wages for federal taxes. So, she was entitled to a refund of $212.

EXERCISE YOUR SKILLS

Find each dependent's standard deduction.

	Earned Income	Unearned Income	Standard Deduction		Earned Income	Unearned Income	Standard Deduction
24.	$250	$100		**25.**	$1,950	$600	
26.	$4,875	$300		**27.**	$5,175	$0	

Use tax tables found in this lesson to solve these problems. Assume that each person claims the standard deduction and that each person was listed on their parents' return as a dependent.

Todd earned $2,406 in wages as a gardener and $148 in interest. His employer withheld $180 in income taxes from his wages.

28. What is Todd's taxable income?

29. How much refund will Todd receive?

Mike Duvalier worked during the summer to earn $1,600 to help pay college expenses. From these earnings, $170 in withholding taxes were deducted. Mike also earned $330 for yard work. No withholding taxes were deducted from these earnings. Find his

30. Taxable income.

31. Tax refund.

Sylvia Wells earned $4,256 last year in part-time wages and $174 in interest. Her employer withheld $450 for income taxes.

32. What is Sylvia's taxable income?

33. How much money will she receive as a refund?

34. Last summer, Otis Wilson earned $2,855. His employer withheld $240 of his wages for income taxes. What is the amount of Otis' tax refund?

35. Luisa Guzman worked part time last year while attending college and earned $4,288. The total withholding taxes she paid were $450. Luisa also earned $78 in interest and $16 in dividends. How much tax refund should she receive?

36. Sue Bent's parents subtracted $2,500 from their adjusted gross income when they listed her as an exemption on their tax return. Sue's taxable income was $148. She paid $277 in withholding taxes. What amount should she expect as a tax refund?

37. ◆ Write a paragraph explaining why it is important for students to file a tax return. Also explain why it is not a good idea to have a large refund at the end of a year.

MIXED REVIEW

1. Find 12% of $250.
2. Find $\frac{1}{8}$% of $4,800.
3. 360 m is what percent less than 480 m?
4. 625 grams is what percent of 5 kilograms?
5. What amount is 250% greater than $50?
6. Ruby Johnson earned $3,312 for college expenses last summer. Her employer deducted $456 in withholding taxes. Ruby also earned $38 in interest from her checking account. Her father listed her as a dependent on his tax return. Use the tax tables found in this lesson to find the tax refund she should receive.
7. The Chou's adjusted gross income last year was $32,908. Their itemized deductions were $4,808. They claimed 3 exemptions of $2,500 each. What was their taxable income?
8. Janice Ortiz sells computer software. The average price of the software is $28. Her average commission is $3.50. What is her average rate of commission?
9. $35 is what part greater than $28?
10. 432×0.001
11. What percent of $90 is $270?
12. $9.053 \div 0.001$
13. Vespa County's tax rate is $38.20 per $1,000 of assessed value. What tax must you pay on a property valued at $64,000, which is assessed at 50% of its value?

Wendy bought these gifts for her spouse: a shirt for $19.79, socks for $4.95, and a belt for $28.99. There was a sales tax of 5% on the purchase.

14. Estimate the total bill.
15. What was the exact bill?
16. Maude James is single with a taxable income of $22,315. Her employer withheld $3,796 from her wages for income tax during the year. Using the tax tables in this lesson, find how much her refund should be.

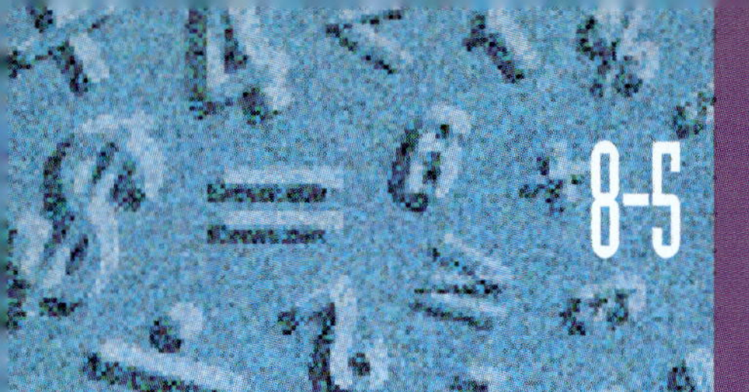

8-5 STATE AND CITY INCOME TAXES

OBJECTIVES

In this lesson, you will learn to

- *calculate state and city income taxes using a graduated tax rate table.*

Most states and some cities tax the incomes of their residents.

◆ What are the major uses of city, village, or town taxes in your area? Make a list of the services that your city, village, or town provides with its tax money. Make a similar list of services provided by your state. Rate the importance of each of the uses to you. Place a "1" next to the tax use that is most important to you, a "2" next to the next more important use, and so on.

WARM UP

a. $12\frac{1}{2} \div 16\frac{1}{4}$

b. 16 m = ? mm

c. Find $\frac{1}{3}$% of 426.

d. Change 125.6% to a decimal.

e. Round 42,489 to the nearest hundred.

STATE AND CITY INCOME TAX

States and cities usually tax citizens at lower rates than the federal income tax. Some states and cities calculate income tax as a percent of federal taxable income. Some use a fixed tax rate no matter how much taxable income a person has. Others use a graduated tax rate like the federal government. That is, the tax rate gets higher as taxable income gets larger. A portion of a graduated tax rate schedule that might be used by a state is shown in Illustration 8-5.1.

For taxable income Over —	But not over —	The tax is —
$ -0-	$5,000	2% of taxable income
5,000	10,000	$100 plus 3% of taxable income over $5,000
10,000	15,000	250 plus 4% of taxable income over 10,000
15,000	20,000	550 plus 5% of taxable income over 15,000
20,000	25,000	1,200 plus 6% of taxable income over 20,000
25,000	30,000	1,550 plus 7% of taxable income over 25,000
30,000	35,000	1,950 plus 8% of taxable income over 30,000
35,000	40,000	2,400 plus 9% of taxable income over 35,000
40,000	45,000	2,900 plus 10% of taxable income over 40,000

Illustration 8-5.1. Partial Graduated Income Tax Table

EXAMPLE Your taxable income last year was $21,600. What was your state income tax?

SOLUTION Using Illustration 8-5.1, find your tax this way:
Look at the tax column for $21,600. (over $20,000; under $25,000)

$1,200	Tax on taxable income up to $20,000
$21,600 − $20,000 = $1,600	Taxable income over $20,000
0.06	Tax rate on income over $20,000
$1,600 × 0.06 = $96	Tax on income over $20,000
$1,200 + $96 = $1,296	Total state income tax

EXERCISE YOUR SKILLS

Use the tax rates in Illustration 8-5.1 to solve problems 1–16.

1. Olive Budde's taxable income last year was $15,798. What was her state income tax?
2. Ana Orestes' state income tax return shows taxable income of $35,470. What is the state income tax on that amount?
3. The Kimura's income on which they must pay state tax is $41,548. What is their state tax?
4. Wynne Odonjo's income subject to state income tax is $12,892. What is his state income tax?

Mr. and Mrs. Simon's gross income for a year was $57,722. From that amount they subtracted $13,230 in adjustments, deductions, and exemptions to find their taxable income.

5. What was their taxable income?
6. What was their state tax?

Amy Rosco is a florist's assistant. Last year, her employer withheld $420 from her wages for state income tax. When she prepared her tax return, Amy showed gross income of $12,400 less $3,000 in deductions and $2,500 in exemptions.

7. What was Amy's taxable income?
8. What was her correct state tax for the year?
9. How much refund was due her?
10. In addition to federal and state income taxes, Rob Traub also has to pay a city income tax. The city income tax rate is $2\frac{1}{2}\%$ of his taxable income. If his taxable income is $18,345, what is his city income tax?

Gil Hark has taxable income of $16,900. He pays a city income tax of 1% on taxable income in addition to state and federal taxes.

11. What is Gil's city tax?
12. What is his state tax?
13. What is Gil's total city and state tax?
14. Carsville charges its residents an income tax of $\frac{1}{2}\%$ of their taxable income. Diana Schmitt lives in Carsville and has taxable income of $23,616. What is her income tax?
15. Francis Lange pays city tax of 2% on taxable income, in addition to state income tax. Her taxable income last year was $22,896. What was her total state and city income tax?
16. Harry Rogers pays a city income tax of $2\frac{1}{4}\%$ on his taxable income of $42,834. In addition, he pays both state and federal income taxes on the same taxable income. If his federal tax last year was $8,231, what was the total of his federal, state, and city income taxes last year?
17. ◆ Look back at your list of local services from the beginning of the lesson. Contact persons in local government to find the source of the tax revenue. Does your city or state have an income tax? If so, use that information to write a word problem.

PUT IT ALL TOGETHER

John and Lisa Rodriguez own a home with an assessed value of $74,000. Their property tax rate is $3.22 per $1,000. They live in a state that has a 6.5% sales tax on all items except food and services.

Their adjusted gross income is $42,000 a year. They have two young sons. This year they had deductions from income of $9,867. They live in a state that has an income tax that is 1.4% of their taxable income.

Prepare a chart or other visual representation showing what percent of their gross income goes to the taxes studied in this chapter. What is the total percent of their income that goes to property and income taxes?

What portion of their income do you think is spent on sales tax? Explain how you estimated this amount. (Hint: Consider what you learned about budgeting in Lesson 6-1.)

MIXED REVIEW

1. 15.3×2.4
2. $32.4 \div 2.4$
3. $4\frac{2}{5} + 6\frac{1}{2}$
4. $17\frac{1}{4} - 6\frac{1}{3}$
5. $8\frac{1}{2} \times 5\frac{1}{3}$
6. $5\frac{1}{3} \div 1\frac{1}{3}$
7. The Carr's taxable income last year was $43,780. They paid a state tax of 3.5% and a city tax of 1.25% on that income. What was the total of the state and city taxes they paid?
8. Ty Wilson works on a piece-rate basis. He completed 140 pieces on Monday, 136 on Tuesday, 148 on Wednesday, and 144 on Thursday. He is paid $0.60 for each piece. How many pieces must he complete on Friday so that his earnings for the 5 days will average $84 a day?
9. Halor Corporation paid 3 workers gross wages of $256.80, $287.77, and $297.89 for working 40 hours each during the week of January 26. Find the total amount of FICA taxes withheld from their gross wages. Use a FICA rate of 7.65%.
10. Vicente is paid 5% commission on all sales up to and including $15,000, and 7% on all sales over $15,000 in any month. Last month Vicente's sales were $25,750. What was the amount of his total commission?
11. A mobile home costs $35,800. The home is estimated to depreciate 6% each year. How much will the home depreciate in 5 years?
12. A truck that sold for $18,500 four years ago is traded in for $7,200 today. What was the average annual depreciation?
13. Last year the Pikanos earned $42,800 from their salaries. They also earned $1,280 in interest and $789 in dividends. The Pikanos paid $6,240 into a retirement fund. What was their adjusted gross income for the year?

EXPLORE TECHNOLOGY

TEMPLATE: Spreadsheet to Find Income Taxes

Will Irons worked at a summer job last year. His parents claimed him as an exemption on their federal income tax return. Will earned $4,178 in wages at a part-time job last year and $126 in interest on his checking account. His employer withheld $550 from his pay for federal income taxes. Will's parents claim him as an exemption on their federal income tax return.

Load ET8 from your Spreadsheet Applications diskette. After you have loaded the template and entered Will's income information into cells B4, B8, and B18, your computer screen should look like the one shown below.

	A	B
1		INCOME TAX WORKSHEET
2		FOR DEPENDENTS
3	INCOME:	
4	Wages	4178
5	Tips	
6	Other Earned Income	
7	Total Earned Income	4178
8	Interest	126
9	Dividends	
10	Other Unearned Income	
11	Total Unearned Income	126
12	Total Income	4304
13		
14	TAX COMPUTATION:	
15	Standard Deduction	
16	Taxable Income	
17	Tax Owed	
18	Withholding Taxes Paid	550
19	Tax or Refund Due	

1. Enter the amount Will can claim as a standard deduction in cell B15. The spreadsheet will calculate Will's taxable income.
2. Use the tax table on page 326 to find the tax Will owes. The spreadsheet will calculate his refund or additional tax due.
3. Formulas within the spreadsheet automatically calculate the value for cells B7, B11, B12, B16, and B19. Write a sentence explaining what the spreadsheet is doing in each of those cells.
4. What would Will's taxable income have been if his earned income last year was $3,800?
5. How much income tax would Will owe if he had earned $4,278 in wages, $89 in interest, and $115 in dividends?

Note: You may want to personalize this exercise by replacing Will's income and tax information with your own, or a friend's, income information.

6. How could you modify this spreadsheet so that it could be used to calculate Will's parents' tax?

Design Your Own

If you regularly perform the same kinds of calculations, you should set up a spreadsheet with formulas to handle the repetitive work. It may take a little effort to plan the layout of such a spreadsheet, but it can save a lot of time in the end.

You work in a state income tax office. As part of your training you've been told to create a spreadsheet to verify state income tax. The income tax is based on the graduated tax table shown in Illustration 8-5.1 on page 331.

Imagine what your spreadsheet would look like. It should include a tax due row or column. It should also include a row or column for each of the income tax ranges shown in Illustration 8-5.1. Write a formula to enter in the tax due column that uses cell references only. Test the formula by calculating the tax due on taxable income of $21,600. The example in the book tells you that the tax is $1,296. If you did not get that answer, check your work and correct your formula. When you get the correct tax on your test data, COPY your formula to the succeeding rows (or columns) so that the spreadsheet can calculate tax due for each of the following taxable incomes:

$4,000	$7,200	$12,000	$15,095	$23,298
$27,289	$32,459	$36,000	$43,000	

Practice with the different currency formatting options in your software before printing out your final copy. The formula should follow a pattern similar to the one started on the background spreadsheet. If your formulas are correct, the tax due for each item should match those on the front spreadsheet.

	A	B	C	D	E	F
1	Income	St. Range	End Range	Min. Tax	Tax Rate	Total Tax Due
2	$21,600	$20,000	$25,000	$1,200	.06	((A2 – B2) * E2) + D2
3	$4,000	0	5000	$0	.02	((A3 – B3) * E3) + D3
					.03	((A4 – B4) * E4) + D4
					.04	
					.05	
					.06	
					.07	
					.08	
					.09	
					.10	

	A	B	C	D	E	F
1	Income	St. Range	End Range	Min. Tax	Tax Rate	Total Tax Due
2	$21,600	$20,000	$25,000	$1,200	.06	$1,296.00
3	$4,000	0	5000	$0	.02	$80.00
4	$7,200	5000	10000	$100	.03	$166.00
5	$12,000	10000	15000	$250	.04	$330.00
6	$15,095	15000	20000	$550	.05	$554.75
7	$23,298	20000	25000	$1,200	.06	$1,397.88
8	$27,289	25000	30000	$1,550	.07	$1,710.23
9	$32,459	30000	35000	$1,950	.08	$2,146.72
10	$36,000	35000	40000	$2,400	.09	$2,490.00
11	$43,000	40000	45000	$2,900	.10	$3,200.00

CHAPTER 8 SUMMARY AND REVIEW

VOCABULARY REVIEW

adjusted gross income *8-4*
assessed value *8-1*
dependent *8-4*
disability benefit *8-3*
earned income *8-4*
estimated tax *8-4*
exemption *8-4*
federal income tax *8-4*
FICA tax *8-3*
graduated tax rate *8-5*
income tax return *8-4*
Medicare *8-3*
property tax *8-1*
sales tax *8-2*
self-employment tax *8-3*
Social Security *8-3*
standard deduction *8-4*
unearned income *8-4*

MATH SKILLS REVIEW

8-1

Compute. Round answers to four decimal places.

1. $20,444 ÷ $34,255
2. $100,000 ÷ $952,340
3. $2.34 × 1,000
4. $900 × 10
5. 0.015 × 100

8-2

6. Find 6% of $205.43, to the nearest cent.
7. Find 12.5% of $5,900,000 to the nearest cent.

APPLICATIONS

8-1
- (Assessed Value ÷ Units per 100 or 1,000) × Tax Rate = Property Tax Due
- Change a tax rate in mills or cents per dollar to a rate in dollars by dividing the number of mills by 1,000 or the number of cents by 100, or moving the decimal place to the left 3 or 2 places, respectively.
- Assessed Value × (Tax Rate Per $1 ÷ 1,0000 or 100) = Tax Due
- Find the decimal tax rate for a district by dividing the amount to be raised through property taxes by the total assessed property value of all property.
- Find the tax on property by multiplying the assessed value by the decimal tax rate.

8. Find the tax on property assessed at $45,900 if the tax rate is $3.045 per $100.

9. A property tax rate is 32.8 mills per dollar. What is the tax rate in dollars?

10. A fire district must raise $985,000 from taxes. The assessed value of property in district is $35,890,000. Find the decimal tax rate needed, to 4 decimal places.

8-2 ◆ Find sales taxes when some items on not taxable by subtracting the nontaxable items from the total and multiplying the result by the sales tax rate.

11. A computer disk is priced at $1.25. What is the sales tax, using a 5% sales tax table?

12. A store sold a new antennae for $56 and installed it for $27. There was a 5.2% sales tax on goods but not on services. Find the total cost.

8-3 ◆ Find FICA taxes by multiplying wages and salaries subject to the taxes by the social security and Medicare tax rates.

◆ Find self-employment taxes by multiplying net income from self-employment by the combined employee and employer FICA tax rate.

◆ Estimate social security benefits by finding your average annual covered income in a benefits table.

◆ Estimate social security benefits for retiring early or late by multiplying the monthly benefit by the early retirement and late retirement percent.

13. Taylor earned $468.92 for the week of February 6th. If the social security tax rate was 6.2% and the Medicare tax rate 1.45%, how much did he pay for those taxes?

14. Jose estimates that his average annual earnings will be $50,000 when he retires at age 65. He estimates that his approximate monthly benefit at that age will be $1,166. What will his approximate yearly benefit be?

15. If Jose (in Exercise 14) retires at age 66, he will receive about 8% more per month in benefits. Find the monthly amount of the benefits, to the nearest dollar.

8-4 ◆ Find adjusted gross income by subtracting adjustments to income from gross income.

◆ Find taxable income by subtracting deductions and exemptions from adjusted gross income.

◆ Use a tax table to find tax due.

◆ Find the tax refund or tax due by subtracting income tax withheld from tax due.

◆ Find the taxable income for a dependent by subtracting the amount of earned income up to the amount of the standard deduction from adjusted gross income.

16. Ricky's adjusted gross income last year was $35,900. He claimed itemized deductions of $5,200 and 2 exemptions at $2,500 each. What was his taxable income?

17. Don and Karla are married and file a joint tax return. Their taxable income last year was $22,173. Using a tax table, what is their tax?

18. Jeanne was claimed as a dependent on her parent's income tax return. She earned $1,700 in wages at a summer job and $139 in interest last year. If the standard deduction is $4,000, what was the standard deduction amount she could claim?

8-5 ◆ Find state and city income taxes using a table or by multiplying taxable income by the income tax rate.

19. Tom's taxable income last year was $31,500. Using a tax table, what is his state income tax?

20. Donnell lives in a state that has a 4% state income tax and a city that has a 2% city income tax on his taxable income. Donnell's taxable income last year was $39,200. What were his state, city, and total state and city income taxes?

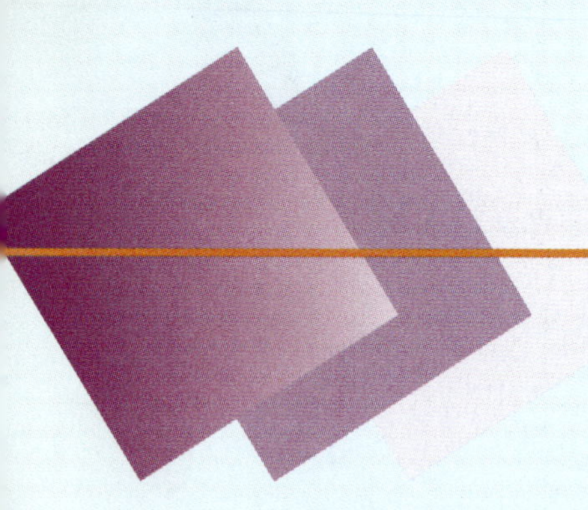

WORKPLACE KNOWHOW: **PARTICIPATES AS A TEAM MEMBER**

Being able to work with other people to accomplish common objectives and goals is a very important skill. Many companies are organizing as teams to get more work done quickly. Team work skills include the ability to contribute ideas, build on members strengths, and be responsible for achieving the team goal. Reread the career profile that began this chapter. That profile discusses how teams work and are evaluated in a warehouse distribution center. What do you like about the team approach discussed? Is there anything that you do not like about it?

Assignment Working as a group, identify a goal that your class would like to achieve. You do not have to plan on achieving the goal. Simply identify one and define a plan for how the goal could be achieved. What team rules would you put into place?

When you are done, write a report describing how you determined what goal to strive for and how the goal could be accomplished. Discuss whether or not the class demonstrated an ability to work as a team on this assignment. Use specific examples to support your opinion.

Work-Skills Portfolio Update your assessment of your computation skills as appropriate. You may wish to add the 5% Sales Tax Table, found in the workbook to your portfolio to show you know how to read and use a sales tax table. Consider including some of your writing assignments from this chapter. If you completed the Integrated Project in the Student Workbook, you may also wish to include the answer sheets to show that you know how to compute FICA taxes.

CHAPTER 8 TEST

Math Skills

1. \$45.67 + \$15.07 + \$23.77
2. $2\frac{1}{5} \times 3\frac{3}{4}$
3. $\frac{1}{6}$ less than \$360 = ?
4. $20.5 \div 1.5$
5. What amount is 25% greater than \$280?

Applications

6. Laura's property is assessed at \$43,700. Her tax rate is \$24.90 per \$1,000 of assessed value. Find her property tax.
7. Ted's house and lot are worth \$74,000. His property is assessed for 40% of its value. The property tax rate in Ted's town is \$3.45 per \$100 of assessed value. What is Ted's property tax?
8. A snowmobile is priced at \$1,890. There is a 5% state and 2% city sales tax. Find the total cost of the snowmobile?
9. Millville needs to raise \$3,557,600 in revenue next year. It expects to receive \$160,000 in license fees and raise the rest in property taxes. The town has property with a total assessed value of \$274,000,000. What tax rate should they charge per \$1,000?
10. Last year Edgar earned \$35,600 as a self-employed consultant. What self-employment tax did he pay, at 15.3%?
11. Emercon paid its employees \$213,500 in wages and salaries in January. Find Emercon's share of FICA taxes, at 7.65%.
12. Eve had gross income of \$34,200. She also had adjustments to income of \$3,000, itemized deductions of \$5,700, and 2 exemptions at \$2,500. What was Eve's taxable income?
13. Tom had a taxable income of \$18,600 last year. His employer withheld \$706 in state withholding taxes from his pay. If Tom's state has a 3% state income tax on taxable income, what amount should Tom receive as a refund?
14. If Karen retires from work at 65, she will be entitled to receive \$460 a month from social security. If she retires at 64, she will receive 93.3% of that amount. How much will she receive a month if she retires at 64?
15. Vince had an earned income of \$1,250 last year from part-time work. His father claimed him as an exemption on his tax return. If Vince's employer withheld \$96 from his wages for withholding taxes, how much should Vince get back as a refund on his federal income taxes?

RADIOLOGIC Technologist

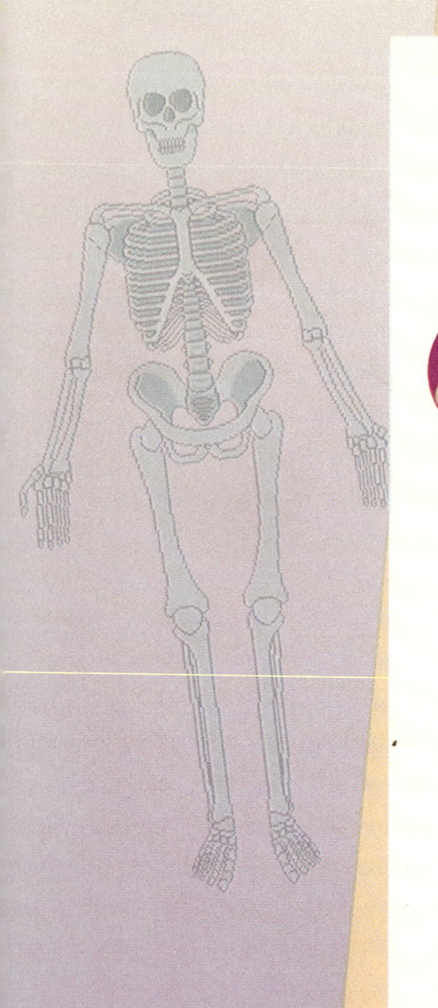

do you have a desire to contribute to someone's health and well-being? Can you remain calm and focused in tense situations? Those are critical qualifications for a health care professional.

Malcolm explored several different types of health careers. He decided on the x-ray field because of the salary potential and because he liked the idea that he could help physicians diagnose and treat medical problems.

Becoming a Radiologic Technologist usually requires a two-year program at either a technical college or community college with "hands-on" training at a hospital. After graduation, Malcolm passed the test to become a Registered Radiologic Technician (RT). That certification, confidence, hard work, and teamwork earned him a full-time job at the hospital where he did his training. RTs may also work in special clinics or private doctors' offices. But, as Malcolm points out, hospitals usually offer higher pay (and a heavier work load).

As an RT, Malcolm helps prepare patients for their x-ray procedures and takes and processes the x-rays. He works closely with radiologists who read the film and help diagnose problems. Malcolm usually works in the

HEALTH CARE

emergency room, although he sometimes works in orthopedics, oncology, and surgery. Some emergency procedures are scary, but there is no time to hesitate in crisis situations.

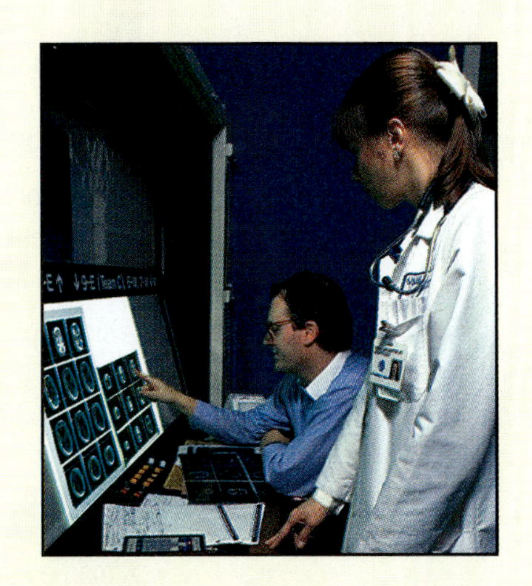

Math (especially percentages and fractions) and science skills are important in training to be an RT. Since these were not Malcolm's strongest subject areas, he admits he had some problems with the course work the first year. A helpful teacher, determination to be successful, extra study, and on-the-job experience all helped him fulfill the math and science requirements. On the job, he uses his knowledge of ratios, angles, measurements, decimals, and averages to position patients and to set the x-ray machines in order to produce reliable films.

Experience and training have been important in Malcolm's professional development. He has learned to do a variety of specialized x-ray procedures, and he takes advantage of technology training offered by the hospital. Newer technologies, such as nuclear medicine, MRT scans, angiography, and sonograms, are replacing more traditional x-rays in some situations. Malcolm aggressively pursues this training because the more skills he has, the more valuable he is to the hospital.

The health care field is booming, and continued growth is expected. A wide variety of technical and service careers exist in the health field. For more information on health care careers, check your local library.

OPPORTUNITIES

9-1 LIFE INSURANCE

OBJECTIVES

In this lesson, you will learn to

- *identify different types of life insurance and their benefits;*
- *calculate life insurance premiums;*
- *identify cash value of a life insurance policy; and*
- *calculate the value of dividends on a policy.*

The main purpose of life insurance is to prevent financial hardship by providing income to people who are dependent on you when you die.

◆ Think about the previous sentence. Then working in a small group, determine how important, on a scale of 1–10, life insurance is in each of the following cases. Explain your ratings.

- A single person who attends school and has no dependents.
- A single person with an aging parent who he or she help support.
- A married couple, both employed, with one young child.
- A married couple with no children. One spouse works full time; the other, part time.
- A married couple with two children. One spouse works; the other takes care of the home and children.
- A widow with two self-supporting adult children.

WARM UP

a. 17.5×0.86

b. $527.859 \div 0.009$

c. $\$4.50 + \$189.00 + \$57.82 + \$0.87$

d. $800.5 - 30.986$

e. Round to the nearest cent: $617.297

f. Round to the nearest thousand: 340,284.2973

g. Find the cost of 180 bolts @ $0.10

LIFE INSURANCE

If your income supports your family, your family will need to replace your income when you die. If you are a homemaker, your surviving spouse may need to pay someone to care for your children and home. In both cases, money is needed to pay funeral costs. Life insurance may also be bought to repay, upon death, debts, such as money that has been borrowed to buy a home or a car. **Life insurance** is the usual way of protecting your family from financial loss when you die. When a person decides to buy life insurance, he or she should determine how much insurance is needed.

Critical Thinking Work in small groups. Look again at the list of situations in the beginning of this lesson. How would you determine how much life insurance protection is needed in each situation? Is it good to buy more life insurance than you need?

Insurance Terminology A life insurance **policy** is the contract between the insured and the insurer. The contract states how much insurance and under what circumstances insurance will be paid. The money paid to an insurance company for life insurance is the **premium.** The person whose life is covered by the policy is the **insured.** The insurance company is the **insurer.** When the insured dies, **death benefits**, usually equal to the face amount of the policy, are paid to the beneficiary. The **beneficiary** is the person named in the policy to receive the death benefits. A life insurance policy may be canceled at any time by the insured. If premiums are not paid, the policy may be canceled by the insurer.

Types of Life Insurance Policies

There are many types of life insurance policies designed to meet different needs and budgets. Three of the most common types are:

- **Term life insurance:** Offers protection for a fixed period of time, such as 1, 5, or 10 years. Insurance can usually be renewed for current face value at a higher premium after the fixed term expires. Term insurance does not include a savings component. Therefore, term insurance is usually less expensive, especially for younger people, than other kinds of insurance.
- **Whole life insurance:** Insures you for a fixed amount for as long as you live. Premiums must be paid for your whole life. In addition to the death benefit, there is a savings component that earns interest and cash value over time. Some companies pay dividends as well. The premium for this type of insurance is usually higher than for term insurance.
- **Limited-payment life insurance:** Gives protection for life, but you pay premiums only for a fixed time, such as 20 years. Monthly payments are higher, and cash values increase at a faster rate.

You will find that individual insurance companies offer slightly different policies. The features, costs, and benefits must be compared to find the policy that is best for your needs.

Life Insurance Premiums

Illustration 9-1.1 shows the premiums an insurance company might charge for each $1,000 units of life insurance purchased. (Note that different rates are given for men and women at different ages. In addition, the rates shown are for nonsmokers.) For example, if you are a female who buys a $15,000 whole life insurance policy at the age of 20, your annual premium would be for 15, $1,000 units. The cost would be calculated as ($15,000 ÷ $1,000) × $6.22 or $93.30.

In addition to the premium, some insurance companies charge a *policy fee* or *administrative fee* to cover the costs of processing an application and issuing a policy.

ANNUAL PREMIUMS PER $1,000 UNITS OF LIFE INSURANCE						
Age of Insured	1-Year Term		Whole Life		20-Payment Life	
	Male	Female	Male	Female	Male	Female
20	$1.39	$1.22	$ 7.07	$ 6.22	$25.29	$23.27
25	1.41	1.24	8.37	7.37	27.77	25.54
30	1.45	1.28	10.16	8.94	30.75	28.29
35	1.62	1.43	12.58	11.07	34.15	31.42
40	1.91	1.68	16.30	14.34	38.89	35.78
45	2.45	2.16	20.02	17.62	44.72	41.14

*Note: Rates shown are for nonsmokers.

Illustration 9-1.1. Sample Life Insurance Premiums Per $1,000 Units

Notice that women pay a lower premium than males do at the same age. This is because women as a group live longer than men. Some states, however, have laws that do not allow insurers to charge different rates for males and females.

EXERCISE YOUR SKILLS

Use the annual premium table in Illustration 9-1.1 to solve these problems. Find the annual premium for each policy.

	Kind of Policy	Age and Sex	Face of Policy	Premium
1.	Whole Life	25, male	$10,000	
2.	1-Year Term	30, female	80,000	
3.	20-Payment Life	40, male	12,000	
4.	Whole Life	45, female	14,000	

5. How much more is the annual premium on a $20,000, 20-payment life policy for a male at age 45 than at age 25?

6. A 20-payment life policy for $9,000 is taken out by a female at age 25. What total amount will she pay in premiums by the end of 20 years?

7. Sam Pontel, age 30, is comparing the total premium cost of a $30,000, 20-payment life policy he may take now and the cost of the same policy at age 35. Find the difference in premium costs over 20 years for this policy at the two age levels.
8. Because he smokes, Bert pays 20% more for life insurance. How much more will Bert pay for $50,000 of 1-year term insurance at age 45 than a nonsmoker would pay at the same age?
9. To the nearest percent, what percent greater is the cost of a whole life policy taken out by a male at age 45 than at age 35?

Mary Crane, age 20, wants to invest no more than $600 a year in life insurance. In even thousands of dollars, what is the largest policy she can buy without spending more than $600 annually on a

10. whole-life policy?
11. 1-year term insurance policy?

LIFE INSURANCE BENEFITS

The main benefit of having life insurance is its protection. Some policies have a savings feature.

Cash Value

Whole life and limited-payment life policies build a cash value after premiums have been paid for a few years. **Cash value** is the money that you get if you cancel the policy. (If you cancel a term policy, you get nothing.) The terms of the policy may give you a choice of using the cash value in these ways:

- **Policy Loan.** You can borrow an amount up to the cash value from the insurance company and still keep insurance coverage. You must repay the money you borrow, but you usually pay a lower rate of interest than other lenders would charge.
- **Paid-Up Life Insurance.** The cash value may be used to make a one-time payment to buy a smaller amount of insurance that covers you until you die. You do not pay any more premiums.
- **Extended Term Life Insurance.** You may trade the cash value for term insurance. You will be covered for the original amount of the insurance for a fixed, shorter period of time.

Table of Cash Values A policy that builds cash value would have a table much like the one shown in Illustration 9-1.2.

Year	Cash/Loan Values per $1,000 Units	Paid-Up Whole Life per $1,000 Units	Extended Term	
			Years	Days
1	$ 0	$ 0	0	0
5	17.80	119	8	274
10	76.09	403	19	115
15	170.72	716	21	80
20	327.50	1,087	22	107
25	564.33	1,599	22	312

Illustration 9-1.2. Sample Cash Value Table

If you had paid premiums on a $10,000 policy for 20 years, you could borrow up to $3,275 on the policy ($10,000 ÷ $1,000 = 10 units; 10 × $327.50 = $3,275). You could also buy $10,870 of paid-up life insurance (10 × $1,087 = $10,870). Or you could buy a $10,000 term insurance policy that would cover you for 22 years and 107 days.

Dividends After you have had a policy for a few years, your insurance company may return part of your premium to you as a **dividend**, which is usually shown on the premium notice. You may deduct the dividend from the premium due or leave the dividend with the company to buy more insurance or to earn interest. The cost of the premium minus the dividends is the net cost of the insurance. The way in which a dividend is used is decided when the policy is written.

EXERCISE YOUR SKILLS

12. How much cash would you get if you canceled a $50,000 policy after paying premiums for five years?

13. You have paid annual premiums of $212 on a $25,000 policy for ten years. What amount could you borrow on your policy?

Alfred Buckley has made payments for 15 years on a $33,000 policy. If he canceled the policy,

14. Estimate the cash payment he could get. Then find the actual cash payment he could get.

15. How much paid-up whole-life insurance could he buy?

16. How many years and days of term coverage could Alfred get?

17. What amount of insurance would Alfred get if he traded the cash value of this policy for extended term insurance?

18. Miriam pays a premium of $12.80 per $1,000 for an $8,000 life insurance policy. Her policy paid a dividend of $18.90, which she uses to reduce her premium. How much should Miriam send to the insurance company when she pays her premium?

19. Ben paid annual premiums on a $25,000 whole life policy at a rate of $17.20 per $1,000. After ten years, he canceled the policy and found that its cash value was $97 per $1,000. Over the ten years, he received dividends of $318.55. For the time Ben had the policy, find the net cost of the insurance.

20. ◆ Three different types of life insurance were discussed in this lesson. Return to the list of insurance situations presented at the start of this lesson. Which type of life insurance, if any, would you recommend in each situation and why?

MIXED REVIEW

1. Write $248\frac{2}{5}\%$ as a decimal.
2. Find $12\frac{1}{2}\%$ of $384.
3. $15.10 increased by 30% of itself is?
4. The regional park system tax rate in Odell County is 2.3 mills per dollar of assessed value. Find the tax to be paid on property assessed at $70,000.
5. Iris earned $2,500 last year at her part-time job. Her parents claimed her as a dependent on their federal income tax return. What taxable income did Iris have last year?
6. You want to borrow money on a $60,000 life insurance policy that you have held for 17 years. The loan value of the policy is $175 per $1,000. What is the maximum amount you can borrow?
7. Roger paid annual premiums on a $35,000 whole life policy at a rate of $14.76 per $1,000. At the end of five years, he canceled the policy and received the cash value of $35 per $1,000. Over the five years, he had received total dividends of $62, which was kept in the account. What was the net cost of this policy?
8. Letitia's check register balance on March 31 was $764.20. In making a reconciliation statement, she found that a check for $19 was incorrectly recorded in the register as $91; and she had no record in her register of a service charge of $4.80, earned interest of $1.21, and a deposit of $78.34. What was her correct check register balance?

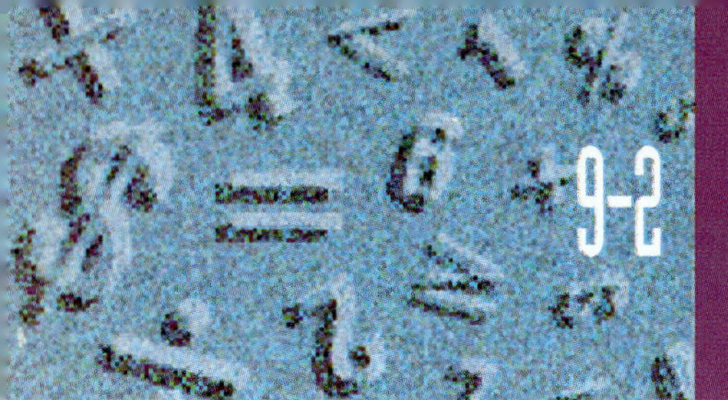

9-2 HEALTH INSURANCE

OBJECTIVES

In this lesson, you will learn to

- *identify different types of health insurance coverage;*
- *calculate medical expenses; and*
- *calculate what health insurance will and will not pay for.*

How many friends or relatives in your age group have had surgery, have been hospitalized, or have needed long-term medical care?

◆ What would you estimate to be the chances of your having a serious medical problem? Considering your current age, do you need health insurance? Are your health insurance needs different from those whom we term the elderly?

WARM UP

a. Find 80% of $4,500.

b. $\frac{5}{8}$ is ?% of 0.625?

c. $15,000 − ? = $6,000

d. ? × 0.76 = 1.14

e. Find 80.5 increased by 25% of itself.

f. What percent of $120 is $54?

g. Round 850,379 to the nearest hundred.

HEALTH INSURANCE

When a person has serious health problems, the cost of medical care can add up to thousands of dollars. Instead of using their savings or current income to pay health care bills, most people rely on health insurance. The insurance company then pays most of the cost of certain kinds of health care up to a specifed maximum. **Health insurance,** like other insurance, protects against financial loss.

Employers often provide *group health insurance* for their employees and their families. In most cases, the employee pays for part of the cost of the *group policy.* If you are not covered by a group policy, you may buy *individual health insurance* for yourself and your family, which is usually more expensive than group coverage.

Older and disabled persons have health insurance through the Medicare programs of the federal government. They may also buy additional private health insurance coverage. State governments provide Medicaid health insurance to people with low incomes, regardless of age.

Kinds of Health Insurance

The three basic kinds of health insurance are **basic health coverage**, **major medical insurance**, and **disability insurance.** Other kinds of health insurance policies include supplemental insurance policies, long-term health care policies, and dread disease policies.

Basic health coverage, usually offered as an insurance package, provides:

Hospitalization benefits, which help pay expenses of a hospital stay, such as hospital room, medicine, lab tests, X-rays, operating room.

Surgical benefits, which cover the fees of doctors who do surgery or who help with surgery in or out of a hospital.

Medical insurance, which pays the fees of other doctors who see you in or out of the hospital, as well as some other medical expenses, such as physical therapy.

Major medical insurance usually supplements basic health coverage. It is designed to help pay the hospital costs and surgical, medical, or other health care expenses due to a major illness or an injury. Often basic and major medical insurance policies are combined into one comprehensive health package.

Disability income insurance replaces part of the income you lose if you are unable to work for an extended period of time because of illness or accident. Usually coverage starts after a specified period, such as 30 days.

HEALTH INSURANCE PLANS

All health insurance plans have policies that state in writing your maximum coverage or benefits. The maximums determine how long each service will be provided or give a maximum dollar amount of coverage. Three common types of health insurance plans are:

- **Traditional plans** that place no restrictions on the doctors or hospitals you choose for medical care. However, you may have to pay a fixed dollar amount or a percentage of all medical bills, depending on the terms of the policy.
- **Health Maintenance Organizations** (HMOs). In an HMO, a small fee is usually charged each time you use a service. You must get all your medical care from doctors, hospitals, laboratories, and drugstores who participate in the HMO.
- **Preferred Provider Plans (PPOs)** combine features of traditional plans and HMOs. A PPO has a network of recommended medical care providers, known as **preferred providers,** who have agreed to take a specified amount of money for their services. PPOs also allow customers to choose a doctor or hospital not in the network. However, customers pay a higher part of the medical service charge when they do not use a preferred provider.

Limits of Coverage A limit is usually set on the number of days of hospital care that are paid by insurance. A limit may also be set on the amount of money paid for a hospital room per day, for doctors' fees, or for the total cost of a health service. You must pay for the cost of health services above the limits.

EXERCISE YOUR SKILLS

Lela Zerkin's hospital bill for 6 days was $1,900. Her surgeon's bill was $1,250. Lela's insurance covered $1,740 of the hospital bill and $1,175 of the surgeon's bill. How much was paid by

1. the insurance company? **2.** Lela?

Gerald Brinson's insurance pays all hospital charges except for a maximum limit of $140 a day for a hospital room. Gerald stayed in a hospital for 9 days and had a private room that cost $210 a day. During his stay, other hospital charges amounted to $738. How much was paid by

3. the insurance company? **4.** Gerald?

Despana Stavros needed hospital care after an accident. Her medical bills for 15 days of care were hospital room, $3,045; X-rays, $590; medicine, $260; operating room, $520; surgeon, $950. Despana's insurance paid $190 a day for the room, $545 for X-rays, $900 to the surgeon, and the full amount of all other charges. Despana paid the amount not covered by her insurance.

5. What was the total of Despana's medical bills?

How much was paid by

6. the insurance company? **7.** Despana?

Ed has disability income insurance that pays 70% of his regular wages if he can't work because of illness or injury. His insurance starts after he has missed 30 days of work. Due to an accident at home, Ed has lost 39 days of work. His total medical expenses were $3,200. If Ed's regular daily wage is $148,

8. what estimated amount of disability income will he get?

9. what actual amount of disability income will he get?

MAJOR MEDICAL INSURANCE PAYMENTS

Many major medical insurance policies require you to pay a part of health care expenses. In a policy with a **deductible** feature, you have to pay a fixed amount each time you get health services. For example, you may have to pay for the first $10 of every X-ray charge. The insurance company pays the rest.

In a policy with a **coinsurance** feature, the insured and the insurer share the cost of health care above the deductible amount. For example, if your policy has an 80% coinsurance feature, the insurance company pays 80% of a health care bill after subtracting the deductible amount. You pay the other 20% of the bill plus the deductible amount.

EXAMPLE Costella has a major medical insurance policy with a $500 deductible feature and an 80% coinsurance feature. Her policy covers the hospital, surgical, and medical expenses of $14,500 that Costella has been charged after being injured in an accident. How much will be paid by the insurance company? By Costella?

SOLUTION

Total covered expenses	$14,500
Less: deductible amount	− 500
Balance to be shared	$14,000
Company's share: 80% of $14,000 balance =	$11,200
Costella's share: 20% of $14,000 balance =	$2,800
Deductible amount	+ 500
Costella's amount to pay	$3,300

EXERCISE YOUR SKILLS

Your major medical policy has an $800 deductible feature and a 90% coinsurance feature. You are injured in an accident and your health care bills amount to $36,000.

10. What amount will be paid by your insurance company?

11. What amount will you pay?

Molly Denard had 6 X-rays taken at a total cost of $360. Under her major medical coverage, the insurance company paid 80% of the cost of X-rays after a $10 deductible fee for *each* X-ray.

12. What was the company's share of the cost of the X-rays?

13. What was Molly's share of the cost?

Adele and Jim Emmet's major medical policy pays 90% of covered expenses for each of them in any year. A $500 deductible feature applies to *each* person's claim. Last year the Emmets made two medical claims. Adele's claim was for $530; Jim's claim was for $950. The insurance company did not allow $70 of Adele's claim as a covered expense. What amount did the insurance company pay for

14. Adele's covered expenses? **15.** Jim's covered expenses?

16. Howard Laird's injury required lengthy hospital and medical care. The fees of his doctors were $8,700 and covered at 100% by his major medical policy. His hospital expenses were $34,460, and the policy covered 90% of the hospital bills beyond a $250 deductible. After Howard left the hospital, a physical therapist made 30 visits to his home at $75 a visit. Howard's policy paid 70% of the therapy bills. Of the total expenses, what amount did Howard have to pay?

Daniel Sparks received medical care in the emergency room of a hospital. His bill for this care included these items: emergency room use, $200; doctor's fee, $85; lab tests, $90; medical supplies, $18. For emergency care, Daniel's major medical coverage had a 70% coinsurance feature with a deductible of $25 for doctors' fees.

17. How much did the insurance company pay?
18. How much did Daniel pay?

Ivy Gould was hospitalized for 17 days. Her total bill for medical care was $24,490. Ivy's major medical coverage pays for 85% of medical expenses above a $750 deductible. Ivy has a disability income insurance policy with another company that pays her $65 each day that she is hospitalized.

19. How much of the bill for medical care does Ivy owe after the insurance company pays its share?
20. After using the disability insurance to pay her medical bill, how much will Ivy still owe?
21. ◆ The costs of health care and insurance have gone up sharply in recent years. Make a list of suggestions for reducing health insurance costs.

MIXED REVIEW

1. $6.65 is what percent greater than $5.32?
2. 36.2 increased by 75% of itself is?
3. $192 is what percent less than $288?
4. $55,600 decreased by 8% of itself is?
5. Sabrina Wooten canceled her $25,000 life insurance policy after 8 years and took the cash value of $62 per $1,000. The annual premiums on the policy were $350. While the policy was in effect, she received a total of $187.40 in dividends. What was the net cost of the policy for the 8 years?
6. John Stroble had a lot that was 20 m by 50 m. He used $\frac{1}{4}$ of the lot for a garden. He planted flowers in $\frac{1}{2}$ of his garden. How many square meters are planted in flowers?

Nannette Frey worked 48 hours last week. She earned $12.52 per hour for the first 37.5 hours. For time over 37.5 hours, she earned time-and-a-half. Deductions of $152.24 were taken out of her paycheck.

7. Find Nannette's gross pay for the week.
8. Find her net pay for the week.

9-3 PROPERTY INSURANCE

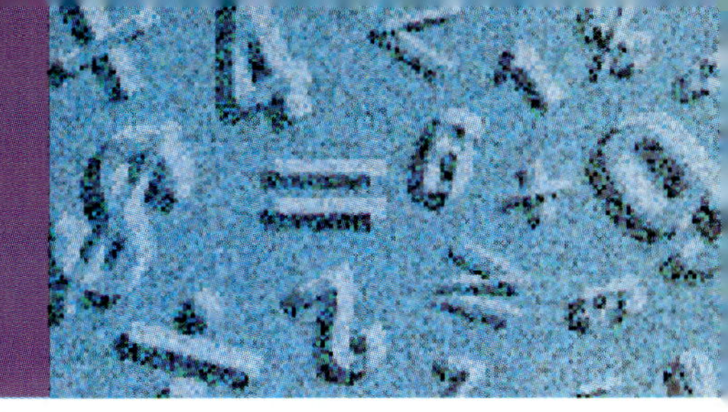

By insuring property, such as a home, you protect yourself from risks that go along with home ownership. Fire, theft, and lawsuits by persons who are injured on your property are some of the risks that may cause you to lose money.

♦ When you take out a mortgage loan to buy a house, the lender will require you to prove that you have insured your house. Why? Would it be simpler to rent an apartment than buy a home to avoid the risks and insurance costs of home ownership?

OBJECTIVES

In this lesson, you will learn to find

- *premiums for property insurance;*
- *amounts collected on insurance claims; and*
- *proportional shares covered by companies.*

WARM *UP*

a. \$7.20 is what percent greater than \$5.76?
b. 42.4 increased by 62.5% of itself is ?
c. \$168 is what percent less than \$224?
d. $990 \div 22{,}000$
e. Find 6% of \$228.47.

PROPERTY INSURANCE POLICIES

A policy that covers your home and protects you against other risks is called **homeowners insurance.** Basic homeowners insurance covers these items:

- *Dwelling*, the home in which you live.
- *Other structures,* such as a garage.
- *Personal property,* which includes the contents of a home, such as clothing, furniture, rugs, and many other items.
- *Additional living expense,* which pays for the extra costs of living when you cannot use your own home because of damage—typically 20% of the amount of the policy.
- *Personal liability,* which protects you in case of lawsuits by persons injured on your property.
- *Medical payments to others,* but not to you or your family, for medical expenses in case of injury on your property.

The amount for which your home is insured determines the protection for other categories. For example if your home is insured for \$60,000, personal property is covered for 50% of that amount, or $30,000; 20% is typical for covering additional living expenses.

Homeowner policies may provide other options as well. For example, a policy may insure personal property when you are away from home. Coverage for contents off the premises is usually 10% of the amount of the policy.

Renters or Tenants Policies If you rent a house or an apartment, you can buy a **renters policy** that provides nearly the same coverage as a homeowners policy except for loss of the building itself. Annual premiums for a renters policy are based on the amount of insurance on the contents of your home. The table in Illustration 9-3.1 shows the annual premium charged by one company for a renters policy.

Replacement Cost Policies Many people insure their property under **replacement cost policies.** With this type of policy, the insurance company will pay the cost of replacing your property at current prices. For example, suppose a leather chair that cost $600 is destroyed by fire. The insurer will pay for a replacement chair that now costs $900 even though the cost is higher than the original purchase price.

Before issuing this type of policy, insurers usually require a survey and inspection of the property. Also, the property must be insured for 100% of its current replacement value with annual adjustments for inflation. Premiums for this type of policy run 10–15% higher than a standard policy because of the extra protection it offers.

Premiums

The premiums you pay depend on many things, such as how much and what kind of coverage you buy, how your house or apartment is built, and where it is located. For example, the premium rates for a brick house near a fire department will be less than for a house made of wood that is far from a fire department.

Property insurance rates are usually based on $100 units of insurance.

Note: *Premium charges are rounded to the nearest dollar.*

EXAMPLE Dave Linz insured his house for $79,000 at an annual rate of $0.51 per $100. Find his premium.

SOLUTION

$79,000 ÷ $100 = 790	Find the number of $100 units.
790 × $0.51 = $402.90	Multiply to find premium.
Premium = $403	Round to the nearest dollar.

EXERCISE YOUR SKILLS

1. Find the premium, to the nearest dollar, for one year for a $51,000 policy at $0.38 per $100.
2. Find the premium, to the nearest dollar, for one year for a $25,000 policy at $0.68 per $100.

The Jan Clark family moved to a new home in the same town. Their new house had the same value as their old home, $189,500. Because the new home is located more than 1,000 feet from a fire hydrant, their homeowners insurance rate increased by $0.09 per $100.

3. What estimated amount more will the Clarks pay each year?
4. What actual amount more will their insurance cost?

5. Esther insured her home for $45,000. The rate was $0.87 per $100. What was the annual premium of this policy?

Bill wants to insure his home for $56,000. The Hob Company's rate is $0.75 per $100, while the Dex Company's rate is $0.68 per $100.

6. What annual premium would be charged by Hob?
7. What annual premium would be charged by Dex?
8. What is the difference between the annual premiums charged by the two companies?
9. Consuela insures her home for $60,000 and pays insurance at a rate of $0.50 per $100. Because she has installed smoke detectors and dead-bolt locks and has a fire extinguisher, her insurance company will give her a 2% discount on annual premiums. Find her annual premium.

Use the table in Illustration 9-3.1 to solve the following problems. Find the annual premium, the amount of off-premises coverage, and the amount of living expense coverage for each apartment.

	Amount of Policy	Distance from Fire Station	Annual Premium	Off-Premises Coverage (10%)	Living Expense Coverage (20%)
10.	$10,000	8 miles			
11.	$16,000	2 miles			
12.	$ 5,000	12 miles			
13.	$12,000	1 mile			

Amount	Distance from Fire Station	
Coverage on Contents	Less Than 5 Miles	5 Miles or More
$ 5,000 or less	$ 61	$ 70
8,000	75	86
10,000	82	94
12,000	94	108
14,000	106	121
16,000	114	130

Illustration 9-3.1. Renters Policy Annual Premiums

CONTROLLING THE COST OF PROPERTY INSURANCE

Deductible Policies Insurance policies can be written for different deductible amounts. With a $100 **deductible policy,** you are responsible for the first $100 of the loss. Discounts may be earned by taking a policy with a higher deductible.

Coinsurance: Coverage Equal to 80% of Property Value Another way to limit the cost of insurance is to purchase a coinsurance policy. With a coinsurance policy, the property owner purchases insurance up to a stated percent of the value of the property. This is usually 80% of the property's value. If you have a property valued at $50,000 and insure it at 80%, the coinsurance coverage is for $40,000. A coinsurance policy must be enough to pay for any mortgage obligations.

EXERCISE YOUR SKILLS

14. Anita Bergman had a $100 deductible homeowners policy. Her home computer valued at $1,200 was stolen. How much did Anita collect from the insurance company?

15. Ned Ryan had a fire in his apartment that destroyed his sofa worth $900. His clothing and drapes, worth $2,800, were damaged by smoke and had to be thrown away. Ned's insurance company paid for the loss less a $250 deductible. How much did the company pay?

16. Ricardo Mentos could purchase a homeowners policy from Bristol Insurance Company for $487 with a $100 deductible clause or for $414 with a $250 deductible clause. How much will Ricardo save by buying the policy with $250 deductible?

17. By increasing the deductible amount on her property insurance, Rita Malzone saved $85 on her annual premium. If her annual premium was $340, what percent of the premium did she save?

Your home is valued at $120,000. Your insurance rate is $0.75 per $100. How much does it cost to insure the house for

18. 100% of its value?

19. 80% of its value?

20. How much more is a 90% coinsurance policy than an 80% coinsurance policy?

SHORT-TERM POLICIES AND CANCELLATIONS

An insurance policy for less than a year is called a **short-term policy.** The premium is found by taking a percent of the annual premium.

Short-term rates are also used to find how much will be refunded if *you* cancel your policy. For example, if you cancel a one-year policy at the end of 98 days, the company will keep 75% of the premium and refund 25% to you. If the one-year premium was $120, your refund would be 25% of $120, or $30.

Days Policy Is in Effect	Percent of Premium Charged
1 Day to 30 Days	25%
31 Days to 90 Days	50%
91 Days to 180 Days	75%
More than 180 Days	100%

Illustration 9-3.2. Sample Short-Term Rates

PROBLEM SOLVING

Use a Venn Diagram

Venn diagrams can be used to show relationships. For instance, an insurance company gave 2% discount on premiums for any 1, 4% discount for any 2, and an 8% discount for all the following:

- dead-bolt locks on doors
- fire extinguisher in kitchen
- smoke alarms

This could be shown as a **Venn diagram.**

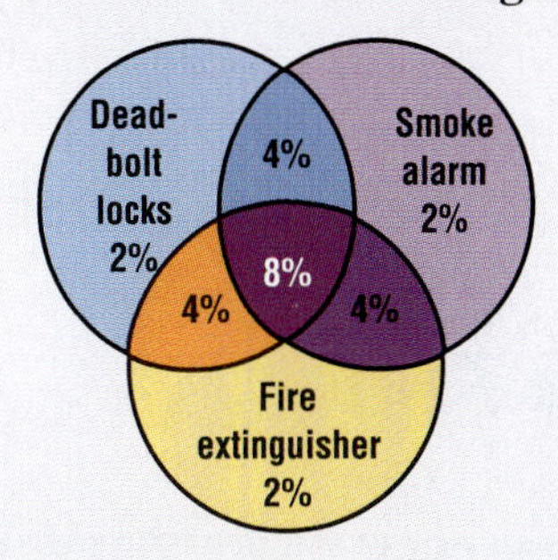

Policies Canceled by an Insurance Company If an insurance company cancels your policy, the refund must be calculated on a *pro rata basis*. This means that the amount of the refund must be in proportion to the amount of time left on the policy.

For example, suppose you pay a premium of $140 on a one-year renters policy and the company cancels the policy at the end of 73 days. The unexpired time for which the premium must be refunded is 365 − 73, or 292 days. The percent that the unexpired time is of one year is 292 ÷ 365, or 80%. The refund is 80% of $140 , or $112.

EXERCISE YOUR SKILLS

Draw Venn diagrams to show relationships for Exercises 21–22.

21. One company rates customers based on being within five miles of a fire station and within 500 feet of a fire hydrant.

22. Another company gives customers the following discounts: more than one smoke detector, 1%; exterior motion-sensor lights, 1%; dead bolts on all doors, 1%; all three, 5%.

23. Millicent Crane paid $0.62 per $100 for $20,000 worth of renters insurance. She canceled the policy after keeping it for 120 days, or about 4 months. How much was her refund?

24. Ronald Green canceled a $4,000 property insurance policy on his sailboat at the end of summer. He had the policy for 89 days after paying $2.45 in premiums per $100 of value. Find his refund.

PROPERTY INSURANCE CLAIMS

If your property is damaged by fire, you have to file a claim with your insurance company. The company will send an adjuster to look at the property and decide on the amount of loss.

Collecting for Loss under a Basic Policy If you have a basic policy, the company will pay the full amount of the loss up to the face value of the policy. It will not pay more than the amount of your policy. If your policy contains a deductible, the insurance company pays the full amount less the deductible up to the face value of the policy.

EXAMPLE Your policy has a face value of \$20,000 with a \$1,000 deductible. How much will the company pay if your loss is \$8,500? \$20,500? \$24,000?

SOLUTIONS \$8,500 − \$1,000 = \$7,500 Amount Paid
\$20,500 − \$1,000 = \$19,500 Amount Paid
\$24,000 − \$1,000 = \$23,000, which is greater than the face value. For a \$24,000 loss, the insurance company pays the face value, \$20,000.

Dividing Loss among Several Companies Some insurance companies may not want to risk issuing policies for large amounts of money. In some circumstances, an owner may have to insure his or her property with more than one company. When a loss occurs, the loss is shared among the companies in proportion to the amount of each policy.

EXAMPLE A warehouse is insured by the United Ins. Company for \$3,000,000 and by the Northern Ins. Company for \$5,000,000. How much should each company pay on a fire loss of \$64,000?

SOLUTION \$3,000,000 + \$5,000,000 = \$8,000,000 total insurance

$\frac{\$3,000,000}{\$8,000,000} = \frac{3}{8}$ United's portion

$\frac{\$5,000,000}{\$8,000,000} = \frac{5}{8}$ Northern's portion

$\frac{3}{8} \times \$64,000 = \$24,000$ Amount paid by United

$\frac{5}{8} \times \$64,000 = \underline{\$40,000}$ Amount paid by Northern

\$64,000 Total

Collecting a Loss When You Have a Coinsurance Policy As described earlier, with a coinsurance policy, the property owner purchases insurance for a stated percentage of the property's value. As long as an insurer carries insurance for the agreed percent, the insurance company will pay claims up to the face value of the policy. If a building with $50,000 value is insured under an 80% coinsurance claim for $40,000, all damage up to $40,000 will be paid. Damages over $40,000 will not be paid.

If the face value of a policy is less than 80% (or the agreed upon coinsurance percent), an insurance company will pay damages up to the fractional part of the actual coverage of the required insurance. The formula used is:

$$\text{Amount Paid by Insurance Company} = \frac{\text{Face Value of Policy}}{\text{Required Amount of Coinsurance}} \times \text{Amount of Loss}$$

If a building with a value of $50,000 and a coinsurance agreement of 80% ($40,000) was actually insured for $24,000 and sustained damage of $7,200, the amount paid by the insurance company would be $4,320. The payment is calculated this way:

$$\frac{\$24{,}000}{\$40{,}000} \times \$7{,}200 = \$4{,}320$$

Critical Thinking Why do you think coinsurance policies are less expensive?

EXERCISE YOUR SKILLS

Find out how much an insurance company will pay for each loss under a basic policy.

25. Property is insured for $18,000 with a $500 deductible. A loss of $18,750 occurs.

26. Property is insured for $105,000 with a $1,000 deductible. A loss of $7,800 occurs.

27. Property is insured for $31,000 with a $1,000 deductible. A loss of $42,000 occurs.

Find each insurance company's share of the fire loss.

		Amount Carried by		Amount Paid by	
	Fire Loss	**Amera Co.**	**Genie Co.**	**Amera Co.**	**Genie Co.**
28.	$ 60,000	$ 2,500,000	$ 5,000,000		
29.	$ 20,000	$12,000,000	None		
30.	$540,000	$20,000,000	$30,000,000		

Find how much an insurance company will pay in each problem.

	Face of Policy	Amount of Loss	Value of Property	Coinsurance Clause	Amount Paid
31.	45,000	$ 8,000	$75,000	80%	
32.	63,000	$ 5,600	$90,000	80%	
33.	40,000	$13,200	$60,000	80%	
34.	45,000	$ 7,000	$62,500	80%	
35.	49,000	$22,500	$70,000	90%	

36. After paying $3,000 for a new roof, Joe Duu figures his house to be worth $80,000. He insures it for $48,000 under a policy containing an 80% coinsurance clause. If a fire causes a $7,600 loss to the house, how much does Joe receive?

37. An auto repair shop valued at $240,000 is insured for $128,000. The insurance policy contains an 80% coinsurance clause. What amount must the insurer pay in case of a fire loss of $3,879?

38. ♦ Make a list of questions you would ask before buying a home insurance policy.

MIXED REVIEW

1. What percent of $1.20 is $0.54?
2. Find 20% of $77.
3. $13 is what part less than $104?
4. $\frac{1}{8} = ?\%$
5. Max insures his house for $34,700. If the rate is $0.47 per $100, find the amount of the premium.
6. The annual premium on a whole-life policy for a 30-year-old male is $10.45 per $1,000 and $7.58 for a female at the same age. What is a male's annual premium on a $24,000 policy?
7. The tax rate in Briggs City is $34.70 per $1,000. If property worth $76,000 is assessed for 60% of its value, what will be the amount of the tax bill?
8. Erik began a trip with a full 64-liter tank of gas. During the trip, he bought 195 liters.When he arrived home, he had 21 liters left. How many liters of gas were used on the trip?
9. An insurance company canceled the homeowners policy of Brian Worthy because he abandoned his residence. The policy cost $73 and was in effect for 43 days. What was the amount of refund.

9-4 AUTOMOBILE INSURANCE

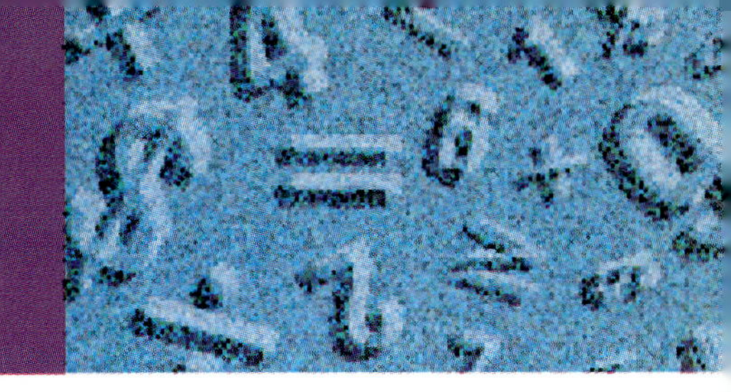

Like other kinds of insurance, automobile insurance is purchased to prevent your having to use personal income and assets from paying costs associated with theft, accident, or other damage.

♦ What factors do insurance companies consider when setting auto insurance rates? Are there factors that especially relate to young people that they should consider? What suggestions do you have for cutting the cost of auto insurance?

OBJECTIVES

In this lesson, you will learn to

- *calculate premiums for various kinds of automobile insurance.*

WARM UP

a. 2.5 × \$356

b. \$65 + \$12 + \$430 + 34

c. Find the cost of 567 @ 1¢.

d. What percent of \$64 is \$80?

e. \$650 is what part greater than \$500?

f. \$54.24 minus 25% of itself is what amount?

AUTOMOBILE INSURANCE

There are four basic types of insurance for motor vehicles that protect you against the risk of financial loss:

- **Bodily injury** Covers your liability for injury to others.
- **Property damage** Covers damage to other people's property, including their vehicles.
- **Collision** Covers damage to your own motor vehicle.
- **Comprehensive damage** Covers damage or loss to your vehicle from fire, theft, vandalism, hail, and other causes.

In some states, bodily injury and property damage coverage are combined into one minimum limit. The limit applies regardless of whether there is injury to one or more persons or whether there is damage to property of others in a single accident.

In addition to the basic types of insurance coverage, states may require additional coverage for which you will be charged. For example, a car owner may be required to buy **uninsured motorists insurance,** which protects against damage to the car or injury to persons in the car caused by a driver who carries no insurance.

Automobile Insurance Premiums

Premiums for automobile insurance may vary from state to state and within a state. Premiums are usually higher in large cities than in small cities and rural areas. Premiums may be higher on cars used for business than those used for pleasure. Premiums may be higher for drivers under 25 years of age than for those over 25.

Type of Insurance	Limits	Annual Premiums for: Pleasure Use Only	Driving to Work	Business
Bodily Injury	$20/40,000	$ 82	$ 90	$118
	25/50,000	98	108	140
	50/100,000	114	126	162
Property Damage	$10,000	$10	$12	$16
	25,000	12	14	18
	50,000	14	16	20
Collision	$100 deductible	$550	$612	$796
	250 deductible	386	428	556
	500 deductible	302	336	436
Comprehensive	$ 50 deductible	$ 92	$101	$132
	100 deductible	68	76	98

Illustration 9-4.1. Sample Annual Automobile Insurance Premiums

EXERCISE YOUR SKILLS

Use Illustration 9-4.1 to find the cost of insurance. If a coverage is not given, assume it is one of these standard coverages: bodily injury, $20/40,000; property damage, $10,000; collision, $100 deductible; comprehensive, $50 deductible.

1. What is the total premium for standard insurance coverage on a car driven to work?
2. On a truck he drives to work, Nolan carries bodily injury insurance of $50/100,000 and $250 deductible on collision. Other coverage is standard. Find his annual premium.
3. Because of his three speeding tickets, Lew Marsh has to pay a premium of 2.2 times the rate for standard coverage. He uses his car to drive to work. What is his premium for one year?

Reva Wilson uses her truck for business and insures the truck with standard coverage.

4. What total annual premium does she pay?
5. If she took the highest deductibles, what amount could she save annually on her truck insurance?
6. Julia Santos owns two cars. One car, used for pleasure only, is insured at standard coverage. Her business car is insured for the greatest amount of bodily injury and property damage coverage. Because she insures both cars with the same company, she gets a 10% discount on her total premium. Find the premium for insuring both cars for one year.
7. A truck used on a farm is insured in the Driving-to-Work category. It has standard coverage with highest deductibles. Since the truck is seldom driven outside the farm, it can be insured for 70% of the usual rate. Find the annual premium.

Olivia McVey is a 16-year-old owner of a car driven for pleasure. Because of her age she must pay four times the usual rate.

8. What annual premium must she pay with standard coverage?
9. To reduce the amount she must pay, she is considering not covering her car for collision and comprehensive damage. What would be her annual premium with this reduced coverage?

PUT IT ALL TOGETHER

10. ♦ What do all of the types of insurance studied in this chapter have in common? What is the distinguishing feature of whole life and limited payment life insurance?

MIXED REVIEW

1. Divide: $9\frac{1}{2}$ by $\frac{1}{4}$
2. Write 45.78% as a decimal.
3. Add: 34.89, 133.9, 4.275, and 1.47
4. Round: $23,896.73 to the nearest thousand
5. Kay Pollard bought a car for $9,502. Four years later she got $2,850 for the car as a trade-in allowance. What was the average annual depreciation of the car?
6. Lloyd Conner's taxable income for last year was $17,200. The state in which he lives charges a tax rate of 2.3% on taxable income. His city tax is 0.5% of taxable income. What is the total of his state and city income tax?
7. The tax rate for Millridge is $4.778 per $100. What is the tax on property in Millridge assessed at $60,000?

EXPLORE TECHNOLOGY

TEMPLATE: Spreadsheet for Figuring Insurance Loss Payments

Sarah Hulmich's house, worth $90,000, is insured for $60,000 under an insurance policy that contains an 80% coinsurance clause. She recently filed a property loss claim for $1,200 with her insurer.

To find out the amount of property loss that Sarah's insurance company will pay, load ET9 from your applications diskette. Your screen should look like the one shown below.

	A	B	C	D	E
1	Finding Amount of Loss Paid by Insurer				
2					
3	Value of Property				
4	Insurance Carried				
5	Required Coinsurance Rate (%)				
6	Amount of Loss				
7					
8	Required Insurance				
9	Amount of Loss Insurer Pays				
10					

Enter the data in the last paragraph to fill in the information for cells B3, B4, B5, and B6.

Answer these questions.

1. To meet coinsurance requirements, for what amount must Sarah insure her house? What formula is used to determine the answer?
2. How much greater or less was the amount of insurance carried than the amount of insurance required? What formula is used to find the answer?
3. Of the total loss, what amount, if any, did the insurance company pay? What formula is used to determine the answer?
4. Of the total loss, what amount, if any, was not covered?

Design Your Own

5. You work in an insurance company. Your job is to calculate payments on co-insurance claims. You set up the following spreadsheet template to increase your productivity. What formulas do you enter into cells C1 and F1?

	A	B	C	D	E	F
1	Property	Coinsurance	Required	Policy	Amount of	Insurance
2	Value	Percentage	Insurance	Face Value	Loss	Payment
3	$75,000	.80		$45,000	$8,000	
4	$90,000	.80		$63,000	$5,600	

Assume that you work in the personnel office of a small company that provides major medical insurance for employees. The major medical policy pays 85% of all bills after a $1,000 deductible is paid. Recently, several employees asked you to help them determine what portion of medical expenses would be paid by the insurance company. You decide to create a spreadsheet that will allow you to answer their questions by simply plugging in the specified information.

Your spreadsheet template looks like the one shown below.

	A	B
1	Finding Amount of Loss Paid By Insurer	
2		
3	Major Medical Expenses	
4	Less Deductible Amount	
5	Balance to be Shared	
6		
7	% Policy Pays	
8	Medical Costs Insurer Pays	
9	Remainder	
10	Plus Deductible Amount	
11	Medical Costs Insured Pays	

6. What formula do you put into cell B5?
7. What formula do you put into cell B8?
8. What formula do you put into cell B9?
9. You should enter B4 in cell B10. Why not put the dollar amount?
10. What formula do you put in cell B11?
11. Think of another formula you could put in cell B11 that would give you the correct answer.

Use the following information to see if your formulas are accurate. An employee has $50,000 in major medical expenses after a serious accident. What does the insurance company pay? How much does the insured have to pay? If your spreadsheet does not show the insurer pays $41,650 and the employee pays $8,350, check your formulas.

Based on the cash payment records from Unit 1 and on what you have learned in this unit, create a budget for yourself. There should be at least three columns: Budget Item, Annual Amount, Monthly Amount. In the first row, record your income annually and monthly. For each line item in your budget, give an annual and a monthly amount.

When you are done, do some "what if" calculations.

12. What happens to your budget if you have to add or increase the cost of automobile insurance by $400 a year?
13. Assume you have to contribute $75 a quarter to health care. How does that impact your budget?
14. You decide you want to save 10% of your monthly income. How do you calculate that?

CHAPTER 9 SUMMARY AND REVIEW

VOCABULARY REVIEW

beneficiary *9-1*
bodily injury *9-4*
cash value *9-1*
coinsurance *9-2, 9-3*
collision insurance *9-4*
comprehensive damage *9-4*
deductible *9-2, 9-3*
dividends *9-1*
extended term life *9-1*
health insurance *9-2*
HMO *9-2*
limited payment life *9-1*
policy loan *9-1*
PPO *9-2*
premium *9-1, 9-3, 9-4*
property damage *9-3*
renters policy *9-3*
replacement cost *9-3*
short-term policy *9-3*
term life insurance *9-1*
whole life insurance *9-1*

MATH SKILLS REVIEW

9-1

1. $12,000 ÷ $1,000
2. 24 × $8.53
3. $25,750 ÷ $1,000
4. $289.54 × 10

9-3

5. $109,000 ÷ $100
6. 825 × $0.65

9-4

7. $53,050 − $2,500
8. $\frac{3}{8}$ × $56,000

APPLICATIONS

9-1
- Divide the amount of insurance desired by the unit measure by which it is sold to find the total units purchased.
- Find the total annual premium by multiplying the premium per unit by the number of units to be purchased.

9. Use Illustration 9-1.1 to find the annual premium for a 20-year-old male for $150,000 of one-year term insurance?
10. Use Illustration 9-1.1 to solve this problem. Krystal, a 30-year-old female who smokes, wishes to purchase a $25,000, 20-payment life insurance policy. Smokers pay 80% more than nonsmokers for the same policy. How much more will Krystal pay for the policy over the 20 years than would a nonsmoker?
11. Muriel pays a premium of $14.34 per $1,000 for $50,000 of whole life insurance. The policy pays an annual dividend of $65.80. Muriel chose to apply the dividend to the annual premium. How much money should she send the insurance company?

Use Illustration 9-1.2. Glenn wants to cancel his $80,000 policy after paying for 20 years. If he cancels the policy,

12. What cash payment will he get? How much paid up whole life could Glenn purchase?

13. For what period, in years and days, could Glenn get extended term life insurance at no extra cost? What amount of extended term insurance would Glenn get?

9-2
- Subtract the deductible from the cost of health care to find the amount eligible for payment by a health insurance carrier.
- Multiply the eligible amount by the coinsurance rate to determine the amount to be paid by the health insurer.

14. Wilbur Lentz spent five days in a hospital. His bills were hospital, $2,580; doctor, $900. His insurance paid $2,125 of the hospital bill and $575 of the doctor's bill. What amount did Wilbur pay?

Ruth Morgan's health insurance has a $250 deductible applied individually for each family member. It also has an 80% coinsurance feature. Health care costs for the Morgan family last year were Ruth, $200; her husband, $950; her daughter, $310. What amount did the insurer pay last year for:

15. Ruth? **16.** Her husband? **17.** Her daughter?

9-3
- Calculate the annual premium for property insurance by multiplying the number of insured units by the rate per unit. Round premiums to the nearest whole dollar.
- Subtract the deductible from property loss to find the amount eligible to be paid by the insurer.
- Refund for Policies Cancelled by Insured = Annual Premium − (Short Term Rate × Annual Premium)
- Refund for Policies Cancelled by the Insurer = Number of Unexpired Dates in Policy ÷ 365 × Annual Premium
- Multiply the loss by the proportion of insurance carried by each insurer to find amounts paid by multiple insurers.
- Coinsurance Requirement = Value of Property × Coinsurance Rate
- Insurer's Loss Payment = Loss × (Face Value of the Policy/Coinsurance Requirement)

18. What is the premium for one year of homeowners insurance for a $72,000 policy at $0.47 per $100?

19. Eddie Wilson canceled his property insurance policy at the end of 146 days. To what refund is Eddie entitled if the original premium was $412? Use Illustration 9-3.2.

20. A tire store was insured for $195,000 by the Wells Group and $455,000 by Regal Insurance. The store was vandalized for a loss of $8,900. What amount of the loss was paid by Wells? by Regal?

9-4 ◆ Calculate car insurance premiums by finding coverage for each type of insurance with its rate and adding all rates.

21. Use Illustration 9-4.1 to find the total premium on a car that is driven to work with standard coverage except for a $500 deductible for collision?

22. Because of his poor driving record, Terrance Primeau, must pay 3.1 times the standard rate for auto insurance and use the car only for driving to work. If Terrance wants standard coverage for this type of use, what will be his annual insurance premium?

23. To cut his premium, Terrance may choose not to cover his car with collision and comprehensive coverage. If he does so, what will his premium be for one year?

WORKPLACE KNOWHOW: **DECISION MAKING**

Decision Making skills include the abilities to identify goals and constraints, generate and evaluate alternatives, consider risks, and choose the best alternative. Decision making skills are **transferable work skills.** That means they are skills that can be applied to a variety of work situations.

Assignment Prepare a brief written or visual report that demonstrates how each of the decision making skills identified would be applied to the selection of an insurance policy. Then identify an area where you have applied decision making skills. What do you consider to be your decision making strengths? weaknesses?

Work-Skills Portfolio In this chapter, you had the opportunity to use a variety of work skills. These included information management (evaluating information to make recommendations regarding insurance needs) money management, including developing a budget and making recommendations for controlling insurance costs. You developed speadsheet templates for office work in the Explore Technology lesson. You may want to include evaluations and samples of these skills in your portfolio. You may also want to include the completed project from your workbook. If you completed the project, your answers will have demonstrated an ability to select correct data from tables and to use data in a practical way. Your answers also indicate that you understand the concepts of different forms of insurance.

CHAPTER 9 TEST

Math Skills

1. $\$125,000 \div \$1,000$
2. $1,080 \times \$0.39$
3. $45 \times \$34.18$
4. $\frac{1}{3} \times \$62,100$
5. $115\% \times \$68$
6. $\$87 + \$18 + \$356 + \$73$
7. $90\% \times \$1,800$
8. $2.8 \times \$731$
9. $\$1,857 - \$500$
10. $0.75 \times \$1,500$

Applications

11. Eloise Hoekstra has a life insurance policy for $15,000 and pays a premium rate of $22.18 per $1,000 annually. The insurance company paid a dividend of $49.80 this year. Find Eloise's premium for the year she deducts the dividend.
12. The cash and loan value of a $30,000 insurance policy held ten years is $82.04 per $1,000. How much cash would the policy owner get if the policy was canceled after premiums were paid for ten years?
13. Your major medical policy has a $500 deductible feature and a 90% coinsurance feature. Last year you spent $1,600 for medical expenses that are covered under your policy. When you file a claim with your insurer for these medical expenses, how much should you expect the insurer to pay?
14. Wendy insured her $60,000 property for 75% of its value. She paid an annual rate of $0.64 per $100 for homeowners insurance. What annual premium did she pay for the coverage?
15. A renter has $5,000 worth of insurance coverage on the contents of her apartment. With a $100 deductible, the annual premium is $80. If she takes a $300 deductible, the annual premium will be reduced by 20%. What would be her annual premium if she takes the higher deductible?
16. Property is insured against fire by the Mutual Company for $2,400,000 and by the Freedom Company for $1,200,000. If the two companies agree on a settlement of $48,000 for fire damages, for what amount will the Freedom Company be liable?
17. Deshawn Zeck's disability income insurance pays 60% of his regular daily wage for each day he is absent from work after being ill or injured. His disability insurance begins after 45 days of absence. Deshawn missed 86 consecutive days of work due to an injury. If his daily pay is $86, to what amount of disability insurance is Deshawn entitled to receive?
18. A VCR worth $250 and a television set worth $890, both new, were stolen from Kirk Spencer. Kirk has a property insurance policy with a $400 deductible. What amount of the loss will the insurance company pay?

UNIT 2 TEST: CHAPTERS 5–9

♦ Choose the best definition for each term.

7-2 **1.** depreciation
6-1 **2.** budget
7-1 **3.** interest
9-1 **4.** premium
5-1 **5.** meter
8-1 **6.** property tax
9-4 **7.** collision insurance

a. The basic unit of length or distance in the metric system.
b. Basic measure of weight in the metric system.
c. A plan for spending income.
d. The cost of using a lender's money to make a purchase.
e. A loss in the value of property caused by aging and use.
f. Taxes collected on real estate.
g. Money paid to an insurance company for life insurance.
h. Covers damage to an owner's car or truck.

♦ Choose the best answer.

5-2 **8.** 100 mm = ? km
a. 10
b. 0.000100
c. 0.1
d. 1 000

6-1 **9.** Charles budgeted for the following monthly expenses. Which is an example of a fixed expense?
a. utility bill
b. mortgage
c. personal expenses
d. transportation expenses

9-3 **10.** Esther insured her home for $45,000. The rate was $0.87 per $100. What was the annual premium of this policy?
a. $372.84
b. $492.25
c. $391.50
d. $380

♦ Solve each problem.

5-3 **11.** 205 g × 32 kg = ? kg

5-3 **12.** 32 L = ? cL

5-2 **13.** 97 km^2 = ? ha

6-2 **14.** Find the estimated and exact extensions: 7 m tubing @ $3.89.

6-3 **15.** 375 ties @ $78.93 per 100 = ?

6-7 **16.** 1,500 watts @ 9.5 cents per KWH = ?

7-4 **17.** The monthly leasing charge for a luxury car is $190. For all miles driven yearly over 15,000, $0.12 a mile is charged. What is the total cost for one year of a car driven 18,700 miles?

7-3 **18.** Arlene Norris bought a car for $8,675 and sold it $7\frac{1}{2}$ years later for $725. What was the annual rate of depreciation?

5-3 **19.** A car used 14 gallons (54 L) of gas to travel 448 miles (756 km). How many miles (km) did the car average per gallon (per L)?

8-2 **20.** A leather coat costs $210 plus 8% sales tax. What is the total cost of the coat?

8-5 **21.** You earn a yearly salary of $23,500. You pay a city tax rate of 1.25% and a state tax rate of 3.5%. What is the total of your yearly city and state taxes?

8-1 **22.** Property tax is assessed at a rate of $7.30 per $100 of assessed value. Find the property tax on property assessed at $44,000.

9-2 **23.** Teresa has a major medical policy that covers 80% of her medical expenses after a $100 deductible is paid. Teresa breaks her leg, incurring medical bills totaling $1,375. What is Teresa's share of the cost?

9-3 **24.** Chi-Luan insured his house for $45,000. The policy required him to carry insurance of at least 90% of the house's current value of $90,000. A fire caused loss of $10,800. What amount did his insurance company pay?

9-3 **25.** A new video camera worth $825 and a new stereo worth $590 were stolen from John. John has property insurance with a $500 deductible. What amount of the loss will the insurance company pay?

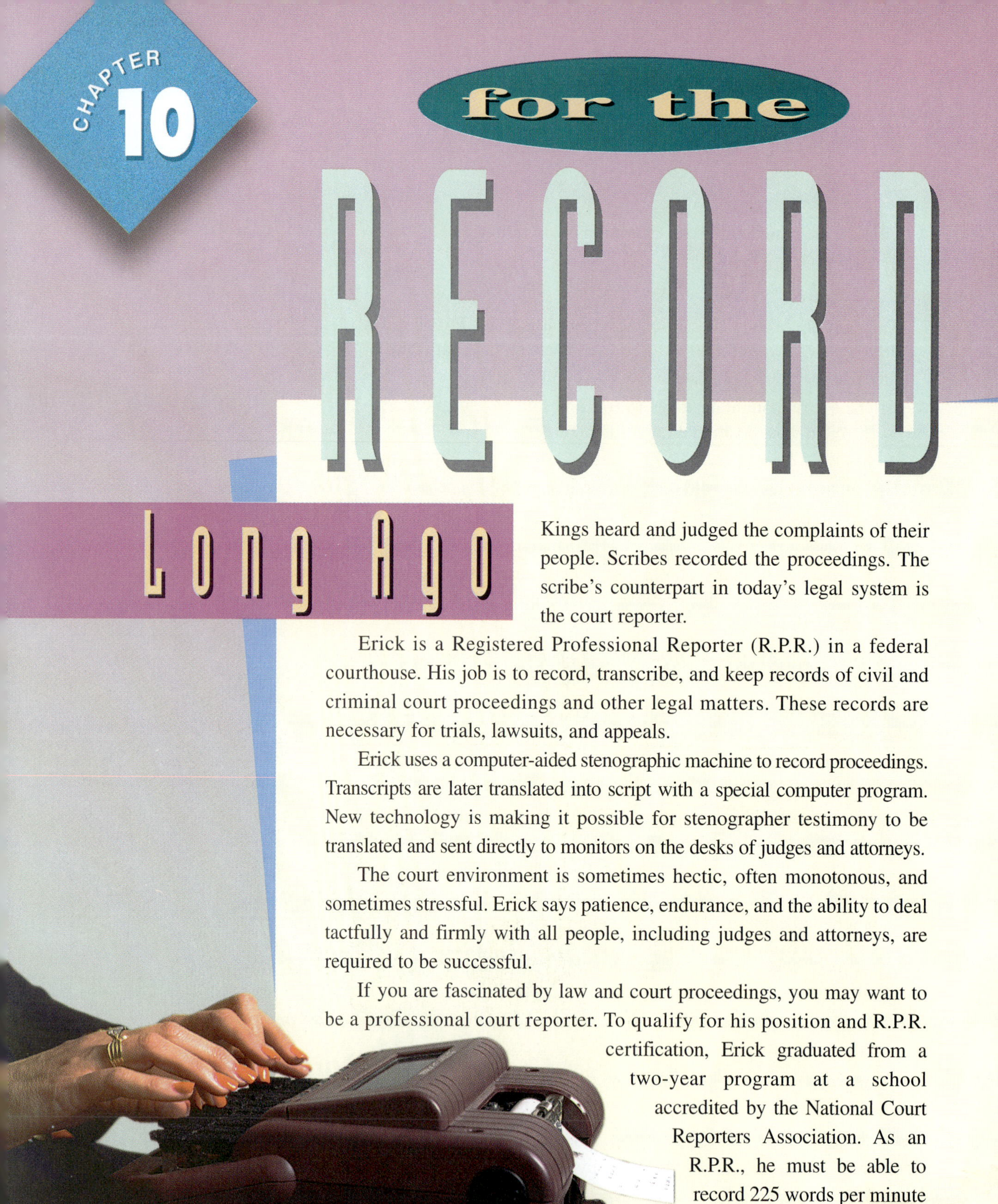

CHAPTER 10

for the RECORD

Long Ago

Kings heard and judged the complaints of their people. Scribes recorded the proceedings. The scribe's counterpart in today's legal system is the court reporter.

Erick is a Registered Professional Reporter (R.P.R.) in a federal courthouse. His job is to record, transcribe, and keep records of civil and criminal court proceedings and other legal matters. These records are necessary for trials, lawsuits, and appeals.

Erick uses a computer-aided stenographic machine to record proceedings. Transcripts are later translated into script with a special computer program. New technology is making it possible for stenographer testimony to be translated and sent directly to monitors on the desks of judges and attorneys.

The court environment is sometimes hectic, often monotonous, and sometimes stressful. Erick says patience, endurance, and the ability to deal tactfully and firmly with all people, including judges and attorneys, are required to be successful.

If you are fascinated by law and court proceedings, you may want to be a professional court reporter. To qualify for his position and R.P.R. certification, Erick graduated from a two-year program at a school accredited by the National Court Reporters Association. As an R.P.R., he must be able to record 225 words per minute with a 96 percent accuracy

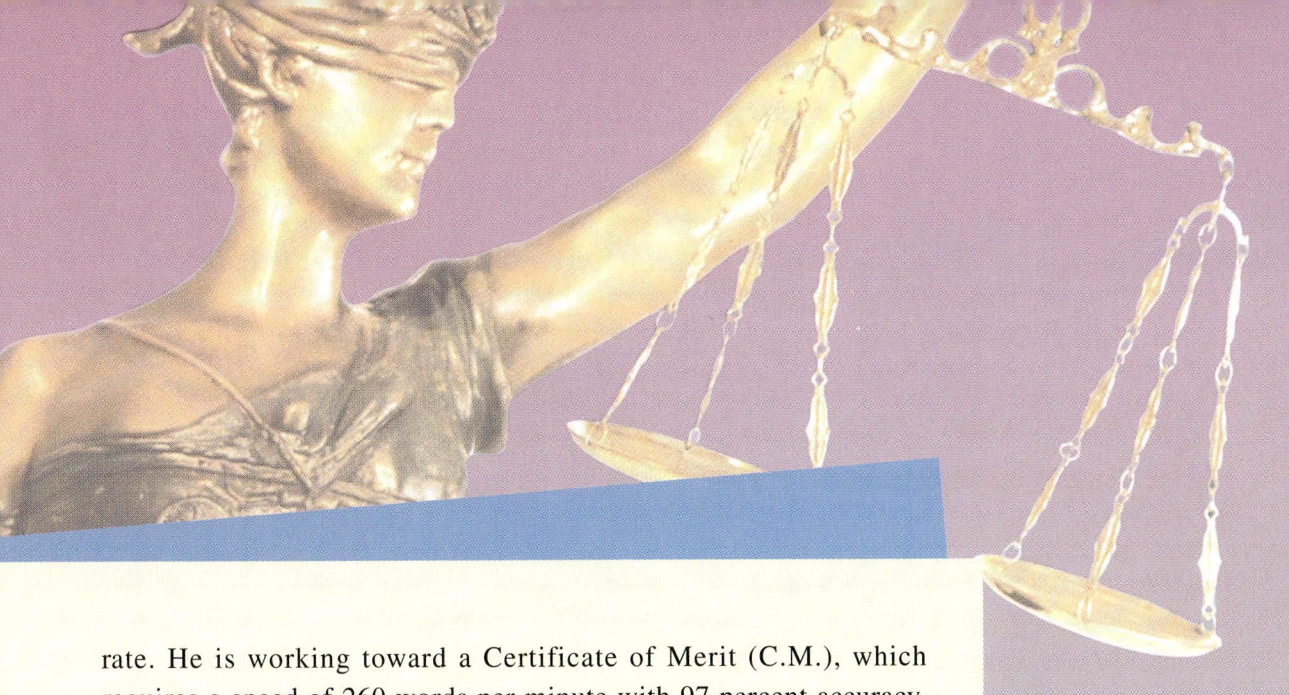

rate. He is working toward a Certificate of Merit (C.M.), which requires a speed of 260 words per minute with 97 percent accuracy. Continued education is required to stay certified.

Several kinds of work opportunities exist for court reporters. Certified reporters work in local, state, and federal courthouses. Their income is steady and benefits are provided. Free-lance reporters may work for law firms, usually out of court, taking depositions. Their work is flexible but not always steady, and benefits are not provided.

Court reporters may also transcribe news shows and sporting events "on the spot" (as words are spoken) to hearing-impaired viewers. These reporters must be able to distinguish homonyms (for example, weather or whether) within microseconds.

Math skills are not a job requirement; but they are helpful for preparing invoices for work performed, keeping business records, and preparing tax returns.

The outlook for court reporting as a career is good. Replacing reporters with recording devices has not been successful. The importance of accurate records ensures that court reporters will be in demand for years to come.

10-1 PASSBOOK SAVINGS ACCOUNTS

OBJECTIVES

In this lesson, you will learn to calculate

- *interest paid on a savings account;*
- *compound interest for simple situations; and*
- *the effective rate of interest.*

Many people put money into savings accounts at banks, credit unions, and savings and loan associations.

◆ Make a list of at least three reasons why you think people put money into savings accounts.

WARM UP

a. Find 6% of $1,200.

b. $\frac{1}{4} \times \$32.80$

c. $0.001 \times \$1,800$

d. 30 m divided by 100 cm = ?

e. Write 0.375 as a percent.

f. Divide $7,000 by $320. Express your answer as a percent, rounded to the nearest tenth.

SAVINGS ACCOUNTS

One reason people open savings accounts is because it is a way of keeping their money safe. Money can be put into or taken out of a savings account easily. Another reason people put money into savings accounts is that they can earn interest on their money. **Interest** is money paid to an individual or institution for the privilege of using their money.

To open an account, you must fill out a **signature card** with your name, address, signature, and Social Security number. This identifies you as the owner of the account. You must also put money into the account. The bank agrees to pay you interest on your money, usually four times a year.

As with a checking account, you may deposit money into or withdraw money from your account. The teller may give you a **receipt,** which is an official record of the transaction. Your transactions may be recorded in a passbook or reported on a statement (usually mailed quarterly).

Critical Thinking In Chapter 7, you learned that when you borrow money for a mortgage loan, you must pay interest to the lending institution. Why do you think banks pay interest to you for putting money into a savings account?

Interest on Savings Accounts

Simple interest is often found on the balance of the account at the end of each quarter. The interest is paid on the first day of the next quarter, or on January 2, April 1, July 1, and October 1. (Since January 1 is a holiday, interest is paid on January 2.) Sometimes interest is found and paid twice a year, or in semiannual periods (six months, or one-half year).

To find the simple interest for any period, first find the interest for a full year. Then multiply that amount by the fraction of a year, such as $\frac{1}{4}$ or $\frac{1}{2}$, for which you want to find interest. For example, if interest is paid at the rate of $5\frac{1}{2}\%$ a year, the interest on \$400.60 paid **semiannually** (twice a year) would be calculated as

$0.055 \times \$400.60 = \$22.033$ interest for one year

$\frac{1}{2} \times \$22.033 = \$11.016$ or \$11.02 interest for semiannual period

Notice that in this type of calculation, you round only the final answer to the nearest cent.

Compounding Interest At the end of each interest period, the interest due is calculated and added to the previous balance in the passbook. The new balance then becomes the principal on which interest is calculated for the next period, if no deposits or withdrawals are made. When you calculate interest and add it to the old principal to make a new principal on which you calculate interest for the next period, you are **compounding interest.** Most banks pay interest compounded either daily or quarterly.

EXAMPLE On January 2, Peter Monroe deposited \$800 in a savings account that pays 6% interest, compounded quarterly. He made no other deposits or withdrawals. If interest is calculated and paid on April 1 and July 1, find the account balance on July 2.

SOLUTION

$\$800 \times 6\% \times \frac{1}{4} = \$12$	Interest for first quarter
$\$800 + \$12 = \$812$	New principal
$\$812 \times 6\% \times \frac{1}{4} = \$12.18$	Interest for second quarter
$\$812 + \$12.18 = \$824.18$	New principal

The account balance on July 2 was \$824.18.

Critical Thinking Explain the difference between simple interest and compound interest.

PROBLEM SOLVING

Use a Formula

When solving problems, you can use a formula.
For instance, the formula for interest is $I = prt$, where I is amount of interest, p is principal amount, r is annual percentage rate, and t is time in years. Study the example below to see how the formula is used.

EXAMPLE If interest is paid at the rate of 6% a year, what is the interest on $270 for one quarterly period?

SOLUTION $p \times r \times t = I$

$\$270 \times 0.06 \times \frac{1}{4} = \$4.05$ interest for one quarter

Critical Thinking Solve the example in the Problem Solving box using the procedure followed for finding simple interest on page 353. What advantages are there to using the formula?

EXERCISE YOUR SKILLS

Find the final balance in each account.

1. April 2, opening deposit, $600; April 17, deposit, $400; May 7, withdrawal, $200; June 21, withdrawal, $300; July 1, interest, $6.82.
2. January 2, opening deposit, $984.31; April 1, interest, $14.29; April 9, withdrawal, $250; May 30, withdrawal, $97; July 1, interest, $9.86; October 1, interest, $10.01.

Find the quarterly simple interest.

3. $300 at 5% a year
4. $750 at 6% a year
5. $217.66 at 4.5% a year
6. $1,400 at $5\frac{3}{4}$% a year

Your bank pays 6% annual interest compounded quarterly on January 2, April 1, July 1, and October 1. You deposited $840 on January 2 and made no other deposits or withdrawals.

7. Find your savings account balance on October 1.
8. How much interest did you earn for these nine months?
9. Jane Eason made a deposit of $1,400 to her savings account on July 1. For the next year she made no other deposits or withdrawals. Annual interest of 5% a year is compounded quarterly. Find the estimated and actual interest earned by Jane by July 1 of the next year.

In solving the following exercises, compound interest is paid quarterly on January 2, April 1, July 1, and October 1. Be sure to read each problem carefully.

10. Lila Mays deposited $400 on April 1 to open an account at the Hiland Savings Bank. The bank pays interest at 6% a year. On July 1, Lila deposited $300 more to her account. She made no other deposits or withdrawals. Find her balance on October 1. *Hint: For July 1, find the interest on $400 before adding the July 1 deposit of $300.*

Some banks pay interest only on the minimum or smallest balance on deposit during an interest period. Ted Garza had a balance of $682 in such an account on July 1. Annual interest is 5% compounded quarterly. He withdrew $170 on August 17 and deposited $200 on September 12.

11. What was Ted's minimum balance during the quarter?

12. How much interest was he paid on October 1?

13. How much did Ted have on deposit on October 1?

MORE ON COMPOUND INTEREST

You have learned to compound interest quarterly and semiannually. Interest may also be compounded annually or daily. Regardless of how interest is earned, the total money in the savings account at the end of the last interest period is called the **compound amount,** assuming that no deposits or withdrawals have been made. The total interest earned, called **compound interest,** is the difference between the original principal and the compound amount.

For example, assume an $800 deposit was kept on deposit for two quarters with interest compounded quarterly at an annual rate of 6%. Assuming no money had been withdrawn, the compound amount would be $824.18. The compound interest is $24.18.

Compound Interest Tables

When you calculate compound interest for several interest periods, you can use a compound interest table like the one in Illustration 10-1.1. The table shows the value of one dollar ($1) after it is compounded for various interest rates and periods.

The table has two parts. The top part is used to calculate annual interest. It may also be used to calculate quarterly and semiannual interest. The bottom part is used to calculate daily interest.

Interest Periods		$1\frac{1}{4}\%$	$1\frac{1}{2}\%$	2%	$2\frac{1}{2}\%$	4%	5%	6%
Annual	1	1.012500	1.015000	1.020000	1.025000	1.040000	1.050000	1.060000
	2	1.025156	1.030225	1.040400	1.050625	1.081600	1.102500	1.123600
	3	1.037971	1.045678	1.061208	1.076891	1.124864	1.157625	1.191016
	4	1.050945	1.061364	1.082432	1.103813	1.169859	1.215506	1.262477
	5	1.064082	1.077284	1.104081	1.131408	1.216653	1.276282	1.338226
	6	1.077383	1.093443	1.126162	1.159693	1.265319	1.340096	1.418519
	7	1.090850	1.109845	1.148686	1.188686	1.315932	1.407100	1.503630
	8	1.104486	1.126493	1.171659	1.218403	1.368569	1.477455	1.593848
	9	1.118292	1.143390	1.195093	1.248863	1.423312	1.551328	1.689479
	10	1.132270	1.160541	1.218994	1.280084	1.480244	1.628895	1.790848
	11	1.146424	1.177949	1.243374	1.312086	1.539453	1.710339	1.898299
	12	1.160754	1.195618	1.268242	1.344888	1.601032	1.795856	2.012197
Daily	30	—	—	—	1.002087	1.003338	1.004175	1.005012
	90	—	—	—	1.006274	1.010048	1.012578	1.015112
	180	—	—	—	1.012587	1.020196	1.025313	1.030452
	365	—	—	—	1.025688	1.041379	1.051998	1.062716

Illustration 10-1.1. Compound Interest Table for $1

Calculating Compound Interest Study the examples below for ways to calculate interest.

EXAMPLES To the nearest cent, find the compound amount for a:

a. $300 deposit at 6% for 5 years

b. $900 deposit at $2\frac{1}{2}\%$ for 8 years

c. $300 deposit at 5% for 30 days

d. $450 deposit at 6% for 90 days

SOLUTIONS

a. 300 × $1.338226 = $401.467800, or $401.47

b. 900 × $1.218403 = $1,096.562700, or $1,096.56

c. 300 × $1.004175 = $301.252500, or $301.25

d. 450 × $1.015112 = $456.800400, or $456.80

Critical Thinking Describe how the table is used to find the interest in each example. Why are the amounts in the table not rounded to the nearest cent since all answers are to be rounded to the nearest cent?

Calculating Quarterly and Semiannual Interest.

When interest is compounded quarterly, you use the multiplier in the table for *four times* the number of annual periods and *one fourth* the rate. When interest is compounded semiannually, you use the multiplier for *twice* the number of annual periods and *one half* the rate.

EXAMPLES

a. Find the amount of $1 compounded quarterly for 3 years at 8%.

b. Find the amount of $1 compounded semiannually for $1\frac{1}{2}$ years at 12%.

SOLUTIONS

a. Use the table for 12 periods (4×3 periods) at 2% ($8\% \times \frac{1}{4}$), or $1.268242.

b. Use the table for 3 periods ($2 \times 1\frac{1}{2}$ periods) at 6% ($12\% \times \frac{1}{2}$), or $1.191016.

EXERCISE YOUR SKILLS

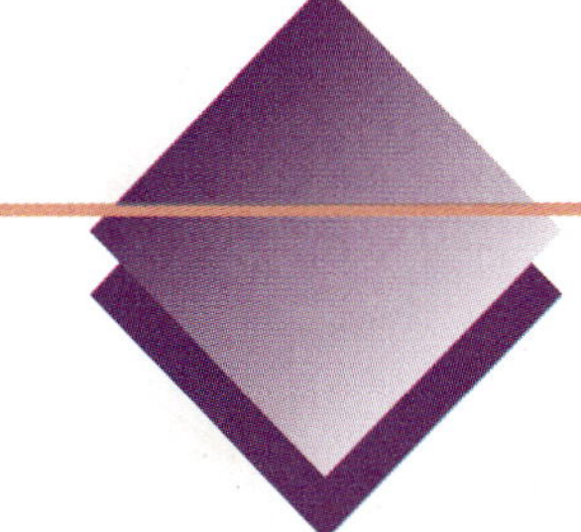

Use the table in Illustration 10-1.1.

	Principal	Rate	Time	Compounded	Compound Amount	Compound Interest
14.	$1,000	5%	4 years	Annually		
15.	700	6%	11 years	Annually		
16.	900	4%	90 days	Daily		
17.	1,000	6%	3 years	Quarterly		
18.	600	8%	2 years	Quarterly		

Calculating the Effective Rate of Interest

The **effective rate of interest** is the rate you actually earn by keeping your money on deposit for one year. The annual rate and the effective rate you earn can be different.

$$\frac{\textbf{Amount of Interest Earned for One Year}}{\textbf{Amount of Money on Deposit}} = \textbf{Effective Rate of Interest}$$

EXAMPLE Find the effective rate of interest for a bank that pays 6% annually compounded quarterly.

SOLUTION A $1,000 deposit will earn interest of $61.36 for one year. (Use the compound interest table.)

$$\frac{\$61.36}{\$1{,}000} = 0.06136 \text{ or } 6.136\% \text{ effective rate of interest}$$

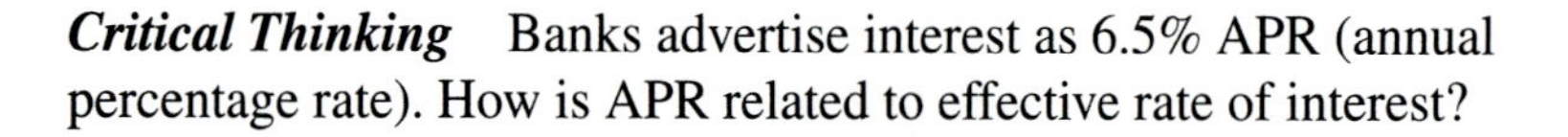

Critical Thinking Banks advertise interest as 6.5% APR (annual percentage rate). How is APR related to effective rate of interest?

EXERCISE YOUR SKILLS

19. You have $400 in an account that yields an effective interest rate of 4.75% per year. How much interest will you earn in one year?

20. The Coles Savings Bank pays 6.2% interest compounded annually. The Drew Savings Bank pays 6% interest compounded daily. Assume a deposit of $2,000 is made in each bank and kept there for one year, or 365 days. At which bank will the higher effective rate of interest be paid? What is the effective rate paid by the higher-paying bank, to the nearest hundredth percent?

For each deposit, find the amount of interest earned for *one year* and the effective rate of interest to the nearest hundredth of a percent.

	Deposit	Annual Rate	Compounded	Interest Earned	Effective Rate of Interest
21.	$1,000	5%	Quarterly		
22.	800	6%	Quarterly		
23.	1,400	10%	Semiannually		
24.	1,810	3%	Semiannually		

25. ♦ Visit or bring in advertisements from at least 3 local banks. Find the annual interest rates for regular savings deposits of $500, $5,000, and $50,000. Determine the annual effective rate for each amount. What patterns do you see? Write a paragraph identifying which bank you would prefer to put each amount of money in and why.

MIXED REVIEW

1. Find $\frac{2}{5}$ of $170.

2. Find $1\frac{1}{2}$% of $3,200.

3. $490 is what percent of $560?

4. $32.80 increased by 60% of itself is?

5. $1.50 is what percent greater than $1.20?

6. A real estate agent sold a piece of land for $1,245,000. The agent charged a $6\frac{1}{2}$% commission. What net proceeds did the seller receive?

7. On July 1, Nell Griffith deposited $4,000 in a savings account that pays 7% interest per year, compounded quarterly. Interest was added to her account on October 1 and January 2. There were no other deposits or withdrawals. What total amount did Nell have on deposit on January 2?

10-2 SPECIAL SAVINGS ACCOUNTS

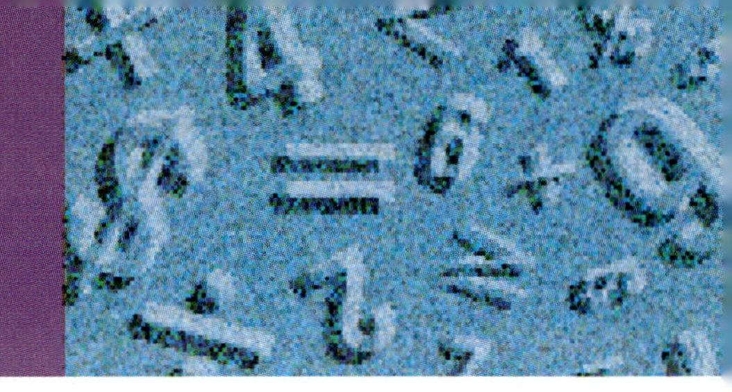

Banks often pay a higher interest rate on savings accounts to customers who agree to keep their money on deposit for a period of time, such as a year, and not make any withdrawals.

◆ Why do banks encourage people to use such savings accounts? Make a list of reasons why banks can afford to pay higher-than-usual interest rates.

OBJECTIVES

In this lesson, you will learn to

- *compare time-deposit accounts to passbook savings accounts; and*
- *calculate penalties on time-deposit accounts.*

WARM *UP*

a. Find 150% of $2,500.

b. Find 1.25% of $2,540.

c. 348 is what percent of 580?

d. $29.40 increased by 20% of itself is what amount?

e. $4.50 is what percent greater than $3.60?

f. 1.8 kg of meat at $6.25 per kg is what amount?

g. Round 330,099 to the nearest thousand.

SPECIAL SAVINGS ACCOUNTS

In addition to regular passbook or statement savings accounts, many banks also offer special savings accounts for long-term savers. The rates paid on these special savings accounts vary, but all banks pay a higher rate of interest on them than on regular savings accounts. Two of the special savings accounts are **time-deposit accounts** and **money market accounts.**

Time-Deposit Accounts

The **time-deposit account** is also known as a **certificate of deposit** or a **savings certificate.** Deposits in time-deposit accounts are usually insured.

Some government rules apply to time-deposit accounts. In addition, most banks require depositors to meet certain requirements. While the specific requirements vary, the following are typical.

- They must deposit a minimum amount. This may be \$500, \$1,000, \$5,000, or \$10,000. More than the minimum may be deposited.
- They must leave the money on deposit for a minimum time. The time may be three months, six months, or from one to ten years. The minimum time may also be called the **term.** The date that marks the end of the term is the **maturity date.**
- They must pay a penalty if money is withdrawn from the account before the end of the term of the deposit. The penalty involves the loss of interest for a certain period of time calculated on the amount of money withdrawn.

Penalties on Time-Deposit Accounts Banks usually charge a penalty for withdrawing money early from a time-deposit account. The penalty is calculated on the money withdrawn from the time deposit account before the end of its term.

EXAMPLE Henrietta Fenwick invested \$3,000 in a 4-year time-deposit account that paid 8% interest compounded annually. She withdrew \$500 before the end of the 4 years. The penalty for early withdrawal was 6 months' interest. What was the amount of the penalty?

SOLUTION $\$500 \times 8\% \times \frac{1}{2} = \$20$ In this case, the penalty applies only to the money withdrawn from the account. Therefore, to calculate the penalty, use the interest formula to figure the simple interest on that amount.

Each bank sets its own penalty for early withdrawals from time-deposit accounts. The penalty usually varies with the term of the time deposit. For example, a 1-year time deposit may carry a penalty of 3 months' interest; a 5-year time deposit may have a 12 months' interest penalty.

It is possible for the penalty to be greater than the interest earned. When that happens, the difference is deducted from the money the depositor is withdrawing to pay the penalty.

Comparing Savings Accounts There are several factors to consider when deciding what kind of a savings account to put your money in. These include how quickly you want to be able to withdraw money, account requirements, and potential interest. You can compare the earning potential of different types of accounts by calculating the interest that would be earned in each type of account over the same time period.

EXERCISE YOUR SKILLS

A $3,000 deposit in a passbook savings account earns interest of $41.25 in 3 months. A 3-month certificate account pays interest at the rate of 6.2% a year calculated and paid each quarter.

1. How much interest is earned on the certificate in 3 months?
2. How much more interest is earned in 3 months by investing in the certificate rather than the passbook account?
3. Your credit union offers a 6-month time-deposit account that pays interest at the rate of 5.8% a year. You plan to deposit $2,000 in this account for 6 months and then reinvest the principal and interest for another 6 months at the same rate. How much interest will you earn in a year?
4. Your bank compounds interest quarterly at 3.5% a year on passbook savings. It also pays 5.45% annual interest on time deposits. How much more would you earn in a year on $7,000 invested in a time deposit rather than a passbook account?

Eli Sarissian has $9,000 on deposit in a 1-year Super Saver Time Deposit account. His contract states that he will be penalized 3 months' interest if he withdraws any part of his deposit before the end of 1 year. Find the penalty if Eli withdraws these amounts early and the penalty is calculated at 7% a year:

5. $2,000
6. $5,000

Tillie Figrel has earned $540 in interest on a $24,000 6-month Money Management Certificate of Deposit she has held for 3 months. She needs the $24,000 now for a down payment on a home. Tillie will have to pay a penalty of 2 months' interest at 9% a year on the $24,000 if she withdraws the money early. *Hint: Interest earned minus penalty equals net interest.*

7. What is the amount of the penalty for early withdrawal?
8. What net interest will she receive after paying the penalty?
9. Earl Comito invested $4,000 in a time-deposit account. After one month, he earned $23.33 in interest and withdrew all of the money. The bank charged a penalty of 6 months' interest at 7%. How much did he receive?

Money Market Accounts

Like time-deposit accounts, **money market accounts** are special savings accounts that offer higher interest rates than regular accounts. If opened in a bank, they are also usually ensured up to $100,000.

- A minimum balance must be kept in the account for the time specified.
- The interest rate paid is fixed for short periods of time, such as 1 month, 3 months, or 6 months.
- A small number of checks may be written against the account.

Banks usually pay a higher interest rate for larger minimum balances. Money may be withdrawn as long as the minimum balance is maintained. If the minimum balance is not kept, a lower interest rate will be paid.

EXERCISE YOUR SKILLS

Gertrude Olkin deposited $3,000 into a money market account. For the first month, her account earned interest at the annual rate of 3.27%. The interest was not added to the account balance. For the next month, interest was paid at the annual rate of 3.3%. Find the amount of interest she was paid for the

10. First month.

11. Second month.

12. For the first quarter of the year, the North Hill Savings & Loan Association paid a rate of 5.43% on Money Market Investment accounts with balances of $1,000 to $5,000. For money market accounts with a balance of $5,001 to $10,000, North Hill added an extra 0.14% to the rate paid. An extra 0.12% was added for money market account balances greater than $10,000. Giancomo Pardo's money market account had a balance of $8,400 during the first quarter. What amount of interest was he paid for the quarter?

13. Write a comparison of the advantages and disadvantages of time-deposit, money market, and regular savings accounts.

14. ◆ Make a list of factors you consider important in choosing a bank in which to save your money. Group your factors into economic and non-economic reasons.

MIXED REVIEW

1. 5 000 mg ÷ 2 = ? g

2. A room 7 m by 5 m = ? m^2

3. $480 is what percent greater than $320?

4. $0.90 is what percent less than $1.20?

5. Alan Daniels had a garden that measures 5.2 meters by 3.5 meters. Find the area of his garden.

6. The Oak Savings Bank compounds interest semiannually at 3.14% a year on passbook savings accounts. It also pays 6.35% annual interest on time-deposit accounts. How much more would you earn in one year if you deposited $6,000 into a time-deposit account instead of a passbook account?

7. Gilbert Zer has a $7,500 savings certificate that has earned $225 in interest. If he withdraws the $7,500 before the certificate term ends, he will pay a penalty of 1 month's interest at 6% annual percentage rate. What net interest will he earn on his investment if he withdraws his money early and pays the penalty?

10-3 PROMISSORY NOTES

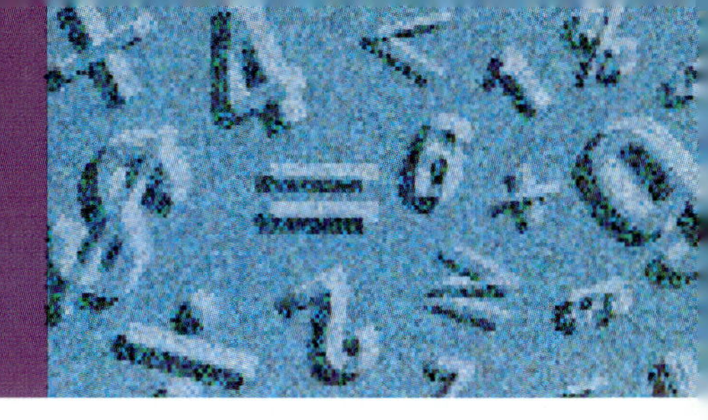

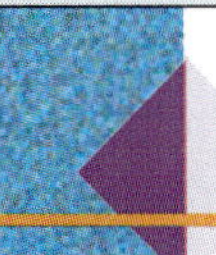

A bank wants to charge you 15% interest on a personal loan for your next vacation. The same bank will charge you only 11% interest on a loan for your next car.

◆ On a sheet of paper, make a list of reasons why you think a bank might charge more interest on a vacation loan than a car loan. Can you think of any reasons you could give a banker to convince them to charge you the same rate on the vacation loan?

OBJECTIVES

In this lesson, you will learn to find

- *interest and amount due on interest-bearing notes and discounted notes for one year;*
- *interest rates for periods other than one year; and*
- *the rate of interest charged on a note.*

WARM UP

a. Find 5% of $1,450.

b. Find 8.5% of $560.

c. Write 1.7 as a percent.

d. $28 is what percent of $700?

e. Write $4\frac{1}{2}\%$ as a decimal.

f. What is 35% of 56 m?

PROMISSORY NOTES

When you borrow money, you usually sign a **promissory note.** This note is your written promise, or IOU, that you will repay the money to the lender on a certain date. Usually you also have to pay for using the lender's money. That cost is called **interest.** Interest is like paying rent to use someone else's money. A note that requires you to pay interest is called an **interest-bearing note.**

Lenders may require a borrower to deposit or pledge property as security for a loan. This property is called **collateral.** Types of collateral that are often used to secure loans are cars, stocks, bonds, and life insurance. Many lenders offer **home equity loans** to home owners. **Home equity** is the owner's stake in a home. It is the difference between what the home could be sold for and what is owed on it. To obtain a home equity loan, the borrower pledges the equity in the home as collateral for the loan. If the loan is not repaid, the lender can seize the collateral and sell it to get the borrowed money back.

Using an Interest-Bearing Note On April 4, 1996, Bess Baker borrowed $3,500 from her bank to buy a used car, which she used as collateral for the loan. Bess signed a **promissory note.** The amount borrowed, $3,500, is the **face,** or **principal.** The day the note was signed, April 4, 1996, is the **date of the note.** The time for which the money was borrowed, one year, is called the **time.** The date the money must be repaid, April 4, 1998, is the **due date,** or **maturity date.** The rate of interest that Bess must pay, 12%, is called the **rate of interest.**

Finding the Interest on a Loan for One Year The formula $I = prt$ is used to find interest owed on a loan.

EXAMPLE Find the amount of interest Bess must pay. Then find the amount she must repay at the end of 2 years.

SOLUTION

$I = prt$	Use the formula
$I = \$3{,}500 \times 0.12 \times 2$	Substitute known values
$= \$840.00$	Interest paid
$\$3{,}500 + \$840 = \$4{,}340$	Amount due at end of 2 years

Using the Interest Formula to Find Interest on a Loan for a Period Other Than One Year Interest rate is always stated as a rate per year. The interest you pay on a loan is proportional to the time for which you borrow the money. For example, for $\frac{1}{4}$ of a year, the interest is $\frac{1}{4}$ of one year's interest. You can calculate interest for periods other than one year with the interest formula. For example,

Principal	×	Rate	×	Time in Years	=	Interest
$7500	×	0.12	×	$\frac{1}{4}$ yr	=	$225
$7500	×	0.12	×	$3\frac{1}{2}$ yrs	=	$3,150
$7500	×	0.12	×	5 years	=	$4,500

Critical Thinking Look at the chart. What relationship do you see between interest and time?

EXERCISE YOUR SKILLS

Find the interest and the amount due at maturity for each note.

	Face of Note	Time	Rate	Interest	Amount Due at Maturity
1.	300	3 yr	12%		
2.	250	3 mo	14%		
3.	720	$2\frac{1}{4}$ yr	$8\frac{1}{2}$%		

4. To finance the remodeling of his basement, Al borrowed $16,200 on an 18-month home equity loan. He signed a promissory note bearing interest at $10\frac{1}{2}$%. What total amount did Al pay on the due date?

5. Carlotta borrowed $2,000 for a vacation trip to Europe. The promissory note she signed was for 3 months at $15\frac{1}{4}$% interest. How much did Carlotta have to pay when the note came due?

EXACT INTEREST AND BANKER'S INTEREST

In earlier lessons, you have dealt with years as whole numbers or simple fractions like $\frac{1}{2}$ or $\frac{1}{4}$. Sometimes in calculating interest for time less than one year, other numbers are used. When the time is shown in days, interest is calculated by either the **exact interest method** or by the **banker's interest method.** Exact interest uses a 365-day year. Banker's interest uses a 360-day year.

The *exact interest method* is used by the United States government and by many banks and other businesses. To find exact interest, you use a year of 365 days. So you show the time as a fraction with 365 as the denominator. For example, you would show 83 days as $\frac{83}{365}$. The exact interest on $1,000 at 6% for 83 days would be

$$P \times R \times T = I$$

$$\$1{,}000 \times 0.06 \times \frac{83}{365} = \$\overset{10}{\cancel{1{,}000}} \times \frac{6}{\underset{1}{\cancel{100}}} \times \frac{83}{365} =$$

$$\overset{12}{\cancel{60}} \times \frac{83}{\underset{73}{\cancel{365}}} = \frac{996}{73} = \$13.6438 \text{ or } \$13.64\text{, to the nearest cent.}$$

The *banker's interest method* is used in place of the exact interest method by some businesses. With this method of finding interest, a year has only 360 days. The 360-day year has 12 months of 30 days each and is known as the *banker's year*. Of course, there really is no such year. It is used because it is easier to calculate with than a 365-day year. Banker's interest on $1,000 at 6% for 83 days would be

$$P \times R \times T = I$$

$$\$1{,}000 \times 0.06 \times \frac{83}{360} = \$\overset{1}{\cancel{60}} \times \frac{83}{\underset{6}{\cancel{360}}} = \frac{83}{6} = 13\frac{5}{6} = \$13.8333$$

or $13.83, to the nearest cent.

Finding the Rate of Interest

If you know the principal and the amount of interest for one year, you can find the rate of interest by dividing the interest by the principal.

$$\textbf{Rate of Interest} = \frac{\textbf{Interest for One Year}}{\textbf{Principal}}$$

If the interest given in the problem is not for a year, you must first find how much the interest would be for one year.

EXAMPLE Ella Stein paid $30 interest on a loan of $1,000 for 3 months. Find the rate of interest she paid.

SOLUTION 3 months $= \frac{1}{4}$ year

$\$30 \times 4 = \$120$	Annual interest
$\frac{\$120}{\$1{,}000} = 0.12$ or 12%	Rate of interest

EXERCISE YOUR SKILLS

First find the exact interest to the nearest cent. Then find the banker's interest to the nearest cent.

6. $960 @ 15% for 270 days

7. $1,800 @ 15% for 36 days

8. $1,200 @ 8% for 240 days

9. $2,400 @ 9% for 45 days

10. $650 @ 6% for 146 days

11. $1,450 @ 7% for 100 days

Larry Wang borrowed $5,000 for 180 days. He paid exact interest at an annual rate of 12%.

12. Estimate the interest Larry owed.

13. What is the exact amount of interest he had to pay?

14. What total amount did he have to repay?

Rob Tovich signed a 180-day note for $1,825. He repaid the loan when due with interest at the annual rate of 12% using a banker's year.

15. How much interest did Rob pay?

16. What total amount did he pay?

17. Tabatha Kerr borrowed $2,400 for 6 months and paid $132 interest. What rate of interest did she pay?

18. Evan Cole paid $19.50 in interest on a loan of $2,600 for 1 month. What rate of interest did he pay?

19. Ali Ahmed borrowed $6,000 and paid $945 in exact interest when the loan came due $1\frac{1}{2}$ years later. What rate of interest did Ali pay?

20. Laura Weiss borrowed $1,500 for 21 months. The total interest she paid was $315. What rate of exact interest did Laura pay?

DISCOUNTED NOTES

Banks and other lenders may lend money to businesses and people for short periods of time, such as 30, 60, or 90 days. These loans are called short-term loans.

Discounting a Noninterest-Bearing Note

When a bank makes a short-term loan, it may require the borrower to sign a note and pay the interest when the loan is made. When interest is collected in advance this way, it is known as a **bank discount.** Because the interest is paid in advance, the note itself does not show any interest rate, and it is called a **noninterest-bearing note.**

The bank collects the bank discount by deducting it from the face of the note. The amount the borrower gets is the face of the note less the discount. When the loan is due, only the face of the note is paid. Obtaining a loan in this way is known as discounting a note.

The percent of discount charged by the bank is called the **rate of discount.** The amount of money that the borrower gets is called the **proceeds.**

EXAMPLE The Palatine State Bank discounted a $6,500 note for Daniel A. Witt at 14% interest for 3 months. Find the proceeds of the note that Daniel receives.

SOLUTION $I = prt$
$I = \$6{,}500 \times 0.14 \times \frac{1}{4} = \$227.50$
Face amount − Interest = Proceeds
$\$6{,}500 - \$227.50 = \$6{,}272.50$
Daniel received $6,272.50

Note that three months later, Daniel would repay $6,500.

Real Rate of Interest When you discount your own noninterest-bearing note, you pay a rate of interest based on the face of the note. However, you do not get the full face amount because interest is deducted in advance. To find the real rate of interest you paid, you must divide the interest paid by the amount you actually received.

For example, if you borrow $100 for one year on a noninterest-bearing note at 10%, you will pay $10 in interest. If you discount that note, however, you have really borrowed only $90 because the interest is deducted in advance. The rate of interest you paid is not $\frac{\$10}{\$100}$. It is $\frac{\$10}{\$90}$. This is a real rate of 11.1%, rounded to the nearest tenth of a percent.

EXAMPLE Find the real rate of interest on Daniel A. Witt's note in the example above.

SOLUTION 3 months = $\frac{1}{4}$ of a year
$\$227.50 \times 4 = \$910$ Interest for 1 year
$\frac{\$910}{\$6{,}272.50} = 0.1451$ or 14.5% Real interest rate

So Daniel actually paid a 14.5% rate of interest.

EXERCISE YOUR SKILLS

Find the proceeds of each note. Each note was discounted on the same day as the date of the note.

	Date of Note	Face	Time	Discount Rate	Proceeds
21.	May 14	$2,000	2 months	9%	
22.	Sept. 3	5,900	3 months	15%	
23.	Feb.22	9,000	4 months	10%	
24.	Mar. 6	12,000	6 months	12%	
25.	Dec.15	500	1 month	8%	

On July 22, the bank discounted Al Klein's 10%, 3-month noninterest-bearing $995 note.

26. Estimate the bank discount.

27. What were the actual proceeds of the note?

PUT IT ALL TOGETHER

◆ On September 23, Eversville Bank discounted Triad Corporation's note for $10,000. The note was dated September 23, due in 3 months, and it carried no interest. The rate of discount was 15%.

28. What was the amount of the bank discount?

29. What proceeds did Triad Corporation receive?

30. What real rate of interest, to the nearest tenth of a percent, did Triad Corporation pay?

The bank discounted Rita Esposito's 6-month noninterest-bearing note for $4,200. The rate of discount was 13%.

31. What was the amount of the bank discount?

32. What proceeds did Rita receive?

33. How much did Rita pay her bank on the maturity date?

34. What real rate of interest, to the nearest tenth of a percent, did Rita pay?

MIXED REVIEW

1. $532.2 + 35.25 + 0.26$
2. 3.52×0.525
3. Write $9\frac{7}{8}\%$ as a decimal.
4. \$90 is $7\frac{1}{2}\%$ of what amount?
5. \$10.64 is 5% less than what amount?
6. Jules began the day with \$75. During the day he received \$25 for mowing a lawn and spent \$1.25 for gas, \$5 for a movie, \$25 for a present, and \$3.75 for lunch. How much money did Jules have at the end of the day?
7. Ellery Clarke's sales for 5 months were \$27,508, \$29,686, \$29,130, \$27,980, and \$28,649. What must be his sales next month if he wants his monthly average to be \$28,000?
8. Karl Obenhous, a secretary, is paid a yearly salary of \$15,860. This is equal to how much a week?
9. Edith and Betty share the monthly costs of their apartment in the ratio of 2 to 3, respectively. The total cost of their apartment last year, including rent, heat, and power, was \$9,300. What was Betty's share of the costs?
10. Henry's sales last week totaled \$15,000. His gross earnings for the week were \$650, of which \$200 was salary and the rest was commission. What percent commission was Henry paid?
11. A collection agency collected 75% of an outstanding bill of \$580 and charged 35% commission. What were the net proceeds for the original lender?
12. On May 31, Eve Day's balances were checkbook, \$339.11 and bank statement, \$394.62. A service charge of \$1.74 had not been deducted in the checkbook. Checks outstanding were #24, \$41.32; #25, \$3.18; #27, \$12.75. Prepare a reconciliation statement for Eve.
13. A nut mixture is made up of peanuts and cashews in the ratio of 3 to 2, respectively. How many pounds of each are needed to make a 45-lb. mixture?
14. Alexandra drove 503 miles in 10.5 hours. What was her average speed?

10-4 INTEREST AND DATE TABLES

OBJECTIVES

In this lesson, you will learn to

- *calculate interest using simple interest tables; and*
- *find the maturity date of a note.*

You started a marathon race at 9:30 A.M. and finished at 3:18 P.M. What was your **elapsed** time in minutes for running the race?

◆ Explain the steps you took to determine elapsed time. What does "elapsed time" mean? Why do you need to find the elapsed time? Work with a partner. Have one person name a starting time or date and an ending time or date and the other find the elapsed time. Exchange roles and repeat.

WARM UP

a. Write $\frac{3}{5}$ as a percent.

b. Write 4.28% as a decimal.

c. Find $112\frac{1}{2}\%$ of $200,000.

d. $600 is what percent of $1,500?

e. 12% more than $248 is what number?

f. $156 less 6% of itself is what number?

INTEREST TABLES

Most banks use computers or specially programmed calculators to find interest and calculate payment amounts on loans. However, some banks use interest tables like the one shown in Illustration 10-4.1 as a quick reference chart. To calculate interest when you do not have the technology available, you can use the tables as well.

The table shows the interest on $100 for a 365-day year. To find the interest on any amount of money using the table, follow these steps:

- Find the interest on $100 for the interest rate and days you want. Do this by reading down the rate column to the point across from the time you want.
- Find the number of hundreds of dollars in the principal by dividing the principal by $100. Do this by moving the decimal in the principal two places to the left.
- Multiply the interest for $100 by the number of hundreds in the principal.

SIMPLE INTEREST TABLE

Interest on $100 for a 365-Day Year

Time (Days)	8%	8½%	9%	9½%	10%	10½%	11%	11½%	12%	12½%
1	0.0219	0.0233	0.0247	0.0260	0.0274	0.0288	0.0301	0.0315	0.0329	0.0342
2	0.0438	0.0466	0.0493	0.0521	0.0548	0.0575	0.0603	0.0630	0.0658	0.0685
3	0.0658	0.0699	0.0740	0.0781	0.0822	0.0863	0.0904	0.0945	0.0986	0.1027
4	0.0877	0.0932	0.0986	0.1041	0.1096	0.1151	0.1205	0.1260	0.1315	0.1370
5	0.1096	0.1164	0.1233	0.1301	0.1370	0.1438	0.1507	0.1575	0.1644	0.1712
6	0.1315	0.1397	0.1479	0.1562	0.1644	0.1726	0.1808	0.1890	0.1973	0.2055
7	0.1534	0.1630	0.1726	0.1822	0.1918	0.2014	0.2110	0.2205	0.2301	0.2397
8	0.1753	0.1863	0.1973	0.2082	0.2192	0.2301	0.2411	0.2521	0.2630	0.2740
9	0.1973	0.2096	0.2219	0.2342	0.2466	0.2589	0.2712	0.2836	0.2959	0.3082
10	0.2192	0.2329	0.2466	0.2603	0.2740	0.2877	0.3014	0.3151	0.3288	0.3425
11	0.2411	0.2562	0.2712	0.2863	0.3014	0.3164	0.3315	0.3466	0.3616	0.3767
12	0.2630	0.2795	0.2959	0.3123	0.3288	0.3452	0.3616	0.3781	0.3945	0.4110
13	0.2849	0.3027	0.3205	0.3384	0.3562	0.3740	0.3918	0.4096	0.4274	0.4452
14	0.3068	0.3260	0.3452	0.3644	0.3836	0.4027	0.4219	0.4411	0.4603	0.4795
15	0.3288	0.3493	0.3699	0.3904	0.4110	0.4315	0.4521	0.4726	0.4932	0.5137
16	0.3507	0.3726	0.3945	0.4164	0.4384	0.4603	0.4822	0.5041	0.5260	0.5479
17	0.3726	0.3959	0.4192	0.4425	0.4658	0.4890	0.5123	0.5356	0.5589	0.5822
18	0.3945	0.4192	0.4438	0.4685	0.4932	0.5178	0.5425	0.5671	0.5918	0.6164
19	0.4164	0.4425	0.4685	0.4945	0.5205	0.5466	0.5726	0.5986	0.6247	0.6507
20	0.4384	0.4658	0.4932	0.5205	0.5479	0.5753	0.6027	0.6301	0.6575	0.6849
21	0.4603	0.4890	0.5178	0.5466	0.5753	0.6041	0.6329	0.6616	0.6904	0.7192
22	0.4822	0.5123	0.5425	0.5726	0.6027	0.6329	0.6630	0.6932	0.7233	0.7534
23	0.5041	0.5356	0.5671	0.5986	0.6301	0.6616	0.6932	0.7247	0.7562	0.7877
24	0.5260	0.5589	0.5918	0.6247	0.6575	0.6904	0.7233	0.7562	0.7890	0.8219
25	0.5479	0.5822	0.6164	0.6507	0.6849	0.7192	0.7534	0.7877	0.8219	0.8562
26	0.5699	0.6055	0.6411	0.6767	0.7123	0.7479	0.7836	0.8192	0.8548	0.8904
27	0.5918	0.6288	0.6658	0.7027	0.7397	0.7767	0.8137	0.8507	0.8877	0.9247
28	0.6137	0.6521	0.6904	0.7288	0.7671	0.8055	0.8438	0.8822	0.9206	0.9589
29	0.6356	0.6753	0.7151	0.7548	0.7945	0.8342	0.8740	0.9137	0.9534	0.9932
30	0.6575	0.6986	0.7397	0.7808	0.8219	0.8630	0.9041	0.9452	0.9863	1.0274
31	0.6795	0.7219	0.7644	0.8068	0.8493	0.8918	0.9342	0.9767	1.0192	1.0616

Illustration 10-4.1. Simple Interest Table

EXAMPLE Find the interest on $650 for 20 days at 10%.

SOLUTION

$ 0.5479	Interest on $100 for 20 days at 10%
× 6.5	Principal, $650, divided by $100
$3.56135	Interest on $100 × number of hundreds

The interest is $3.56. Round to the nearest cent.

When the number of days you want is not shown in the table, you must combine multipliers to get the number.

EXAMPLE Find the interest on $250 for 45 days at 8%.

SOLUTION

$0.6575	Interest on $100 for 30 days at 8%
+0.3288	Interest on $100 for 15 days at 8%
$0.9863	Interest on $100 for 45 days at 8%
× 2.5	Principal, $250, divided by $100
$2.4675	Interest on $100 × number of hundreds
The interest is $2.47.	Round to the nearest cent.

Interest for a rate not shown on the table can be found in much the same way. For example, the interest on $100 @ $18\frac{1}{2}$% for 20 days is the sum of the amount for 9% ($0.4932) and the amount for $9\frac{1}{2}$% ($0.5205).

EXERCISE YOUR SKILLS

Use Illustration 10-4.1 to find the interest to the nearest cent.

1. $800 @ 10% for 10 days
2. $400 @ 12% for 15 days
3. $3,250 @ 9% for 25 days
4. $75 @ 12% for 20 days
5. $300 @ $9\frac{1}{2}$% for 45 days
6. $500 @ 12% for 60 days
7. $750 @ 9% for 120 days
8. $400 @ 19% for 10 days
9. $2,500 @ 20% for 30 days
10. $900 @ 18% for 16 days

Juanita Torres borrowed $700 on a note for 60 days with interest at 12%.

11. Using Illustration 10-4.1, what interest did she pay?
12. What total amount did she owe when the note was due?

You need to borrow $500 for 20 days to help pay a $750 bill due on May 6. You can borrow the money at 17%.

13. What will be the interest using a banker's year?
14. What will be the interest using Illustration 10-4.1?

DUE DATES

A promissory note usually shows on its face the due date, or maturity date, of the note. In this section, you will learn how to find the due dates of notes. You will also learn how to find the number of days between the date of a note and its maturity date.

Finding the Due Date When the Time Is in Months When the time of a note is in months, you find the due date by counting that number of months forward from the date of the note. The due date is the same day in the month you stop with as the date of the note. If the date is the last day of the month, the note is due on the last date of the month in which the note is due. For instance, a 1-month note issued on January 31 would be due February 28 (or February 29 in a leap year).

Finding the Due Date When the Time Is in Days The time of a note may be shown in days. In this case, you can find the due date by counting, from the date of the note, the indicated number of days.

EXAMPLE Find the maturity date of a 90-day note dated May 28.

SOLUTION

	90 days	Length of note
May	− 3 days	Days left in May
	87 days	Subtract
June	− 30 days	Days in June
	57 days	Subtract
July	− 31 days	Days in July
	26 days	Subtract
August 26		Maturity date

Finding the Number of Days between Two Dates Sometimes you may need to find the number of days between two dates.

EXAMPLE Find the number of days from June 14 to August 23.

SOLUTION

June 14 to June 30	16 days
July 1 to July 31	31 days
August 1 to 23	+ 23 days
Total	70 days

EXERCISE YOUR SKILLS

Find the due date for each note. Assume February has 28 days.

	Date Issued	Time		Date Issued	Time
15.	February 12	1 month	**16.**	November 14	2 months
17.	March 31	4 months	**18.**	March 31	3 months
19.	March 5	30 days	**20.**	January 29	60 days
21.	January 30	45 days	**22.**	April 14	75 days
23.	December 28	80 days	**24.**	June 9	120 days

Find the number of days from

25. January 5 to March 12

26. November 12 to January 29

27. May 6 to August 22

28. September 6 to January 4

29. February 23 to May 5

30. July 29 to August 8

PUT IT ALL TOGETHER

♦ On March 5, Suba Jain signed a note for $5,000 at 14% exact interest. She paid the note on June 3.

31. For how many days was interest due?

32. What amount of interest did Suba owe?

33. What was the total amount due on June 3?

34. On July 13, Tony Dimato borrowed $8,000 to buy a new car for $12,500. He signed a note with exact interest at 8%. If he paid 100% of the note off on September 24, what total amount did he owe?

MIXED REVIEW

1. 3.75 equals what percent?

2. $\frac{3}{4}$ equals what percent?

3. $30 is what percent of $250?

4. What amount is $1\frac{1}{4}\%$ of $84?

5. Find the due date of a 3-month note dated August 31.

6. The interest on a loan of $4,300 for 3 months is $129. What is the rate of interest?

The population of Akins this year was 23,205. Last year, it was 21,000.

7. Estimate the percent of increase.

8. What is the exact percent of increase?

9. A 14-oz box of Vigor hot cereal costs $0.77. A $1\frac{1}{8}$-lb box of Wakup hot cereal costs $1.08. Which brand costs less per pound and how much less?

10. Edith bought fertilizer for her lawn. Her lot was 25 m wide and 50 m long. Her house took up 100 m^2 of the lot. If one bag of fertilizer covered 350 m^2, how many whole bags did Edith need?

11. On July 1, Ken Van's balances were bank statement, $456.70 and check register, $333.25. The bank statement included a $1.75 service charge and $1.45 in earned interest. The following checks were outstanding: #678, $56; #682, $23; #683, $44.75. Complete a bank reconciliation statement for Ken.

12. A store sells shirts for $17.80 each, or 3 for $49.95. How much would you save by buying 3 shirts at a time instead of 3 shirts one at a time?

10-5 INSTALLMENT BUYING

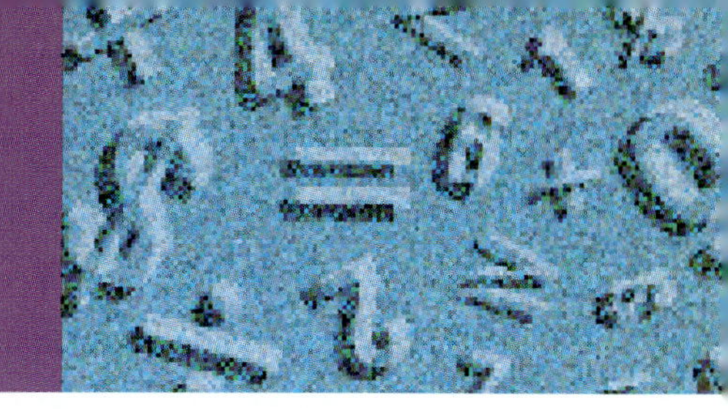

"**B**ut I need a car to get to work, and I need to work to earn money to buy a car!" exclaimed Tracy.

◆ Discuss Tracy's problem and how it might be solved. Work in small groups. Brainstorm ideas that might help Tracy solve the dilemma.

OBJECTIVES

In this lesson, you will learn to

- *find the installment price and finance charge on installment loans;*
- *find monthly payments; and*
- *calculate the annual percentage rate (APR) on a loan.*

WARM UP

a. 116% of $715 is what amount?

b. What amount is $37\frac{1}{2}\%$ of $120?

c. What is $16\frac{1}{4}\%$ as a decimal?

d. Find the number of days between March 22 and July 2?

e. 35% less than $230 is what amount?

INSTALLMENT BUYING

Boats, cars, TV sets, VCRs, furniture, clothes, and many other items are often bought on an **installment plan.** The installment plan is also called the *time payment plan.* When you buy on the installment plan, you are really borrowing the seller's money and paying it back in part payments.

You may have to make a **down payment** of part of the price at the time you buy something on an installment plan. You may also have to sign an **installment contract** in which you agree to pay the unpaid balance in weekly or monthly payments.

Finding the Installment Price and Finance Charge The installment price is higher than the cash price because the seller adds a **finance charge** to the cash price. This charge pays the seller interest on the money and covers the extra cost of doing business on the installment plan. If there are no other charges, such as insurance, the finance charge is the difference between the installment price and the cash price. A wise buyer compares the costs of installment buying and paying cash.

EXAMPLE A camcorder has a cash price of $1,300. To buy it on an installment plan, you pay $130 down and $75 a month for 18 months. Find the finance charge. By what percent is the installment price greater than the cash price?

SOLUTION

$ 130	Down payment	
+1,350	($75 × 18 months)	Total payments
$1,480	Installment price	
−1,300	Cash price	
$ 180	Finance charge	

$\frac{\$180}{\$1,300} = 0.13846$ or 13.8% Percent greater

Finding the Number of Months and Monthly Payment Sometimes you may know the installment price and down payment and need to find the amount of the monthly payment or the number of months to pay.

EXAMPLE The installment price of a compact disk player is $400. You must pay $40 down and make payments for 20 months. What will be your monthly payments?

SOLUTION

$400	Installment price
− 40	Down payment
$360	Remainder to pay

$360 remainder to pay ÷ 20 months = $18 monthly payment

EXERCISE YOUR SKILLS

Find the installment price and finance charge for each item.

	Item	Cash Price	Installment Terms
1.	Boat	$1,500	$150 down; $67.50 a month for 24 months
2.	Computer	2,500	$500 down; $115 a month for 20 months
3.	Bicycle	250	$25 down; $26.50 a month for 9 months

4. You buy a camera for $50 down and a total installment price of $455. You pay $33.75 per month. For how many months will you have to make payments?

5. You pay $75 down and $8.75 a week for some jewelry. If the total installment price is $250, for how many weeks do you have to make payments?

6. A motorcycle costs $2,300. You pay $500 down and the rest in 18 equal installments. How much will each installment be?
7. A hair dryer that sells for $30 cash may be bought for $3 down and $5.76 a month for 5 months. By what percent is the installment price greater than the cash price?
8. You can buy a watch for $150 cash or pay $30 down and the balance in 12 monthly payments of $11.50. By what percent would your installment price be greater than the cash price?

INSTALLMENT LOANS

Rather than pay a retail store a down payment and monthly payments, you can also obtain an **installment loan** from a bank or credit union. When you borrow on an installment loan, you receive the face amount of the loan. You repay the face amount in installments, usually monthly. One month's interest on the unpaid balance is also paid. If the interest rate on the loan is 1.5% a month, this rate is equal to 18% a year ($1.5 \times 12 = 18$). Sometimes a service charge is added to the cost of the loan.

Repayment Schedule Below is a schedule of payments on a $600 loan. The loan was repaid in six monthly payments of $100 on the principal, plus a 1.5% finance charge on the unpaid monthly balance. The payment decreased as the finance charge decreased.

LOAN REPAYMENT SCHEDULE				
End of -	Unpaid Balance	Finance Charge: 1.5% of Unpaid Balance	Payment on Principal	Total Payment
1st month	$600.00	$ 9.00	$100.00	$109.00
2d month	500.00	7.50	100.00	107.50
3d month	400.00	6.00	100.00	106.00
4th month	300.00	4.50	100.00	104.50
5th month	200.00	3.00	100.00	103.00
6th month	100.00	1.50	100.00	101.50
Totals		$31.50	$600.00	$631.50

Illustration 10-5.1. Installment Loan Repayment Schedule

The finance charge was 1.5% of the unpaid balance. Each payment was the finance charge plus the $100 payment on the principal.

Critical Thinking If the interest rate in Illustration 10-5.1 is 18% per year, why is the interest charge less than $54 ($600 \times 0.18 \times 0.5 = \$54$)?

Level Payment Plans Many lenders calculate payments so that each payment is the same amount. This payment method is called the **level payment plan.**

EXERCISE YOUR SKILLS

For each loan, make a schedule of monthly payments of the finance charges and principal as shown in Illustration 10-5.1.

	Amount Financed	Monthly Repayment Plan Number of Payments	Payment on Principal	Finance Charge on Unpaid Balance	Total Payments
9.	$500	5	$100	1%	
10.	100	10	10	1.5%	
11.	800	10	80	1.5%	

12. A member of a credit union borrows $720 agreeing to repay it in 6 equal monthly payments plus a finance charge of 1.25% a month on the unpaid balance. Find the total finance charge.

Kay borrowed $600 from a finance company and repaid it in 6 monthly payments of $100 plus a finance charge of 1.75% on the unpaid balance. A $6 service charge was due with the first payment.

13. What was the total finance charge on the loan?

14. What was the total amount Kay paid to the finance company?

ANNUAL PERCENTAGE RATE (APR)

To find the rate of interest on a loan for one year, you divide the interest paid in a year by the principal. Finding the rate of finance charges on an installment loan is not as easy. As you know, the cost of borrowing money may include more than interest. It may also include service charges. Also, since you make payments on the loan each month, you are not borrowing the whole principal for the full time of the loan. So the law makes the lender tell the borrower what annual rate is charged on the loan. The **annual percentage rate** (APR) is a percent that shows the ratio of the finance charges to the amount financed.

The easiest way to find the annual percentage rate is to use tables like the ones shown in Illustration 10-5.2. To use the tables, you need to know the number of monthly payments for the loan and the finance charge per $100 of the amount financed. To find the finance charge per $100 of the amount financed, you must first multiply the finance charge by 100. You then divide the product by the amount financed.

$$\textbf{Finance Charge per \$100 of Amount Financed} = \frac{\textbf{Finance Charge} \times \textbf{100}}{\textbf{Amount Financed}}$$

After you have found the finance charge per $100 you can use the tables in Illustration 10-5.2 to find the annual percentage rate.

Number of Payments	Annual Percentage Rate										
	12¾%	13%	13¼%	13 ½%	13¾%	14%	14¼%	14½%	14¾%	15%	15¼%
	Finance Charge per $100 of Amount Financed										
3	2.13	2.17	2.22	2.26	2.30	2.34	2.38	2.43	2.47	2.51	2.55
6	3.75	3.83	3.90	3.97	4.05	4.12	4.20	4.27	4.35	4.42	4.49
9	5.39	5.49	5.60	5.71	5.82	5.92	6.03	6.14	6.25	6.35	6.46
12	7.04	7.18	7.32	7.46	7.60	7.74	7.89	8.03	8.17	8.31	8.45
15	8.71	8.88	9.06	9.23	9.41	9.59	9.76	9.94	10.11	10.29	10.47

Number of Payments	Annual Percentage Rate										
	15½%	15¾%	16%	16¼%	16½%	16¾%	17%	17¼%	17½%	17¾%	18%
	Finance Charge per $100 of Amount Financed										
6	4.57	4.64	4.72	4.79	4.87	4.94	5.02	5.09	5.17	5.24	5.32
12	8.59	8.74	8.88	9.02	9.16	9.30	9.45	9.59	9.73	9.87	10.02

Number of Payments	Annual Percentage Rate										
	18¼%	18½%	18¾%	19%	19¼%	19½%	19¾%	20%	20¼%	20½%	20¾%
	Finance Charge per $100 of Amount Financed										
6	5.39	5.46	5.54	5.61	5.69	5.76	5.84	5.91	5.99	6.06	6.14
12	10.16	10.30	10.44	10.59	10.73	10.87	11.02	11.16	11.31	11.45	11.59

Illustration 10-5.2. Partial APR Tables

EXAMPLE Find the annual percentage rate for a $600 installment loan that must be paid off in 6 months. The finance charge for the loan is $36.

SOLUTION The finance charge on $600 for 6 months was $36.

$$\text{Finance charge per } \$100 = \frac{\$36 \times 100}{\$600} = \$6$$

Using the Annual Percentage Rate Tables, read across the rows for 6 payments until you come to the amount closest to $6. Since $5.99 is the closest amount, use that rate, $20\frac{1}{4}\%$.

EXERCISE YOUR SKILLS

Find the finance charge per $100 of amount financed.

15. $800 borrowed, $78 finance charge

16. $250 borrowed, $41 finance charge

17. $760 borrowed, $89.60 finance charge

Find the annual percentage rate.

18. Finance charge of $5.75 per $100 for 9 payments

19. Finance charge of $2.50 per $100 for 3 payments

20. Finance charge of $8.85 per $100 for 15 payments

Eldridge P. Wisneuski borrowed $1,300 to upgrade his computer system. Eldridge repaid the installment loan in 12 monthly payments of $116 each.

21. What was the finance charge on this loan?

22. What was the finance charge per $100 of amount financed?

23. What was the annual percentage rate?

◆ Angela Barrett borrowed $4,000 and agreed to repay the installment loan in 12 monthly payments. The finance company calculated simple interest of 10.5% on the $4,000. The finance company also charged a $25 service charge.

24. What is the simple interest on $4,000 for 1 year at 10.5%?

25. Find the total amount Angela will pay over the life of the loan.

26. How much will Angela's monthly payments be?

27. What was the finance charge per $100 of amount financed?

28. What was the annual percentage rate?

MIXED REVIEW

1. Find $12\frac{1}{2}\%$ of $32.

2. What amount is 265% of $80?

3. Write the decimal equivalent of $45\frac{1}{4}\%$.

4. Use the formula $I = prt$ to find the simple interest on $1,410 for 1 month at 18% annually.

5. You borrow $700 from a finance company and pay it back in 12 monthly payments of $63.50 each. What is your finance charge on this loan?

6. The cash price of a TV was $120. Dan Herr bought it for $12 down and 12 monthly payments of $9.80. By what percent did the installment price exceed the cash price?

7. Ann bought an audio system on the installment plan for $790. She paid $90 down and the balance in equal monthly installments of $50 each. How many months did it take Ann to pay for the system?

10-6 CREDIT CARDS

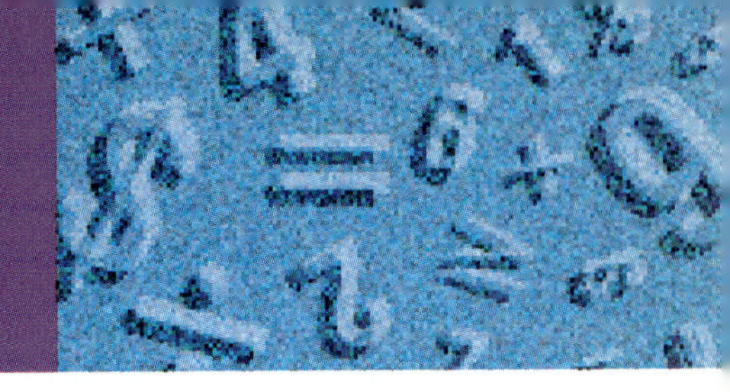

There are many types of credit cards. There are **bank credit cards,** such as MasterCard and Visa, and **travel and entertainment charge cards,** such as American Express and Diners Club. Many car owners use **oil company credit cards** to pay for fuel and automobile repairs. Large department and chain stores often provide customers with **store credit cards** as well.

◆ List some advantages of each type of credit card to consumers and to companies. Then make a list of disadvantages.

OBJECTIVES

In this lesson, you will learn to

- *find net receipts from credit card sales;*
- *find the cost of using a credit card; and*
- *calculate finance charges using the previous balance and average daily balance methods.*

WARM UP

a. Multiply 804.25 by 7.2.

b. Divide 80.006 by 73.4.

c. 6 km + 325 m = ? m

d. Write $105\frac{1}{2}\%$ as a decimal.

e. \$2.45 is what percent greater than \$2.10?

f. What amount is $62\frac{1}{2}\%$ greater than \$44?

CREDIT CARDS

In addition to the variety of types of credit cards, there are also different types of credit available with cards. Two of the most common are 30-day accounts and revolving charge accounts. **30-day charge accounts** require you pay the monthly balance in full each month. **Revolving charge accounts** allow you to pay a portion of the bill or the total bill.

Most credit cards state a maximum amount that can be charged to the card, known as the **credit limit.**

Finding Net Receipts from Credit Card Sales Many businesses do not offer credit to their customers directly. Instead they accept credit cards for the purchase of merchandise or service. There are two major reasons why. When a firm offers credit to a customer through its own store account, it does not receive the money from the sale until the customer pays. The firm also must maintain the credit records. The money from credit card sales, however, is received quickly from the credit card company; and firms do not have to do as much bookkeeping.

When the credit card company pays the firm, it deducts a percentage of each sale to cover the cost of providing the credit service. The percentage varies, but usually it is between 3% and 6% of the total sale.

For example, the Super Credit Card Company charges 5% of total sales for its services. If a sale is made for $400 using their card, the net amount that the store receives is $380.

$400 × 0.05 **$20 deducted by the credit card company**

$400 − $20 **$380 net amount received from the credit card sale**

Because of these fees, some companies offer a discount if you pay cash for merchandise or services.

EXERCISE YOUR SKILLS

Algon, Inc., accepts MidTown credit cards. Algon, Inc., had total credit card sales last year of $32,800,000. The MidTown Card Company charges 4% of sales for its services.

1. How much did Algon pay the credit card company last year?
2. Find Algon's net receipts from credit card sales last year.
3. Last month Tyme, Inc., had $385,000 of AmeriCard charge sales and $268,000 of sales using UniCard credit card. If AmeriCard charges 5% and UniCard charges 3% for services, what were Tyme's net receipts from these companies?

A store estimates that its next year's sales will be $740,000 if it does not take credit cards. If it takes General Card, sales are estimated to be $980,000, $500,000 of which will be made using the card. The General Card Company charges 6% for its services. Using these estimates:

4. What will be the net sales receipts next year with the card?
5. What will be the increase in net sales by using the card?
6. Jennifer Froelich bought a chain saw for $275 and received a 4% discount for paying cash instead of using a credit card. What did Jennifer pay for the saw?

Finding the Cost of Using a Credit Card Credit cards are a convenience. Instead of carrying cash around, you can use a credit card to make purchases. For this convenience, most credit card companies charge an annual membership fee that may be $15 to $50 a year. Other companies may charge a fee for each purchase.

A credit card company usually sends a monthly statement showing purchases, payments, fees charged, and minimum amount due. You are usually given 25 days to pay your statement balance. If you pay the balance in full within that time, there are no additional charges. When you pay your balance in full, you have had free use of the creditor's money for the specified time. An exception to this "free use" of money is if you borrow money using **cash advance slips** or ATM withdrawals. Advance slips are issued by some credit card companies. Interest is charged from the date of the cash advance loan at a *daily* interest rate.

When you do not pay your credit card balance in full, you are using credit cards to borrow money. Credit card companies charge interest to borrowers in the form of a finance charge. Credit card companies may also charge **late fees** for accounts not paid on time.

Like other forms of interest studied in this chapter, the interest rate on credit card companies is based on an **annual percentage rate (APR)** that is calculated monthly. For example, if your credit card company charges an annual percentage rate of 18%, the interest rate on your account will be calculated as 18% ÷ 12 = 1.5% monthly. The monthly finance charges may be calculated using either the previous balance or the average daily balance method.

Previous Balance Method With the previous balance method, interest is charged on the balance in the account on the final billing date of the previous month. If the previous balance was 0, the amount due on the account is the total of all new charges plus any fees. If there is a previous balance, a finance charge will be calculated for that amount. The total amount due is calculated by adding current charges to the previous balance plus interest. For example,

Previous balance:	$125.00
Finance charge based on APR of 18%: $125 × (18% ÷ 12) =	$ 1.88
New charges:	$100.00
New account balance:	$226.88

Average Daily Balance Method When this method is used, the interest rate is applied to the **average daily balance** of charges in the account over the course of the billing period. For example, Judy's credit card charges a monthly interest rate of 2%. The amount of money charged to her account on November 8 was $400. During the billing period (November 8–December 7) she made the following charges: November 15, $75 and November 25, $180. On November 18, she made a payment of $91.

You can use a chart or spreadsheet, like the one on the next page, to calculate her daily average balance. A **Transaction** is either a payment (subtracted from the balance) or a new charge (added to the balance). The **Balance at end of Day** is the previous balance plus or minus any payments or new charges. The **Number of Days** is the number of days that balance remains in effect. The **Sum of Daily Balances** is the balance multiplied by the number of days.

Dates	Transaction	Balance at end of Day	Number of Days	Sum of Daily Balances
11/8	+400	$400	1	$400
11/9–11/14	0	$400	6	$2,400
11/15	+$75	$475	1	$475
11/16–11/17	0	$475	2	$950
11/18	−91	$384	1	$384
11/19–11/24	0	$384	6	$2,304
11/25	+$180	$564	1	$564
11/25–12/7	0	$564	13	$7,332
Total			31	$14,809

To calculate the daily average balance, divide the total of the Sum of Daily Balances by the total Number of Days. Then apply the monthly interest rate to that amount to find the finance charge.

Average daily balance: $14,809 ÷ 31 = $477.71
Finance charge (2% × $477.71) = $9.55
Ending balance: $487.26

Remember, you are only charged a finance charge when the previous month's bill was not paid in full.

EXERCISE YOUR SKILLS

7. Angelo's credit card statement balance was $460. He did not pay by the due date, so the credit card firm added a finance charge of 1.5%. How much was the finance charge?
8. Laura's annual credit card membership fee is $25. She also pays 70¢ per purchase. If she made 90 purchases last year, what was the cost of Laura's card for the year?
9. SecurCard has an annual fee of $20 and a finance charge of 1.65% per month on the unpaid balance. In May, SecurCard charged Mark the annual fee and a finance charge on his $275 unpaid balance. Find the total on Mark's monthly statement.
10. Ivy pays a $25 annual membership fee and an 18% finance charge on unpaid balances to Global Card. Belinda pays a $15 annual membership and a 20% finance charge on unpaid balances to Major Card. Both have an average monthly unpaid balance of $40. Who pays the most in yearly membership fees and finance charges? How much more per year?
11. Fess borrowed $300 for 20 days on his credit card using a cash advance. The company charged a daily finance charge of 0.0628%. What was the amount of the finance charge?

12. Bo borrowed $700 for 35 days from her credit card company using a cash advance check. The daily finance charge was 0.0453%. What was the finance charge on Bo's loan?
13. Ali borrowed $500 on a cash advance for 45 days on his credit card. The finance charge rate was 0.0512% per day. What total finance charge did Ali pay?
14. CardCo charges a daily finance charge of 0.0722% on all cash advances. How much would your finance charge be if you borrowed $200 for 60 days using the cash advance?
15. Write a word problem that requires the average daily balance method be used to calculate the finance charge. Exchange problems with someone else. Use a chart like that shown in the lesson to find the average daily balance, finance charge, and new balance on an account.
16. ♦ Having a credit card carries with it much responsibility. Investigate what happens when a credit card account is not paid promptly. Find out what is meant by a credit rating.

PUT IT ALL TOGETHER

17. Think about what you have learned about interest in this chapter. What common elements are there between interest paid on savings accounts and interest owed on loans?
18. Assume you have $7,500 earning interest in a 3-year time-deposit savings account at the rate of $7\frac{1}{2}\%$. You want to buy a car for $7,500. You can get a 3-year car loan with no money down at a 4% interest rate. Using what you have learned in this chapter, explain why it will be less expensive in this case to borrow the money.

MIXED REVIEW

1. Add $8\frac{1}{2}$ and $1\frac{3}{8}$.
2. $45 \text{ m} \div 2.4 \text{ cm} = ? \text{ cm}$
3. What is the due date of a 90-day note dated May 5?
4. LeBlanc's accepts Premier credit cards. LeBlanc's must pay Premier 3.5% on all credit card sales. If LeBlanc's sells $47,800 this month on the credit card, what net amount will they receive from Premier?
5. Midwest Card charges a $22 annual membership fee and a finance charge of 1.3% a month on all unpaid balances. In May, Midwest charged you the membership fee and a finance charge on the $450 unpaid balance in your account. What was the total of your May statement?
6. Ben borrowed $360 from a finance company and repaid the loan in 18 monthly payments of $25.40 each. What was the finance charge?
7. Dee bought a garden tiller on the installment plan for a total cost of $450. She paid $100 down and the rest in monthly installments of $14 each. How many months did it take Dee to pay for the tiller?

EXPLORE TECHNOLOGY

TEMPLATE: Spreadsheet to Compare Savings Options

Jose Hernandez wants to compare the interest earned on a passbook savings account and a time deposit account. Using a spreadsheet will allow him to adjust savings amounts and rates easily and see the interest from both accounts.

Load ET 10 from your applications diskette. Enter the data in blue below (cells B2–B5 and C2–C5) into the spreadsheet as shown below. When you do, the spreadsheet will calculate the total amount of the deposit at the end of the period, the interest earned, and the difference between interest earned on passbook and time deposit accounts.

The data represents a six-month deposit of $15,000 into two accounts. The passbook account pays 6% interest compounded quarterly with money on deposit for two interest periods. The time deposit account pays 8% interest, compounded semiannually, with money on deposit for one interest period.

	A	B	C
1	**Account Information**	**Passbook**	**Time Deposit**
2	Interest Rate (%)	6.00%	8.00%
3	Times Compounded in Year	4	2
4	Periods on Deposit	2	1
5	Amount of Deposit	15000.00	15000.00
6			
7	Compound Amount 1	225	600
8	Compound Amount 2	228.38	
9	Compound Amount 3		
10	Compound Amount 4		
11	Total Compount Int.	453.38	600.00
12	Compound Amount	15453.38	15600
13	Difference	−146.63	146.63

Answer these questions about your completed savings plan comparison record.

1. Why does the compound interest increase for the passbook account?
2. Change the time deposit account interest rate to 6%, the same as the passbook rate. In which account would Jose earn more interest? How much more?

Design Your Own

Jay Isen plans to borrow money to buy a used boat. Lake Marine Bank offers him a 24-month loan of $120 per month.

Create a spreadsheet like the one below that will allow Jay to calculate the rate of interest for the boat. Include a row as shown to check that the rate of interest is accurate. Enter the appropriate formulas using cell references whenever possible.

	A	B	C	D
1		**Lake Marine**		
2	Original Principal	2500		
3	Monthly Payment	120		
4	Number of Payments	24		
5	Finance Charge			
6	Exact Rate of Interest			
7	Check Exact Rate of Interest			

3. What finance charges did Jay pay the bank? What formula did you use to calculate this charge?
4. What was the rate of interest for the loan? What formula did you use to calculate the rate of interest?
5. What formula did you use to verify the rate of interest?

Jay decides to look around to compare loan rates. Port Charles Bank offers him an 18-month loan of $150 per month. Lake Blue Line Bank offers him a loan with monthly payments of $135 for two years. Redesign your spreadsheet so that it looks like the one shown below. Enter the information about all three loans on the revised spreadsheet. When you are done, answer the following questions.

	A	B	C	D	E	F	G	
1	**Bank**	**Principal**	**Mo. Pay**	**Number**	**Total**	**Finance**	**Rate of**	
2				**Payments**	**Cost**	**Charge**	**Interest**	
3	Lake Marine							
4	Port Charles							
5	Blue Line							

6. Which bank offers the lowest rate of interest? the highest?
7. Given all the terms, which loan costs the least?
8. Which spreadsheet design did you find easier to work with? Why?

CHAPTER 10 SUMMARY AND REVIEW

VOCABULARY REVIEW

annual percentage rate (APR) *10-5*
bank discount *10-3*
banker's interest *10-3*
collateral *10-3*
compound interest *10-1*
date of note *10-3*
discounted notes *10-3*
due date *10-3*
effective rate of interest *10-1*
elapsed time *10-4*
face value *10-3*
finance charge *10-5*
installment loans *10-5*
interest *10-1*
money market accounts *10-2*
promissory note *10-3*
rate of interest *10-3*
savings accounts *10-1*
signature card *10-1*
time-deposit accounts *10-2*

MATH SKILLS REVIEW

10-1

1. $2{,}400 \times 0.05$
2. $500 \times 0.12 \times 2\frac{1}{2}$

10-2

3. Tell which amount is greater: \$4,080 or \$4,079.98.

10-3

4. $\$4{,}500 + \$450$
5. $\frac{\$120}{\$1{,}800} = ?\%$

APPLICATIONS

10-1
- Add deposit and interest amounts and subtract withdrawn amounts to find savings account balances.
- For simple interest, use the formula $I = prt$, where I is interest, p is principal, r is rate, and t is time in years.
- To find compound interest, add the interest for the first compounding period to the account balance; then use the new balance as the principal. Repeat the process as needed.
- To find interest using a compound interest table, multiply the value of \$1 by (a) the rate and interest period by the principal when figuring annual or daily interest; (b) $\frac{1}{4}$ the rate and 4 times the annual periods when figuring quarterly interest; and (c) $\frac{1}{2}$ the rate and twice the annual periods for semiannual interest.
- Divide interest earned for one year by the amount of money on deposit to find the effective rate of interest.

6. You have \$350 in a savings account that pays 4% annual simple interest. Find the interest for 1 quarter.
7. On January 2, Tim deposited \$1,500 in a 3.8% savings account compounded quarterly. How much interest would be earned on October 1 if a deposit of \$700 was made on July 1?

8. Find the effective rate of interest (to the nearest tenth) on a $600 deposit at 5% compounded semiannually.

10-2 ◆ Time deposits, like money markets, pay more interest than passbook savings but have penalties for early withdrawal.

◆ Compare time and deposit passbook accounts by finding the difference between the interest amounts earned by both accounts.

9. You have two passbook accounts of $2500 each that pay 2.5% interest compounded quarterly. By combining the two accounts, you can deposit the money into a money market account that pays 3.6% annual interest compounded quarterly. How much more interest will you earn in three months?

10. Web put $8,000 in a 5-year 7% savings certificate. After keeping the money on deposit for 6 months, he withdrew the money. The penalty for early withdrawal is 9 months' interest at 7%. How much is the penalty?

10-3 ◆ Use $I = prt$ to find interest. For fractional parts of a year, multiply principal times rate times a fraction. For exact interest, use a 365-day year. For banker's interest, use a 360-day year.

◆ Find bank discount by multiplying the face of the note by the discount rate. The proceeds of the note equals the face of the note minus interest.

◆ Find the effective rate of interest on a discounted note by dividing the amount of discount by the proceeds.

11. Find the amount due at maturity on a $2,400 loan for $1\frac{1}{2}$ years at 4%.

12. Leroy borrowed $4,800 for 6 months and paid $264 exact interest. What rate of exact interest did Leroy pay?

13. Vi discounted her 3-month noninterest-bearing note for $1,200 dated May 1 on the same day. The rate of bank discount was 8%. What proceeds did Vi receive?

10-4 ◆ Find interest using an interest table.

◆ Find the number of days between two dates using a table or arithmetically.

What is the due date of an 80-day note dated

14. September 9?

15. December 27?

10-5 ◆ Find the total finance charge for an installment loan by subtracting the original loan amount from the sum of the monthly payments or by adding service charges and interest.

◆ Find the annual percentage rate (APR) by finding the finance charge per $100 of the amount financed in a table.

16. Diedra borrowed $800 and repaid the installment loan in four monthly payments of $200 plus a finance charge of 1.5% on the unpaid balance. What was the total finance charge?

17. Ned borrowed $1,000 and repaid it in six monthly payments of $180 each. What was the finance charge?

18. You buy a TV for an installment price of $600 putting $150 down and paying the rest over 12 months. What is the amount of the monthly payment?

10-6 ◆ Find the finance charge on an unpaid credit card balance by multiplying the unpaid balance by the finance charge rate. Card company may use previous balance or unpaid balance method.

19. May's average monthly unpaid balance on her credit card statement is $300. Her credit card company charges a 1.65% finance charge per month on unpaid balances and also an annual membership fee of $25. Find her total annual costs.

WORKPLACE KNOWHOW: MANAGING TIME

Time management skills include the ability to select relevant activities, set priorities, and create a plan and schedule for achieving goals.

Assignment This chapter has been about the relationship between time and money. Think about some of your financial goals. Select one. Write down some steps you can take to achieve your goal. Based on those steps, how long do you think it will take to achieve your goal?

Work Skills Portfolio Update You may wish to include an assessment of your understanding of the basic interest formula and application of percents. How does your assessment of percent skills compare to your self-assessment after you finished Chapter 4? You may wish to include the completed monthly interest and payment chart from the integrated project in the student workbook. You may also wish to include the spreadsheet you created to compare loan options in the Explore Technology lesson as examples of your understanding of interest formulas.

CHAPTER 10 TEST

Math Skills

1. $\$114.72 \div 4$
2. $\$4{,}000 \times 7\% \times \frac{1}{2}$
3. $700 \times \$1.157625$
4. $\$850 + \$125 - \$218$
5. $\$8{,}154.20 - \$8{,}050$
6. $\$600 \times 5.43\% \times \frac{1}{4}$

Applications

7. You have $2,000 in a 4% savings account compounded quarterly. How much will you have in the account after two quarters if no other deposits or withdrawals are made?

The value of $1 at 4.6% annual interest compounded daily for 90 days = $1.011565, and 180 days = $1.023263. Using these values, find the interest a $700 deposit earns in

8. 90 days
9. 180 days
10. A deposit of $400 earns $15.37 interest when an annual rate of 3.72% is compounded daily. To the nearest hundredth of a percent, what is the effective rate of interest?
11. A penalty of 9 months' interest at a rate of 7% on the entire account balance is charged for early withdrawals from time-deposit accounts. Find the penalty for early withdrawal from a time-deposit account with a balance of $3,500?
12. What is the due date of a 60-day note dated October 12.
13. John signs a three-month note for $1,000 bearing exact interest at 15% per year. How much must he repay after three months?
14. The interest on a six-month loan for $10,000 was $550. What annual rate of interest was charged?
15. Juanita borrowed $500 and paid 15 monthly payments of $36 each. What was the finance charge per $100 of the amount financed?
16. You can buy a TV for $650 cash or pay $50 down and $108 per month for six months. By what percent is the installment price greater than the cash price, to the nearest whole percent?
17. Silvia buys a $550 tool set. If she pays cash, the store gives her a 4.5% discount. How much will she save by paying cash?
18. Kris has an unpaid balance in his credit card account of $520 for a total of 25 days. If his credit card company charges a daily finance charge of 0.084%, what total finance charge in dollars will Kris pay?
19. What are the net proceeds of a four-month discounted note for $3,600 discounted at 12% exact interest?

Chapter 11

Start the Presses

Looking for a Job

where you can see the results of your efforts in a finished product? If so, you might enjoy a career in the printing industry. Marcus works at a company that specializes in printing four-color magazines and books. He says the printing industry offers a variety of career choices, good pay, and opportunities for promotion.

One job critical to the printing process is that of press operator. Press operators set up the printing presses and monitor printed output to make sure the printed pages look good. The job is critical because the costs of press

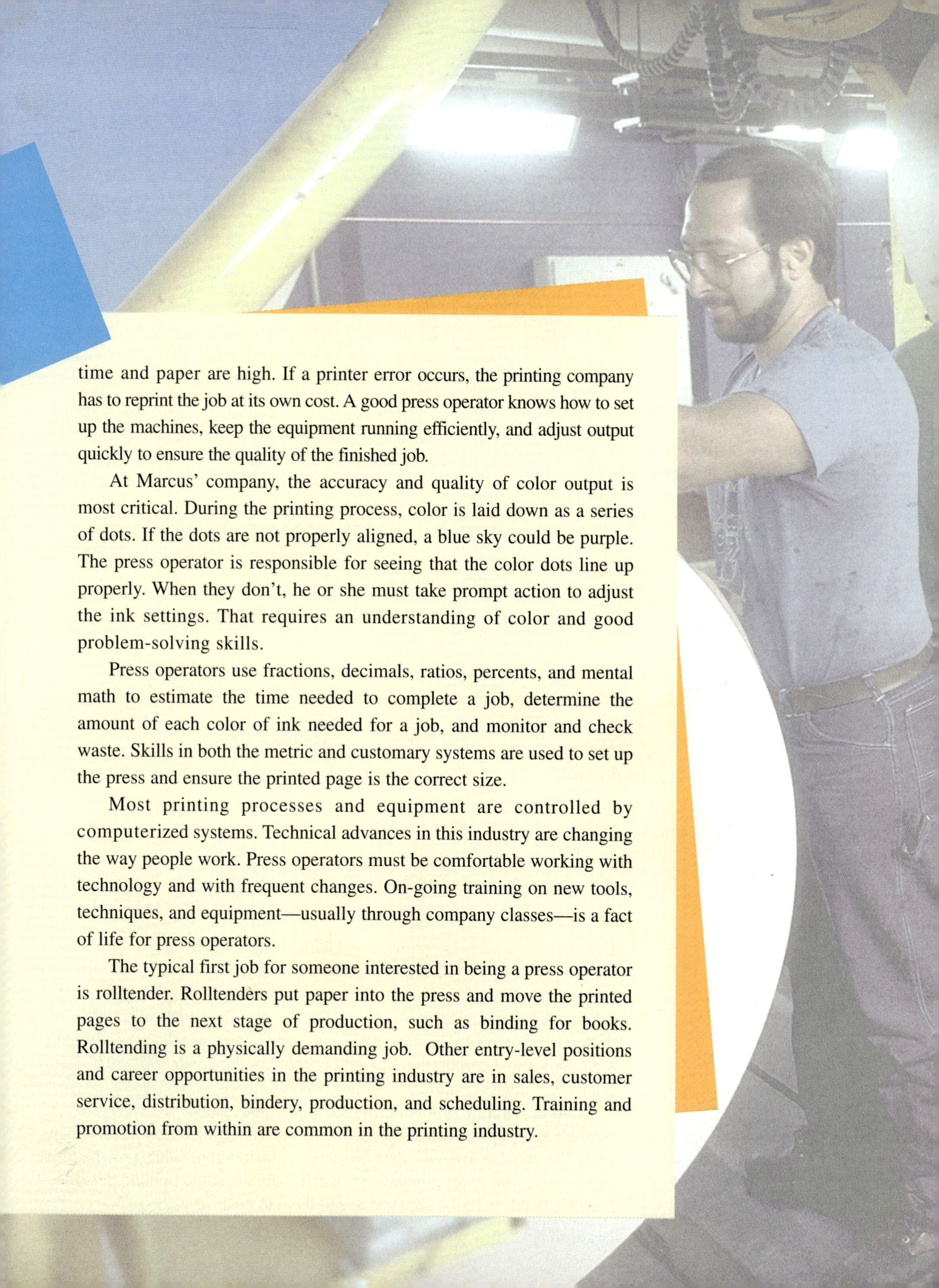

time and paper are high. If a printer error occurs, the printing company has to reprint the job at its own cost. A good press operator knows how to set up the machines, keep the equipment running efficiently, and adjust output quickly to ensure the quality of the finished job.

At Marcus' company, the accuracy and quality of color output is most critical. During the printing process, color is laid down as a series of dots. If the dots are not properly aligned, a blue sky could be purple. The press operator is responsible for seeing that the color dots line up properly. When they don't, he or she must take prompt action to adjust the ink settings. That requires an understanding of color and good problem-solving skills.

Press operators use fractions, decimals, ratios, percents, and mental math to estimate the time needed to complete a job, determine the amount of each color of ink needed for a job, and monitor and check waste. Skills in both the metric and customary systems are used to set up the press and ensure the printed page is the correct size.

Most printing processes and equipment are controlled by computerized systems. Technical advances in this industry are changing the way people work. Press operators must be comfortable working with technology and with frequent changes. On-going training on new tools, techniques, and equipment—usually through company classes—is a fact of life for press operators.

The typical first job for someone interested in being a press operator is rolltender. Rolltenders put paper into the press and move the printed pages to the next stage of production, such as binding for books. Rolltending is a physically demanding job. Other entry-level positions and career opportunities in the printing industry are in sales, customer service, distribution, bindery, production, and scheduling. Training and promotion from within are common in the printing industry.

11-1 INVESTING IN BONDS

OBJECTIVES

In this lesson, you will learn to find

- *the market price of a bond;*
- *total investment in bonds, including commissions and other costs;*
- *the cost and redemption value of savings bonds; and calculate*
- *interest on bonds; and*
- *the rate of income from investments in bonds.*

Financial experts usually advise keeping 3–6 months living expenses in a bank savings account. They recommend putting additional savings dollars in other income producing investments.

◆ Make a list of 3–5 goals that will require some money. For example, do you want to take a three-month vacation overseas? buy a house? retire at 40? Select one of your goals. How much money do you think it will take to achieve your goal? Calculate how long it will take you to reach that dollar goal if you put $1,000 dollars into a savings account that pays 5% interest compounded quarterly. Do you know of any way to make money "grow" faster?

WARM UP

a. $0.815 \times \$36{,}000$

b. $558 \div 31{,}000 = ?\%$

c. Write $\frac{1}{2}\%$ as a decimal.

d. $0.1 \times 1 \text{ km} = ? \text{ m}$

e. $21.60 is what percent less than $36?

f. $56.24 increased by $\frac{1}{8}$ of itself is what amount?

INVESTING IN BONDS

Organizations, such as corporations, federal, state, and local governments, and even school districts, may borrow large amounts of money to pay for major projects, such as renovating factories, paving roads, or building new schools. The amounts they borrow are often in the hundreds of millions of dollars. To borrow the money they issue bonds. Persons who are willing to lend money to a corporation buy one or more of the bonds. Buyers of the bonds are called *bondholders* of the corporation.

A **bond** is a form of long-term promissory note. Bonds are a written promise to repay the loaned money on the due date, which may be anywhere from 5 to 30 years from the date of issuance. In addition, interest on the money borrowed is paid to the bondholder, usually semiannually. Bondholders may keep the bonds until the due date or sell them to other investors.

Corporations often use their property as collateral for the money they borrow. If the loan is not repaid, the bondholders may take over the corporation's land, buildings, or equipment. This guarantee is made between the corporation and a bank or trust company, called the *trustee.* The trustee is appointed by the corporation to represent the bondholders as a group in their dealings with the corporation.

Bonds are usually issued with a **face value** of $1,000, which is printed on the front of the bond. Face value is the amount of money that the issuer agrees to pay the bondholder on the due date. Bonds may be issued with other face values, such as $500, $5,000, or $10,000. Illustration 11-1.1 is an example of a bond with a face value of $1,000.

Critical Thinking Since bonds are backed by the collateral of a company, does that mean there is no risk involved in purchasing a bond?

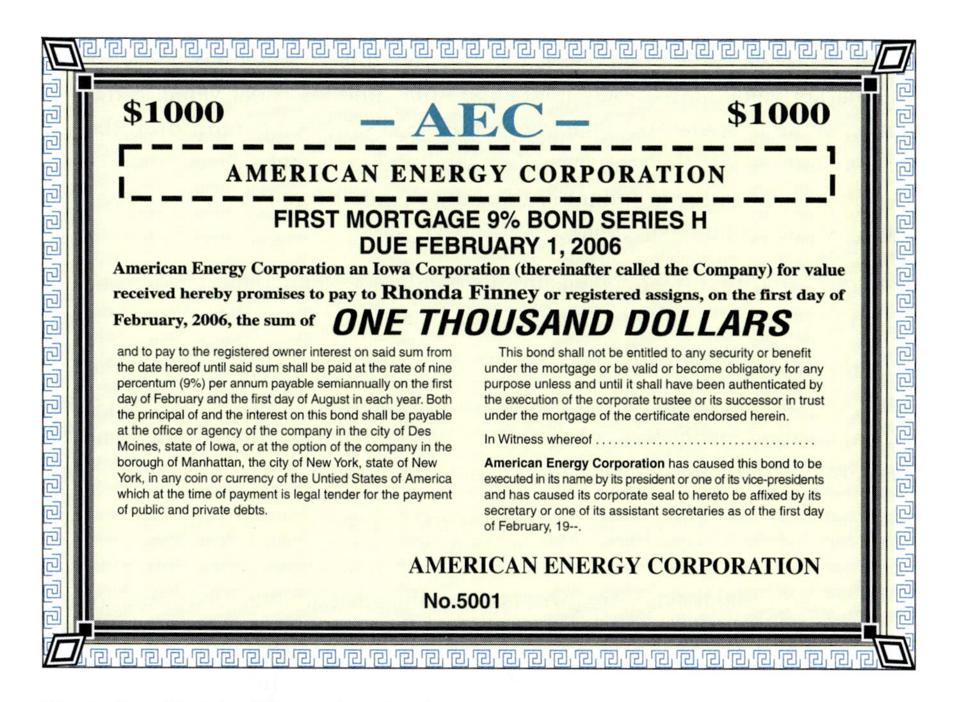

$1000 – AEC – $1000

AMERICAN ENERGY CORPORATION

FIRST MORTGAGE 9% BOND SERIES H
DUE FEBRUARY 1, 2006

American Energy Corporation an Iowa Corporation (thereinafter called the Company) for value received hereby promises to pay to Rhonda Finney or registered assigns, on the first day of February, 2006, the sum of ONE THOUSAND DOLLARS

and to pay to the registered owner interest on said sum from the date hereof until said sum shall be paid at the rate of nine percentum (9%) per annum payable semiannually on the first day of February and the first day of August in each year. Both the principal of and the interest on this bond shall be payable at the office or agency of the company in the city of Des Moines, state of Iowa, or at the option of the company in the borough of Manhattan, the city of New York, state of New York, in any coin or currency of the Untied States of America which at the time of payment is legal tender for the payment of public and private debts.

This bond shall not be entitled to any security or benefit under the mortgage or be valid or become obligatory for any purpose unless and until it shall have been authenticated by the execution of the corporate trustee or its successor in trust under the mortgage of the certificate endorsed herein.

In Witness whereof ....

American Energy Corporation has caused this bond to be executed in its name by its president or one of its vice-presidents and has caused its corporate seal to hereto be affixed by its secretary or one of its assistant secretaries as of the first day of February, 19--.

AMERICAN ENERGY CORPORATION

No.5001

Illustration 11-1.1. Corporation Bond

Bondholders—the people who invest in bonds—do so because the interest paid is usually higher than that on savings accounts. While bonds provide a relatively safe investment option, investing in bonds is riskier than depositing money in a bank, and requires more study. For example, these are some of the details you must consider before you buy bonds: price and interest rates, tax laws, ability to resell the bonds, earnings record and credit history of the issuer, and business conditions.

Investment Banking Houses

When an organization decides to issue bonds, it does so for a large amount of money. For example, the bond shown for the American Energy Corporation might be just one of 100,000 issued to borrow $1,000,000. The entire issue is often sold to an *investment banking house.* The banking house then sells the bonds to investors at a slight increase in price over what the banking house paid for them.

For example, a corporation may issue 100,000 bonds, each with a face value of $1,000. The banking house may buy them from the corporation at $993 each and sell them at $1,000 each.

Par Value and Market Value Bonds have two kinds of value: par value and market value. The **par value** of a bond is the same as its face value, or the amount that is printed on the face of the bond. It is the amount that the borrower promises to pay the bondholder on the due date.

The **market value** of a bond is the price at which the bond is being sold. The market value may or may not be the same amount as par value. If the market value is greater than par value, the bond sells at a **premium.** If the market value is less than the par value, the bond is sold at a **discount.**

The amount of the premium or discount is the difference between the market value and the par value.

Critical Thinking If a bond ensures its bondholder a guaranteed price at maturity date, why could the market value be less or greater than the par value?

How the Market Price of a Bond is Quoted The market price, or market value, of a bond is quoted as a percent of the par value. For example, a price quotation of 97 means 97% of the par value. The market price of the bond is found by multiplying the par value by the percent.

EXAMPLE The banking house of Miller & Row offers American Energy Corporation bonds at 97. What is the price of one of the corporation's $1,000 bonds? Are these bonds selling at a discount or a premium?

SOLUTION

$0.97 \times \$1,000 = \$970$	Price of one $1,000 bond
Selling at a discount.	The price is less than the face value of the bond.

If the price of a $500 par value bond is 102, the market price is 102% of $500, or $510. This bond is selling at a premium.

Finding the Total Investment in Bonds When You Know the Market Price and the Number of Bonds Bought To find the total investment in bonds, you must find the market price of one bond and multiply by the number of bonds bought.

EXAMPLE Lenore Walters bought four $1,000 Zontech Company bonds at 108. What was her total investment in Zontech bonds?

SOLUTION

$108\% = 1.08$	
$1.08 \times \$1,000 = \$1,080$	market price of 1 bond
$4 \times \$1,080 = \$4,320$	total investment

EXERCISE YOUR SKILLS

Mental Math State the market price, in dollars and cents, of one $1,000 bond at each quoted price. Also state whether the bond is selling at a discount or at a premium.

1. 98

2. 93

3. 106

4. 101

5. $97\frac{1}{4}$

6. $105\frac{3}{4}$

7. $104\frac{1}{8}$

8. $85\frac{7}{8}$

Find the amount of money invested in each of the bond purchases.

Forlund County, $1,000 bonds:

9. 3 @ 92 **10.** 7 @ 89 **11.** 9 @ 107

Northeast Power Corporation, $1,000 bonds:

12. 6 @ 98 **13.** 5 @ 95 **14.** 8 @ 104

Rowland Community Schools, $500 bonds:

15. 4 @ 112 **16.** 8 @ 90 **17.** 12 @ $78\frac{3}{4}$

SELLING BONDS

After a bond issue has been sold by the investment banking house, anyone who wants to buy the bonds must buy them from a bondholder who wants to sell their bonds before the maturity date. Likewise, a bondholder who wants to sell must sell to someone who wants to buy bonds.

Using a Broker to Buy and Sell Bonds Bonds are seldom sold directly by the seller to the buyer. Usually, they are handled through a **broker,** who is a dealer in stocks and bonds.

Each buyer and seller places an order with a broker to buy or sell a particular bond. Each broker then sends the order to one of the exchanges, such as the New York Stock Exchange. In the exchanges, orders to buy and sell are received from all over the world. The actual buying and selling takes place when the buy and sell orders are matched.

Broker's Commission When bonds are bought and sold through a broker, the buyer is charged a **broker's commission** or *brokerage fee.* The commission may be a fixed amount per bond, such as $30 per bond. It may also be a variable amount based on the number of bonds purchased, such as $20 per bond for buying 1 to 5 bonds, and $15 each for buying 6–15 bonds. The commission may also be a per transaction charge plus a fixed amount per bond.

Finding the Total Investment in Bonds When the Market Price, Commission, and Number of Bonds Are Known When bonds are purchased through a broker, the amount of the investment is the market price of the bonds plus the broker's commission.

EXAMPLE Through a broker, Esteban Velazquez bought 3, \$1,000 Tinwell Company bonds at 96 plus \$20 commission per bond. What was the amount of his investment?

SOLUTION

$0.96 \times \$1{,}000 = \$960$	market price of 1 bond
$3 \times \$960 = \$2{,}880$	market price of 3 bonds
$3 \times \$20 = \$60$	commission on 3 bonds
$\$2{,}880 + \$60 = \$2{,}940$	total investment

Daily Bond Quotations

Each day's bond sales on the exchanges are shown the next day on the financial pages of many city newspapers. An example of what a daily bond quotation from the New York Exchange might look like is shown in Illustration 11-1.2.

Reading the Bond Quotation Table The sales of Arnold Power Company bonds for a day are shown on the first line of the table.

At the right of the name, the "$9\frac{1}{4}$" means that the bonds pay an interest rate of $9\frac{1}{4}\%$. The "07" means that the bonds are due in the year 2007.

The number of bonds sold, 19, appears in the Volume column. In the Close column, the "96" is the last price at which a sale was made that day.

The "$+\frac{1}{2}$" in the Net Change column means that the last price for the day was $\frac{1}{2}$ higher than yesterday's closing price. So, the closing price yesterday must have been $95\frac{1}{2}$, or \$955.

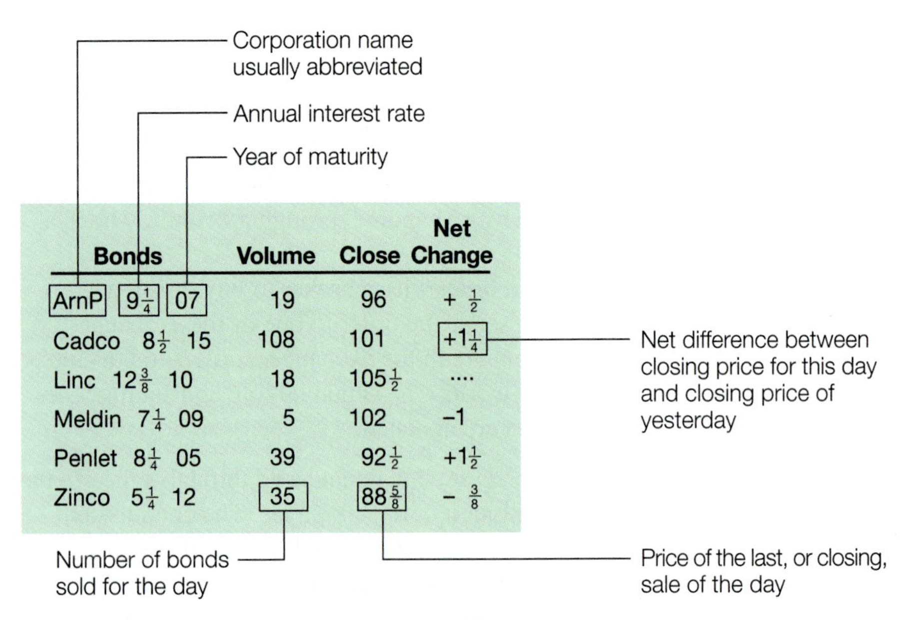

Bonds	Volume	Close	Net Change
ArnP $9\frac{1}{4}$ 07	19	96	$+\frac{1}{2}$
Cadco $8\frac{1}{2}$ 15	108	101	$+1\frac{1}{4}$
Linc $12\frac{3}{8}$ 10	18	$105\frac{1}{2}$	....
Meldin $7\frac{1}{4}$ 09	5	102	-1
Penlet $8\frac{1}{4}$ 05	39	$92\frac{1}{2}$	$+1\frac{1}{2}$
Zinco $5\frac{1}{4}$ 12	35	$88\frac{5}{8}$	$-\frac{3}{8}$

Illustration 11-1.2. Daily Corporation Bond Quotations

EXERCISE YOUR SKILLS

Find the total investment in each of the following bond purchases. The par value of each bond is \$1,000. The commission in each case is \$15 per bond.

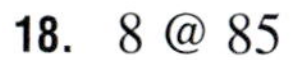

18. 8 @ 85 **19.** 5 @ 92 **20.** 4 @ 103

21. 10 @ $93\frac{5}{8}$ **22.** 16 @ $112\frac{3}{4}$ **23.** 7 @ $99\frac{1}{2}$

24. Adam Edlund invested in 14 bonds with a par or face value of \$1,000 each. He paid \$1,035 for each bond, and the brokerage fee was \$12 per bond. How much did Adam invest?

25. Simone Mullins bought 8, \$1,000 bonds, but she paid only \$730 for each of them. If \$17 commission was charged per bond, what was her total investment in the bonds?

26. Otto Bawol bought 4, \$1,000 bonds at $87\frac{1}{2}$. Two days ago the price of the bonds was 85. The brokerage fee was \$35 for the transaction and \$7.50 for each bond. What was the amount of Otto's investment in the bonds?

Use the information in Illustration 11-1.2 for these problems.

27. What annual rate of interest is paid on the Meldin bonds listed in the bond quotation table?

28. In what year are the Zinco bonds due for payment?

29. What was the last, or closing, price, in dollars, of Penlet bonds for the day shown in the table?

30. How many Linc bonds were sold for the day?

31. What was the last price, in dollars, of Cadco bonds on the day before the day shown in the table?

UNITED STATES SAVINGS BONDS

The United States Government borrows money in a variety of ways. One of the most popular ways is the **Series EE savings bonds.** These bonds are issued in denominations of \$50, \$75, \$100, \$200, \$500, \$1,000, \$5,000, and \$10,000. The **denomination** is the amount printed on the face of the savings bond.

When you buy a savings bond, you pay one-half of its denomination. So, a \$500 savings bond costs \$250, a \$1,000 bond costs \$500, and so on. Bonds may be purchased at most banks and through payroll savings plans offered by employers.

The rate of interest paid on savings bonds, currently about 5.25%, changes twice a year. The savings bond rate is calculated as a percentage of the Treasury Bill rate, which is the rate the federal government pays when it borrows money.

Redeeming Savings Bonds When redeeming a bond, you exchange the bond for money. The amount you get, which includes the original purchase price and interest, is called the **redemption value.** Illustration 11-1.3 shows a table of redemption values for a $50 savings bond. Since redemption values change from time to time, the amount you get when redeeming a bond may differ from amounts in the table. To find the redemption value of a $100 bond, multiply by two (2 × $50), and so on for larger denominations.

End of Year	Redemption Value	End of Year	Redemption Value
1	$26.08	5	$38.38
2	28.14	6	41.66
3	30.30	7	45.42
4	32.94	8	49.32

Illustration 11-1.3. Redemption Table for $50 Savings Bond

EXAMPLE Cindy buys a $100 savings bond and holds it for 6 years. Find: (a) The bond's redemption value and (b) the interest earned on the bond.

SOLUTION

a. $\$100 \div \$50 = 2$ equivalent number of $50 bond units

$41.66 redemption value of $50 bond

× 2

$83.32 redemption value of $100 bond

b. $83.32 redemption value of bond

$\$100 \times \frac{1}{2} = \$50.00$ purchase price of bond

$\$83.32 - \$50 = \$33.32$ interest earned on bond

EXERCISE YOUR SKILLS

When necessary, use the redemption values from Illustration 11-1.3 to solve the problems.

32. What is the purchase price of a $50 savings bond? a $75 savings bond?

33. If you bought eight $500 savings bonds, what total amount would you pay for this purchase?

34. Jean Hertz redeemed 3, $50 savings bonds after holding them for 4 years. What total amount did she receive?

35. Merilou Hogan has a $200 savings bond. What amount will she receive if she redeems the bond at the end of 2 years?

36. Wesley Henneck owns ten $50 savings bonds. How much interest will he have earned if he redeems the bonds after 7 years?

BOND INCOME

When you invest in bonds, the corporation or governmental organization that issues the bonds makes interest payments to you. These payments are your income from the bonds.

Bond Interest

The interest rate of a bond is based on the par value. Since the par value is the principal, the interest formula is:

Par Value × Rate × Time = Interest

EXAMPLE Find the interest for one year on a $1,000 par value, 8% bond.

SOLUTION $1,000 × 0.08 × 1 = $80, interest for 1 year
If the interest is paid semiannually, the amount of each interest payment for this bond would be $40.
$\$1{,}000 \times 0.08 \times \frac{1}{2} = \$40$, semiannual interest

Income from an Investment in Bonds To find your income from bonds, you first find the interest you receive for one bond. Then multiply that result by the number of bonds you own.

EXAMPLE Find the annual income from five $1,000 bonds paying 13% interest.

SOLUTION

0.13 × $1,000 = $130	interest on 1 bond
5 × $130 = $650	income from 5 bonds

Finding the Rate of Income on Investment Before you buy bonds, you should find out what rate of income you will receive. This will help you compare bonds with other types of investments.

You can find your rate of income by dividing the annual income in interest by the total investment in bonds. By dividing in this way, you are comparing the part with the whole. Your annual income from interest is the part. The amount you invested in bonds is the whole with which you compare the part.

$$\textbf{Rate of Income} = \frac{\textbf{Annual Income}}{\textbf{Investment}}$$

Rate of income is often called **yield.** You say your *yield* on an investment is 8.2%, or an Alpha Cable Company bond *yields* 8.2%.

EXAMPLE What is the rate of income, or yield, on a $1,000, 8% Alpha Cable Company bond priced at 94 plus $40 commission? Round to the nearest tenth of a percent.

SOLUTION

0.08 × $1,000 = $80	annual income
0.94 × $1,000 = $940	amount invested
$940 + $40 = $980	total investment
$\frac{\$80}{\$980} \approx 00.081632 = 8.2\%$	rate of income

EXERCISE YOUR SKILLS

Find the annual income in each problem.

37. 4, $1,000, 9% bonds

38. 16, $1,000, 5% bonds

39. 7, $1,000, 6% bonds

40. 12, $500, 8.25% bonds

41. 8, $500, 7% bonds

42. 10, $1,000, $6\frac{1}{8}$% bonds

43. 16, $1,000, 7.25% bonds

44. 3, $1,000, 8.5% bonds

45. What is the annual income from four $1,000, $8\frac{1}{8}$% bonds?

46. You own 23 bonds with a par value of $1,000 each and paying 7.75% interest. Find your annual income from these bonds.

47. How much is each interest payment on a $500, 9% bond if the interest is paid semiannually on June 1 and December 1?

48. What estimated and actual annual interest would you get from 8, $1,000 par value bonds that pay $6\frac{7}{8}$% interest?

For each of these bond investments, find the total investment in the bonds and the total annual income from the investment.

	Bonds Owned	Par Value per Bond	Price Paid	Commission per Bond	Total Investment	Interest Rate	Annual Income
49.	5	$1,000	87	$10.00		8%	
50.	10	1,000	$92\frac{1}{2}$	7.50		12%	
51.	30	1,000	$105\frac{1}{4}$	5.00		$10\frac{1}{2}$%	
52.	45	1,000	96	5.00		7.6%	

For each bond, find the rate of income on the investment to the nearest tenth of a percent.

	Par Value	Interest Rate	Price Paid	Commission	Rate of Income
53.	$1,000	10%	91	$25	
54.	1,000	8.5%	74	35	
55.	1,000	$9\frac{1}{2}$%	112	25	
56.	500	7.8%	62	25	

57. A $1,000 bond, paying 8% interest, was bought at $68\frac{1}{2}$, plus $30 commission. Find the rate of income on investment.

58. A $1,000, 11% bond was bought at 124. The previous day the bond sold at 120. Find the rate of income on the investment.

Note: In the following exercises, when commission is not stated, it is included in the price. When the par value is not given, assume it is $1,000.

59. What is the rate of income on a 12% bond bought at 108?

60. What is the yield on a 9% bond bought at 98?

61. Art Products Company $9\frac{1}{2}\%$ bonds can be purchased at 107. How much money must be invested in the bonds to produce an annual income of $1,520?

Al Horwath bought 7 Nole Company $1,000 par value, 13% bonds at 150 plus a commission charge of $10 per bond. Interest on these bonds is paid semiannually on February 1 and August 1.

62. What semiannual interest payment will Al receive from this investment in bonds?

63. What amount must be invested in Bloom County 7% bonds at 80 in order to earn an annual income of $2,800?

Louis Bard can buy 9% mining company bonds at 84, or 7% oil company bonds at 68.

64. Which bond pays the higher yield?

65. What is the rate earned by the bond with the higher yield?

66. ◆ Contact your local bank. Find the current rates for savings bonds. If you hold a $1,000 savings bond for 5 years, what will its redemption value be?

MIXED REVIEW

1. Find 4.6% of $39,000.
2. What is 0.25% of $200?
3. Write 5 as a percent.
4. Find $12\frac{1}{2}\%$ of $64.
5. 12 km + 325 m = ? m
6. 2 500 mL = ? liters
7. A total of 6 bonds are purchased at 107 plus $12.50 commission per bond. What is the total investment in bonds?
8. Kelly Lind borrowed $5,100. Find the due date and the amount due at maturity on this 75-day, 8% note, dated May 16. Use a 360-day year for this note.
9. On July 1, Bob deposited $740 in a bank account that paid 5% interest per year, compounded semiannually. Interest was added on January 2 and July 1. Find his balance on July 1 of the next year if he made no other deposits or withdrawals.
10. You write a check for two thousand, seven dollars and six cents. Write that amount in numbers.

11-2 INVESTING IN STOCK

OBJECTIVES

In this lesson, you will learn to

- *calculate the total cost of purchasing stocks;*
- *read a stock table;*
- *calculate the yield on stock investments;*
- *calculate annual dividends from stocks; and*
- *find the proceeds from the sale of stock.*

Companies often issue stock to raise money to help reach their goals. For example, a company with a successful product may need money to expand. A new company may need money to help pay for the designing and testing of a product that it hopes will be successful in the marketplace.

◆ Consider the two companies just described. In which of these companies would you be willing to invest your money? Why?

WARM *UP*

a. $24,000 × 0.2815

b. $18 is what percent of $72?

c. 2 kg = ? g

d. What amount is 6.5% less than $30?

e. $1,782 is what percent smaller than $1,800?

INVESTING IN STOCKS

Investors who buy the shares of a corporation are called the **shareholders** or *stockholders* of a company. Each shareholder gets a **stock certificate** that shows on its face the number of shares it represents. Each certificate is registered with the corporation in the owner's name. Stockholders may buy and sell their shares on a stock exchange. A sample stock certificate is shown in Illustration 11-2.1.

The money stockholders invest becomes a permanent part of the corporation. Unlike bondholders' investments, the money does not have to be repaid. As part owners, however, stockholders have a right to share in the profits of the corporation. Their projected gain is not fixed as it is with bondholders. The profits that the corporation distributes to its shareholders are called **dividends.** Dividends are usually paid quarterly. They may also be paid semiannually or annually.

COMMON STOCK

78
NUMBER

CAPITAL STOCK
$5,000.00
50 SHARES PAR VALUE $100 EACH

50
SHARES

Martin Leasing Company

This Certifies that Alex Preskow is the owner of Fifty - - - - - - - - - - - - - - - - - - - Shares of the Capital Stock of **Martin Leasing Company,** transferable only on the Books of the Corporation by the said owner, in person or by duly authorized Attorney, upon the surrender of this Certificate properly endorsed.

In Witness Whereof We have hereunto set our hands and affixed the seal of the Company at 9:50 a.m. this 10th day of April 19 - -

Charles Baker
SECRETARY

Christine Poloka
PRESIDENT

SHARES $100 EACH

Illustration 11-2.1. Stock Certificate

Critical Thinking Explain the difference between a bond holder and a stockholder of a corporation.

Selling a Stock Issue An issue of stock is sold in much the same way as an issue of bonds. The whole issue is usually sold to an investment banking house. The banking house then sells the shares to investors at a somewhat higher price than it paid for them.

After an issue of stock has been sold the first time, investors buy and sell the shares at any price they agree on.

The price at which a stock is sold is called the **market price** or *market value.* The market price of a stock depends largely on what investors think about the future earnings of the company. The market price changes often.

How the Market Price of a Stock is Quoted The trade of 1,000 shares of Weylin International stock may be posted as shown in Illustration 11-2.2. The price at which a stock is sold is its market price. The market price, or market value, of a share of stock is usually quoted in dollars per share. For example, a quotation of "$58\frac{1}{2}$" means that the price of one share is $58.50.

WI... $58\frac{1}{2}$... 1000

Illustration 11-2.2. Stock Trade

Stockbroker's Commission After a stock issue has been sold through an investment banking house, purchases and sales of stock are made through brokers in the same way that bonds are bought and sold. The broker charges a commission or brokerage fee for the service.

The amount of a broker's commission depends on the price of the stock and the number of shares bought. A special type of broker, called *a discount broker,* charges lower commissions but also often gives less service and advice to customers.

Finding the Total Cost of a Stock Purchase

When you buy stock through a broker, the total cost of the stock is the market price of the stock plus the broker's commission.

Market Price + Commission = Total Cost

EXAMPLE Richard Lim bought 300 shares of Crestone stock at 31. The broker charged him \$97 commission. Find the total cost of the stock.

SOLUTION

$300 \times \$31 =$	\$9,300	market price
	97	broker's commission
$\$9,300 + \$97 =$	\$9,397	total cost

Daily Stock Quotations

Each day's sales on the stock exchanges are shown the next day on the financial pages of many large city newspapers. Some examples of quotations from a stock exchange for one day are shown in Illustration 11-2.3.

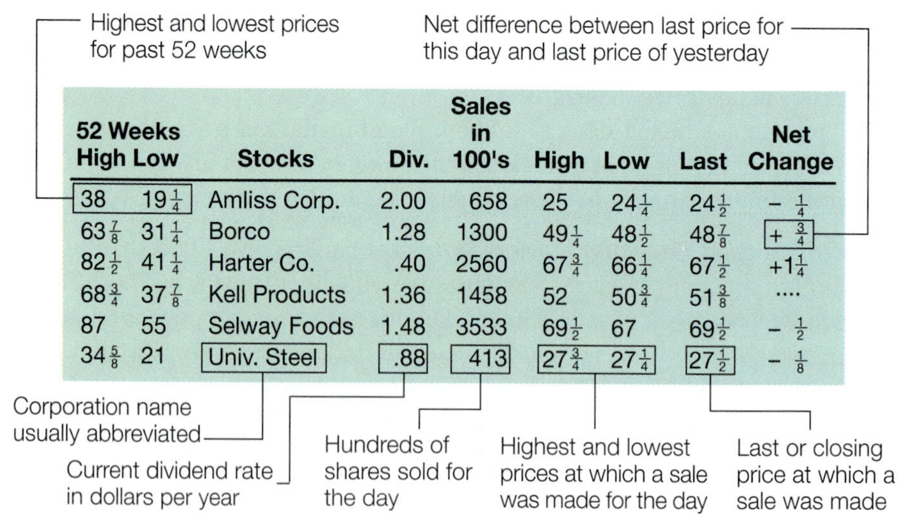

52 Weeks High	Low	Stocks	Div.	Sales in 100's	High	Low	Last	Net Change
38	$19\frac{1}{4}$	Amliss Corp.	2.00	658	25	$24\frac{1}{4}$	$24\frac{1}{2}$	$-\frac{1}{4}$
$63\frac{7}{8}$	$31\frac{1}{4}$	Borco	1.28	1300	$49\frac{1}{4}$	$48\frac{1}{2}$	$48\frac{7}{8}$	$+\frac{3}{4}$
$82\frac{1}{2}$	$41\frac{1}{4}$	Harter Co.	.40	2560	$67\frac{3}{4}$	$66\frac{1}{4}$	$67\frac{1}{2}$	$+1\frac{1}{4}$
$68\frac{3}{4}$	$37\frac{7}{8}$	Kell Products	1.36	1458	52	$50\frac{3}{4}$	$51\frac{3}{8}$	....
87	55	Selway Foods	1.48	3533	$69\frac{1}{2}$	67	$69\frac{1}{2}$	$-\frac{1}{2}$
$34\frac{5}{8}$	21	Univ. Steel	.88	413	$27\frac{3}{4}$	$27\frac{1}{4}$	$27\frac{1}{2}$	$-\frac{1}{8}$

Illustration 11-2.3. Corporation Stock Quotations

Reading the Stock Table The sales of Amliss Corporation are shown in the first quotation on the table. The figures at the left of the company name show the highest price paid, \$38, and the lowest price paid, \$19.25, for the stock during the past 52 weeks.

At the right of the company name, the "2.00" is what Amliss is paying its shareholders as a dividend for each share ($2).

The "658" in the Sales in 100s column means that 65,800 shares were sold during the day.

The High column shows the highest price paid for the stock during the day ($25). The Low column shows the lowest price paid for the stock during the day ($24.25). The Last column shows the price of the last, or closing, sale of the day ($24.50).

The "$-\frac{1}{4}$" in the Net Change column means that the last price of the day was $\frac{1}{4}$ of $1 or $0.25 lower than the last price yesterday. The last price yesterday must have been $24.75, or $24.50 + $0.25.

EXERCISE YOUR SKILLS

Find the total cost of each stock purchase.

	Number of Shares	Market Name of Stock	Price	Commission	Total Cost
1.	100	Arcell	28	$67.60	
2.	200	Concorp	$20\frac{5}{8}$	86.81	
3.	100	Dryer Glass	$15\frac{1}{4}$	50.33	
4.	50	Gen. Importers	$64\frac{3}{8}$	73.67	
5.	20	IMX	$109\frac{7}{8}$	58.86	
6.	80	Montenet	$30\frac{1}{2}$	62.38	
7.	400	Nortox	4	46.80	
8.	140	Strand	$36\frac{5}{8}$	101.09	
9.	30	Tole Instrum.	$45\frac{1}{4}$	41.22	
10.	150	XOE Computer	$55\frac{1}{4}$	240.60	

11. Diane was charged a commission of $50 to buy 20 shares of Stone & Kerr stock at 62. What was her total investment in the stock?

12. A discount broker offers a 40% discount off the $80 commission charged by another broker to handle the purchase of 30 shares of Eaton at $70\frac{1}{2}$. How much could a buyer save by purchasing from the discount broker?

Use Illustration 11-2.3 to answer each of the following.

13. During the past 52 weeks, what stock in the table sold at the highest price of all?

14. What stock sold at the lowest price during the last 52 weeks?

15. What stock paid the highest dividend? What amount was paid?

16. What stock paid the lowest dividend? What amount was paid?

17. How many shares of Borco were sold?

18. How many shares of Universal Steel were sold?

19. What stock from the table showed a net change of exactly $0.75 for the day?
20. What stock had the greatest net change for the day?
21. For what stock was the last price also the highest price for the day?
22. What stock had the same closing price as the day before?

STOCK INCOME

Shareholders earn income on their investments in stock through **stock dividends.** They may receive dividends for as long as they own stock. The amount of dividends paid by a company usually depends on how much profit the company makes.

Classes of Stock

Many corporations issue two classes of stock—common stock and preferred stock. Both types represent ownership. They differ, though, in how they share in profits that are paid out as dividends.

Preferred stock has its dividend set by the corporation at the time it is issued. The dividend is set as a specific rate, such as 5%, or $5 a share. **Common stock** is the ordinary stock of a corporation. Common stock does not have a set dividend. There is no guarantee that dividends will be paid to either class of stockholder. When dividends are paid, they go first to shareholders of preferred stock.

Dividends When a corporation decides to distribute profits to the shareholders, it *declares a dividend.* The dividend may be shown either as a percent of the par value of the stock or as an amount of money per share. For stock with no par value, the dividend is always declared as an amount per share.

For example, suppose a corporation declares a dividend of 3.4% of the par value of $100. The amount of the dividend on each share is 3.4% of $100, or $3.40. The corporation may declare either a dividend of 3.4% or a dividend of $3.40 a share.

Income from Dividends When a corporation pays dividends on stock that you own, you earn income. To find the total amount of the dividend payment you get, you multiply the dividend on one share by the number of shares you own. Your yearly income is the total of the dividends you get in one year.

EXAMPLE Myrna Wolf owns 70 shares of Antel Computer common stock, par value $100. If a dividend of 4.5% is declared, how much should Myna get in dividends?

SOLUTION Find the dividend on one share. Multiply to find the amount on all shares.

$0.045 \times \$100 = \$4.50 \qquad 70 \times \$4.50 = \$315$

PROBLEM SOLVING

Find a Pattern Some problems are simplified if you can find a pattern and use it to solve problems or answer questions.

Suppose you own some stock and are thinking of selling it. You could look for patterns that might help you make a decision. You have tracked its selling price and dividends. You find these results:

	Selling Price	Dividends
Jan. 1	$25\frac{1}{8}$	2.1%
April 1	$22\frac{1}{2}$	1.4%
July 1	$20\frac{7}{8}$	0.5%
Oct. 1	$18\frac{3}{8}$	3.2%
Jan. 1	19	3%

What patterns do you see? Would you have sold the stock in July? If today is January 2, would you sell the stock? Explain your reasoning.

Dividends are sometimes calculated and paid quarterly (four times a year) or semiannually (twice a year).

EXAMPLE What yearly income will Roy and Jean Zern get from an investment in 60 shares of Heritage Furniture stock if a dividend of $0.47 is paid quarterly?

SOLUTION $60 \times \$0.47 = \$28.20$ dividend for 1 quarter
$4 \times \$28.20 = \$112.80$ yearly income

Finding the Yield on a Stock Investment The yield, or rate of income, received from an investment is found by dividing the annual income from the investment by the amount invested.

For stocks, the investment is the total cost of the stock, including any expenses or commission paid in obtaining the stock. The income is the amount of annual dividends. The yield is equal to the annual dividends divided by the total cost and is stated as a percent.

$$\textbf{Yield} = \frac{\textbf{Annual Dividends}}{\textbf{Total Cost of Stock}}$$

In the example above, if the Zerns had paid a total of $2,400 for their stock, and earned annual dividends of $112.80, their annual yield would be

$$\text{Annual Yield} = \frac{\$112.80}{\$2{,}400}$$

$$\text{Annual Yield} = 0.047 \text{ or } 4.7\%$$

EXERCISE YOUR SKILLS

23. Nora owns a share of stock that cost \$18 and pays a quarterly dividend of $\$0.31\frac{1}{2}$. Find the yield for that stock.

Find the total annual dividend received by each shareholder.

	Shares Owned	Par Value per Share	Annual Dividend Rate	Dividend
24.	100	\$100	6%, annually	
25.	80	100	9%, annually	
26.	82	50	$7\frac{1}{2}$%, annually	
27.	50	—	\$0.65 per share, quarterly	
28.	160	—	1.70 per share, quarterly	

Find the yield on the investment in each of the following stocks. Round all answers to the nearest tenth of a percent.

	Total Cost per Share	Par Value per Share	Dividend Rate	Dividend Payable	Yield
29.	\$130	\$100	8%	Annually	
30.	60	100	5%	Annually	
31.	90	—	\$0.57 per share	Quarterly	
32.	56	—	1.30 per share	Quarterly	
33.	12	—	0.17 per share	Quarterly	

Ada Garza bought 50 shares of stock at \$47.375 a share. She paid a commission of \$62.43 for this purchase. The stock pays an annual dividend of 6% on a par value of \$50.

34. What total dividend does Ada get from the 50 shares?

35. What yield does Ada earn on this investment?

Regal Paint Company stock sells for \$28 and pays an annual dividend of 2.7% on a par value of \$100. Esteem Company has a market price of $8\frac{1}{4}$ and pays a quarterly dividend of \$0.18.

36. Which stock earns a higher yield?

37. How much higher?

Page Alarm Company stock pays a regular annual dividend of \$2.50 a share.

38. How many shares must you buy to get an annual income of \$1,000 from the investment?

39. What total investment will you make in the stock if you buy it at 56 and pay \$82 per 100 shares for commission?

A stock paying an annual dividend of \$1.40 a share sells at 34.

40. How many shares of stock must you buy at that price to get an annual dividend income of \$308?

41. What total investment, without calculating commissions, would you have to make to earn the dividend income you want?

PROFIT AND LOSS ON STOCK INVESTMENTS

When shareholders sell their stock, they either make a profit, break even, or take a loss. To find out how they did on their investments, the investors must consider the cost of the stock, its selling price, and the expenses of selling.

Finding the Proceeds from the Sale of Stock

When you sell stock, you must pay a commission. You may also pay other charges such as a handling or service fee and a Securities and Exchange Commission (SEC) fee. Sometimes the state in which you sell stock charges a *transfer tax*. When you buy stock, you do not pay a transfer tax or an SEC fee.

When you sell stock through a broker, you get the market price less the commission and all other charges (service fee, SEC fee, transfer tax). The amount you receive is called the *net proceeds*.

Market Price − (Commission + other charges) = Net Proceeds

EXAMPLE Find the net proceeds from the sale of 50 shares of stock at 30 with commission and other charges of $62.

SOLUTION

50 × \$30 = \$1,500	market price
\$1,500 − \$62 = \$1,438	net proceeds

Finding the Profit or Loss on the Sale of Stock Your profit or loss on a sale of stock is the difference between the total cost of purchasing the stock and the net proceeds. If the amount of the net proceeds is greater than your total cost, you have a profit. If it is less than your total cost, the result is a loss.

Net Proceeds − Total Cost = Profit
Total Cost − Net Proceeds = Loss

EXERCISE YOUR SKILLS

Find the net proceeds from the sales of the stock below.

	Name of Stock	Shares Sold	Selling Price	Commission & Other Fees	Net Proceeds
42.	Zenwell Mining	100	34	\$61.34	
43.	Melork Properties	50	$12\frac{1}{4}$	48.22	
44.	Dryson Chemical	230	81	145.57	
45.	Harbridge Hospital	80	$28\frac{1}{8}$	52.68	

Toni Ulworth sold 100 shares of stock at 192. Her total expenses from the sale of the stock were \$50.11.

46. What were her estimated net proceeds from the sale?

47. What were her actual net proceeds from the sale?

Find the profit or loss in each of these sales, labeling each loss.

	Name of Stock	Shares Traded	Selling Price	Commission & Other Fees	Total Cost of Purchase	Profit/ Loss
48.	UDC Polymers	100	42	\$88.04	\$3,935.56	
49.	King Theaters	80	$8\frac{1}{2}$	50.73	558.42	
50.	Sun Graphics	400	17	97.39	7,003.67	
51.	Com-Design	75	$28\frac{1}{4}$	57.58	2,840.31	
52.	Euro Shipping	110	37	86.47	3,200.75	

53. Gerardo bought 60 shares of stock for a total cost of \$2,985. He later sold the stock at \$51 a share and had selling expenses of \$94. What was his profit or loss from this sale of stock?

Finding Total Gain

Besides any profit you may make on the sale of stock, you may also earn dividends on stock you own. To find your **total gain** from owning stock, you must combine the profit or loss from the sale of stock with the dividends received.

EXAMPLE You bought 40 shares of XYZ stock for a total cost of \$3,757 and kept the stock for 3 years during which time you received semiannual dividends of \$3.10 per share. You sold the stock and received net proceeds of \$4,211. Find the total dividends and total gain from owning and selling the stock.

SOLUTION $40 \times 3 \times 2 \times \$3.10 =$ \$744.00 total dividends.
$\$4{,}211 + \$744 - \$3{,}757 = \$1{,}198$ total gain.

EXPLANATION
Add all income received from stock ownership (dividend and sale proceeds) to find gross income. Subtract total investment in the stock to find total gain.

54. Darian Woods bought 75 shares of stock for a total cost of $1,298. After receiving 6 quarterly dividends of $0.27 per share, he sold the stock and received net proceeds of $1,194. What was his total gain from owning and selling the stock.

55. Jeanette sold 62 shares of Wexor Van Lines at a profit of $345 and 120 shares of Ark Adhesives at a loss of $490. Before she sold the stock, she received 3 quarterly dividends of $0.90 a share on the Wexor Van Lines stock. Ark Adhesives paid no dividends. What was Jeannette's total gain from owning and selling both stocks?

56. ♦ Suppose you had $1,000 to invest in the stock market. Choose two or more stocks from a current stock market report and invest in them. Track your purchases for 1 week. Did the value of your investment change in the week? Compare your results with other students' results.

MIXED REVIEW

1. 7.004×0.035
2. $\frac{2}{3}$ of $9.51
3. $37,000 × 2.38
4. $5{,}724 \div 12{,}000$
5. Find $1{,}800 \div 10.7$, rounded to the nearest tenth.
6. $798.72 is what percent of $12,800?
7. 84 increased by 22.5% of itself is ?
8. How many m^2 are in a lot 22 m by 40 m?
9. Charles bought 120 shares of 7% preferred stock with a par value of $150. What is his annual dividend income?
10. Ava borrows $900 and agrees to repay the loan in 36 payments of $31 each. How much is the finance charge on the loan?
11. Dominic Avtar wants to buy enough $1,000, 7% bonds to have an annual income of $1,610. How many bonds must he buy?
12. Janet Zurn sold 100 shares of Entinol stock at $32\frac{1}{2}$. By selling the stock through a discount broker, she saved 20% of the $88.20 commission charged by another broker. What were Janet's net proceeds from the sale of this stock?
13. What is the rate of income on a share of preferred stock that costs $90 and pays a semiannual dividend of $3.60 a share?

11-3 INVESTING IN MUTUAL FUNDS

OBJECTIVES

In this lesson, you will learn to

- *calculate the number of shares purchased;*
- *find amount of and rate of commission; and*
- *calculate profit or loss from investments in mutual funds.*

The trading cards of some first-year baseball players may become fairly valuable in future years. This usually happens when the player pictured becomes one of the superstars of the game.

◆ Assume you collect baseball cards. Write a strategy you would use to have a good chance of ending up with the trading card of a future superstar? What are the possible pitfalls of your strategy?

WARM *UP*

a. 8.005×0.046

b. \$2,500 + \$6.35 + \$82.01

c. Write $\frac{3}{5}$ as a percent.

d. Estimate the product of $\$0.19 \times 190$.

e. \$820 is what percent of \$16,400?

f. Find $1,200 \div 2.1$, rounded to the nearest tenth.

INVESTING IN MUTUAL FUNDS

If you make an investment in the stock of only one company, you can make a good profit if the company does well. If the company is not profitable, however, you may lose all or part of your investment. You may be able to lessen your risk of making an unwise investment, by spreading your investment dollars among several companies. **Mutual funds** are investment companies designed to make it easy for a single investor to do that.

Nature of Mutual Funds Mutual fund investment companies use the money from their investors to buy stock in a number of companies. By investing in a large number of companies, the mutual fund increases its chances of buying stocks that will be profitable. There are many different kinds of mutual fund companies. Some have aggressive profit goals, others have more moderate growth goals. Some mutual funds specialize in certain kinds of stock; others purchase stock from a wide variety of organizations.

When selecting a mutual fund to invest in, as with other investment options, you should do some research before investing your money. You want to select a mutual fund based on their investment requirements, financial goals, stock portfolio, and performance record.

Mutual fund shares are traded based on net asset values. You can calculate **net assets** by finding the total value of the mutual fund's investments and subtracting any money owed to others. The **net asset value,** or NAV, is found by dividing the net assets by the number of shares held by stock holders. For example, a fund with net assets of $10,000,000 and 500,000 shares issued will have a net asset value of $20 ($10,000,000 ÷ 500,000).

The net asset value of mutual funds is calculated daily and is published on the financial pages of daily newspapers in a form similar to that shown in Illustration 11-3.1. Following the fund name, the net asset value appears in the NAV column.

Fund Name	NAV	Offer Price
Denton Fund	10.03	10.96
Frontier Growth	39.80	41.03
Kaner High Yield	10.58	11.48
Mark International	6.37	N.L.
Newsome Growth	17.87	N.L.
Payne Hall New Hz.	9.27	N.L.
Randall Explorer	17.54	19.17
Walker A-1 Fund	12.21	N.L.

Illustration 11-3.1. Mutual Fund Quotations

Types of Mutual Funds Two types of mutual funds are listed in Illustration 11-3.1: **no-load funds** and **load funds.** The term load means the same as commission. *No-load funds* are sold without a commission being charged. The no-load funds are identified by the abbreviation "N.L." in the Offer Price column.

Some of the funds shown in Illustration 11-3.1 have two numbers or prices following their names. These funds are called *load funds.* When you buy load funds, you pay the amount shown in the Offer Price column, which includes a commission charge.

Finding the Total Investment in Mutual Fund Shares The mutual fund investment is calculated in two different ways, depending on the type of fund being purchased. For load funds, the total investment is found by multiplying the **offer price** by the number of shares bought. For example, in Illustration 11-3.1 the offer price of the Kaner High Yield fund is 11.48. So, the cost of 200 shares is $2,296 (200 × $11.48).

For no-load funds, the NAV is multiplied by the number of shares to find total investment. In Illustration 11-3.1, you can see that Newsome Growth has a NAV of 17.87. The total investment in 100 shares would be $1,787 (100 × $17.87).

Finding the Number of Shares Purchased Many people who buy mutual fund shares invest a certain dollar amount, such as $500, instead of buying a certain number of shares. To find the number of shares purchased, divide the amount of the investment by the price of a mutual fund share. The price of a no-load fund is its net asset value. The price of a load fund is its offer price.

Suppose $500 is invested in a no-load mutual fund with a net asset value of 6.37. The number of shares bought is 78.493 ($500 ÷ $6.37). The number of shares is calculated to three decimal places to assure that the entire dollar amount is invested.

Finding the Rate of Commission When you buy no-load funds, you are not charged for commission. For load funds, the amount of commission is the difference between the net asset value and the offer price. To find the rate of commission on the purchase of load funds, divide the amount of commission by the offer price.

Offer Price − Net Asset Value = Commission
Commission ÷ Offer Price = Rate of Commission

EXAMPLE What is the rate of commission, to the nearest percent, on Frontier Growth funds with a net asset value of 39.80 and an offer price of 41.03?

SOLUTION $41.03 − $39.80 = $1.23 amount of commission
$1.23 ÷ $41.03 ≈ 0.02997 ≈ 3% rate of commission

Profit or Loss from Mutual Fund Redemption

When shares are *redeemed,* or sold back to the mutual fund company, the investor is paid the net asset value. The proceeds from the sale are found by multiplying the net asset value by the number of shares redeemed.

Proceeds = Number of Shares × Net Asset Value

The profit or loss from owning mutual fund shares is calculated by finding the difference between the proceeds and the total amount invested.

Proceeds − Amount of investment = Profit
Amount of Investment − Proceeds = Loss

EXAMPLE Find the proceeds from the sale of 200 shares of a mutual fund with a net asset value of 12.30. Find the amount of the profit or loss if the total investment in the 200 shares is $1,950.

SOLUTION $200 \times \$12.30 = \$2{,}460$ proceeds
$\$2{,}460 - \$1{,}950 = \$510$ profit

EXERCISE YOUR SKILLS

Using the information in Illustration 11-3.1, find the total investment in the mutual fund purchases.

	Mutual Fund	Number Shares	Total Investment
1.	Payne Hall New Horizon	40	
2.	Mark International	132	
3.	Denton Fund	80	
4.	Randall Explorer	260	

5. Amy Lester bought these Pacific Trends, no-load mutual fund shares on different days: 50 shares, NAV 13.56; 120 shares, NAV 13.84; 90 shares, NAV 14.05. What was her total investment in Pacific Trends?
6. Foster A-2, a no-load fund, has a net asset value of 16.82. Foster Growth, a load fund, is quoted with a net asset value of 7.50 and an offer price of 8.03. What total investment would be made if 100 shares each of Foster A-2 and Foster Growth were purchased?

Calculate shares to the nearest thousandth share.

7. Today's NAV of the Ark-Stone Group, a no-load fund, is 6.73. At that price, how many shares can be bought for $8,000?
8. Antoine LeBaron has $2,500 to invest in Udrell Technology, a no-load fund with a current NAV of 16.20. How many shares of Udrell can Antoine purchase?
9. Rose invested $5,000 in a load fund that has a current NAV of 16.50 and offer price of 17.34. How many shares did she buy?

Cecile owns 804.726 shares of Lorgen New Vision, a load mutual fund. She is investing another $500 in Lorgen shares which are quoted today with a net asset value, 22.60; offer price, 24.75.

10. Estimate the number of shares Cecile can buy today.
11. What is the actual number of shares she can buy?
12. After this purchase, how many shares of Lorgen will she own?
13. Stenwell Growth Fund is quoted at 6.08 NAV, 6.24 offer price. To the nearest tenth of a percent, what commission rate is charged?
14. Your broker quotes Leland Air Fund at these prices: NAV, 14.06; offer price, 14.80. Find the rate of commission.

Find the amount of commission and the rate of commission, to the nearest tenth, for each of the mutual fund shares.

	Mutual Fund	NAV	Offer Price	Amount of Commission	Rate of Commission
15.	Pentel Government	6.29	6.42		
16.	Zoll Energy Fund	12.78	13.44		
17.	Allen Asia Fund	17.75	19.40		
18.	Will Technology	8.58	8.98		

A commission of $0.78 is charged for a mutual fund with an offer price of 10.04.

19. What is the actual rate, to the nearest tenth percent?

The Dart Oil Fund has a current NAV of 11.87. Elmo redeemed the 320 Dart shares that he had bought for a total cost of $3,198.

20. What were his proceeds?

21. Find his profit or loss.

The Odgen Fund had a NAV of 7.94 and an offer price of 8.10 on the day that Herb Mills invested $4,000 in the fund. Six months later, the fund was quoted at these prices when Herb redeemed his investment in the fund: NAV, 8.11; offer price, 8.28.

22. What actual number of shares did Herb buy?

23. What proceeds did he receive?

24. What amount of profit or loss did Herb's investment make?

25. ◆ When investing in sports trading cards, there is a danger that the card may be damaged in some way. Such damage may make a $1,000 card worth less than $100. Speculate about any dangers that are associated with buying mutual funds. Compare your list with other classmates.

MIXED REVIEW

1. 500 is what part greater than 400?

2. Estimate the cost of 3.1 lb at $18.75.

3. 400 grams is what percent of a kilogram?

4. Find the cost of 1,300 lb @ $82 per cwt.

5. Write $2\frac{2}{15}$ as a decimal, rounded to the nearest tenth.

The Boswenth Americas Fund trades at a net asset value of 17.44 and an offer price of 18.23.

6. What commission amount is paid per share?

7. To the nearest tenth of a percent, find the rate of commission.

8. Mark bought 350 shares of Alcan Fund for $7,192.50. Two years later he sold his Alcan shares at a 21.75 NAV. What was his profit or loss on this investment?

11-4 INVESTING IN REAL ESTATE

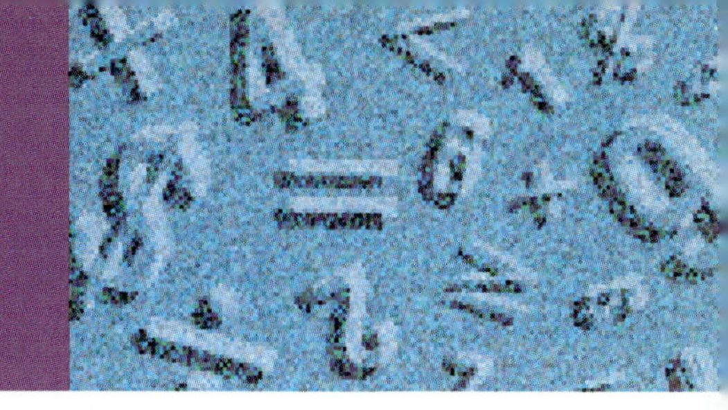

Real estate is an attractive investment option to many people. Some people invest in single or multiple-family dwellings; others invest in office buildings. Still others may invest in land for farming or mining.

◆ Assume you want to buy a small apartment building. What problems do you expect to have by being a landlord? Can you think of ways to prevent or solve those problems? Will you have to work full time at being a landlord? Based on your answers to these questions, what do you think are some of the advantages and disadvantages of investing in real estate?

OBJECTIVES

In this lesson, you will learn to

- *calculate net income and rate of return from a real estate investment;*
- *find the monthly rent that should be charged; and*
- *calculate and use rate of return on investment.*

WARM UP

a. \$15.75 ÷ 10.5

b. $\frac{1}{2} \times \frac{3}{4}$

c. Find 3% of \$108,000.

d. Write 0.051 as a percent.

e. What amount is 40% more than \$35.50?

f. Find \$400 ÷ \$5,500, to the nearest tenth of a percent.

INVESTING IN REAL ESTATE

When you invest in real estate, the rent you receive from others is your gross income from the investment. Your net income is the amount that is left after you have paid all the expenses of owning the property.

Even though you may collect rent monthly, you usually calculate your income and expenses on an annual basis. For example, assume that you collect rent of \$700 a month and have annual expenses of \$7,600. Your gross income for the year is 12 × \$700, or \$8,400. Your net income for the year is \$8,400 − \$7,600 or \$800.

Expenses of Owning Rental Property As Illustration 11-4.1 shows, only part of the money you collect as rent is profit. The rest of the money is usually used to pay for taxes, repairs, and insurance. Generally, you will also have to pay interest on a mortgage loan.

Since rental property consists of buildings that wear out because of age and use, the real estate investor will also have to calculate depreciation.

Illustration 11-4.1. Where Rental Money Goes

Finding the Annual Net Income from a Real Estate Investment The annual net income is the amount left after deducting the annual expenses from the annual rental income.

Annual Rental Income − Annual Expenses = Annual Net Income

EXAMPLE Lori Nalin bought a house for \$50,000. She made a \$10,000 cash down payment and got a \$40,000 mortgage for the balance of the purchase price. She rented the house to a tenant for \$500 a month. Her annual payments for taxes, repairs, insurance, interest, depreciation, and other expenses totaled \$4,700. What annual net income did she earn?

SOLUTION

$12 \times \$500 = \$6,000$	annual rental income
$\$6,000 - \$4,700 = \$1,300$	annual net income

Finding the Rate of Return on a Real Estate Investment The rate of return, or yield or rate of income, is based on the cash investment in the property. It is found by dividing the annual net income by the cash investment.

$$\textbf{Rate of Return} = \frac{\textbf{Annual Net Income}}{\textbf{Cash Investment}}$$

The rate of return for Lori Nalin's house is calculated this way:

$$\text{Rate of Return} = \frac{\$1{,}300}{\$10{,}000} \qquad \text{Rate of Return} = 0.13 \text{ or } 13\%$$

EXERCISE YOUR SKILLS

Find the annual net income for each real estate owner.

	Monthly Rent Income	Annual Expenses: Taxes	Repairs	Insurance	Interest	Depreciation	Other	Annual Net Income
1.	$590	$1,720	$470	$500	$2,100	$1,300	$50	
2.	310	950	360	380	1,100	560	40	
3.	840	2,800	210	590	3,000	2,050	90	
4.	540	1,630	840	360	1,870	1,200	75	

5. Marvin bought a condominium and lot for $72,000. He paid $18,000 cash and got a mortgage for the balance. He rented the condominium to a tenant for $950 a month. For the first year, Marvin's expenses were: mortgage interest, $4,800; 3% depreciation on the house valued at $58,000; taxes, repairs, insurance, and other expenses, $3,750. What was his net income for the year?

6. Nicole bought an eight-unit apartment building for $170,000. During the first year of ownership, she received $460 a month for the rent of *each apartment unit.* Her expenses for the year were: mortgage interest, $20,400; 2% depreciation on the building valued at $130,000; taxes, repairs, insurance, and other expenses, $14,900. Find her net income for the year.

Find the rate of return, to the nearest tenth of a percent, that these owners will get on their cash investments.

	Cash Investment	Monthly Rental Income	Annual Expenses: Interest on Mortgage	Other	Rate of Income
7.	$21,000	$500	$2,400	$1,900	
8.	38,000	780	1,600	3,300	
9.	46,000	1,800	9,600	9,300	
10.	12,000	390	2,200	1,450	
11.	31,000	640	2,200	2,150	

12. Lester Collins took out a $62,000 mortgage on a 2-family house after making a down payment of $11,000 as a cash investment. During the first year, he rented one unit at $560 a month and the other unit at $580 a month. For the year, he paid $8,200 in mortgage interest and $4,400 in other expenses. To the nearest tenth of a percent, what rate of income did Lester earn on his cash investment?

For $78,000, Bridgette can buy an office building that rents for $920 a month. Taxes, insurance, and repair expenses average $3,900 annually. Depreciation is estimated at $1,200 a year. Bridgette plans to pay $78,000 cash for the property. To the nearest tenth of a percent, find the

13. Rate of income she will make on her cash investment.
14. Rate of income she will make on her cash investment by charging $1,050 monthly rent and keeping expenses the same.
15. Kim Ruiz bought a 6-unit apartment house for $70,000 and made a cash down payment of $20,000. The first year, she rented each of the 6 apartments at $290 a month. Her expenses for the year were: mortgage interest, $9,800; depreciation at 4% of the house's value of $55,000; taxes, insurance, and other expenses, $6,300. Find Kim's rate of income on her cash investment, to the nearest percent.
16. Zack Eason made a $7,000 down payment on a house that cost $24,000. He rented the house at $490 monthly for the first year. During the year he had these expenses: taxes, $990; insurance, $270; interest, $2,400; repairs, $900; depreciation at 2% of the cost of the house. What was Zack's rate of income on his cash investment?

REAL ESTATE INCOME

When you invest in real estate, you should set a goal for the rate of income that you want to earn on your cash investment. The rent you charge must cover the net income that you want to earn and all expenses of owning the property.

Determining What to Charge for Rent To determine how much to charge for rent, you must first determine annual expenses and an amount you want to earn as net income on your investment.

$$\text{Annual Net Income} + \text{Annual Expenses} = \text{Annual Rental Income}$$

$$\frac{\text{Annual Rental Income}}{12} = \text{Monthly Rent}$$

EXAMPLE Ezra wants to earn 15% annual net income on his $10,000 cash investment in property. His annual expenses of owning the property are $2,700. What monthly rent must Ezra charge?

SOLUTION

$0.15 \times \$10,000 = \$1,500$	desired annual net income
$\$1,500 + \$2,700 = \$4,200$	annual rental income
$\$4,200 \div 12 = \$350$	monthly rent

Critical Thinking What if property around Ezra rents for $250. Should he still charge $350? Why or why not? What if surrounding property rented for $1,000.

Capital Investment and Expenses If you plan to be a property owner, you will spend money on capital investments and expenses. ***Capital investment*** is the amount of the original investment plus anything else spent for improvements that make the property more valuable. Adding a room to your house and building a garage are examples of capital investment.

Money spent for repairing and replacing broken items does not increase the value of property, but it is an expense of owning the property. Expenses are deducted from the gross income of property to find net income. The money paid out returns the property to its original condition. Such expenses might include repainting the house, replacing a broken sidewalk, and repairing leaking faucets.

EXERCISE YOUR SKILLS

Find the monthly rent the owner must charge.

	Cash Investment	Desired Annual Net Income	Annual Expenses	Monthly Rent
17.	$20,000	14% of investment	$3,500	
18.	8,000	7% of investment	2,560	
19.	45,000	12% of investment	9,120	
20.	36,000	9% of investment	8,400	

Elvira Needham bought two lots for $12,000 at a tax sale. She estimates that her yearly expenses of owning these lots will be $540. A nearby factory wants to use the lots for parking trucks overnight. To earn a 15% rate of income on her investment,

21. What annual rent should Elvira charge?

22. What estimated monthly rent should she charge?

23. What actual monthly rent should Elvira charge?

Brian Mahaney bought a vacant warehouse with a $12,000 cash down payment. He estimates that expenses will be $4,320 for the first year. Brian's goal is to earn a rate of income of 8% on his investment.

24. To reach this goal, what monthly rent must he charge?

25. If Brian's annual expenses go up by 5% in the second year, what monthly rent will he have to charge in the second year to reach his investment goal?

26. Cynthia Drexel bought a 5-year old condominium for $68,000 and immediately spent $2,200 to install a deck. She spent $1,600 to paint the interior and make minor repairs. What was her capital investment in the condominium?

Marv can buy a house for $28,000 cash. To make it more appealing to renters, Marv also has to spend $12,000 to improve the property by adding a room and paving the driveway. Marv estimates the total annual expenses of owning the house to be $2,600.

27. What would be Marv's capital investment in the house?

28. What monthly rent will he have to charge, after the improvements are made, in order to make a net income of 7% on his total capital investment?

29. Rick Weiss plans to buy a vacant 4-unit apartment by paying $26,000 in back taxes due the city. To meet current building and safety codes, he will have to spend $32,000. Rick calculates that each unit can be rented for $280 a month and that total annual expenses for all units will be $8,700. If he makes this capital investment, what rate of income will Rick earn, to the nearest tenth of a percent?

30. ♦ Consult a local real estate professional. In your area, what is the expected rate of return on real estate investments?

PUT IT ALL TOGETHER

31. In the first lesson of this chapter, you estimated how long it would take to pay for a goal with an initial investment of $1,000 in a bank savings account. In succeeding lessons, you estimated how a similar dollar investment in various investment vehicles might increase or decrease. Write the conclusions you have reached about risks involved in investments versus the risk involved in keeping money in a relatively low growth savings account.

MIXED REVIEW

1. $48 \div \frac{3}{4}$
2. Find 115% of 60.
3. $\frac{5}{12} - \frac{1}{6}$
4. 3.8 L − 250 mL = ? L
5. 48 is what percent of 150?
6. Find 2% less than $25.50.
7. The annual expenses on a house that costs $60,000 cash are estimated to be $3,900. What monthly rent must be charged to earn a net income of $7\frac{1}{2}\%$ on the cash investment?
8. Susan bought a vacant building and lot for $56,000. During the first year she spent $22,000 to divide the building into smaller spaces. In the first year, her cash expenses of ownership were $8,350. Depreciation was calculated at 2% on the $42,000 original worth of the building alone. What was Susan's capital investment?
9. Jessica withdrew $2,400 from a time-certificate account that paid 8% annual interest. Because she withdrew her money before the end of the term, the bank charged a penalty of 2 months' interest. What was the amount of the penalty?

TEMPLATE: Spreadsheet to Monitor Personal Investments

The following problems involve three different ways in which you can invest your own money and hopefully realize a profit. When you load ET11 from your applications diskette, you can make changes to the template as you consider the possibilities.

Template ET11A: **Spreadsheet for Finding Rental Rate of Income**
Enter the following information into the appropriate cell on template ET11A. The data represents investments in rental property that costs $50,000 and rents for $890 per month. In addition, you have made the following list of estimated annual expenses of ownership: $1,620 for taxes; $430 for repairs; $500 for insurance; $2,300 for interest; $1,100 for depreciation; and $100 for miscellaneous expenses.

When completed, your spreadsheet should look like the one shown below.

	A	B
1	Finding Rental Income	
2		
3	Cash Investment	$50,000.00
4	Monthly Rent	$890.00
5	Annual Expenses:	
6	Taxes	$1,620.00
7	Repairs	$430.00
8	Insurance	$500.00
9	Interest	$2,300.00
10	Depreciation	$1,100.00
11	Misc.	$100.00
12		
13	Annual Results	
14	Gross Rental Income	$10,680.00
15	Less Expenses	$6,050.00
16	Net Income (Loss)	$4,630.00
17	Rate of Income (%)	9.26

1. What formula was used to calculate each of the annual results in B14 – B17.

Template ET11B: **Spreadsheet to Compare Bond Investment Yields**
You have come into possession of two corporation bonds and wish to compare the annual income earned from each bond as well as their yields based on current prices. Load ET11B from your Spreadsheet Applications diskette. Enter the data shown in blue for Bond I (in column B) and Bond II (in column C).

	A	B	C
1	Finding Bond Investment Yields		
2		Bond 1	Bond 2
3	Number of Bonds	7	12
4	Par Value	$1,000	$500
5	Bond Interest Rate (%)	8.20%	9.75%
6	Market Price	.975	1.03125
7	Commission per Bond	$6.00	$7.50
8			
9	Total Investment	$6,867.00	$6,277.50
10	Annual Income	$574.00	$585.00
11	Annual Yield (%)	8.36%	9.32%

2. What formulas were used to calculate the bond totals in cells C9, C10, and C11?

***Template ET11C:* Spreadsheet to Calculate Stock Purchase Profit/Loss**
Two years ago you bought 100 shares at a total cost of $2,000. You just sold the stock at $34.50 a share, paying selling costs of $78. During the two years you owned the stock, you received eight quarterly (every three months) dividend payments of $0.55 for each share.

Load ET11C from the diskette. Complete the spreadsheet template with the information given above. Your completed spreadsheet should look like this:

3. What formula was used to calculate each of the summary information in Cells B14 – B17.

	A	B
1	Stock Ownership Analysis	
2		
3	Stock Purchase Price	$2,000.00
4		
5	Sales Data:	
6	Number of Shares	100
7	Selling Price/Share	$34.50
8	Selling Costs	$78.00
9		
10	Dividend Data:	
11	Dividend Amount	$0.55
12	No. of Payments	8
13		
14	Net Proceeds	$3,372.00
15	Net Profit (Loss)	$1,372.00
16	Dividends Received	$440.00
17	Net Gain (Loss)	$1,812.00

Design Your Own

How would you compare on a single spreadsheet which of the previous investments provided the most annual income?

1. What piece of information do you need to compare the investments?

2. Complete on paper or on computer a spreadsheet to compare investments. The spreadsheet might look something like the one below. Whatever it looks like, it should allow you to compare the annual income you received from rental, bond, and stock purchases.

	A	B	C	D	E
1		Bond 1	Bond 2	Rental	Stock
2	Total Investment				
3	Annual Income				
4	Yield (or Rate of Income)				

3. Which investment yielded the highest annual return?

CHAPTER 11 SUMMARY AND REVIEW

VOCABULARY REVIEW

annual net income *11-1*
bond *11-1*
broker *11-1*
capital investment *11-4*
commission *11-1*
common stock *11-2*
denomination *11-1*
discount *11-1*
dividends *11-2*
face value *11-1*
load funds *11-3*
market price *11-2*
market value *11-1*
mutual funds *11-3*
net asset value *11-3*
net proceeds *11-2*
no load funds *11-3*
offer price *11-3*
par value *11-1*
preferred stock *11-2*
premium *11-1*
rate of income *11-1*
rate of return *11-4*
redemption value *11-1*
savings bond *11-1*
shareholders *11-2*
stock *11-2*
stock certificate *11-3*

MATH SKILLS

11-1

1. $850 × 8
2. 7 × $987.50
3. $100 ÷ 2
4. $1,000 × 0.075 × 0.5
5. Find $13\frac{1}{2}$% of $1,000

11-3

Round to the nearest tenth of a percent.

6. $1.34 ÷ $40.92
7. $0.18 ÷ $8.98

APPLICATIONS

11-1
- Calculate the market price of a bond by multiplying the par value of the bond by the price quotation expressed as a percent.
- Use a redemption table to find the amount to be received when a savings bond is redeemed.
- Find the interest on a bond by multiplying the bond's par value by the rate and time.
- Calculate the rate of income or yield by dividing the annual income received by the total investment in bonds.

8. Sue bought 6, $1,000 Ezy Co. bonds at 94 plus a transaction fee of $37 and commission of $7 per bond. What was the amount of her investment?

Elio buys a $500 savings bond and holds it for 5 years. Use Illustration 11-1.3 to find the bond's

9. redemption value. **10.** interest earned.

11. Marc owns 15, $500 corporation bonds that pay 6.5% interest. Find his total income from this investment.

12. Karen bought 6, $1,000, 8% bonds at 102. Commission paid was $7 per bond. What yield does Karen receive on her bond investment, to the nearest tenth percent?

11-2 ♦ Find the total cost of a stock purchase by finding the market price of the stock and adding the commission.

♦ Find total payment for a dividend shown as a percent by multiplying it by par value then multiply by number of shares.

♦ Calculate the rate of income or yield on stock investment by dividing the annual dividends by the total cost of stock.

13. Teri Walker bought 400 shares of M-X Corp. stock at 34. The broker charged a commission of $195. Find the total cost.

Refer to Kell Products in Illustration 11-2.3. Find the stock's

14. lowest price in last 52 weeks.

15. annual dividend paid.

16. You own stock that cost $56 a share and pays a quarterly dividend of $0.28 a share. What is the rate of income?

17. Marge Nole bought 200 shares of Tropic Juice Company at $12\frac{1}{2}$ and paid a commission of $88. She later sold the stock at $15\frac{3}{4}$, paying a commission of $96 and other charges of $4.25. Find Marge's profit or loss on the stock sale.

11-3 ♦ Find the total investment in mutual fund shares by multiplying the number of shares times the price paid.

♦ Find rate of commission on load funds by subtracting net asset value from offer price, then dividing the result by offer price.

18. How many shares of a no-load mutual fund with a net asset value of 12.50 may be purchased if $6,000 is invested?

19. A mutual fund with a net asset value of 8.36 is sold at an offer price of 8.65. Find the commission, to nearest tenth.

11-4 ♦ Find the annual net income from a real estate investment by subtracting annual expenses from annual rental income.

♦ Find the yield on a real estate investment by dividing the annual net income by the cash investment in the property.

♦ Calculate monthly rent to be charged by dividing the expected annual rental income by the annual number of months rented.

20. Angie bought a $62,000 house for $7,000 down. She rented it for $950 a month. Her expenses: interest, $5,800; 2% per year depreciation; other expenses, $2,900. For the year, find her net income.
21. Find Angie's rate of income on investment, to nearest tenth of a percent.
22. Maynard Cosgo wants to earn 12% annual net income on a $8,000 cash investment. He estimates annual expenses to be $7,500. What monthly rent must Maynard charge?
23. Jane Seewick bought an abandoned house for $12,000. She spent $4,000 to add a driveway, $3,000 to repair the roof, $4,500 to repair interior walls, and $200 to repair plumbing. What was Jane's capital investment in the house?

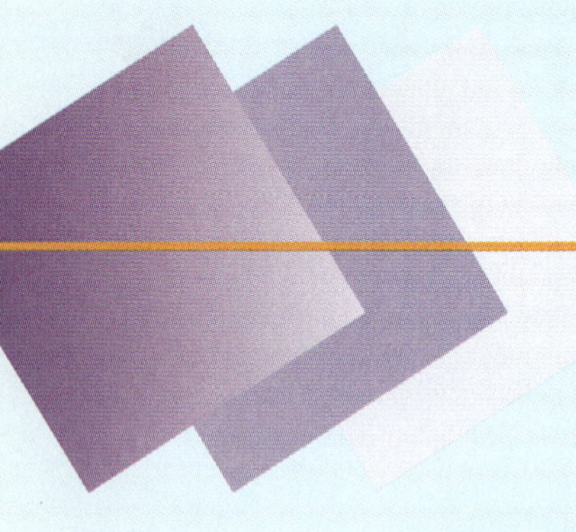

WORKPLACE KNOWHOW: **ACQUIRE AND EVALUATE INFORMATION**

The ability to obtain and evaluate data is an important workplace and life skill. For example, in this chapter, you have learned about different kinds of investment vehicles. Before investing in a company, you would find out if it was making money or losing money. In the same way you can investigate company's or career fields to determine if you might be interested in working in that field.

Assignment Work in small teams. Reread the profile of the printing press operator at the start of this chapter. What does it tell you about the printing industry? Suppose you want to know more about this industry and companies working within it. Make a list of questions you have and how you would go about gathering information. Have each team report their lists to the class.

Work Skills Portfolio Include your assessment of your knowledge of and interest in the material learned in this chapter. If you completed the integrated project in the student workbook, include your answers to investment questions as an indication of your skills. You may also want to note how you feel about the subject. Do you think you would enjoy working in an investment company or enjoy managing your own investments? Consider how you feel about taking financial risks. You may also want to include an assessment of your research skills and a description of what you learned from the WorkPlace KnowHow project.

CHAPTER 11 TEST

Math Skills

1. $0.965 \times \$500$
2. 500 @ \$0.12
3. $\$250 \div 2$
4. $2.5\% \times \$96{,}000$
5. $3.6\% \times \$100$
6. $6 \times \$0.235 \times 100$
7. 100 shares at $28\frac{1}{2}$
8. $\$9{,}816 \div 12$
9. $\$0.37 \div \$8.20$, to nearest tenth percent
10. Find $\$500 \div \$10{,}000$ to the nearest percent.

Applications

11. Dale Wade bought 12, \$500 par value bonds at 96. Commissions and other expenses were \$10.50 per bond. Find the total cost.
12. Find the yield on a \$1,000 par value, 12% bond bought at 120.
13. Mark Sun bought 40 shares of stock at $47\frac{3}{4}$, plus commission of \$82.50. What was the total cost of the purchase?
14. Helen owns 80 shares of ElEm common stock paying a quarterly dividend of \$0.82 per share and 140 shares of ElEm preferred stock, \$100 par value, paying an annual dividend of 6.8%. Find the annual dividend she gets from common stock. Preferred stock.

Beverly Goraco bought 150 shares of a stock at a total cost of \$4,672. She kept the stock for 2 years, during which time she received semiannual dividends of \$1.30 a share. She then sold the 150 shares at $33\frac{3}{4}$ and was charged a commission and other costs of \$85.

15. What was her profit or loss from the sale of the stock?
16. What was her total gain from owning and selling the stock?
17. Henry invested \$4,500 in a no-load mutual fund with a net asset value of 8.72. How many shares did he buy?

Arlon Maxwell made a \$29,000 down payment on a resort condominium that sold for \$85,000. His average weekly rental income will be \$400, while his total annual expenses will be \$15,800. Find his

18. annual net income.
19. rate of income earned, to the nearest tenth of a percent.
20. Sue Blakely paid \$24,000 for 5 acres of land. Her annual expenses of owning the land are \$900. Sue wants to earn an annual rate of 25% on her investment and cover all her expenses for the year, what monthly rent should she ask for?

UNIT 3 TEST: CHAPTERS 10–11

♦ Choose the best definition for each term.

10-2 **1.** term
10-5 **2.** finance charge
11-1 **3.** market value
10-1 **4.** interest
11-1 **5.** bond
10-5 **6.** annual percentage rate (APR)
10-3 **7.** collateral

a. Money paid to an individual or institution for the privilege of using their money.
b. Minimum amount of time money must be on deposit in a time-deposit account.
c. Special savings accounts that offer higher interest rates than regular accounts.
d. Property pledged as security for a loan.
e. A charge that pays the seller interest on money and covers the extra cost of doing business on the installment plan.
f. A ratio of the finance charges to the amount financed.
g. Maximum amount that can be charged to a credit card.
h. The price at which a bond is being sold.
i. Investors who buy the shares of a corporation.
j. A written promise to repay loaned money on the due date, which may be anywhere from 5 to 30 years from the date of issuance.

♦ Choose the best answer.

10-1 **8.** $5\frac{1}{2}\%$ interest on $600.00, paid semiannually, is

a. $33 **b.** $66
c. $16.50 **d.** $8.25

10-2 **9.** A $4,000 money-market account earns monthly interest of 2.7%. Find the monthly interest paid.

a. $108 **b.** $1,080
c. $400 **d.** $120

10-3 **10.** Find the amount due at maturity for the following note: Face of note 600; Time $4\frac{3}{4}$ year; Rate 11%

a. $913.50 **b.** $600
c. $666 **d.** $313.50

11-1 **11.** You buy five $1,000 bonds at .89. What is your investment?

a. $5,000 **b.** $4,450
c. $4,500 **d.** $890

11-2 **12.** 420 shares of stock at 28, with broker's commission of $78, equals ? total investment.

a. $11,760 **b.** $11,732
c. $13,944 **d.** $11,838

◆ Solve each problem.

Find the banker's interest to the nearest cent.

		Principal	Rate	Time	Interest
10-3	**13.**	$6,000	$8\frac{1}{2}$%	$3\frac{1}{2}$ yr	
10-3	**14.**	$17,400	14%	1 yr	
10-3	**15.**	$23,000	10%	6 mo	

10-5 **16.** Katherine bought a new living room sofa for $1,400. She must pay $140 down and make payments for 36 months. What will be her monthly payments?

10-5 **17.** Cash price: $750. Installment terms: $75 down; $66 a month for 12 months. Find the installment price and finance charge.

10-6 **18.** A credit card company charges $4\frac{1}{2}$% of total sales for its services. A customer charges $375 at a clothing store. What net amount does the clothing store receive, to the nearest cent?

10-2 **19.** Jefferson invested $2,500 in a 3-year time-deposit account that paid 7% interest, compounded annually. He withdrew $600 before the end of the 3 years. The penalty for early withdrawal was 6 months' interest. What was the amount of the penalty?

11-3 **20.** Find the rate of commission on mutual funds with a net asset value of 42.37 and an offer price of 44.63.

11-2 **21.** The annual yield on stock purchased for $3,700, earning annual dividends of $164.20, would be ?

11-1 **22.** Find the number of shares of mutual funds bought for $800 at 4.83.

11-4 **23.** Find the rate of return, to the nearest tenth of a percent given the following:

Cash investment $32,000 Monthly rental income $620
Interest on mortgage $3,300 Other Expenses $1,800

11-4 **24.** Karen wants to earn 15% annual net income on a $15,000 property investment. What monthly rent must Karen charge, if her annual expenses of owning the property total $2,310?

10-1 **25.** On April 1, Sarah deposited $1,200 in a savings account that pays 5% interest, compounded quarterly. She made no other deposits or withdrawals. Interest is calculated and paid on July 1 and October 1. Find the account balance on October 2.

CHAPTER 12

a Sound Career

DO YOU LIKE TO listen to music all day? Play with sound? Work with technology? Those are a few of the qualities needed by a sound engineer.

Mathhias is a sound-recording engineer and owner of a studio that creates soundtracks for radio and TV commercials, independent films, and corporate presentation and training videos. His job is to create soundtracks that are "just what the customer wants."

To those interested in a career as a sound engineer, he says the first step is to develop the technical skills needed to operate and maintain computer sound equipment.

The engineers at Mathhias' company use a digital sound system that records multiple sound tracks on a single computer disk. This allows them to move sound on one track, without changing the complete recording. The process is similar to moving paragraphs around in a word processing program.

In addition to technical skills, which can be learned on the job or in a technical program, sound engineers must

- **understand the physics of sound** and the dynamics of how sound signals are translated "from Point A to Point B" onto the tape or disk.
- **have "an ear for sound"** The ability to listen and recognize mistakes, sound distortions, and what works or doesn't is a skill that comes from experience.

Mathhias uses these math skills all the time.

- **Decimals to measure time and lengths of sound in hundredths of a second.** A 60-second radio commercial must be 60 seconds exactly; a sound is heard for a precise time, such as 5.12 seconds.
- **Equations and estimations** to determine exactly when and how far apart sounds will be heard on a recording.
- **Ratios and percents** to compare the amount of time voices or sounds are heard on a recording.

Opportunities for sound engineers exist in recording studios, concert halls, theaters, TV and radio studios, film and video production companies, and corporate communications departments. Mathhias warns that the recording industry is competitive. "Enthusiasm, a desire to do the work, and persistence are absolutely essential. Wherever you work, observe everything and learn all you can. When the opportunity presents itself, jump in and prove yourself."

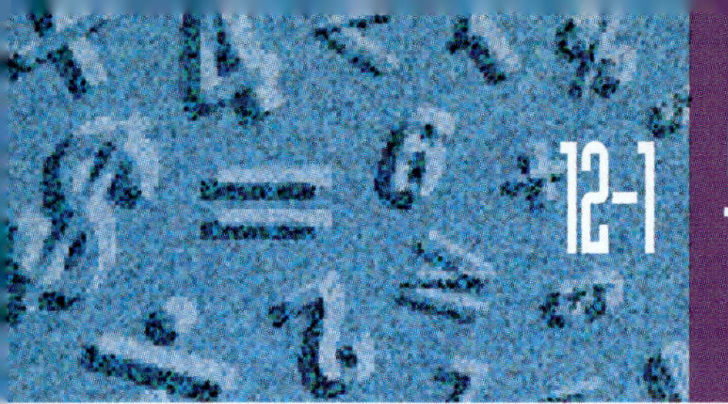

12-1 MEASURES OF CENTRAL TENDENCY

OBJECTIVES

In this lesson, you will learn to

- *find the mean, median, mode, and range for a set of data; and*
- *interpret and complete a frequency table.*

You are at Lake La Donna. Is it safe to dive from any bluff into the lake? Why or why not?

◆ Work in small groups. Write as many examples as you can think of for the word *average*.

LAKE LA DONNA
Average Depth: 9 1/2 ft

WARM UP

a. Change $\frac{1}{4}$% to a decimal. **b.** Multiply 48 by $37.50.

c. Find 106 divided by 0.52 rounded to the nearest tenth.

d. How many days are there between June 5 and September 8?

Tell whether each number is even or odd:

e. 7 **f.** 2 **g.** 16 **h.** 49 **i.** 2,001

MEASURES OF CENTRAL TENDENCY

Earlier in this course you found the average pay of employees. In this section, you will study three types of averages called the mean, median, and mode. As a group, these averages are called **measures of central tendency.**

Mean

The **mean,** or **arithmetic average,** is found by adding a group of numbers and dividing the sum by the number of items added. The mean is the best-known and most used measure of central tendency. The group of numbers is sometimes referred to as the **data.** *Data* is plural for *datum.*

EXAMPLE During the third quarter, the Surnell Company made these profits: July, \$38,000; August, \$52,000; September, \$33,000. Find the mean profit per month.

SOLUTION \$38,000 + \$52,000 + \$33,000 = \$123,000 Total profit

\$123,000 ÷ 3 = \$41,000 Mean profit per month

Median

The **median** is the middle number in a set of data that is arranged in either ascending or descending order. One-half of the numbers will be on either side of the median.

To find the median, arrange the numbers in order. If the group has an odd amount of numbers, the number in the middle is the median. If the group has an even amount of numbers, add the two middle numbers and divide by 2 to find the median.

EXAMPLES Find the median for each set of data.

a. 12, 18, 17, 10, 15, 23, 12, 23, 9

b. 9, 0, 1, 9, 5, 4, 9, 2

SOLUTIONS **a.** There are 9 numbers, and 9 is an odd number.

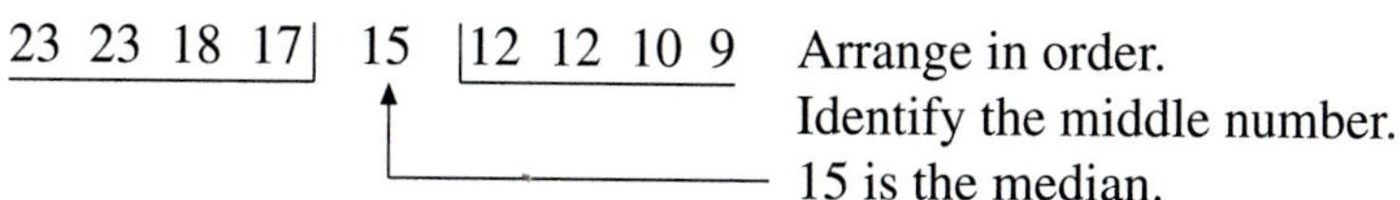

b. There are 8 numbers, and 8 is an even number.

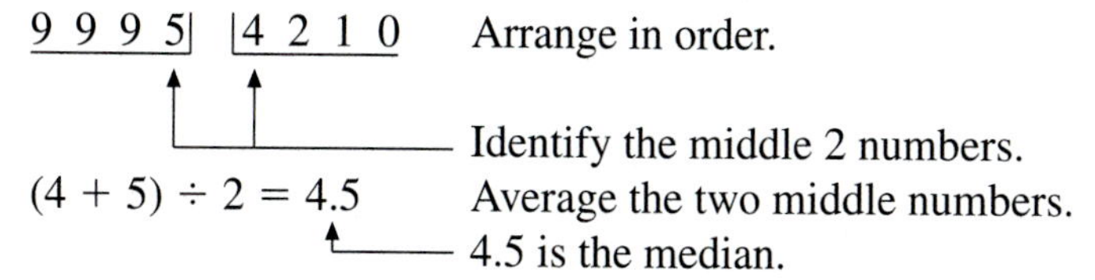

Mode

The **mode** is the number that occurs most frequently in a group of numbers arranged in order. There may be no mode or more than one mode in a set of data.

EXAMPLES Find the mode for the other examples on this page.

SOLUTIONS **Mean Example:** \$52,000, \$38,000, \$33,000 No Mode

Median Example a: 23, 23, 18, 17, 15, 12, 12, 10, 9
23 and 12 both appear twice. 23 and 12 are the modes.

Median Example b: 9, 9, 9, 5, 4, 2, 1, 0
9 appears three times. 9 is the mode.

Range

As you work with measures of central tendency, you may also be asked to find the range. The **range** is the difference between the highest and lowest numbers in a set of data. In the first example on the previous page (for the mean) the range was 52,000 − 33,000 or 19,000. For median examples, the range for example a was 14; and for example b, the range was 9.

EXERCISE YOUR SKILLS

For each set of data, find the mean, median, mode, and range.
Round to the nearest tenth.

1. 17, 12, 18, 22, 12, 11, 23

2. 16, 25, 19, 29, 18, 25

3. 2,184; 8,105; 19,238; 9,053

4. 0.2, 0.18, 1.5, 0.7, 0.2, 0.7, 0.9, 1.3

Workers at a small business were paid these amounts per hour: $8; $8.25; $8.45; $8.25; $8; $8.40; $8.20; $8; $8.10. Find the

5. mean

6. median

7. mode

8. range

Apartments in a certain area rented for these monthly rates: $420; $410; $405; $435; $455; $425; $450; $410; $400; $350; $320; $420; $440; $425; $410; $450; $430. Find the

9. mean

10. median

11. mode

12. range

FREQUENCY DISTRIBUTION

When you work with larger sets of data, it may be difficult to list them in order. Making a **frequency distribution** is one way of arranging numbers. To make a frequency distribution table, list all the data. Every time a piece of data is used, make a vertical mark next to it. Make every fifth mark a slash to group the marks in sets of five. This makes counting (tallying) easier. From the frequency distribution table, you can calculate the mean, median, mode, and range.

EXAMPLE During a 17-day period, daily sales of tires were: 28, 23, 20, 28, 28, 24, 29, 20, 28, 21, 24, 28, 20, 29, 28, 23, 24. Make a frequency distribution table. Then find the mean, median, mode, and range.

SOLUTION Frequency Distribution

29	//
28	~~////~~ /
24	///
23	//
21	/
20	///

Mean: Multiply and add. Then divide.

$(29\times2)+(28\times6)+(24\times3)+(23\times2)+(21\times1)+(20\times3) = 425$

$425 \div 17 = 25$

Median: Find the middle number, in this case the ninth number, or **24.**

Mode: Most frequent number, 28.

Range: $29 - 20 = 9$

EXERCISE YOUR SKILLS

For Exercises 13–16, make a frequency distribution table. Then find the mean, median, mode, and range.

13. In a 10-day period, a club sold these numbers of magazine subscriptions: 25, 22, 24, 26, 24, 22, 25, 22, 20, 20.
14. The reports generated by a computer per day were: 28, 19, 28, 21, 24, 26, 28, 17, 19, 20.
15. These numbers of surveys per hour were completed by a researcher during interviews at a shopping mall: 15, 14, 11, 16, 18, 14, 17, 12, 9, 10, 12, 14, 16, 11, 17, 10.
16. During a recent month, a shop had these daily numbers of customers: 14, 21, 16, 18, 14, 20, 19, 20, 14, 27, 17, 22, 18, 14, 21, 14, 19, 20, 14, 27, 19, 12, 14.
17. During a year, the Mel-Tran Company bought 15,000 bolts at \$0.016; 20,000 bolts at \$0.0154; and 5,000 bolts at \$0.018. Make a frequency distribution table. Then find the price paid per bolt, stated as the mean, the median, and the mode. Round answers to the nearest thousandths.
18. ◆ Survey 40 students as to how long (to the nearest $\frac{1}{2}$ hour) they spent on homework last night. Make a frequency table. Then find the mean, median, mode, and range of the data.

MIXED REVIEW

1. Estimate \$23.99 ÷ 12
2. Estimate \$5,100,000 ÷ 25
3. Round 0.1896 to the nearest hundredth.
4. For 6 days this week, the numbers of defective frying pans produced by a factory were: 210, 225, 206, 214, 180, 225. Find the mean, median, and mode of defective frying pans.

During a 6-week period, a truck was driven these numbers of miles weekly: 1,250; 1,469; 781; 960; 1,351; 1,076.

5. Shown as a mean, what average number of miles was the truck driven weekly rounded to the nearest mile?
6. What was the range of weekly miles driven?

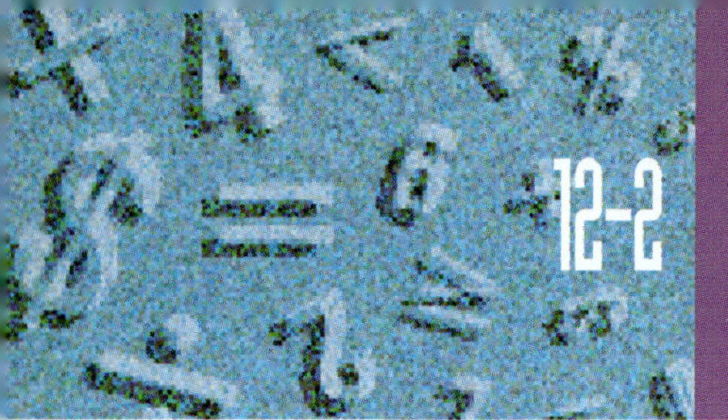

12-2 PROBABILITY

OBJECTIVES

In this lesson, you will learn to

- *find the probability of simple events;*
- *find experimental probability; and*
- *draw a diagram to determine probability.*

Suppose you ask your teacher whether there will be a quiz tomorrow and he says "probably." Do you think you will have a quiz? Is it certain you will have a quiz? What if your teacher said that he would toss a coin three times. If the results were heads each time, you would have a quiz. Now do you think you will have a quiz?

◆ Flip a coin 3 times. Record the results. Combine your results with all class members. How many times did three heads come up?

WARM *UP*

a. Estimate 487×21

b. Estimate $3{,}016 \div 18$

c. $\frac{1}{2} \times 9 \times \frac{2}{3}$

d. Find $\frac{2}{5}$ of 550

e. Write 0.34% as a decimal

f. Find $9.5 \div 2.4$. Round the quotient to the nearest hundredth.

PROBABILITY

Probability is a way of mathematically predicting the chance that an event will occur. This section deals with the basic concepts of probability and its everyday and business uses.

Chance Events You have two marbles that are alike in every way except color. The marbles, one blue and one white, are placed in a bag. Suppose you are asked to pick one of the marbles from the bag without looking. You have no way of knowing which marble is blue and which marble is white. So, you have no way of picking one rather than the other. This means that the marble you pick will be a *random choice,* and the outcome will be a **chance event.**

Probability of a Chance Event With this bag of marbles, what are the chances that your outcome will be to pick the blue marble? Two outcomes are possible: (1) you pick the blue marble, or (2) you pick the white marble.

Since your pick is a random choice, one outcome is just as likely as the other. The outcomes are *equally likely*. The chance that you will pick the blue marble is 1 out of 2, which may be shown as the fraction $\frac{1}{2}$. The probability of the event that you will pick the blue marble is $\frac{1}{2}$. You may also say that the probability is 0.50, or 50%.

Now suppose that the bag holds 6 blue marbles, 3 white marbles, and 1 red marble, making a total of 10 marbles. Again, the marbles are alike except for color. What are the chances of your picking a blue marble? Since there are 6 blue marbles, there are 6 chances out of 10 that you will pick a blue marble. So, the probability of the event that you will pick a blue marble is $\frac{6}{10}$, $\frac{3}{5}$, 0.6, or 60%.

Since 4 of the marbles are not blue, there are 4 chances out of 10 that you will not pick a blue marble. So, the probability of your not picking a blue marble is $\frac{4}{10}$, or $\frac{2}{5}$.

Certain or Impossible Events If a bag holds nothing but 5 yellow marbles, the chances of picking a yellow marble are 5 out of 5, which is $\frac{5}{5}$ or 1. In this case picking a yellow marble is an event that is certain to happen, because no other outcome is possible. Any event that is certain has a probability of 1.

From a bag holding nothing but yellow marbles, the event that you will pick a white marble is an impossible outcome. The chances of picking a white marble are 0 out of 5, which is $\frac{0}{5}$, or 0. Any event that is impossible has a probability of 0.

Probability and Large Numbers The probability of an event occurring tells you about how many times you can expect the event to happen in a large number of tries. For example, if a perfectly balanced coin is flipped, it is as likely to come up heads as it is tails. So, you can say that the probability of its coming up heads is $\frac{1}{2}$. Such a coin is often called a "fair" coin.

This does not mean that out of 40 flips you are sure to get 20 heads. You may get several more or several less than 20. What it does mean is if you flip the coin many times, say 1,000 times, the number of heads should not be far from $\frac{1}{2}$ of 1,000. The longer you keep flipping, the nearer the number of heads should come to $\frac{1}{2}$ the number of flips. This is called *the principle of large numbers.* It is one of the main concepts of probability. For a large number of tries of the same kind, you can predict the outcome with only a small amount of error.

EXAMPLE You toss a standard six-sided die once. What is the probability of throwing a 5? If you toss the die 900 times, how many times would you expect to get a 5?

SOLUTION Six sides, only one has a 5: Probability of a 5 $= \frac{1}{6}$
Expected number of 5s in 900 rolls is $\frac{1}{6} \times 900$, or 150

EXERCISE YOUR SKILLS

1. If a fair coin is flipped and falls freely, it is as likely to turn up heads as tails. What is the probability of tails?

A bag holds 3 plastic disks, 1 green, 1 red, and 1 yellow. One disk is picked at random.

2. What is the probability that it will be green?
3. What is the probability that it will not be green?

Five black socks and four blue socks are in a drawer. You pick one sock without looking.

4. What is the probability that it will be black?
5. What is the probability that it will not be black?
6. What is the probability that it will be either black or blue?
7. What is the probability that it will be white?

Forty balls, numbered from 1 to 40, are placed in a drum. What is the probability of your picking a ball that is

8. the number 4?
9. even numbered?
10. has the digit 5 on it?

A box of paper clips holds 20 red clips, 30 blue clips, and 50 yellow clips. One clip is picked at random. What is the probability that it will be

11. red?
12. blue?
13. green?
14. not blue?
15. either blue or yellow?

16. The probability of an event is $\frac{1}{5}$. About how many times would the event happen in 6,000 tries?

PROBLEM SOLVING

Make a Diagram

You are to flip a coin and toss a die to determine your move in a game. Heads you move forward; tails, backward. The die tells you how many spaces to move. To score 10 points, you need to move 3 spaces forward or 5 spaces backward. What is the probability that you will score 10 points on your next turn?

To solve the problem, you can draw a diagram that shows possible results.

coin	die	result
	1	h, 1
	2	h, 2
heads	3	h, 3
	4	h, 4
	5	h, 5
	6	h, 6
	1	t, 1
	2	t, 2
tails	3	t, 3
	4	t, 4
	5	t, 5
	6	t, 6

The diagram shows there are 12 possible results. Two are the desired results, so the probability is $\frac{2}{12}$ or $\frac{1}{6}$.

Four cards marked 1, 2, 3, and 4 are put into a hat.

17. You pick two cards. What is the probability that the sum of the two cards is 5? Make a diagram.

18. What is the probability that the sum of the two cards is 10?

19. If one card is picked at random, what is the probability that the card picked will be the 4 card?

20. A card is picked and then put back 2,400 times. About how many times would the card picked be the 3 card?

Experimental Probability In the examples so far you could calculate the probability of the event from the description of the conditions. In many cases, this information may not be known, and the probability must be based on experiment or experience.

For example, if a coin is worn and out of balance, heads and tails are not equally likely when the coin is flipped. The probabilities for this coin must be found by experiment; that is, we must flip the coin and record the results. Suppose you got 827 heads and 173 tails in 1,000 flips. You could estimate the chances of a head as 0.827. The chances of tossing a tail are 0.173. For future flips of this coin, you can predict that about 82.7% will be heads and about 17.3% will be tails.

Quality control departments often test a few samples of the products they make to see what percent are up to standard and what percent are not. If these samples are chosen by chance, they are called **random samples.** Suppose 100 random samples of staplers are tested and 3 were defective. This represents 3% of the 100 samples. On the basis of this test, you estimate that about 3% of the whole batch of staplers from which the samples were taken will be defective. In a batch of 500 of these staplers, the number of defective staplers will be about 3% of 500, or 15.

Probability Based on Experience Life insurance companies use records of births and deaths to make *mortality tables* that show how many persons in a sample group of 100,000 live babies have reached certain ages. The numbers at each age have been rounded to the nearest thousand to make calculating easier.

Age	Number Living
0	100,000
10	97,000
20	96,000
30	95,000
40	93,000
50	90,000
60	82,000
70	66,000
80	39,000

Illustration 12-2.1. Sample Mortality Table

From the table, you can calculate the probability that a person will reach a particular age. For example, 93,000 of the sample of 100,000 people born reached the age of 40. So at birth, the probability of living to be 40 is $93{,}000 \div 100{,}000$, or 93%. Of the 93,000 who were living at age 40, 82,000 were still living at age 60. So, the probability that a person who is 40 will live to be 60 is $82{,}000 \div 93{,}000$, or 0.88, to the nearest hundredth.

EXERCISE YOUR SKILLS

21. In a factory, 400 random samples of computer chips were tested. Twelve chips were defective. If 1,200 of these chips are made, about how many will be defective?

During the last 365 days, a weather forecaster's predictions have been right 292 days. They have been wrong 73 days.

22. What is the probability that the forecaster's prediction of tomorrow's weather will be right?

23. Over the next 30 days, about how many times would you expect the forecast to be right?

Refer to the mortality table. What is the probability that a person born today will

24. live to be 20? **25.** not live to be 20?

To the nearest hundredth, what is the probability that a person 10 years old will live to be age

26. 50? **27.** 60?

28. 70? **29.** 80?

30. ♦ Random sampling is often done for surveys of different kinds. There are different ways of obtaining a random sample, such as cluster sampling, systematic sampling, and convenience sampling. Work in small groups to find out how each of these types of sampling are completed.

MIXED REVIEW

1. What is $\frac{1}{3}$ of 2,016? **2.** Find 4.2% of 5,000.

3. \$45.78 + \$332.98 − \$189.00 + \$23.81 − \$1.56

4. What amount is 20% greater than \$48?

5. For 5 months, Ektron kept track of the number of consecutive days their copy machine worked before it needed service. The days between service calls were 15, 22, 19, 34, 40, 5, 12. Find the mean number of days between service calls.

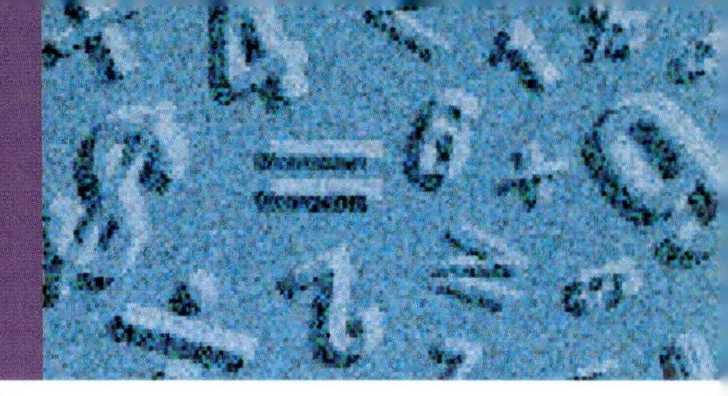

12-3 BAR AND LINE GRAPHS

A newspaper included the graph below in an article that claimed a dramatic rise in crime in a city between this month and last month. Do you think the graph is misleading? If yes, in what ways?

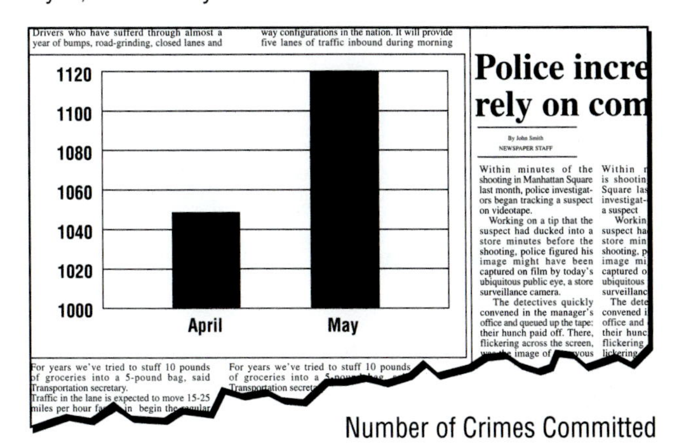

Number of Crimes Committed

OBJECTIVES

In this lesson, you will learn to interpret and use

- *bar graphs; and*
- *line graphs.*

WARM UP

a. What is $\frac{1}{6}$ of \$468.30?

b. Find $87\frac{1}{2}\%$ of \$2,480

c. \$3 + \$58.94 − \$21.01 + \$0.45

d. Find 30% less than \$570

e. Write $\frac{1}{10}\%$ as a decimal

f. Write 2.5 as a percent

BAR AND LINE GRAPHS

Business firms use graphs frequently to show data about their companies or industries. Graph often show facts and trends more clearly than do numbers in tables. Graphs may be produced by hand or by computer programs that use data stored on disk files.

Vertical Bar Graphs The **vertical bar graph** in Illustration 12-3.1, with bars running up and down, shows the daily sales of The Home Center for a week. There is one bar for each day, Monday through Saturday. The height of each bar shows the sales for each day. The scale for measuring the bars is on the left side of the graph.

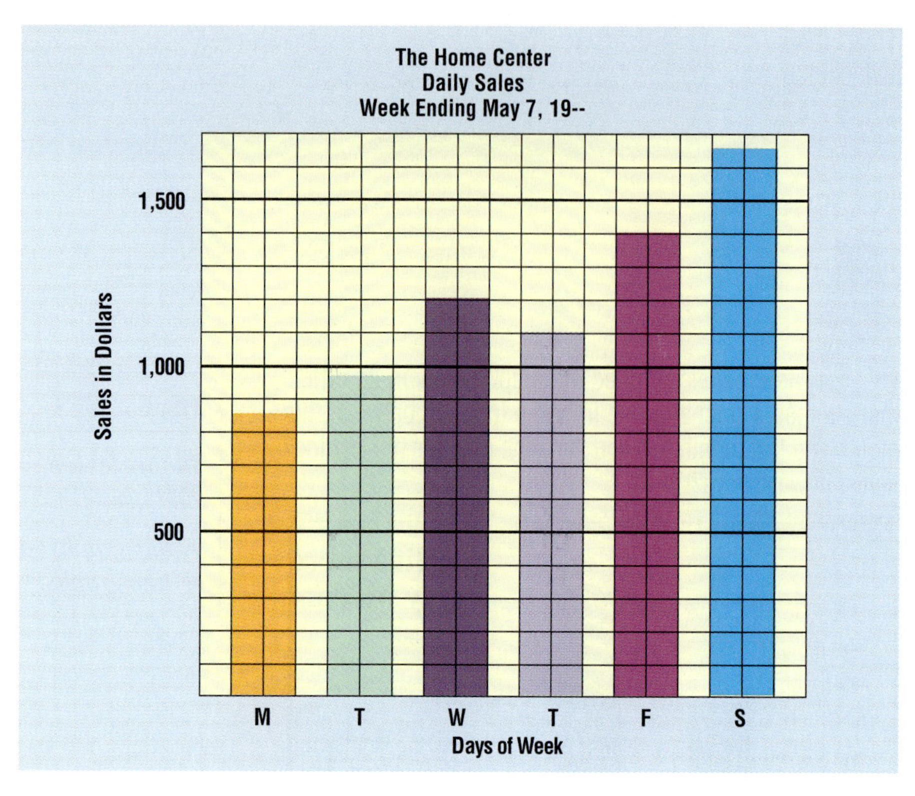

Illustration 12-3.1. Vertical Bar Graph

Each vertical block on the graph equals $100 of sales. The lines for $500 and multiples of $500 are labeled. The height of each bar is correct to the nearest $50. For example, sales for Monday were $842.20. That amount was rounded to the nearest $50, or $850. The top of the Monday bar was then put halfway between the $800 and $900 lines.

The heading or title of the graph shows the company name, identifies the data shown, and gives the time period that the graph represents.

Horizontal Bar Graphs The **horizontal bar graph** in Illustration 12-3.2, with bars running left to right, shows the sales for four types of goods sold by The Home Center for the quarter ending June 30. It looks like the vertical bar graph except that the bars are horizontal.

Each horizontal block on the graph equals $2,000. The amounts for each bar were rounded to the nearest $1,000 before the graph was made. For example, sales of electrical goods totaled $27,890 but are shown as $28,000.

Critical Thinking Refer to the illustrations in the text. What makes a good graph? Identify some characteristics of good graphs.

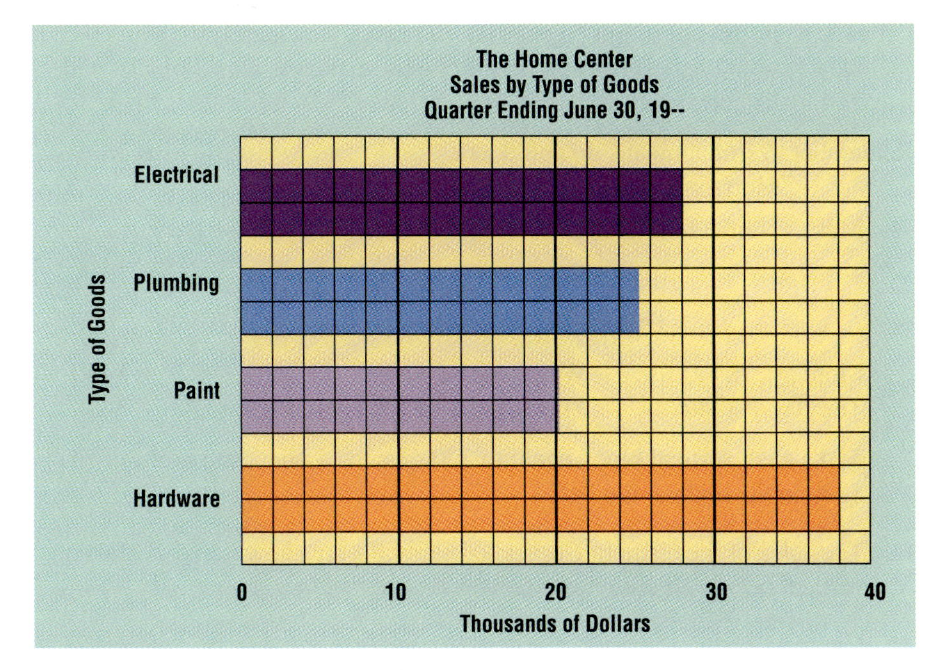

Illustration 12-3.2. Horizontal Bar Graph

EXERCISE YOUR SKILLS

The vertical bar graph in Illustration 12-3.1 shows daily sales to the nearest $50. Use that graph to answer each of the following questions.

1. On which day were sales the largest?
2. On which day were sales the smallest?
3. On which days did sales exceed $1,300?
4. On which 2 days was the difference between sales the largest?

The horizontal bar graph in Illustration 12-3.2 shows sales of four types of goods sold by The Home Center for the quarter ending June 30. Use that graph to answer each of the following questions.

5. What was the approximate dollar amount of hardware sales during the quarter?
6. Which type of good had the lowest sales during the quarter?
7. For which 2 types of goods were the sales most nearly the same during the quarter?
8. How much greater were the sales of hardware than the sales of paint?

For each exercise, use graph paper with 10 blocks, or squares, to the inch. Include a heading for each graph. In all graphs, make the bars 2 blocks wide; leave 2 blocks between bars.

Remember: Vertical means up and down.
Horizontal means across.

9. Make a vertical bar graph showing these facts:

The Candle Shop
Daily Sales, February 7–12, 19—

Monday	$440	Wednesday	$380	Friday	$460
Tuesday	$340	Thursday	$360	Saturday	$480

Make each vertical block equal to $20 of sales. Label each multiple of $100 in the scale at the left.

10. The sales of the Camera Corner for 6 months are shown below. Make a vertical bar graph with these facts:

Camera Corner
Sales for July–December, 19—

July	$26,000	September	$28,000	November	$32,000
August	$18,000	October	$22,000	December	$46,000

Make each vertical block equal to $2,000 of sales. Label each multiple of $10,000 in the vertical scale.

11. Make a horizontal bar graph with these facts:

Do-It-Yourself Remodeling
Sales, November 16–21, 19—

Paneling	$2,600	Floor Tile	$1,100
Doors	$2,800	Sinks	$1,800
Ceramic Tile	$1,300	Cabinets	$2,200

Make each horizontal block equal to $100 of sales. Mark the scale at the bottom of the graph to show each multiple of $1,000.

12. Make a horizontal bar graph showing last year's sales in 6 selling areas of the Protect-All Company.

Home Alarms	$110,365.20	Locks	$ 58,358.70
Car Alarms	$125,720.10	Steel Doors	$101,395.19
Garage Door Openers	$ 96,359.22	Repairs	$116,682.44

Make each block on the horizontal scale equal to $2,000. Round each sales calculate to the nearest $2,000 before entering it into the graph.

Line Graphs The **line graph** in Illustration 12-3.3 shows the sales of The Home Center by months. The time scale runs from left to right and is at the bottom of the graph. The dollar scale runs from bottom to top and is at the left.

The monthly sales were rounded to the nearest $1,000. The line was made by first placing dots showing each month's sales. The dots were then joined by drawing a line with a ruler.

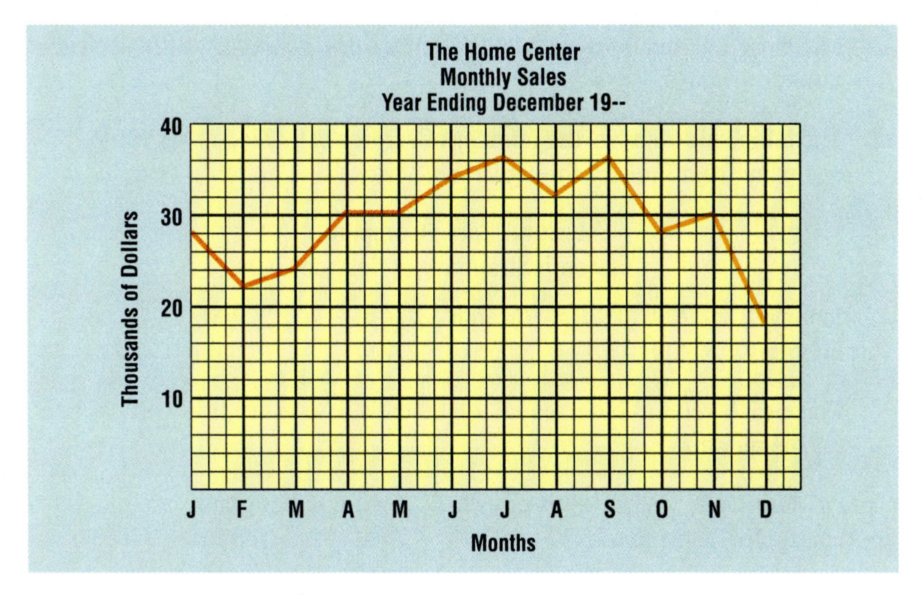

Illustration 12-3.3. Line Graph

EXERCISE YOUR SKILLS

Illustration 12-3.3 shows the monthly sales of The Home Center to the nearest $1,000. Sales for January were about $28,000; for February, about $22,000; and so on. Use the line graph to answer these questions.

13. What were the approximate sales for each month from September through December?

14. In which month were sales the lowest?

15. In which months did sales fall below $24,000?

16. In which month(s) did sales exceed $35,000?

17. Between which 2 consecutive months did sales increase the most? decrease the most?

18. Show these facts on a line graph:

The Poster Shop
Sales for May 22–28, 19—

Sunday	$420	Wednesday	$230	Friday	$350
Monday	$270	Thursday	$320	Saturday	$460
Tuesday	$260				

Make each vertical block equal to $10 of sales. Use every fifth vertical line for the days.

19. The table below shows the number of service calls made by a cable company from January through June. Make a line graph.

Merlund Cable Company
6-month Summary of Service Calls

January	2,200	March	2,700	May	2,100
February	2,900	April	2,400	June	1,900

Make each vertical block equal to 100 calls. Use every fifth vertical line for the months.

Critical Thinking What kinds of uses do you think bar graphs are most appropriate for? Line graphs?

20. ♦ Make up your own data that could be shown on a line or bar graph. Create an accurate graph. Then create a graph that misrepresents or distorts the data in some way.

MIXED REVIEW

1. $56 is what part greater than $48?
2. 23.8 L + 1.85 L + 390 mL = ? L
3. $266.60 is what percent less than $310?
4. 104 m^2 ÷ 13 m = ? m
5. A vertical bar graph is drawn on graph paper that has 8 blocks to an inch. If each block is equal to $200, what is the greatest amount that can be marked on a graph that is 2 inches high?
6. A salesperson made the following number of telephone calls to customers last week: Monday, 24; Tuesday, 16; Wednesday, 21; Thursday, 16; Friday, 22. Show the mean, median, and mode for the number of calls in the week.

12-4 CIRCLE AND RECTANGLE GRAPHS

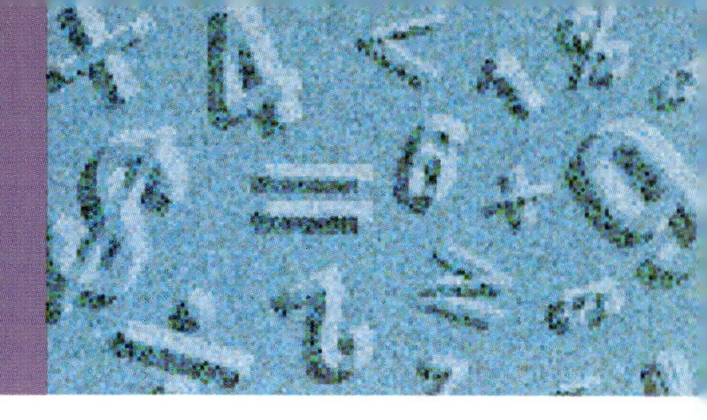

A small town allocates its tax dollars in this way: county expenses, 12%; township expenses, 2%; road and bridge expenses, 2%; town expenses, 25%; community college expenses, 4%; school district expenses, 55%. Would you use a line graph to show this data? Why or why not?

◆ What other type of graph could you use? Describe or draw a graph you have seen in the news that could be used to show this information.

OBJECTIVES

In this lesson, you will learn to

- *interpret and make circle graphs; and*
- *interpret and make rectangle graphs.*

WARM UP

a. Find $\frac{1}{8}$ of $50.56.

b. Find 36% less than $486.

c. Express the product of $4\frac{1}{2}$ and $3\frac{1}{4}$ as a decimal rounded to the nearest hundredth.

d. Find the exact interest on a $4,500 loan at $8\frac{1}{2}$% for 120 days.

CIRCLE GRAPHS

Bar graphs are frequently used to compare quantities to each other. Line graphs show change over time. **Circle graphs** are used to show how parts relate to the whole (100%) and to each other.

The full circle of 360 degrees (360°) is the whole, or 100%. Each part is shown by a sector in the circle graph. The circle graph in Illustration 12-4.1 shows in percents how Tricia budgets her income.

Tricia earns $250 per month from her after-school job. She has prepared this budget.

In making the circle graph, the dollar amounts were first changed to percents, then to degrees, rounded to the nearest whole number.

Tricia's Budget

Entertainment	$ 50
Clothes/Personal	50
Meals	30
School	20
Miscellaneous	25
Savings	75
Total	$250

Budget Category	Amount	Expressed as Percent	Degrees in Sector
Entertainment	$ 50	$\frac{\$50}{\$250} = 0.20$ or 20%	20% of 360° = 72°
Clothes/Personal	$ 50	$\frac{\$50}{\$250} = 0.20$ or 20%	20% of 360° = 72°
Meals	$ 30	$\frac{\$30}{\$250} = 0.12$ or 12%	12% of 360° = 43°
School	$ 20	$\frac{\$20}{\$250} = 0.08$ or 8%	8% of 360° = 29°
Miscellaneous	$ 25	$\frac{\$25}{\$250} = 0.10$ or 10%	10% of 360° = 36°
Savings	$ 75	$\frac{\$75}{\$250} = 0.30$ or 30%	30% of 360° = 108°
TOTALS	$250	$\frac{\$250}{\$250} = 1.00$ or 100%	100% of 360° = 360°

To make a circle graph, draw a circle with a compass. Be sure to mark the center of the circle. Use a protractor to draw the central angles for each sector.

Central Angles

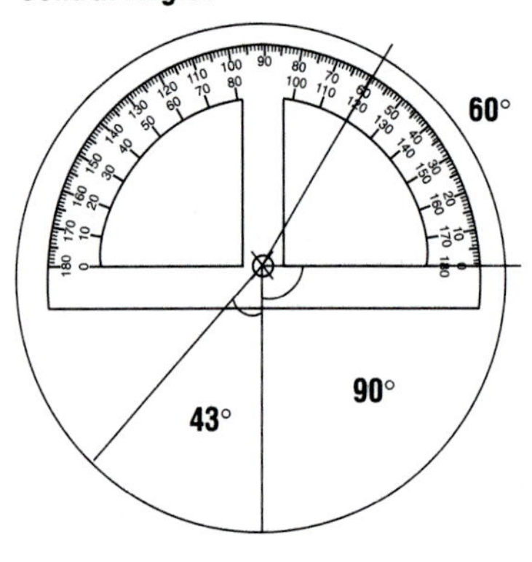

Tricia's Budget

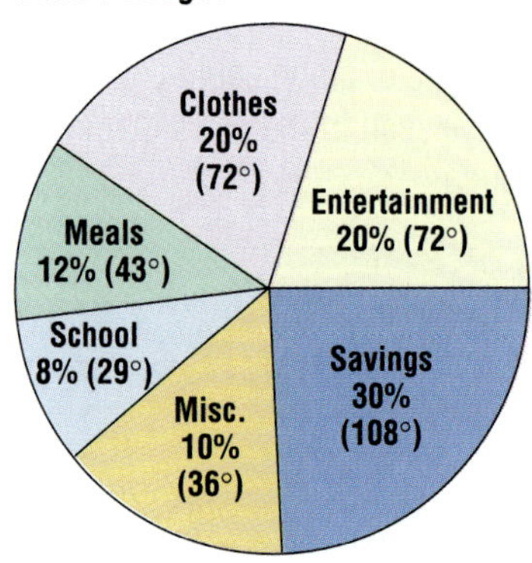

Illustration 12-4.1. Circle Graph

EXERCISE YOUR SKILLS

Last year's sales in the 4 departments of the Redi-Solutions Computer Company were as follows: Hardware, 50%; Software, 30%; Training, 15%; and Repair, 5%.

1. On a circle graph, how many degrees should be used to show the sales of each department?
2. Make a circle graph showing the percent for each department.

The Teal Company spent $12,000 on advertising as shown.

Newspaper	$3,600	Circulars	$3,000
Radio	2,400	Mailings	1,800
Yearbook ads	600	Miscellaneous	600

3. Give the percent of advertising the Teal Company spent on each type.
4. Make a circle graph for the Teal Company's advertising costs.

5. Robolink Company calculates that the $50,000 cost of building a small robot and installing it at the customer's site is divided into these parts: raw materials, $5,000; direct labor, $12,500; overhead, $15,000; delivery and testing; $7,500; training, $10,000. Make a circle graph showing these parts as percents of the total cost of the robot.

RECTANGLE GRAPHS

A **rectangle graph** is a single vertical or horizontal rectangular bar that is divided into sections. The entire rectangle represents the whole, or 100%. The sections show the parts of the whole. A rectangle graph shows how the parts relate to each other and to the whole.

Each part is proportional in size to the whole. For example, suppose a 5 inch bar represents $1,000. To show a part equal to $400 you would mark off two inches, of the rectangle. ($\frac{\$400}{\$1,000} = \frac{2}{5}$, $\frac{2}{5} \times 5$ inches = 2 inches)

Rectangle graphs are most often used to show dollars or percents. In the *vertical rectangular graph* in Illustration 12-4.2, the whole rectangle represents Tri-Co's corporate income of $2,000,000. Parts of the rectangle, also shown in dollars, represent each division's contribution to the corporate income.

Critical Thinking How are the heights of the parts for the graph in Illustration 12-4.2 found? What if the income had been: Alpha, $200,000; Beta, $900,000; and Gamma, $900,000?

The *horizontal rectangular graph* in Illustration 12-4.3 shows the same information as in Illustration 12-4.2. However, data is presented as percents.

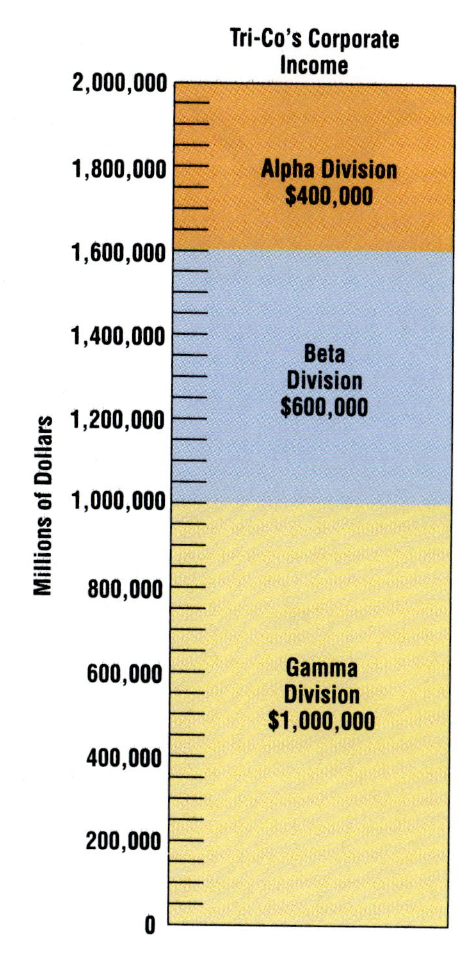

Illustration 12-4.2. Vertical Rectangle Graph

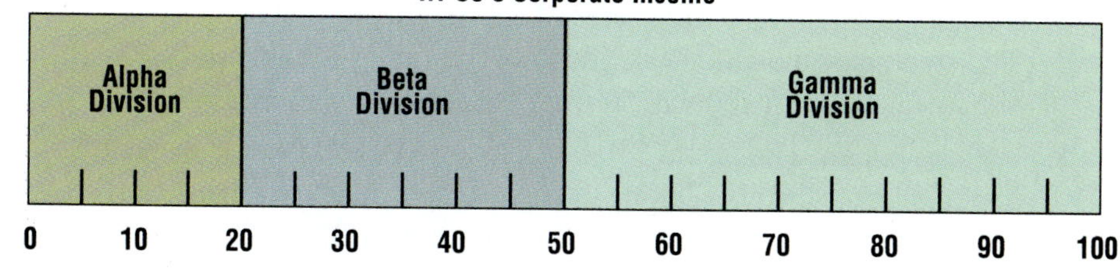

Illustration 12-4.3. Horizontal Rectangle Graph

EXERCISE YOUR SKILLS

6. How could you change Illustration 12-4.3 so that the graph shows the amount of total sales.

MusicLand Sales for November

Old Records $5,000	Cassettes $10,000	CDs $35,000

Use the rectangle graph above for Exercises 7–10.

7. What information does the rectangle graph show?
8. What were the total sales for the month?
9. What percent of sales came from CDs?
10. Make a vertical rectangle graph that shows the same information in percents. Be sure your graph indicates total amount of sales.
11. ♦ Find data in a newspaper or magazine or on television that could be displayed in a circle or rectangular graph. Make the graph. Be sure it is properly labeled.

MIXED REVIEW

1. Divide 78 by $2\frac{1}{2}$.
2. What part less than $618 is $515?
3. Find the exact interest at 10% on $7,500 for 4 months.
4. Find the number of days from August 19 to November 8.
5. What is the tax rate on $1,000 if the rate is 23.8 mills on $1.
6. An annual salary of $16,380 is equivalent to what monthly salary?
7. Selena Krelin bought 150 shares of Crane Tractor preferred stock. The stock paid a quarterly dividend of 3% on a par value of $100. What was the annual dividend from the stock?
8. Richard bought a house for $60,000, paying $12,000 in cash and getting a mortgage for the rest. During the first year, he rented the house for $600 a month. His expenses were mortgage interest, $4,300; taxes, $1,260; other expenses, $910. What was the rate of income on Richard's original cash investment to the nearest tenth of a percent?
9. Lila Harris is paid a salary of $250 a week and a commission of 2% on net sales. Last week her gross sales were $4,720 and customer returns were $112. How much did Lila earn last week?
10. On November 30, Darrell's check register had a balance of $852.07. His bank statement balance on the same date was $682.56. Checks outstanding were #259, $56.71; #261, $33.78. A slip enclosed with the statement showed that the bank charged his account $260 for the November car loan payment due the bank. Prepare a reconciliation statement.

12-5 ECONOMIC STATISTICS

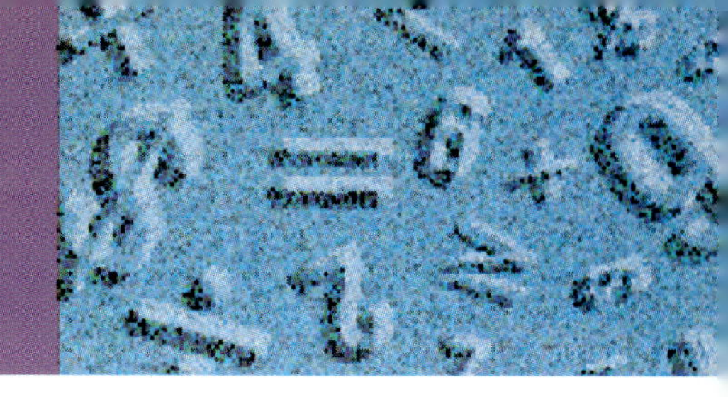

THE LAW OF STATISTICS: If the statistics do not support your viewpoint, you obviously need more statistics!

◆ Work in small groups. Choose an issue that has two distinct viewpoints. Then make statements to support each viewpoint. On what do you base your statements?

OBJECTIVES

In this lesson, you will learn to

- *find rate of inflation;*
- *find purchasing power of a dollar;*
- *read and use consumer price index; and*
- *read and use unemployment tables.*

WARM *UP*

a. Multiply 45.3 by 1,000

b. Divide 7.05 by 1,000

c. Find the number of days between May 6 and July 3

d. Find the tax rate per $100 if the rate is 31.7 mills on $1

ECONOMIC STATISTICS

Government agencies publish many statistics to provide a measure of how well the economy is doing. These measures also tell us how the economy affects businesses, families, and individuals. The statistics we hear most often deal with inflation and unemployment.

Inflation

For consumers, business firms, and the government, **inflation** means that the prices of goods and services they buy are rising. The U.S. Department of Commerce keeps track of prices and publishes a report that tells how much inflation has occurred within the past year. A calculate in the report, called the *rate of inflation,* shows the percent increase in prices from the previous year. Illustration 12-5.1 shows the annual rate of inflation for a ten-year period.

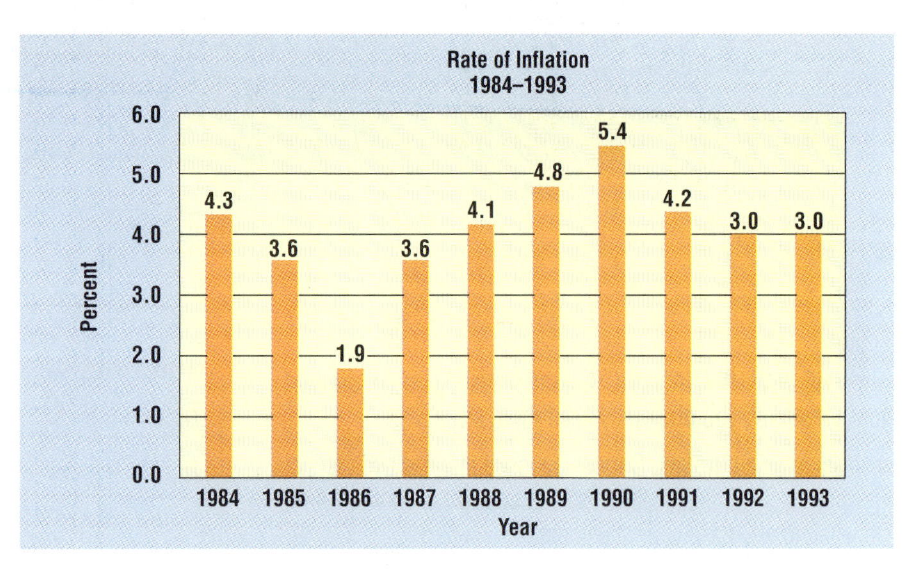

Illustration 12-5.1. Annual Rates of Inflation

Purchasing Power of the Dollar When inflation occurs, each dollar buys less than it did in the past. The **purchasing power of the dollar** is a measure of how much a dollar may now buy compared to what it could have bought during some base period. The **base period** is a period in time with which comparisons are made.

Purchasing Power of the Dollar 1984–1993	
Year	Purchasing Power
1982-1984	$1.000
1985	0.928
1986	0.913
1987	0.880
1988	0.846
1989	0.807
1990	0.766
1991	0.734
1992	0.713
1993	0.692

Illustration 12-5.2. Annual Purchasing Power, or Worth, of the Dollar

Illustration 12-5.2 shows the purchasing power of the dollar from 1982–1993. In the base period, 1982–1984, the dollar was worth its full value of $1.00. In 1985, the dollar was worth $0.928 compared to the base value. The purchasing power of the dollar dropped $0.072 in one year ($1.00 – $0.928). This means that a 1985 dollar could buy only about 93¢ worth of the same goods that could have been bought in the base period. The 1985 dollar is worth less because of inflation, and it would take about $1.07 to buy the same goods that could have been bought for $1 in the base period.

EXERCISE YOUR SKILLS

Use the tables provided to answer the following questions.

1. What was the rate of inflation for 1990?
2. For the years shown, which year had the lowest rate of inflation? What was the rate?
3. In which year was the highest rate of inflation recorded, and what was the rate?
4. During which two consecutive years was the total rate of inflation the greatest? What was the total rate?

For the 10-year period from 1984 through 1993,

5. What was the total rate of inflation?
6. What was the average rate of inflation to the nearest tenth percent?
7. Comparing 1990 to the base period, what was the purchasing power of the dollar in 1990?
8. Comparing 1990 to the base period, by what amount did the purchasing power of the dollar drop?

What amount did the purchasing power of the dollar drop from the base period through

9. 1987?
10. 1991?
11. 1993?
12. In what year did the purchasing power of the dollar drop the greatest amount? What was the amount?
13. What was the first year in which the purchasing power of the dollar dropped to less than $\frac{3}{4}$ of its value in the base period?
14. During which 2-year period did the purchasing power of the dollar show the least amount of change? What was the amount of change?

Consumer Price Index You have already seen how a base period is used to compare the purchasing power of the dollar. Another measure of prices that uses a base period is the Consumer Price Index. The **Consumer Price Index** (CPI) uses a single number, called an index number, to report how much the prices of goods and services typically bought by consumers have changed in comparison to a base period. The index number for the base period is always equal to 100.

Illustration 12-5.3 shows index calculates for various categories of consumer goods and services. The "all items" column gives an average number considering all categories and is the number commonly used when referring to the CPI.

The CPI may be expressed in several ways. For example, the CPI for 1991 is 136.2. This means that the cost of goods in 1991 was 136.2% of their cost in the base period. The percent increase in prices from the base period to 1991 is 36.2% (136.2 – 100.0). Looking at the relationship in another way, it cost $136.20 in 1991 to buy the same goods for which you would have paid $100 in the base period.

Historical Report — Consumer Price Index, 1984–1993 Categories of Goods and Services							
Year	**CPI All Items**	**Energy**	**Food**	**Shelter**	**Apparel and Upkeep**	**Transpor-tation**	**Medical Care**
1982–84	100.0	100.0	100.0	100.0	100.0	100.0	100.0
1985	107.6	101.6	105.6	109.8	105.0	106.4	113.5
1987	113.6	88.6	113.5	121.3	110.6	105.4	130.1
1989	124.0	94.3	125.1	132.8	118.6	114.1	149.3
1991	136.2	102.5	136.3	146.3	128.7	123.8	177.0
1993	144.5	104.2	140.9	155.7	130.4	130.4	201.4

Illustration 12-5.3. Consumer Price Index for Selected Years

Unemployment Rate The **unemployment rate** tells the percentage of the total labor force that is not working. The **labor force** consists of all people who are willing to work and who either have a job or are looking for a job.

Illustration 12-5.4 shows the unemployment rate for different persons for one month as estimated by the U.S. Department of Labor.

1993 Unemployment Rates by Race, Sex, and Age	Rate for All...	Rate for 16–19 Year-Old...
Workers	6.8	
Male Workers	7.1	20.4
Female Workers	6.5	17.4
White Workers	6.0	16.2
Black Workers	12.9	38.9
Hispanic Workers	10.6	26.2

Illustration 12-5.4. Unemployment by Various Categories

EXERCISE YOUR SKILLS

Use the appropriate table to answer the following.

15. What was the CPI for all items for 1993?

16. By what percent did the CPI increase from the base period to 1993?

17. Which category of goods and services shows the smallest price increase between the base period and 1993?

18. Which 2 categories of consumer goods and services had price increases greater than those of the CPI from the base period to 1993?

By what percent did prices for shelter increase from

19. The base period to 1989?

20. The base period to 1993?

Yvonne earned $10,000 a year and Tom earned $4 an hour in the base period. Since then, they both have received wage increases equal to the increase in the CPI. To the nearest cent, in 1993

21. Yvonne should have been earning what annual wage?

22. Tom should have been earning what hourly wage?

In 1993

23. Which persons had the highest rate of unemployment? What was the rate?

24. For male and female workers, did men or women have a higher unemployment rate? What was the difference in rates?

25. For every 1,000 Hispanic workers, how many would you expect to be unemployed?

26. What type of person has the lowest unemployment rate?

PUT IT ALL TOGETHER

27. ♦ Work in small groups. Choose a topic related to inflation rates, Consumer Price Index, or unemployment rates. Gather *current* data on your topic and prepare a presentation for the class. Your presentation should include graphical presentations of data as well as oral or written summaries.

MIXED REVIEW

1. Multiply 0.056 by 100.

2. Find $\frac{2}{3}$ of $540.

3. Round $45,678 to the nearest $1,000.

4. Divide 3.7 by 2.3, round quotient to the nearest hundredth.

Use Illustration 12-5.3 to compare the Food and the Transportation categories for 1993.

5. Which category of prices increased faster between 1991 and 1993?

6. What was the difference in price increases?

August sales of the Dubois Company were $246,172.99. If August's sales were shown on a horizontal bar graph, what amount would be recorded if the sales were rounded to the nearest

7. $5,000?

8. $10,000?

9. Find the median of this group of numbers: 5, 15, 10, 8, 9, 5, 14, 16, 7, 9, 8.

10. Six cars are parked in a lot. Three of the cars are blue, 2 are white, and 1 is green. What is the probability that the first car leaving the lot will be blue, assuming no other cars enter?

EXPLORE TECHNOLOGY

SPREADSHEET: Comparing Inflation Indicators

In Lesson 12-5 you learned that the purchasing price of the dollar declines as inflation rises. You were also introduced to the Consumer Price Index which shows how prices of specific goods and services have changed in comparison with a given base period. Both measures are indicators of inflation rate. Economists use the CPI to calculate the official rate of inflation that you hear or read about in the news. In this lesson you will compare the inflation rates that can be calculated using both measures.

Load ET12 from your applications diskette. When you do you will see data on the purchasing power of the dollar and the CPI. Do not look at the formulas used in the spreadsheet until you are told to do so.

	A	B	C	D	E
1		Purchasing	Inflation		Inflation
2		Power of	Rate Based	CPI	Rate Based
3	Year	Dollar	on Dollar	Index	on CPI
4	1982–1984	$1.00	1	100	1
5	1985	.928	7.759%	107.60	7.60%
6	1986	.913			
7	1987	.880			
8	1988	.846			
9	1989	.807			
10	1990	.766			
11	1991	.734			

Following is a formula that can be used to calculate the inflation rate based on the purchasing power of the dollar.

$$\frac{\text{(Purchasing Power of the Previous Year} - \text{Purchasing Power of Current Year)}}{\text{Purchasing Power of the Current Year}} \times 100$$

EXAMPLE: Find the inflation rate for 1991 using the information in the Purchasing Power of the Dollar table.

SOLUTION: $0.766 - 0.734 = 0.032$ Subtract purchasing power of the current year (1991) from the purchasing power of the previous year (1990)

$0.032 \div 0.734 \times 100 = 0.0436$ or 4.36% Divide the answer by the purchasing power of the current year, then multiply by 100 to find the inflation rate.

1. Write the above formula so your computer spreadsheet can calculate the inflation rate based on purchasing power. Place the formula in cell C11. Then check the spreadsheet formula in cell C5 of the template. If the pattern of your formula matches that in cell C5, copy the formula from C6 to C10.

To calculate the official rate of inflation, use this formula:

$$\text{Rate of Inflation} = \frac{\text{Most Recent Year's Price Index} - \text{Earlier Year's Price Index}}{\text{Earlier Year's Price Index}} \times 100$$

2. Write this formula so your computer spreadsheet can calculate the rate of inflation for 1991 based on the CPI. Then check your formula with the formula in cell E5 of the template. If it is similar, copy the formula to Cells E6:E10.

Your completed spreadsheet should look like this:

	A	B	C	D	E
1		Purchasing	Inflation		Inflation
2		Power of	Rate Based	CPI	Rate Based
3	Year	Dollar	on Dollar	Index	on CPI
4	1982–1984	$1.00	1	100	1
5	1985	.928	7.759%	107.60	7.60%
6	1986	.913	1.643%	109.60	1.86%
7	1987	.880	3.750%	113.60	3.65%
8	1988	.846	4.019%	118.30	4.14%
9	1989	.807	4.833%	124	4.82%
10	1990	.766	5.352%	130.70	5.40%
11	1991	.734	4.360%	136.20	4.21%

3. Why do you think that to estimate inflation based on the purchasing power of the dollar your base year is the current year and for the CPI the base year is the earlier year?

4. If your spreadsheet has graphing capabilities, consult your documentation or on-line HELP to learn how to graph your spreadsheet. You should be able to create a graph similar to the one shown below.

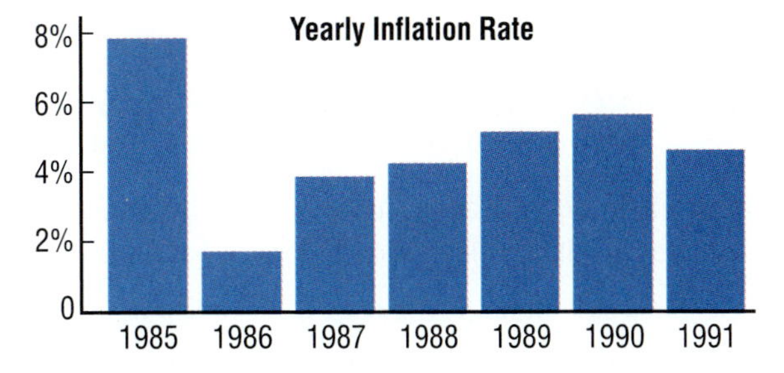

Design Your Own

Create a spreadsheet that reflects the information in Tricia's budget on page 452 of Lesson 12-4. Enter formulas that will calculate budget totals and budget percentages. If your spreadsheet has graphing capabilities, create a circle graph or pie chart. Compare your pie chart to the one shown in the book.

Next enter budget information for yourself or your family into your budget spreadsheet template. If your spreadsheet has graphing capabilities, create a circle graph. If not, add a column to your spreadsheet to calculate the number of degrees in a circle each budget item requires. Then hand draw a circle graph for your budget.

CHAPTER 12 SUMMARY AND REVIEW

VOCABULARY REVIEW

arithmetic average *12-1*
base period *12-5*
certain events *12-2*
chance event *12-2*
circle graph *12-4*
Consumer Price Index (CPI) *12-5*
experimental probability *12-2*
frequency distribution *12-1*
horizontal bar graph *12-3*
impossible event *12-2*
inflation *12-5*
labor force *12-5*
line graph *12-3*
mean *12-1*
median *12-1*
measures of central tendency *12-1*
mode *12-1*
purchasing power of the dollar *12-5*
probability *12-2*
probability based on experience *12-2*
range *12-1*
random samples *12-2*
rectangle graph *12-3*
unemployment rate *12-5*
vertical bar graph *12-3*

MATH SKILLS REVIEW

1. 2,345 ÷ 139 = ?
2. 5,893 − 4,799 = ?
3. 26 ÷ 14 = ?
4. 9,542 × 56% = ?
5. 36% of 360 = ?

APPLICATIONS

12-1
- Find the mean by dividing the sum by the number of items.
- Find the median by finding the middle number in an ordered group of numbers
- Find the mode by finding the most frequently occurring number.
- Find the range by subtracting the lowest number from the highest number in a group of numbers.
- Make a frequency distribution table by arranging the numbers in a group and tallying each number as many times as it occurs in the group

A baseball player made the following hits on a 7-day road trip: day 1, 3; day 2, 1; day 3, 0; day 4, 4; day 5, 3; day 6, 2; day 7, 3.

6. Find the mean number of hits, to the nearest tenth.
7. Find the median number of hits.
8. Find the mode for the number of hits.
9. A piecework employee produced the following number of items in a 10-day period: day 1, 34; day 2, 32; day 3, 33; day 4, 29; day 5, 34; day 6, 35; day 7, 34; day 8, 34; day 9, 30; day 10, 31. Make a frequency distribution table and then find the (a) mean, (b) median, (c) and mode.

12-2 ♦ Find the probability of a chance event occurring by dividing the number of ways the event can happen by the total number of possible outcomes.

♦ Predict the number of times an event will occur by multiplying the number of tries by the probability of the event occurring.

A box of chocolate candies contains 5 with pecans and 7 without any nuts in them. If you picked a candy from the box without looking, what is the probability that you will pick a:

10. Candy with pecans?

11. Candy without pecans?

12. Candy with walnuts?

13. Chocolate candy?

14. Random samples drawn from a factory production line show that 2% of the products are defective. If 3,600 of the products are produced, how many are defective?

12-3 ♦ Interpret bar and line graphs by reading and understanding the scales and values in these graphs.

♦ Construct bar and line graphs on bar graph paper

The horizontal bar graph below shows production at a factory during a year by product. Answer these questions about the graph:

15. Which product had the most sales?

16. Which product had the least sales?

17. What products had sales of less than $80,000,000?

18. What was the amount of sales for rods?

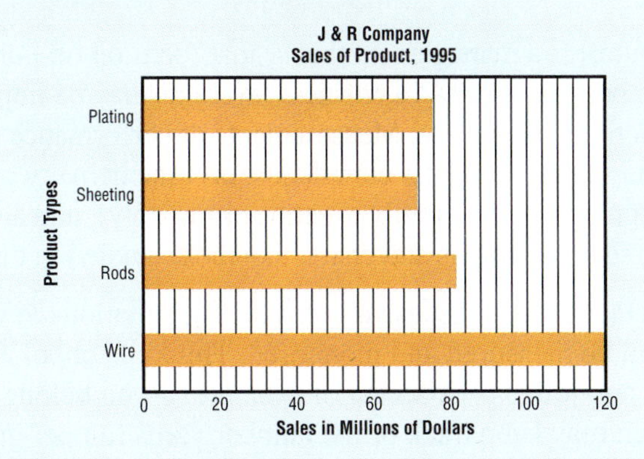

A line graph is drawn on graph paper that has 8 blocks to an inch. If each block is equal to $50:

19. What is the largest number that can be marked on a graph 5 inches high?

20. How many blocks are needed to show an item with a value of $1,250?

12-4 ♦ Interpret rectangle and circle graphs by reading and understanding the scales and values in these graphs.

♦ Construct rectangle and circle graphs.

The rectangle graph below shows the games won, lost, and tied by a college hockey team during a month. Answer these questions about the graph:

21. How many hockey games were played in the month?

22 How many hockey games did the team win? Lose? Tie?

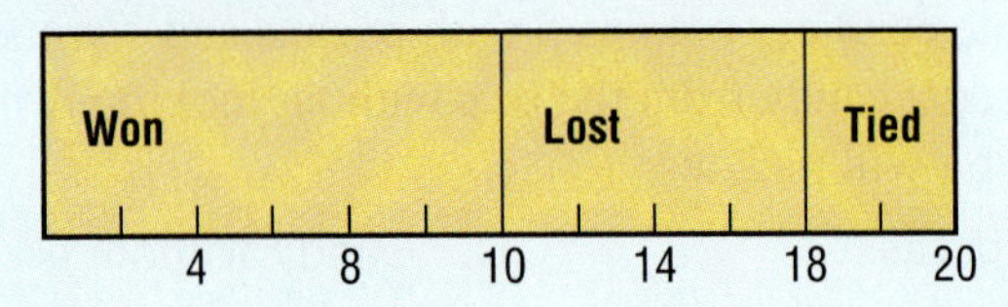

A company had four salespeople: Acker, Berg, Chen, and Devon. Their sales for the month were, respectively, $26,000; $28,000; $31,000; and $35,000. If you were constructing a circle graph:

23. How many degrees would be in the entire graph?

24. How many degrees would be needed to represent Acker's sales?

12-5 ◆ Read and interpret graphs and tables displaying economic statistics.

25. Using Illustration 12-5.3, what category of goods and services rose the most from inflation in 1985? 1987? 1989? 1991? 1993?

WORKPLACE KNOWHOW: **MONITORS, CORRECTS, AND IMPROVES SYSTEM PERFORMANCE**

Total quality control is a management theory that focus on on-going improvement in performance. At a business level this may be improving products, productivity, or service. Measurements of performance over time, such as sales, number of customer complaints, on-time delivery of product, are reported on a regular schedule (daily, weekly, or monthly), and are frequently presented in the form of graphs and charts like those studied in this chapter.

Assignment Have a class discussion of different situations in which performance can be measured and monitored. Think of school activities, personal goals, and job experiences. For example, if you belong to a sports team, your coach may keep track of the number yards run per game. A word processor may be evaluated on words keyed per minute and number of errors.

Choose one of the situations discussed. Identify a performance indicator. Make up data about performance. Then create a chart or graph reflecting your data. Discuss what makes graphs and charts more or less effective and the value of regular reporting.

Work Skills Portfolio Include in your portfolio an assessment of your understanding of the statistical and economic indicators studied, and copies of any presentations you created. If you completed the integrated project, you may also include the graphs you prepared for the management report.

CHAPTER 12 TEST

Math Skills

Use mental calculations when you can.

1. 24% more than $128 = ?
2. Multiply $3\frac{1}{7}$ by $4\frac{2}{3}$.
3. Divide $4,300 by 0.2.
4. Find the exact interest on a loan of $5,300 @ 9% for 60 days.
5. What part less than $350 is $280?

Applications

A computer program counted the number of errors it found each day in data sent by users during the first half of a month. The daily totals of errors were 20, 24, 31, 20, 28, 26, 20, 21, 22, 23, 19, 17, 18, 21, 20. What was the average number of errors per day figured as

6. mean?
7. median?
8. mode?
9. What was the range of the daily error totals in the exercise above?

In a 365-day year, the months of April, June, September, and November have 30 days. February has 28 days; the rest of the months have 31 days. If you select 1 month at random, what is the probability that it will contain

10. Exactly 30 days?
11. More than 30 days?
12. On a vertical bar graph, a line 1-inch long represents $500. What amount of money would be represented on this graph by a line $5\frac{1}{2}$ inches long?
13. You are using graph paper that has 10 blocks to the inch. The dollar scale of a rectangle bar graph you are making is to be from $0 to $100,000. Each block represents $2,000. How many inches will you need to draw the dollar scale?

The circle graph below represents the income made by a shop from towel sales, bedspread sales, and linen sales.

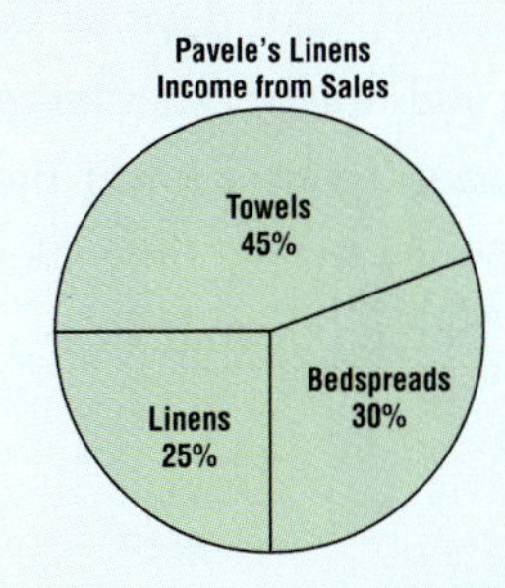

If the income earned by the shop in 1 week was $4,200, how much money was earned from

14. bedspread sales?
15. linen sales?

CHAPTER 13

BUILDING YOUR OWN

DO YOU ENJOY . . . being your own boss? Amber does. She's turned her skills at fixing things and working with tools into a career as a self-employed home repair and remodeling specialist.

Recent projects include building bookcases, installing kitchen cabinets, and reshingling a roof. She also provides "builder's services" (putting up drywall, painting walls, and installing locks) on houses under construction.

Amber enjoys variety, picking her own jobs, and working for herself. When things go wrong, however, she must rely on her own problem-solving skills and creativity.

Being self-employed means having to look for and bid on jobs. Bids tell customers how much a project will cost and when it will be completed. The type, amount, and cost of materials must be specified.

Often Amber has to help clients determine what they want. Being honest, patient, courteous, and persuasive helps. It may take days to estimate the cost of a room addition or other large project. If the work is awarded to a lower bidder, Amber gets no money for the time spent with the customer and on the bid.

FUTURE

Amber sees the cost of the bid effort, the expense of medical and business insurance, and no paid holidays as the major disadvantages of self-employment.

Math skills are constantly used in the building trades. Amber says she uses these math skills all the time:

- **estimation and formulas for bids;**
- **decimals and fractions for accurate measurements;**
- **area and angles for construction; and**
- **ratios to read designs and floor plans**

Some computer skills are necessary since floor plans and design specifications are often created on computers. Amber uses business software to generate invoices and calculate her taxes.

It took Amber time to get the money and experience to start her own business. After vocational school, she worked eight years for a construction company. She got a contractor's license, did home repair jobs on the side, read business books, and built her collection of tools (valued at over $40,000) before going out on her own. Today, Amber's business is mostly word of mouth. Quality work, completed on time and at a fair price, is the key to Amber's success. Amber says, to be successful in your own business you need skill, confidence, and motivation.

13-1 PROFIT AND LOSS

OBJECTIVES

In this lesson you will

- *learn the purpose of an income statement;*
- *distinguish between cost of goods sold and operating costs;*
- *learn the basic formulas for determining profit and loss; and*
- *analyze an income statement.*

Manufacturers make products. Retailers or merchandisers sell those products. Service businesses, such as hair stylists, attorneys, and personal trainers, sell a service as their primary product.

◆ Make a chart with three columns. Label the columns "Retail," "Service," and "Manufacturing." Under each column make a list of all the costs you think are involved in running that business. Look at your list. Are there different kinds of costs involved? How would you distinguish different types of costs?

WARM *UP*

a. \$3,500 + \$6,400 + \$9,500 + 5,400

b. \$25,000 − \$12,700

c. Find 20% of \$50,000.

d. Find 18% of 8,000.

e. Write $\frac{\$2{,}500}{\$15{,}000}$ as a percent, rounded to the nearest tenth of a percent.

FINDING PROFIT AND LOSS FOR A BUSINESS

There are several different kinds of businesses. Manufacturing businesses produce products from raw materials. Merchandisers buy those products and sell them to consumers. Merchandisers may be wholesalers or distributors who sell to other businesses. Or they may be retailers who sell directly to individual consumers. There are also service businesses that provide services, such as hair cuts or lawn care. To stay in business, a company must make a **profit** for its owners. That means a business must bring in more money than it spends.

An **income statement** shows how much money has been earned and spent during a given time period. If more money has been made than spent, there is a profit. If more money has been spent than earned, there is a loss. Income statements are usually prepared monthly, quarterly, and annually. Two income statements are shown in Illustration 13-1.1. One is for a retailer, Lynn's Books. The other is for a manufacturer, Bob's Cabinet Factory. We will use their income statements to identify the costs of doing business and the mathematics necessary to determine profit or loss.

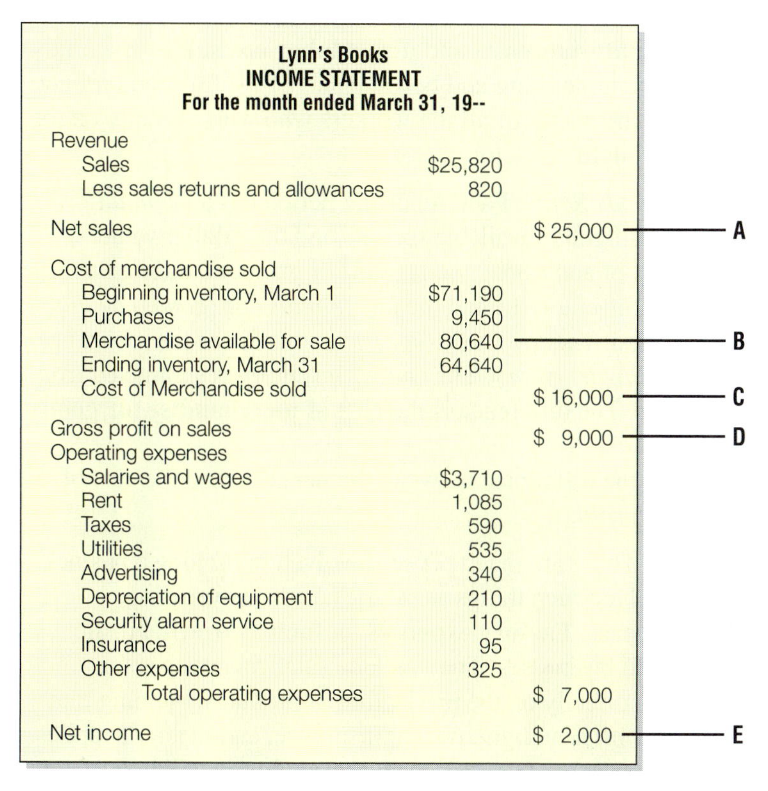

Lynn's Books
INCOME STATEMENT
For the month ended March 31, 19--

Revenue			
Sales	$25,820		
Less sales returns and allowances	820		
Net sales		$ 25,000	A
Cost of merchandise sold			
Beginning inventory, March 1	$71,190		
Purchases	9,450		
Merchandise available for sale	80,640		B
Ending inventory, March 31	64,640		
Cost of Merchandise sold		$ 16,000	C
Gross profit on sales		$ 9,000	D
Operating expenses			
Salaries and wages	$3,710		
Rent	1,085		
Taxes	590		
Utilities	535		
Advertising	340		
Depreciation of equipment	210		
Security alarm service	110		
Insurance	95		
Other expenses	325		
Total operating expenses		$ 7,000	
Net income		$ 2,000	E

A **Net Sales** = Sales – Returns and Allowances

B **Goods Available For Sale** = Beginning Inventory + Purchases

C **Cost of Merchandise Sold** = Goods Available for Sale – Ending Inventory

D **Gross Profit** = Net Sales – Cost of Merchandise Sold

E **Net Income (or Loss)** = Gross Profit – Total Operating Expenses

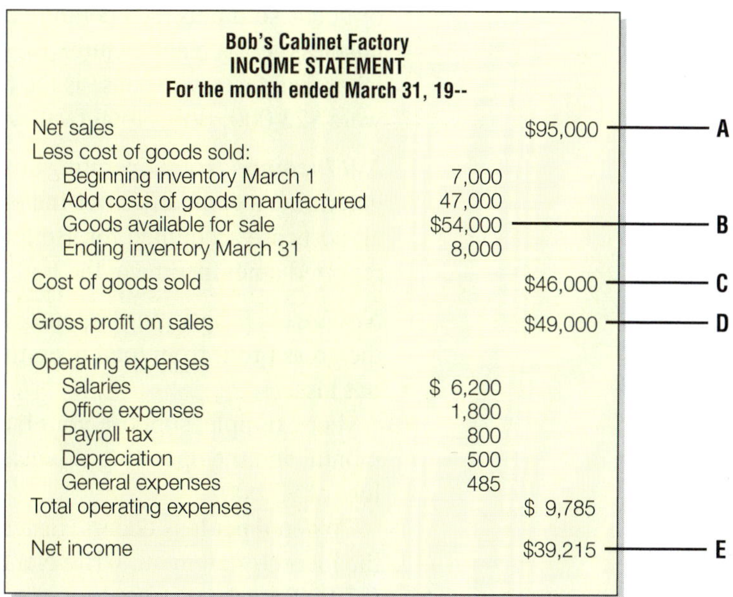

Bob's Cabinet Factory
INCOME STATEMENT
For the month ended March 31, 19--

Net sales		$95,000	A
Less cost of goods sold:			
Beginning inventory March 1	7,000		
Add costs of goods manufactured	47,000		
Goods available for sale	$54,000		B
Ending inventory March 31	8,000		
Cost of goods sold		$46,000	C
Gross profit on sales		$49,000	D
Operating expenses			
Salaries	$ 6,200		
Office expenses	1,800		
Payroll tax	800		
Depreciation	500		
General expenses	485		
Total operating expenses		$ 9,785	
Net income		$39,215	E

Illustration 13-1.1. Income Statements

Sales **Sales,** or *gross sales,* is the total value of the goods sold in a period of time. Some goods may be returned for refunds, and allowances or reductions in price are given for damaged goods. These *sales returns and allowances* decrease sales, so they are subtracted from sales. The amount left is called **net sales *(A).***

Cost of Goods (or merchandise) Sold These are expenses directly related to the cost of the merchandise (or goods) sold. For a retailer, these are the costs of purchasing merchandise. For a manufacturer, the costs of goods sold is the sum of the **raw materials costs** and **direct labor costs.** Raw materials are the items used in manufacturing and become part of the finished product. **Direct labor costs** are the wages of all the workers who work directly on the products as they go through the factory.

Finding the Cost of Goods Sold Each reporting period has a *beginning inventory*, or record of all goods available for sale and their dollar value. Companies add the cost of additional goods made or puchased during the time period to the beginning inventory to find the *merchandise or (goods) available for sale **(B)*** for the period. The inventory on hand at the end of the reporting period is called the *ending inventory*. The ending inventory is subtracted from goods available for sale. The difference is the **cost of merchandise (or goods) sold (C)** for that period.

Gross Profit (D) is the difference between the net sales and the cost of merchandise (or goods) sold.

Operating Expenses These are the expenses involved in running a business. Operating expenses include such things as salaries, taxes, advertising, equipment, and office space. **Factory expenses**, or **factory overhead**, are similar to those for retail businesses. They include salaries or wages for employees not working directly on the production of products, building rent, depreciation of equipment, heat, light, power, insurance, and factory supplies. Operating expenses are usually itemized, then totaled. For a retailer, the cost of goods sold plus the operating costs of a business are the **total costs** of that business. For a manufacturer, the sum of the costs of raw materials, direct labor, and factory expense is the **total factory cost.** Lynn's Books total costs were $23,000; Bob's total factory costs were $55,785.

Net Income Total operating costs are deducted from gross profit to determine a company's **net income *(E).*** This is the actual profit earned after all costs are accounted for. Both Lynn's Books and Bob's Cabinet Factory reported a net income at the bottom of their statements.

Net Loss If operating expenses are greater than gross profit on sales, subtract the gross profit from total operating expenses. The result, or difference, is a **net loss.**

For example, suppose your business had a gross profit of $13,000 for a month, and the operating expenses were $16,000 that month. You had a net loss of $3,000.

When a net loss occurs, instead of writing Net Income at the bottom of the income statement, write Net Loss. Put parentheses around the amount ($3,000) to show it is a loss, or negative amount.

EXERCISE YOUR SKILLS

For Exercises 1–4 write the formula necessary to calculate the item requested. Then show your answer.

1. The Cycle Shop's sales for the first quarter of the year were $152,296. Sales returns and allowances for the quarter were $6,450. What were the net sales for the quarter?
2. On January 1, Gill Hardware had an inventory of merchandise costing $75,070. During the year, merchandise costing $201,500 was bought. Merchandise inventory on December 31 was $50,730. Find the cost of merchandise sold for the year.
3. The Tech Company's records show these amounts for the quarter ending December 31: Finished goods inventory October 1, $32,475; cost of goods produced during the quarter, $82,890; finished goods inventory, December 31, $30,750; net sales for the quarter, $135,720. Find the cost of goods sold and the gross profit on sales for the quarter.

4. The Cap Place collected the following information about their business in May. Total sales: $22,000. Sales returns $1,500. Cost of Merchandise Sold: $4,500. Operating expenses: Advertising, $3,000; Salaries, $3,500; Store Space, $2,000; Depreciation, $325; Other expenses, $890. Create an income statement for The Cap Place. Show Net Sales, Cost of Goods Sold, Gross Profit, Total Operating Expenses, and Net Income (or Loss) for the month.

The factory records of Kargill Enterprises show earnings for last quarter of $1,187,000. Costs for last quarter were raw materials, $478,392.70; direct labor, $610,729.20; supervisory salaries and wages, $56,388.80; rent, $25,825; depreciation and repairs, $31,103.76; power, $15,281.50; factory supplies, $16,950.40; other factory expense, $6,204.86. What was the

5. cost of goods sold?
6. factory overhead?
7. total factory cost?
8. net income or loss?

Critical Thinking Lynn's profit for the time period shown on her income statement was $2,000. Bob's net profit was $39,215. Are they both doing well in their business? Is Bob's business doing better than Lynn's? How do you know?

ANALYZING INCOME STATEMENTS

Businesses use income statements to report income. They can also be used to analyze a business. Percentage analysis shows relationships among items to net sales. Ratio analysis compares items on an income statement. Following is a percentage analysis for Lynn's Books.

Item	Amount	Formula %	Comments
Net sales	$25,000	$25,000 ÷ 25,000 = 100%	Net sales is the base for calculating percentage components. It is always 100%.
Gross Sales	$25,820	$25,820 ÷ 25,000 = 103.28%	Gross Sales may be greater than net sales.
Cost of goods sold	$16,000	$16,000 ÷ 25,000 = 64%	Cost of goods is 64% of net sales.
Gross profit	$9,000	$9,000 ÷ 25,000 = 36%	$0.36 of each sales dollar results in gross profit.
Operating expenses	$7,000	$7,000 ÷ 25,000 = 28%	$0.28 of each sales dollar goes to cover operating costs.
Net income	$2,000	$2,000 ÷ 25,000 = 8%	$0.08 of each sales dollar is pure profit or net income. This is also referred to as the **return on sales.**

The **merchandise turnover rate** is the number of times per year that a store replaces, or turns over, its average stock of merchandise. Different kinds of businesses have different acceptable turnover rates. Businesses that sell durable and expensive goods such as furniture or large appliances often have a low turnover rate, such as 2.4. Businesses that sell perishables such as food often have a high turnover rate.

$$\textbf{Annual Merchandise Turnover Rate} = \frac{\textbf{Cost of Merchandise Sold for the Year}}{\textbf{Average Merchandise Inventory for the Year}}$$

EXAMPLE Calculate the merchandise turnover rate for Lynn's Books based on the March income statement.

solution	
$ 71,190	Merchandise inventory, March 1
+ 64,640	Merchandise inventory, March 31
$135,830	Total of inventories

$\$135{,}830 \div 2 = \$67{,}915$	Average inventory
$\$16{,}000 \div \$67{,}915 = 0.23558$	Cost of Goods Sold for Period ÷ Average Inventory = Monthly rate[1] Round to nearest thousandth.
$12 \times 0.236 = 2.83$	Multiply by 12 for Annual rate. Round to nearest hundredth.

[1]***Note:*** Since Lynn's Books' income statement is for a month, we multiplied the monthly rate by 12 to find the equivalent rate for a year. If the income statement had been for a quarter, we would have multiplied 0.236 by 4 to find the annual rate. **Always show merchandise turnover as an annual rate correct to the nearest hundredth.**

Comparing Income Statements Income statements are most helpful as an analysis tool when statements for similar periods are compared over time. For example, Lynn's Books sales are highest in the month of December. A percentage comparison of key items on her March and December statements is shown below.

	March	%	December	%
Net Sales	$25,000	100%	$75,000	100%
Cost of Goods Sold	$16,000	64%	$48,000	64%
Gross Profit	$9,000	36%	$27,000	36%
Operating Expenses	$7,000	28%	$9,000	12%
Net Income	$2,000	8%	$18,000	24%

Critical Thinking The percentage of sales for cost of goods sold was similar in both months. What accounts for the large percentage increase in net income?

EXERCISE YOUR SKILLS

Prepare a percentage analysis for Bob's Cabinet Factory, using the income statement shown on page 469.

9. Cost of Goods Sold

10. Gross Profit on Sales

11. Operating Expenses

12. Net Income.

13. Find the annual merchandise turn over rate for Bob's Cabinet Factory.

JoAnn Booth's business had net sales of $70,000. The cost of merchandise sold was $42,000, and the gross profit was $28,000.

14. The cost of merchandise sold was what percent of net sales?

15. The gross profit on sales was what percent of net sales?

From July through December, McDougall's Closet Shop took 3 merchandise inventories: $44,000, $94,100, and $62,300. The cost of merchandise sold during the 6-month period was $300,600.

16. What was the store's average merchandise inventory?
17. What was the merchandise turnover rate for the 6 months?
18. Find the equivalent merchandise turnover rate for a year.

♦ The records of the Paulson Company last year showed these data: net sales, $440,000; cost of merchandise sold, $286,000; operating expenses, $105,600.

19. Find the gross profit on sales and the net income for last year.
20. Find the percent of net sales for cost of merchandise sold, gross profit on sales, operating expenses, and net income.

MIXED REVIEW

1. 345×9.346
2. $723{,}632 \div 8$
3. What amount is 150% of $42?
4. $481.60 is what percent less than $560?
5. Write 89.2% as a decimal, rounded to the nearest hundredth.

On net sales of $241,000, Jill had a gross profit of $118,090 and operating expenses of $59,045. What percent of net sales, as an estimated rate and actual rate, was the

6. Gross profit?
7. Net income?
8. For the third quarter of the year, a store's 3 merchandise inventories were $108,400, $87,100, and $155,500. The cost of merchandise sold for the quarter was $292,500. What was the annual merchandise turnover rate?
9. Syl Reiner borrowed $3,000 for 1 month and paid $3,042 on the maturity date. What interest rate did he pay?
10. What is Halina's annual income from 340 shares of stock if the stock pays a regular quarterly dividend of $0.42 per share?
11. Allen Cole made an 840 km car trip. He averaged 12 km per liter of gas. (a) How many liters of gas did he use? (b) At 35¢ a liter, how much did he spend for gas?
12. Brenda Sanders had $1,105.25 in her checking account on Monday morning. During the week, she wrote checks for $14.53, $532.57, and $105.92. She made a deposit of $443.19. Find her checking account balance at the end of the week.

13-2 CALCULATING AND ANALYZING OPERATING COSTS

OBJECTIVES

In this lesson, you will learn to calculate

- *computer costs;*
- *costs of labor;*
- *cost of a unit of office work; and*
- *apportioned factory costs to appropriate departments.*

Operating costs are all expenses not directly involved in the cost of goods, or merchandise sold. That includes such things as advertising, depreciation, labor costs, interest on debts, shipping costs, insurance, legal and professional fees, office expenses, business equipment, and office supplies. As you saw in the income statement, all operating expenses are subtracted from the gross income of a company. They impact profits and should be watched closely.

◆ In this lesson, we will look at three operating expenses: technology, labor, and office space. Which of these costs do you think is the highest? the lowest? What percentage of total operating costs do you think should be allocated to each item?

WARM UP

a. 360 × \$405.20

b. 2.4 + 0.082 + 19

c. 1,250% of 0.4

d. \$8.15 × 40

e. 7 600 m = ? km

f. Estimate 793 ÷ 39

g. Find what percent \$12,000 is of \$66,000.

TECHNOLOGY COSTS

Computer technology is changing rapidly. A company's technology costs may be anything from stand-alone personal computers to networked mainframe systems. There are two major costs when computer systems are purchased. First, there is the purchase price of the system and all its parts. These parts, common to nearly every type of computer system, include:

- The **central processing unit,** or brain of the computer.
- **Storage devices,** such as disk and tape drives, that store computer programs and data.
- **Input devices,** such as computer terminals with keyboards, that are used to enter data into a computer system.
- **Output devices,** such as printers and terminals with display screens, that are used to get data out of the computer system.
- **Software,** or a computer program, which is a set of instructions that tells the computer system what to do.

Second, there is the cost of installing the computer system. On large computer systems, installation costs may include raised flooring, special electrical wiring, cables, air conditioning, and security systems. Even on small computer systems, special cables may be needed to connect computers to each other. Separate electrical outlets and wiring may also be required.

Because large computer systems can be very costly to purchase, many firms lease them. Some firms lease computer systems because leasing lets them upgrade to new models easily.

Software Costs Computer systems consist of **hardware,** such as terminals, disk drives, and printers, and **software,** or computer programs.

There are two basic types of computer programs: *operating systems*, which manage the computer hardware, and *application programs*. Application programs instruct the computer system to process your data the way you want it processed. Examples of application programs are payroll programs, word processing programs, spreadsheet programs, and computer games.

A firm can buy software in a store, or it can develop programs for its own needs. When a firm develops programs itself, it uses the skills of computer programmers. Large computer programs may be made up of thousands of lines of computer instructions, called "lines of code." A good programmer may produce from 20–40 fully tested and finished lines of computer code in one day.

Storing computer software and data is another cost. The capacity of computer storage is often measured in terms of the number of characters, or **bytes,** of data that can be stored. For example, a floppy disk may be able to store 720,000 characters, or bytes, of data (letters, numbers, and other characters). People who use computers call 1,000 bytes of data 1 **kilobyte,** or KB. They call 1,000,000 bytes of data 1 **megabyte,** or 1 MB. 720,000 bytes are 720 Kilobytes or 720K.

EXERCISE YOUR SKILLS

The Accounting Department of Milo Manufacturing Co. just bought 20 personal computer systems. Each personal computer cost $2,200, the printer (used by all systems) cost $600, and the software (used by all) cost $1,200.

1. What was the cost of each system?
2. What was the total cost of the systems to the department?

Daner, Inc., leases 10 large computer printers for $175 each per month.

3. What is the annual cost of the printers to the firm?
4. The Data Processing Department of Davis, Inc., assigned 3 programmers the job of developing a specialized inventory management program. Each programmer was paid $25 an hour and spent 640 hours writing and testing the program. What was the cost of labor used to develop the program?
5. A programmer, paid $24 an hour, took 900 hours to develop a sales management program made up of 3,000 lines of computer instructions. What was the average labor cost per line?

LABOR COSTS

Depending on the type of business, labor expenses can account for as much as 40%–90% of the operating expenses of a business. Some labor expenses are generally required by law, such as payroll, FICA, and unemployment taxes. Other benefits, such as health insurance, vacation, and savings plans, make a company an attractive employer. Illustration 13-2.1 shows the benefits package and costs for one employee at a large company. Note that at this company, some expenses, such as health insurance, are shared by employer and employee. In some companies the employer pays the full amount of insurance. In other companies, employees pay for their own insurance.

Benefit	The Company Pays	You Pay
Health Care	$ 3,912	$ 996
Disability	798	117
Life, Dependent Life, AD&D, and Business Travel Accident Insurance	151	134
Pension Plan	1,326	0
Savings Plan	1,260	2,521
Social Security Tax (FICA)	2,796	2,796
State Unemployment Insurance	249	0
Federal Unemployment Insurance	434	0
Worker's Compensation	42	0
Total Benefit Costs	**$10,968**	**$6,564**

Total Company Benefit Costs Plus Annual Income $47,510

Total Company Benefit Costs $10,968

Total Annual Income $36,542

Total Company Costs for Your Benefits as a Percent of Your Total Annual Income: 30%

The Company also provides other valuable benefits and programs at no cost to you including:

- Holidays;
- Vacation;
- Personal Days;
- Employee Assistance Program;
- Tuition Reimbursement; and
- Bereavement Days.

Illustration 13-2.1 Example of Benefit Costs

EXERCISE YOUR SKILLS

Last year, Intermart, Inc., employed 30 workers, all of whom received at least $7,000 in total wages. The FUTA tax rate was 6.2% on the first $7,000 of each employee's wages. Intermart also had to pay state unemployment taxes of 3.7% on the first $7,000 of each employee's wages.

6. What was the firm's FUTA tax?
7. Estimate then calculate the exact amount of state unemployment taxes due.
8. If Intermart deducts 90% of state unemployment taxes from their FUTA taxes, what is the net FUTA taxes they owe?

Emily is paid $5.50 per hour. Retirement benefits, insurance, and other fringe benefits add another 18% to her wages. If she is paid for 2,000 hours of work during 1 year,

9. How much is she paid in straight wages for the year?
10. What are her total wages for the year, including straight wages and fringe benefits?

OFFICE SPACE

The cost of office space is another, often major, operating expense. The cost of office space may include the cost of renting, lighting, heating, cooling, insuring, cleaning, maintaining, and repairing the space used by the office. If the office is owned instead of rented, the cost may include depreciation, mortgage interest, and real estate taxes.

PROBLEM SOLVING

DRAW A DIAGRAM

When solving a problem, drawing a diagram can be a useful strategy. Kyle is designing the offices for his boss and coworkers. There is a large square room to be used. His boss will require $\frac{1}{4}$ of the space for her office. He will need $\frac{1}{2}$ as much space as his boss. The rest of the room is to be used for 6 workstations. Kyle, the secretary, has drawn this first draft on the new floor plan. What is missing? How would you improve on Kyle's draft to make it more efficient?

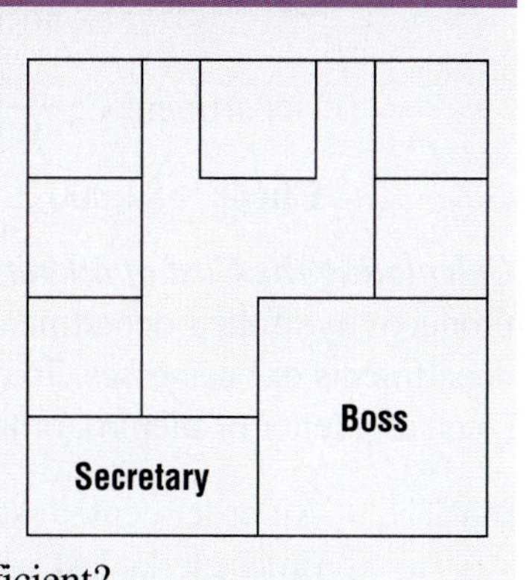

Use the Problem Solving Strategy to solve these exercises.

An office manager wants you to increase the number of work stations that can be put in a space 40 feet wide by 30 feet long. That space now contains 4 rows of 5 workstations each. Each workstation requires a minimum of 30 square feet of space. Aisles must be 3 feet wide. Movable walls can be put in to separate workstations, but you must be able to get to each workstation without going through another.

11. If the office space rents for $12,420 per year, what is the annual rent per workstation with the old arrangement?
12. Draw a diagram to show how you would rearrange the workstations.
13. By what percent is the number of workstations increased in the same space?
14. Find the annual rent per station with your arrangement.

ALLOCATING AND ANALYZING OVERHEAD COSTS

Distributing Expenses Companies may allocate overhead expenses on department budgets. How allocations are made varies from company to company and with the expense. For example, taxes and insurance may be distributed on the basis of the value of the equipment in each department. Cleaning expenses may be distributed on the basis of floor space. Other overhead expenses may be distributed on the basis of number of employees in each department. This kind of distribution is common in factories, where factory expense is often charged to each department.

EXAMPLE A factory pays monthly rent of $20,000. What amount should be charged to each of its 3 departments? Department A has 2,000 ft^2 (square feet) of floor space; Department B, 5,000 ft^2; and Department C, 3,000 ft^2.

SOLUTION 2,000 + 5,000 + 3,000 = 10,000 ft^2 total floor space

Department A: $\frac{2{,}000}{10{,}000}$ or $\frac{1}{5}$ of \$20,000 = \$ 4,000

Department B: $\frac{5{,}000}{10{,}000}$ or $\frac{1}{2}$ of \$20,000 = \$10,000

Department C: $\frac{3{,}000}{10{,}000}$ or $\frac{3}{10}$ of \$20,000 = \$ 6,000

Check $4,000 + $10,000 + $6,000 = $20,000

Calculating the Cost of a Unit of Overhead Managers like to compare the productivity of their department or company with the productivity of other departments or businesses. To do so, they find the cost figure of a single job (such as a letter or memo), or the cost figure for a single worker or workstation.

EXAMPLE An order center estimates that the costs of workstations for 10 order clerks last year were wages and fringe benefits, $138,000; space, $8,590; utilities, $890; depreciation, $13,600; supplies, postage, and telephone, $17,098. Find the average cost of each workstation.

SOLUTION Step 1 Add all operating costs.
$138,000 + 8,590 + 890 + 13,600 + 17,098 = $178,178

Step 2 Divide total cost by the number of workstations
$178,178 ÷ 10 = $17,817.80 average cost per workstation.

EXERCISE YOUR SKILLS

15. In the example shown for calculating the cost of a unit of office space assume the office was open 250 days last year. What was the cost of each workstation per day, to the nearest cent?
16. For the same example, if each clerk usually completed 130 orders each day, what was the cost of each order, to the nearest cent?
17. The 4 departments of Tensor Electronics, Inc., use this floor space: Printers, 3,000 ft^2; Copiers, 2,000 ft^2; Assembly, 1,800 ft^2; Finishing, 1,200 ft^2. The annual maintenance cost of the building is $18,000. The cost is distributed on the basis of the floor space of each department. How much should each department be charged annually?
18. Nestor, Inc., pays its managers $400,000 a year. This expense is charged to the departments on the basis of the number of workers. The number of workers in Department A is 108; Department B, 36; Department C, 12; and Department D, 84. What amount should be charged to each department?
19. An office manager studied 15 workers and found that, on average, each worker spent 1 hour of each 8-hour work day on coffee breaks, personal telephone calls, and other personal business. The average wage of the workers was $52 a day. What was the cost per day to the office of this nonworking time?
20. ◆ Write 3 ways of measuring productivity and costs in your classroom.

MIXED REVIEW

1. Multiply: $16\frac{1}{8}$ by 24
2. Divide: 0.048 by 0.008
3. Write 0.4078 as a percent.
4. Find 12% of $4,350.
5. An insurance office pays a file clerk $5.25 an hour. The clerk can find and pull 95 letters per hour. What would be the labor cost for the clerk to find and pull 3,515 letters?
6. Sarah can buy a VCR for $350 cash, or she can make a down payment of $50 and pay $27.25 for 12 months. How much more is the installment price than the cash price?
7. Stock paying quarterly dividends of $1.80 are bought for a total cost of $90 per share. What rate of income is earned on the investment?
8. Three departments of a store use this floor space: Jewelry, 400 square feet; Household, 1,500 square feet; Paint, 600 square feet. Cooling costs are charged to departments on the basis of floor space. What should be each department's share of summer cooling costs of $1,200?
9. Property worth $72,000 is assessed at 45% of its value. The tax rate is $52.122 per $1,000 of assessed value. What is the tax on the property?

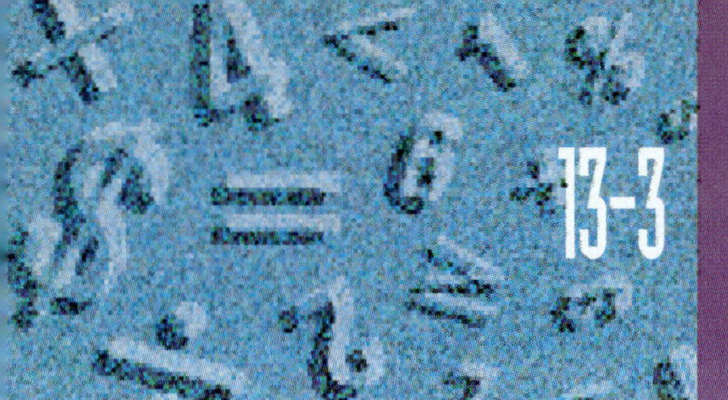

13-3 DEPRECIATION COSTS

OBJECTIVES

In this lesson, you will learn to calculate depreciation using

- *the declining-balance method;*
- *the sum-of-the-years method; and*
- *the modified accelerated cost recovery system method.*

In earlier chapters, you discussed the depreciation of cars and property. Business property, such as equipment, also depreciates in value.

◆ Identify at least 2 reasons why you think business equipment depreciates. How do those reasons compare to the reasons why personal property depreciates?

WARM UP

a. 20% of $18,300

b. $14\frac{1}{2}$% of $50,000

c. $450 is what percent of $9,000?

d. To the nearest tenth of a percent, find $8,500 ÷ $57,500.

e. Find the average of 315, 354, 634, 525, 444, 288, and 331.

DEPRECIATION

The depreciation of assets that have a life of more than one year can be a major operating expense for many kinds of businesses. The calculation of depreciation is regulated by the Internal Revenue Service. They allow depreciation to be calculated in several ways.

STRAIGHT-LINE AND DECLINING-BALANCE METHODS

You have already learned how to calculate depreciation on cars using the *straight-line method.* In that method, the amount of depreciation is the same for each year.

Suppose you bought a car for \$8,500 and planned to keep it five years. You estimated its trade-in value to be \$2,500 five years later. The average annual depreciation would be (\$8,500 -\$2,500) ÷ 5 or \$1,200.

Office and factory equipment usually depreciates more during the early years of use than during later years. Because of this, businesses use depreciation methods that deduct greater amounts in the early years than in later years. Two such methods are the declining-balance method and the sum-of-the-years-digits method.

When calculating depreciation, the term **book value** is used. Book value is the original cost of the asset less the total depreciation to date. For example, if a machine that cost \$280,000 has been depreciated \$40,000, its book value, or current value, is \$240,000 (\$280,000 − \$40,000 = \$240,000).

Declining-Balance Method of Calculating Depreciation

The **declining-balance method** uses a fixed rate of depreciation for each year. Because the rate is applied to a declining or decreasing balance, the amount of depreciation decreases each year.

EXAMPLE A truck costing \$15,000 is estimated to depreciate 20% each year. What is the estimated book value of the truck at the end of the first and second years?

SOLUTION

0.20 × \$15,000 = \$3,000	First year depreciation
\$15,000 − \$3,000 = \$12,000	Book value—End of first year
0.20 × \$12,000 = \$2,400	Second year depreciation
\$12,000 − \$2,400 = \$9,600	Book value—End of second year

EXERCISE YOUR SKILLS

Use the declining-balance method to solve these problems.

1. A machine that cost \$20,000 decreases in value each year at the rate of 10%. What is the book value of the machine at the end of the fourth year?
2. Tuxton Foundry buys a press for \$5,000. The estimated life is 12 years, and the annual depreciation is 8%. How much does the company expect the press to be worth at the end of 6 years?

Copy and complete the table below.

	Property	Original Cost	Rate of Depreciation	Book Value	
				At End of First Year	At End of Third Year
3.	Drill	$1,500	7%		
4.	Press	$8,500	8%		
5.	Painter	$4,000	9%		
6.	Tools	$3,800	12%		
7.	Lift	$7,000	14%		

THE SUM-OF-THE-YEARS-DIGITS METHOD

Another way to calculate depreciation is the **sum-of-the-years-digits method.** This is a variable-rate method. Like the declining-balance method, the sum-of-the-years-digits method provides the greatest amount of depreciation in the first year and smaller amounts of depreciation after that.

For example, if you estimate that you will use a machine for five years, the amount of depreciation is calculated this way:

- Add the years together: $1 + 2 + 3 + 4 + 5 = 15$ years.
- Then depreciate the machine:

  $\frac{5}{15}$ of the total depreciation for the first year.

  $\frac{4}{15}$ of the total depreciation for the second year.

  $\frac{3}{15}$ for the third year.

  $\frac{2}{15}$ for the fourth year.

  $\frac{1}{15}$ for the fifth year.

EXAMPLE A machine costing $10,000 will be used for 5 years, then traded in for $4,000. Find the book value of the machine at the end of the second year.

SOLUTION

$10,000	Original cost	Sum-of-the-years-digits:
− 4,000	Trade-in value	$1 + 2 + 3 + 4 + 5 = 15$
$ 6,000	Total depreciation	

$\frac{5}{15} \times \$6,000 = \$2,000$ First year depreciation

$\$10,000 - \$2,000 = \$8,000$ Book value—End of first year

$\frac{4}{15} \times \$6,000 = \$1,600$ Second year depreciation

$\$8,000 - \$1,600 = \$6,400$ Book value—End of second year

EXERCISE YOUR SKILLS

Use the sum-of-the-years-digits method for these problems.

8. Garmen Products bought a hydraulic winch for $31,700. They plan to use it for 6 years and then trade it for $6,500. Find the book value of the winch at the end of the second year.
9. Diaz Bros. bought a loading machine for $40,000. The firm estimates that the loader will be used for 9 years and then be traded in for $5,000. What will the book value of the loader be at the end of the first and third years?
10. Sheng, Inc., bought a forming machine for $68,500. The company plans to use the machine for 10 years and then sell it. The company estimates the resale value will be $13,500. Find the depreciation for each of the first 3 years.

Tone Grain Farms bought 5 items of equipment, weighing 14 metric tons, for $420,000. The company plans to use the equipment for 16 hours daily for 15 years and then scrap it for no value.

11. What will be the book value of the equipment at the end of 3 years of depreciation?

THE MODIFIED ACCELERATED COST RECOVERY SYSTEM (MACRS)

For federal income tax purposes, the **modified accelerated cost recovery system (MACRS)** must be used to calculate depreciation for most property placed in service after 1986. The MACRS method allows you to claim depreciation over a fixed number of years depending on the class life of the property. The *class life* means how long the Internal Revenue Service (IRS) will let you depreciate the property.

The lives of different types of property are classified into 3, 5, 7, 10, 15, 20, 27.5, or 31.5 years by the Internal Revenue Service (IRS). For example, the IRS puts most office equipment and cars into a 5-year class life.The rate of depreciation to be used for each year of a property's life is set by the IRS and varies with each class life.

You calculate the depreciation deduction for any one year by multiplying the original cost by the rate of depreciation for that year. Trade-in value, or salvage value, is not used in the MACRS method.

The table of depreciation rates for a property with a 5-year class life bought in the middle of the first year and used until the middle of the sixth year is shown in Illustration 13-3.1.

Last Half of	First Year	20.0%
	Second Year	32.0%
	Third Year	19.2%
	Fourth Year	11.5%
	Fifth Year	11.5%
First Half of	Sixth Year	5.8%
	Total	**100.0%**

Illustration 13-3.1. Table of MACRS Rates for 5-Year Property

For example, Anna Velez bought a new business car for $9,000 on July 1. The car has a 5-year class life. For income tax purposes, Anna can claim these amounts of depreciation on the car:

Last Half of	**First Year**	**0.20 × $9,000 =**	**$1,800**
	Second Year	**0.32 × 9,000 =**	**2,880**
	Third Year	**0.192 × 9,000 =**	**1,728**
	Fourth Year	**0.115 × 9,000 =**	**1,035**
	Fifth Year	**0.115 × 9,000 =**	**1,035**
First Half of	**Sixth Year**	**0.058 × 9,000 =**	**522**
	Total Depreciation		**$9,000**

EXERCISE YOUR SKILLS

Use the schedule of depreciation in Illustration 13-3.1.

Burton Concrete Company paid $4,200 for a laser printer that had a class life of 5 years. They bought the printer on July 5.

12. What amount of depreciation was allowed on the machine for each year?

13. What was the book value of the machine at the end of its life?

A machine bought on July 10, costs $45,000 and has a 5-year class life for MACRS depreciation.

14. What total depreciation is allowable for the first year's use?

15. What is the book value of the machine after the third year?

An asset, bought in July, cost $100,000 and has a 5-year recovery life under MACRS.

16. What is the total amount of depreciation allowed for the life of the asset?

17. What is the book value of the asset after the first year's use?

18. What is the book value of the asset after two year's use?

Critical Thinking Another method of depreciation is called the **unit of performance (or production) method.** What do you think this means?

19. ◆ Prepare a study guide that identifies the key facts and formulas for each method of depreciation studied in this lesson. Include any other information that may be helpful to you in completing depreciation exercises.

MIXED REVIEW

1. Multiply $4\frac{1}{2}$ by $3\frac{1}{2}$

2. Divide 534.12 by 100

3. Find exact interest on $600 at 14% for 6 months.

4. Write $\frac{3}{8}$ as a decimal to the nearest hundredth.

5. To make 300 book covers, the costs were: materials, $3,589.12; labor, $6,128.43; factory overhead, $721.31. What was the average cost of each cover, to the nearest cent?

6. Four departments of a factory use this floor space: A, 5,600 square feet; B, 2,400 square feet; C, 4,800 square feet; D, 3,200 square feet. The annual rent of the building, $180,000, is distributed on the basis of floor space. How much should be charged to each department?

7. Lester Williams bought a car for $12,900 and drove it 15,000 miles in the first year. His expenses that year were: depreciation, 20% of the cost of the car; interest at 12% of the cost of the car; gas, oil, insurance, and other expenses, $1,860. Find the operating cost of the car per mile, to the nearest tenth of a cent.

13-4 SHIPPING AND ADVERTISING COSTS

OBJECTIVES

In this lesson you will

- *learn about different mailing and shipping options;*
- *calculate advertising costs for different media; and*
- *calculate the cost and profit produced from a direct mail campaign.*

Suppose you decide to start a small business. How would you let people know about your product or service?

◆ Make a list of 5 ways you could advertise your product or service. If you could only use one of those methods to advertise your product, which one would you choose? Why?

WARM UP

a. \$274 − (\$12 + 15.50)

b. \$475 × 7.5%

c. \$253 ÷ 0.5

d. \$375 × 0.765

e. Find 14.5% of \$105,000

MAILING AND SHIPPING

Businesses use the U.S. Postal Service, private carriers, and shippers to communicate with and service their customers. In addition, businesses may pay shipping and handling costs when they purchase items.

Mailing and shipping costs are generally considered an operating expense of a business. Illustration 13-4.1 shows some of the mailing services available from the U.S. Post Office. Private carriers such as U.P.S. and Federal Express also offer different levels of express and priority delivery services. The cost differences between types of services can be significant. Businesses should monitor their shipping expenses to be sure economical and appropriate service choices are being made. For example, a 3 oz letter could be mailed in any of the following ways:

EXAMPLES

Item	Cost*
3 oz letter shipped first class mail	$0.78
3 oz letter shipped next day express mail	$10.75
3 oz letter included in a third class mailing	$0.226

*Costs in effect December 1995.

Service	Comments
First Class Mail	Use for correspondence, bills, invoices weighing up to 11 ounces (2–5 days to reach destination).
Priority Mail	Use for mail weighing more than 11 oz. Two - three-day service.
Third Class Mail	Usually for large quantity mailings such as direct mail advertising. Cost based on number and weight of pieces. Must mail minimum of 200 pieces or 50 pounds.
Fourth Class Mail	For packages over 1 pound. Packages must be under 70 pounds and be less than 108 inches in combined length and **girth***. Delivery time 2–7 days. Also called Parcel Post.
Business Reply Mail	Often used with advertising mailings to solicit orders from customers. The business pays the return postage fee for the customer. In addition to postage, a per piece service charge and annual permit fee is required.
Express Mail	Material that needs to be delivered the next day. Premium price.

**Girth is the measurement around a package at its thickest part. If a package is 30 inches long, 10 inches wide, and 4 inches deep its girth is 10 + 4 + 10 + 4 or 28 inches.*

Illustration 13-4.1. Some Mailing Services of U.S. Postal Service

Freight is used most often for shipping heavy, bulky goods. Freight shipments may be sent by airplane, truck, train, or ship. Freight agents provide prices on the cost of specific shipments to worldwide locations. Shipping charges are usually quoted on the basis of weight per 100 pounds. For example, a 2,500 pound package is being shipped by freight. Charges are quoted as $37.42 per 100 pounds. The cost of the freight shipment is calculated as $2,500 \div 100 \times \$37.42 = \$935.50$.

Shipping charges vary from vendor to vendor. However, shipping charges are usually based on the size of the item being shipped, the distance to be traveled, and how quickly the package must meet its destination point. Special services, such as insurance, door-to-door delivery, and collect on delivery (C.O.D.) shipments, usually increase the cost of shipping.

EXERCISE YOUR SKILLS

1. You live in New York. An important package must be at a customer's California office tomorrow morning. Use Illustration 13-4.1. What is the best way to ship the package?

Argos Castings, Inc. quotes Barr Foundry a price of $379.85 plus transportation charges of $37.80.

2. What will be the total cost of the casting to Barr Foundry?

Which of the packages below can be sent by parcel post?

	Length	Width	Depth	Weight
3.	23 inches	8 inches	14 inches	16 lb
4.	50 inches	12 inches	7 inches	72 lb
5.	60 inches	15 inches	12 inches	50 lb

6. Merlin Motors ordered a hoist that weighed 1200 pounds. The charge was $21 per 100 pounds. What was the shipping cost.

ADVERTISING COSTS

Producing a great product that is very salable is the goal of most businesses. Once a product is produced, a business must make consumers aware of the product, its benefits, and where they can buy it.

An advertiser must decide on the best way to reach its audience. For example, a toy company introducing a new doll may advertise during Saturday morning children's television. Advertisers want to know an ad's reach and frequency. **Reach** is how many people see or hear an ad. **Frequency** is the number of times an ad is seen or heard.

Television and radio ads usually have wide reach. The cost of radio and television advertising depends on the length of the commercial, the ratings for the show the ad is aired on, the time of the day when the ad is carried, and if it is a national or local ad. Television advertising is expensive. A local station may charge $500 for a 30-second ad run in the afternoon. A 30-second national advertisement run in prime time (5 P.M. to 11 P.M.) may cost $50,000. Radio advertising is relatively inexpensive.

Print advertisements often contain more information than radio and tv ads. Prices of print ads depend on the size and frequency of the ad, as well as on the circulation (reach) of the publication.

Print ads are usually billed by column size or fractional page. Column size is the width and length of an ad. The formula for determining an ad's cost is Number of Inches × Cost Per Column Inch × Number of Column Widths. For example, a newspaper charges $50 per column inch for a one-column width ad. An ad that is 5 column inches long and one column width wide would be billed as 5" × $50 × 1 = $250. An ad in the same paper that is 5 column inches long and three column widths wide would be billed as 5" × $50 × 3 = $750.

Illustration 13-4.2 is an example of an advertising rate card that shows ad costs for a magazine with 60,000 subscribers. The 1X, 3X, etc. refer to the frequency rate, or the number of times the ad will run. To calculate the total cost of an ad, find the base price by matching the size ad with the frequency rate. Then, add any additional charges, such as for preferred position or color. For example a full page, 4-color ad that runs 3 times would cost $2,070 + $1,020 or $3,090 per placement. If the same advertisement appeared on the back cover, the cost would be $2,145 + $1,020 or $3,165. The cost of the full contract (all 3 placements) for that ad on the inside front cover is $3,165 × 3 or $9,495.

ADVERTISING RATES				
SIZE	**1X**	**3X**	**5X**	**12X**
Full page	$2,195	$2,070	$1,945	$1,820
$\frac{2}{3}$ page	$1,450	$1,385	$1,285	$1,220
$\frac{1}{2}$ page	$1,115	$1,055	$995	$940
$\frac{1}{3}$ page	$775	$735	$675	$635
Cover Rate	$2,275	$2,145	$2,000	$1,860
Color Charges: Add to per page rate				
2-color	$275	$255	$240	$225
4-color	$1,040	$1,020	$970	$930

Illustration 13-4.2. Advertising Rate Card

Direct Mail Advertising is advertising sent to you by mail. Magazine subcriptions are frequently advertised in this way.

The success of a direct mail campaign depends on how many of the people mailed to are part of the target audience for the product. For example, a car manufacturer may send a direct mail campaign for a car warranty to everyone who purchased a new car from one of their dealers in the last two months. That would be a targeted advertising campaign. A high response rate could be expected. On average, advertisers report response rates from 2% to 15%.

The cost of direct mail advertising includes the cost of creating and reproducing the advertising material and mailing the material out. Other costs may include return postage for pre-paid mail back cards and postal fees, such as bulk mail permits.

EXAMPLE An advertiser sends out a direct mail campaign of 5,000 pieces. Each piece weighs 0.2087 lb. The bulk mail charge to mail the piece is billed as $0.073 per piece plus $0.465 per pound. There is also an $80 bulk mailing permit fee. The advertiser receives 2,000 replies in the postage-paid envelope he provided. The postage charge for each return order is the cost of a first class letter plus a $0.10 service charge. Finally the advertiser must pay an $85 business reply mail permit fee to the post office. What was the total postage bill for the mailing and responses?

SOLUTION **Step 1: Calculate outgoing bulk mail costs.**

- **Number of Pieces × Cost Per Piece = Total Piece Charge**
  5,000 × $0.073 = $365 total piece charge
- **Total Pieces × Per Piece Weight in Pounds = Total Pounds**
  5,000 × 0.2087 = 1,043.5 pounds
- **Total Pounds × Cost Per Pound = Weight Charge**
  1,043.5 × $0.465 = $485.2275 weight charge
- **Piece Charge + Weight Charge + Bulk Mail Fee = Total Outgoing Mailing Costs.** Round answer to the nearest cent.
  $365 + $485.228 + $80= $930.23, total outgoing mailing costs

Step 2: Calculate return mailing costs.

- **First Class Postage + Per Piece Service Charge = Per Piece Return Postage**
  $0.32 + $0.10 = $0.42 per piece return postage
- **Postage Per Piece × Number Pieces Returned = Total Return Postage**
  $0.42 × 2,000 = $840.00 total return postage
- **Business Reply Mail Fee + Postage = Total Return Mailing Costs**
  $85 + $840 = $925 total return mailing costs

Step 3: Add all charges to find cost.

- **Total Outgoing Mailing Costs + Total Return Mailing Costs = Total Cost of Bulk Mailing**
  $930.23 + $925 = $1,855.23 total mailing costs

Critical Thinking What factors influence the price that can be charged for carrying an advertisement?

EXERCISE YOUR SKILLS

For radio and t.v., assume that 30-second ads cost $\frac{1}{2}$ as much as 60-second ads.

7. A local television station, WZAB, has set these rates for 60-second advertisements: daytime shows, $900; prime-time shows $1,500. Your company is planning an advertising campaign that will include: 8 30-second daytime ads, 6 30-second prime-time ads, and 2 60-second prime-time ads. How much will the adverting campaign cost?
8. For the Super Bowl one year, the rates for a 30-second commercial were $375,000. How much did it cost to run 4 60-second advertisements during that Super Bowl?
9. A local radio station, WXYZ, charges $20 for a 30-second ad and $37.50 for a 60-second ad. Your company wants to run 4 30-second ads and one 60-second ad every day for two weeks. How much will this radio advertising campaign cost for two weeks?

The *Daily Herald*'s weekday advertising costs are $35 per column-inch for 1 to 12 column inches and $32.50 per column-inch for 13-24 column inches. Saturday ads cost 5% more; Sunday ads cost 10% more.

10. You want to place a 14-inch, one column width ad in the *Daily Herald.* Draw the size of the advertisement. How much would the ad cost per week day?
11. A store ad in the Daily Herald is 3 columns wide and 6 inches deep. The ad first appears on Monday and runs for 10 consecutive days. How much does it cost to run the ad for those 10 days?

Use the rate card shown in Illustration 13-4.2.

12. What is the one-time cost of a $\frac{2}{3}$ page, 2C ad with a 1X frequency rate?

13. How much is saved on three $\frac{2}{3}$ page ads, if you agree in advance to a 3X frequency rate.

14. Make up an exercise using the rate card. Solve it. Then exchange your exercise with a classmate's exercise and solve each other's exercise.

PUT IT ALL TOGETHER

♦ Richard, an artist, has a successful part-time business doing charcoal-drawings of infants for $50 each. He wants to see if he can expand this business. After doing some research, he finds he can collect the names of 200 new parents each month from birth announcements in various newspapers in his area. He estimates that he will receive calls from 10% of the families he sends to and that ultimately 5% of everyone he mails to will order a drawing. Because this business is a sideline, Richard assumes his only major operating expense is advertising. His other operating costs for the business amount to $50 a month.

Richard decides to put together a year-long direct-mail advertising campaign. He spends $3,000 to create, copy, and package 2,400 copies of his direct mail package. Each advertising envelope weighs 0.2069 lbs. The third class charges are $0.124 per piece to mail plus $0.687 per total pounds. He also must pay an $80 postage fee to set up a bulk mail account.

15. What is the total cost of creating and mailing the package for the full year? What is the average monthly cost?

16. How many orders does Richard expect to receive from his campaign each month? in the year?

17. Richard's estimate of his expenses and the number of orders for charcoal drawings for the year was correct. Did he earn a profit on this business? Show gross revenue, total expenses, and net income or loss.

18. What is Richard's return on sales?

MIXED REVIEW

1. Multiply 1.2078 by 10; 100; 1,000

2. Divide 36.8 by 10; 100; 1,000

3. A firm employs 446 workers with a total weekly payroll of $178,500. How much FICA taxes must it pay for both its employees and itself if the rate is 7.65%?

4. Naperville's tax rate is $4.633 per $100 of assessed value. Find the tax to be paid on property assessed at $24,000.

5. You insure your $72,000 house for 90% of its value. At a rate of 65 cents per $100, what is your annual premium?

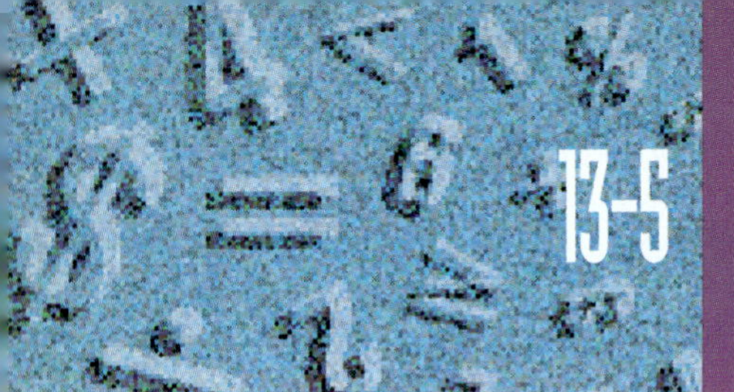

13-5 THE BALANCE SHEET

"But I don't understand why the bank won't lend me the money." complained Mona Shettles. "I always pay my bills as soon as I can!"

◆ Have you ever known someone in Mona's situation? Can you think of some reasons why Mona could not get a bank loan? Make a list of reasons. Compare your list with a classmate's list.

OBJECTIVES

In this lesson you will

- *learn to identify assets, liabilities, and capital or owner's equity;*
- *distinguish LIFO, FIFO, and weighted average inventory costing methods; and*
- *prepare and analyze a balance sheet.*

WARM UP

a. 2.5×18

b. $560 \times 1\frac{5}{8}$

c. $816.29 − $758.47

d. 14¢ × 250

e. Round 1,878.787 to the nearest ten.

f. Solve for N: $50 \div 10 = 6.25 \div N$.

THE BALANCE SHEET

A **balance sheet** shows a company's financial status as of a certain date. A balance sheet is prepared at least once a year. The balance sheet for Lynn's Books as of March 31, 19—, is shown in Illustration 13-5.1.

Notice that the balance sheet is divided into three categories: assets, liabilities, and capital.

Assets are all the things that have value and that are owned by a business. Lynn's assets include her merchandise inventory, her store equipment, and supplies. Money in the form of cash is also considered an asset. Assets may be classified as current or long lived. *Current assets* can be turned into cash or used within a year. Current assets include cash, inventory, and supplies. Long-lived assets, such as machinery, will be used for more than a year.

You were introduced to cost of goods, or merchandise, inventories in Lesson 13-1. An **inventory** is a list of items and their values. Generally, the value of an inventory is based on the cost of the item, not the selling price. The exact value of an inventory is found by taking a *physical inventory* or count of items in stock and multiplying that number by the cost to produce each item. This assumes that the cost of each item in inventory is known. When the exact cost is not known, a business may use one of these commonly accepted inventory cost systems: **first-in first-out (FIFO) method, last-in first-out method (LIFO),** or the **weighted-average method.** These methods are explained in llustration 13-5.2. The value of Lynn's merchandise inventory is $64,640.

Lynn's Books BALANCE SHEET March 31, 19--	
Assets	
Cash	$ 8,300
Merchandise inventory	64,640
Store equipment	16,000
Store supplies	2,100
Total assets	$91,040
Liabilities	
Hubbel Book Co.	$19,000
Acril Supply	11,000
Total liabilities	$30,000
Capital	
Lynn Gordan, Capital	$64,640
Total liabilities and capital	$91,040

Illustration 13-5.1. Balance Sheet for Lynn's Book Place

Liabilities A business often gets some of its assets by buying them on credit and promising to pay later. The persons to whom the money is owed are called the **creditors** of the business. The amounts owed to creditors are the **liabilities** of the business. Like assets, liabilities may be current or long lived. Lynn's liabilities are current. An example of a long-lived liability would be a mortgage.

Capital If all the assets of a business are owned free of debt, the owner's share of the business is equal to the total value of the assets. If there are liabilities, the value of the owner's share, called **capital** or *owner's equity*, is found by subtracting the liabilities from the assets. Since Lynn Gordon is the owner of the bookstore, the capital account is written in her name. The total capital of $60,000 is Lynn's claim against the assets of the business.

Assets − Liabilities = Capital

EXAMPLE Lynn Gordon owns a bookstore called Lynn's Books. The bookstore owes the Hubbel Book Co. $19,000 and Acril Supply $11,000 for merchandise Lynn has bought on credit. Lynn's Books has assets worth $91,040. Find Lynn's capital or owner's equity.

SOLUTION $19,000 + $11,000 = $30,000 total liabilities
$91,040 − $30,000 = $61,040 capital

Following are three methods of finding the value of an ending inventory. Use the inventory report for The Harp Store to see how each works.

The Harp Store Inventory Period: February 1, 1995 – January 31, 1996			
	Units	**Unit $**	**Total $**
February 1, Beginning Inventory	4	$40	$ 160
Purchases			
April 14	10	42	420
June 22	8	44	352
Sept. 6	16	45	720
Nov. 28	12	48	576
Total	50		$2,228
January 31, 1996 Ending Inventory	16		??????

- **First In, First Out (FIFO)** Assumes that goods or merchandise purchased first are used first and the value of the ending inventory is based on the cost of the most recently purchased items. For the chart shown above, the value of inventory based on the FIFO method would be calculated using the most recent purchase prices (November 28 and September 6)

  12 of the 16 units valued at $48 = $576
  4 of the 16 units valued at $45 = 180
  Ending Inventory Value: $756

- **Last In, First Out (LIFO)** Assumes that goods or merchandise purchased last are used first. The value of the ending inventory is based on the cost of goods purchased first. Using this method, the value of the ending inventory on January 31 would be based on the earliest purchase prices (February 1, April 14, and June 22):

  4 units at $40 $160
  10 units at 42 420
  2 units at 44 88
  Ending Inventory Value: $668.00

- **Weighted Average Method** Inventory is priced at the average price per unit of the beginning inventory plus the cost of all purchases during the fiscal year. The ending inventory value using this method is calculated as:

  $2,228 ÷ 50 = $44.56 weighted average per unit
  16 × $44.56 = $712.96 value of inventory on hand

Illustration 13-5.2. Inventory Costing Methods

Source: Century 21 Accounting, 5th edition. South Western Educational Publishing.

EXERCISE YOUR SKILLS

Prepare a balance sheet for each of these problems using Illustration 13-5.1 as a guide. State the amount of capital.

Vince Regis owns a garden shop with these assets: Cash, $2,100; Merchandise, $51,700; Store Equipment, $7,800; Store Supplies, $930. He owes the Grossel Co. $3,175 and the Tipaloy Supply Co. $1,850 for merchandise bought on credit.

1. What are Vince's total assets?
2. What are his total liabilities?
3. What is Vince's capital?

4. Samantha Merks owns a paint store. She has these assets: Cash, $3,850; Merchandise, $42,000; Store Supplies, $520; Store Equipment, $9,200; Delivery Truck, $8,400; Land and Building, $60,000. She owes the State Bank $8,600 and the Logan Manufacturing Company $23,800. What is her capital?

5. David Salamer has a ceramics business with these assets: Cash, $1,400; Accounts Receivable, $1,600; Merchandise, $23,500; Kilns, $3,700; Shelving, $2,100; Land and Buildings, $25,500. He owes Trek Pottery $7,500, Lane Supply Co. $3,300, and Long Lake Bank $9,100. What is David's capital?

6. Karen Carns owns a delivery company. She has 6 employees whose total annual wages are $94,000. On December 31 last year, her assets were: Cash, $2,620; Office Supplies, $2,875; Delivery Equipment, $29,250. On that date, she owed Rondell Office Supply, $580; Steel Oil Company, $2,700; Liberty Finance Co., $17,200; and the Hartley Repair Garage, $1,200. What was Karen's capital?

7. On December 31 of last year, Clyde Slade, a small engine repair shop owner, had these assets and liabilities. Calculate his assets, capital, and total liabilities.

Clyde's Repair Shop
Balance Sheet, December 31, 19—

Cash	$ 2,744	Debts owed to creditors:	
Inventory	40,316	On-Time Distributors	$ 6,230
Shop Supplies	1,650	Do-All Suppliers	2,970
Shop Equipment	18,500	Clyde Slade, Capital	
Total Assets		Total Liabilities and Capital	

8. Jill Grover's business had these assets on December 31, 19—: Cash, $3,842; Accounts Receivable, $4,915; Merchandise Inventory, $97,150; Store Equipment, $21,470; Delivery Equipment, $14,752; Supplies, $1,732. Her liabilities were a debt to the Dren Company of $44,655. What were her assets and capitals?

9. As of March 31, 19—, Bezard Brothers' had these assets: Cash, $15,450; Accounts Receivable, $4,550; Merchandise Inventory, $79,990; Store Equipment, $5,430; Delivery Equipment, $20,450; Supplies, $450. The company owed Roform Supply $5,960; Meldex Inc. $2,670; and Cole Co. $1,010. What were the Bezard Brother's assets, liabilities, capital, and total liability and capital?

10. Saturn Jewelry Suppliers had these assets on June 30, 19—: Cash, $17,643.10; Accounts Receivable, $32,524.65; Merchandise Inventory, $234,631.50; Supplies, $13,398.00; Equipment, $129,463.00, and building, 154,770.00. The company owed $325,000 to the Lern County Bank and $140,000 to New Creations, Inc. What was the company's assets, liability, and capital?

ANALYZING A BALANCE SHEET

As with an income statement, a balance sheet provides owners, bankers and creditors with critical information. Two key financial indicators from a balance sheet are:

Key Indicator	Formula	Explanation
Current Ratio	$\frac{\text{Current Assets}}{\text{Current Liabilities}}$	Indicates a company's ability to pay short term debts. Creditors use it to help determine if they should lend money to a business. Rule of thumb is this ratio should be about 2:1.
Debt to Net Worth	$\frac{\text{Total Liabilities}}{\text{Owner's Equity}}$	Shows amount of money owed to creditors relative to the equity value of a company. A high debt to net worth ratio indicates a company has a substantial amount of debt.

EXERCISE YOUR SKILLS

Use Lynn's balance sheet and income statement, with the formulas just reviewed, to show Lynn's Books :

11. Current ratio

12. Debt to Net Worth Ratio

MIXED REVIEW

1. 42,585 ÷ 36

2. 74% of 348 is ?

3. $6\frac{7}{8} - 2\frac{4}{20}$

4. \$14 increased by $\frac{1}{8}$ of itself is ?

5. Find the number of days from March 28 to June 3.

6. A computer that cost \$3,800 is expected to depreciate 20% each year using the declining balance method. Find the expected value at the end of the second year.

Net sales for The Exotic Bird Store were \$32,000. Operating expenses were \$5,685, and the cost of merchandise sold was \$15,000.

7. What percent of sales was the cost of merchandise?

8. What was the net income or loss for The Exotic Bird Store?

13-6 DISTRIBUTING BUSINESS INCOME

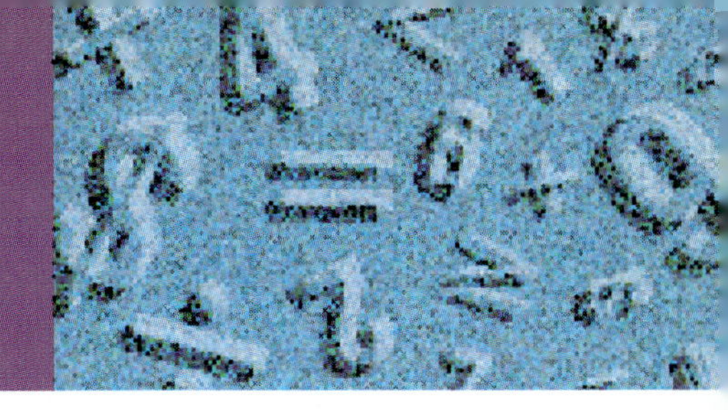

The balance sheet shows how much income a business has. How that income is distributed among owners is determined by the type of business and the agreements governing the business. In this lesson, we will look at how income is distributed in partnerships, corporations, and cooperatives.

◆ Suppose you have a fantastic idea for a new business but no money to get started. You need one or more partners to finance your business and help you operate it. Whom would you choose as a partner? Would you choose a relative or a close friend? How would decisions be made? How would net income be divided? Explain your reasoning for each decision.

OBJECTIVES

In this lesson, you will learn to

- *distribute net income among partners;*
- *find the rate of dividend paid on capital stock;*
- *calculate retained earnings; and*
- *calculate benefits of belonging to a cooperative.*

WARM *UP*

a. $240.60 × 36

b. 80,000 × $0.15

c. 3.25% × $60,000

d. 0.25 × 2.3

e. 9.6 kg equals how many grams?

DISTRIBUTING INCOME AND LOSS IN A PARTNERSHIP

A business owned by one person is called a **sole proprietorship.** A business owned by two or more persons is called a **partnership.** When you form a partnership, you usually sign a *partnership agreement.* This agreement often tells how much money the partners invest in the business and how they will share the net income or net loss. Some ways to distribute income or loss are as follows:

- equally between the partners.
- in proportion to the partners' investments.
- by paying interest to the partners on their investments.
- in a fixed ratio or percent
- by combining two or more of the above methods.

If no way is specified, income or loss is distributed equally.

Net Income Distributed in Proportion to Investments To find how much each partner receives when net income is shared in proportion to investments, show each partner's investment as a fractional part of the total investment. Then, multiply the total partnership net income by each partner's fraction.

EXAMPLE Altmon and Baines invest \$60,000 and \$30,000, respectively, in a partnership. They agree to share net income in proportion to their investments. At the end of the first year, the partnership earns a net income of \$63,000. Find each partner's share of the net income.

SOLUTION Total investment \$90,000

Altmon's investment \$60,000

Altmon's share $\frac{\$60,000}{\$90,000} = \frac{2}{3}$

Baines' investment 30,000

Baines' share $\frac{\$30,000}{\$90,000} = \frac{1}{3}$

Altmon's share $= \frac{2}{3} \times \$63,000 = \$42,000$

Baines' share $= \frac{1}{3} \times \$63,000 = \$21,000$

Check: \$42,000 + \$21,000 = \$63,000 total net income

Critical Thinking When the shares of the partners are written as ratios or as percents, what should the sum of the shares be?

Net Income Distributed by Fixed Ratio or Fixed Percent In some partnerships, net income is shared in a fixed ratio, such as 5 to 4. In this case, the net income is divided into nine equal parts with five parts going to one partner and four parts to the other partner. In other partnerships, each partner's share may be a certain percent of the net income, such as 55% to one partner and 45% to the other.

EXAMPLE Lawrence and Turrell, business partners, agree to divide a net income of \$45,000 in the ratio of 5 to 4. What is each partner's share?

SOLUTION $\frac{5}{9} \times \$45,000 = \$25,000$ Lawrence's share

$\frac{4}{9} \times \$45,000 = \$20,000$ Turrell's share

Critical Thinking In the example above, how were the fractions $\frac{4}{9}$ and $\frac{5}{9}$ arrived at?

Net Income Distributed Equally After Paying Interest on Investments When partners invest different amounts, the partners often are given interest on their investments. Then the rest of the net income is divided equally.

EXAMPLE Kirby and Dubois form a partnership, investing $70,000 and $60,000, respectively. They agree to give each partner 9% interest on the investment and to divide the remainder of the net income equally. The net income for the first year is $60,000. Find each partner's share of the net income.

SOLUTION

Kirby's interest (9% of $70,000)	$ 6,300
Dubois' interest (9% of $60,000)	5,400
Total interest	$11,700

Net income	$60,000
Interest	− 11,700
Remainder	$48,300

$48,300 ÷ 2 = $24,150 partner's share of remainder

	Kirby	Dubois
Interest	$ 6,300	$ 5,400
Remainder	24,150	24,150
Total share	$30,450	$29,550

Check: $30,450 + $29,550 = $60,000 net income

EXERCISE YOUR SKILLS

For Exercises 1–2, the partners distribute income in proportion to their investment. Find the net income each partner will get.

	Investments			
	Partner X	Partner Y	Partner Z	Net Income
1.	$6,000	$18,000		$8,400
2.	$18,000	$27,000	$36,000	$12,618

3. In the partnership of Dujovny and Ching, Dujovny's investment is $25,000 and Ching's is $30,000. Net income is divided in proportion to their investment. Their net loss for the first year is $8,965. What is each partner's share of the loss?

4. The investments of 3 partners are Baird, $12,000; Ehardt, $36,000; and Polley, $48,000. Each year the net income is distributed in proportion to the partners' investments. Last year, the firm's net sales were $480,800. The cost of merchandise sold was $312,800, and the operating expenses were $96,960. The partners estimate that the business could be sold for $250,000. How much of last year's net income did each partner get?

5. Daniels and Multon formed a partnership. Their agreement states that Daniels should receive 40% of any net income and Multon should get 60%. The net income for the first year was $82,500. Find each partner's total share of the net income for the first year.

6. In the Pace-Gibs partnership, Pace invested $35,000 and Gibs invested $42,000. They agreed to divide net income in the ratio of 7 to 4, with Pace receiving the larger share. Last year's net income was $147,400. Find each partner's share.

Find each partner's share of the net income in each problem. Interest is paid on the investments. The rest of the net income is divided equally.

	Investments			Interest on Investment	Net Income for Year
	Partner A	Partner B	Partner C		
7.	$50,000	$30,000		8%	$23,000
8.	$22,000	$37,000	$46,000	12%	$38,100
9.	$82,000	$58,000	$69,000	15%	$59,010

10. Healy and Myers are partners with investments of $50,000 and $75,000, respectively. The partners receive 9% annually on their investments. The rest of the net income is divided equally between them. Last year the firm made a net income of $68,428 and a gross profit on sales of $149,750. What is each partner's total share of this net income?

11. Dunn, Rice, and Hardaway formed a partnership. They invested $9,400, $14,000, and $18,500, respectively. For the first year, their gross profit was $82,370 and their expenses were $68,139. The partners got 7% on their investment, and the rest of the net income was shared equally. What was each partner's total share of the net income for the first year?

DISTRIBUTING CORPORATE AND COOPERATIVE INCOME

Starting a Corporation A **corporation** is a business owned by several people who legally act as one person under a *charter.* The charter is usually granted by the state government and tells what kinds of business the corporation can do. It also tells the par value and number of shares of stock that can be issued.

For example, suppose three people own sole proprietorships that produce exercise equipment. After meeting, they decide to put their companies together to form a corporation that will build and sell a broad line of exercise equipment. Because their companies have about the same value, each person will receive the same share of ownership in the corporation.

The three founders of the corporation obtain a charter from the state to incorporate as the Exerco Company. They then issue $750,000 of capital stock, which consists of 15,000 shares of $50 par value stock. **Capital stock** is the total value of the stock issued by a corporation. The capital stock is distributed to the founders in equal shares, each worth $250,000.

As soon as the corporation charter is issued by the state, the Exerco Company may begin doing business. The three original founders now become shareholders of the Exerco Company.

At the end of the first year, the Exerco Company issues a balance sheet, which is shown in Illustration 13-6.1.

Exerco Company BALANCE SHEET December 31, 19--	
Assets	
Cash	$109,500
Accounts receivable	92,800
Merchandise inventory	230,400
Supplies	8,850
Office equioment	29,800
Delivery equipment	125,650
Land and buildings	345,000
Total assets	$942,000
Liabilities	
Accounts payable	$147,000
Capital	
Capitol stock	$750,000
Retained earnings	45,000
Total liabilities and capital	$942,000

Illustration 13-6.1. Exerco Company Balance Sheet

Retained Earnings and Dividends Regularly, such as each month or year, the net income is figured and recorded in an account called **retained earnings.** In the balance sheet in Illustration 13-6.1, the retained earnings are shown in the capital section as $45,000.

The directors of the corporation then decide how much of the net income is to be distributed to the shareholders and how much is to be kept by the corporation for research, expansion, improvements, or for emergencies. The net income distributed to each shareholder is called a *dividend.* The **rate of dividend** may be shown as a percent or in dollars and cents.

Finding the Rate of Dividend To find the rate of dividend as a percent, divide the dividend by the value of the capital stock. To find the rate in dollars and cents per share, divide the dividend by the number of shares.

$$\frac{\textbf{Dividend}}{\textbf{Value of Capital Stock}} = \textbf{Rate of Dividend (as a percent of capital stock value)}$$

$$\frac{\textbf{Dividend}}{\textbf{Number of Shares}} = \textbf{Rate of Dividend (in dollars and cents per share)}$$

EXAMPLE The Exerco Company, with capital stock of $750,000 (15,000 shares at $50 par value per share) has a net income of $60,000 for the second year. What rate of dividend will be declared if all the net income is to be paid to the shareholders as dividends?

SOLUTIONS $60,000 ÷ $750,000 = 0.08 = 8%, rate as a percent

or

$60,000 ÷ 15,000 = $4, rate in dollars per share

When the capital stock of a corporation includes both common and preferred shares, first find dividends on preferred shares and subtract that amount from net income. The remainder is net income and may be paid as a dividend on common shares.

Finding the Total Dividend and the Amount Kept in the Retained Earnings Account Usually only a part of the net income is distributed to the shareholders. The rest is retained by the corporation.

EXAMPLE The Exerco Company, with capital stock of $750,000, has a net income of $120,000 for the third year of operation. The directors vote to pay a 4% dividend and to keep the rest of the net income in the retained earnings account. Find the total dividend and the amount kept in the retained earnings account.

SOLUTION 4% of $750,000 = $30,000, total dividends paid

$120,000	net income
− 30,000	total dividend
$ 90,000	retained earnings

A **cooperative** is similar to a corporation, except that its customers are usually the shareholders of the business. The net income is distributed among the shareholders in two parts:

- a dividend on the capital stock, which is distributed to the shareholders in proportion to the number of shares they own
- a patronage or customer dividend, which is distributed to the shareholders in proportion to how much they have bought from the cooperative

EXERCISE YOUR SKILLS

12. The Kasco Corporation, which has $900,000 of capital stock, has a net income of $108,000 for 1 year. What rate of dividend, shown as a percent, should be declared if all the net income is to be distributed to the shareholders?

13. A corporation that has 51,000 shares of capital stock paid an annual dividend of $122,400 after making a net profit of $193,000. What rate of dividend was paid in dollars and cents per share?

Maxtel Corp. has 8,000 shares of capital stock. The directors voted a dividend of $4,720. In cents per share, what is the

14. Estimated rate of dividend?

15. Actual rate of dividend?

The Ranswell Corporation has 30,000 shares of $100 par value stock outstanding. Last year the corporation made a net income of $280,000. The directors voted that 75% of the net income was to be paid as dividends.

16. What was the dividend rate per share as a percent?

17. What was the dividend amount in dollars per share?

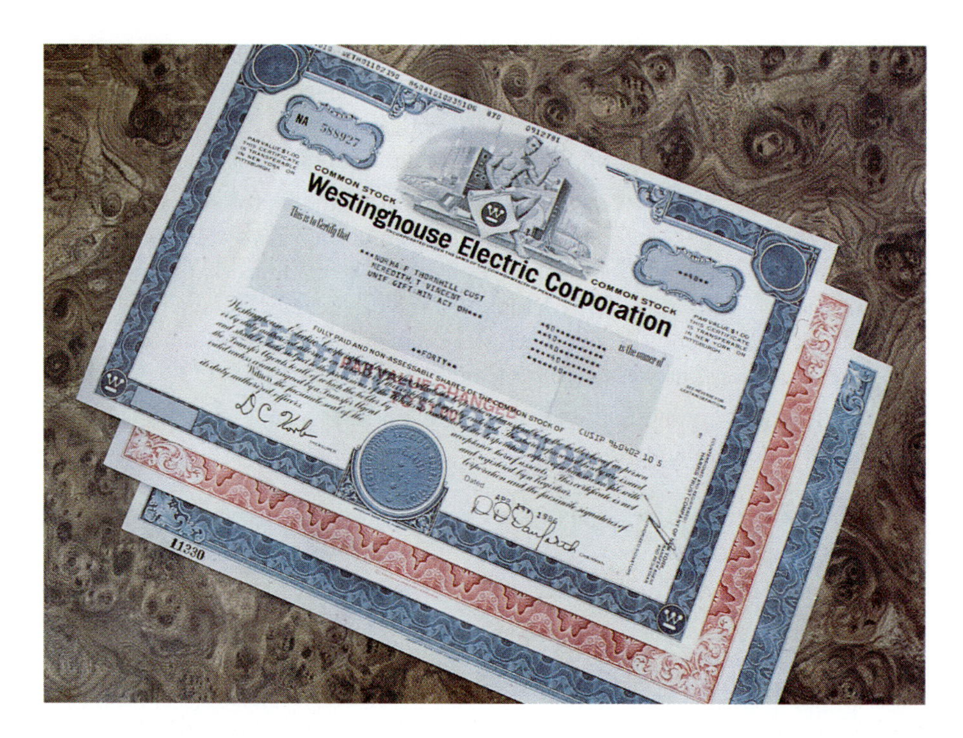

18. The FSK Company has $120,000 of preferred stock and $480,000 of common stock. The dividend rate on the preferred stock is 6%. The company's net income for 1 year is $50,400. What is the largest percent of dividend that can be paid on the common stock?

19. The Devlin Corporation issued 25,000 shares of $100 par value stock. In the current year, the corporation had a net income of $228,400 and declared a 6.75% dividend to shareholders. The rest was kept in the retained earnings account. What amount was kept as retained earnings?

20. A corporation has 175,000 shares of capital stock. It earned a net income of $325,000 during the year. During the year, 4 quarterly dividends of 8 cents a share were paid. The rest of the net income was kept in the retained earnings account. How much is kept in the retained earnings account?

21. A corporation has 5,000 shares of 6% preferred stock (par value $100 per share) and 70,000 shares of common stock ($10 par value per share). The corporation's net income for last year was $132,720. The directors declared a 6% dividend on the preferred stock and a dividend of 40¢ per share on the common. The net income left was kept in retained earnings. How much was kept in the retained earnings account?

22. Katarina Pentek is a member of a buying club organized as a cooperative. At the end of 1 year, the directors of the cooperative declare a 6% dividend on the capital stock and an 8% customer dividend on the sales to the store's customers. Katarina owns 20 shares of the stock, each with a par value of $25. She bought $1,250 from the cooperative during the year. What is the total dividend she should receive from the cooperative?

23. At the end of last year, the Farmers' Coop. declared a 5% dividend on the capital stock and an 8% patronage dividend. Mark owns 30 shares of the capital stock, each with a par value of $60. He bought $1,650 worth of merchandise from the cooperative last year. Find his total dividend.

24. The Penrod family owns 80 shares of stock in a cooperative (par value of each share, $15). They buy most of their groceries, lawn care products, and office supplies from the cooperative. On December 31, the cooperative declared a 9% dividend on the capital stock and a 2% patronage dividend. What is the total dividend the Penrod family should receive if their purchases from the cooperative totaled $8,470?

25. ◆ Write a paragraph that compares and contrasts business corporations and private cooperatives.

MIXED REVIEW

1. Round $107,800 to the nearest $1,000.
2. Find the product of $\frac{1}{3}$ and $588.
3. Estimate the quotient of $789 ÷ 0.82. Then find the quotient, rounded to the nearest cent.
4. The assessed value of property in Grand Heights is $19,520,000. To run the city, $947,227 is to be raised by property taxes. What tax rate per $100 is needed to the nearest tenth of a cent?
5. The basic annual premium on a house insured for $125,000 is $756. The policyholder wants additional personal property insurance of $14,000 to cover the video and sound recording equipment he uses as a hobby. That coverage will cost $1.35 per $100. What total premium will be paid by the owner?
6. The Washburn Corporation, with capital stock of $1,500,000, earned a net income of $206,974 last year. A dividend of 4.5% was paid to the shareholders and the remainder of the net income was put in retained earnings. How much was put in retained earnings?
7. Semtel Corporation has issued 5,250 shares of capital stock at a par value of $100. The corporation earned $84,000 in net income last year and declared all of it as a dividend. What was the rate of dividend, as a percent?

13-7 BANKRUPTCY

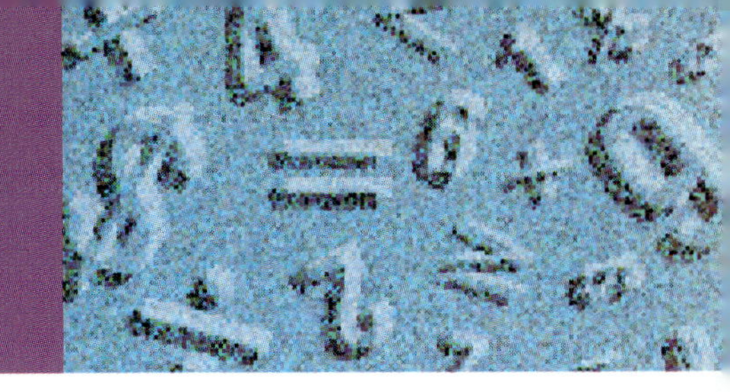

What do you think happens to people who become bankrupt? Do they ever have to repay their debts? If bankrupt persons don't have to repay their debts in full, is it fair to others who make all of their payments?

OBJECTIVES

In this lesson, you will learn to

- *calculate percent and amount of claims that can be paid under bankruptcy.*

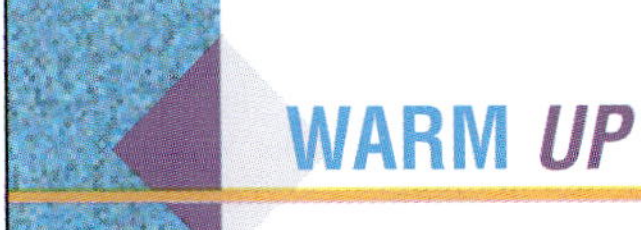

WARM UP

a. 15% of $120,000

b. $86.72 + $9.80 + $12

c. $900 × 8% × 2 years

d. Write 0.0643 as a percent.

e. What part of $500 is $25?

f. Find a 4.6% tax on $33,740 of income.

BANKRUPTCY

When a business keeps operating at a loss, the amounts it owes may become more than its assets are worth. When this happens, the business is *insolvent* and a court may declare it *bankrupt.* The court then appoints a trustee or receiver to sell all the assets and pay the debts.

Paying Debts After selling the assets, the trustee must pay the legal costs of the bankruptcy and any other **preferred claims** that the law says must be paid first. Then the money that is left is paid to the creditors in proportion to their claims. The percent to be paid to each creditor is found by dividing the total cash available for the creditors by the total of all creditors' claims.

$$\frac{\textbf{Cash Available for Creditors}}{\textbf{Total Creditors' Claims}} = \textbf{Percent Paid on Each Claim}$$

For example, if the total cash available for creditors is $5,000 and the total of creditors' claims is $20,000, each creditor will get 25% of the claim ($5,000 ÷ $20,000 = $\frac{1}{4}$ or 25%).

EXAMPLE A hobby shop was declared bankrupt by the court. The shop's assets were sold by the trustee for $12,000. Legal costs of bankruptcy and other preferred claims totaled $7,000. Creditors' claims totaled $20,000.

a. What percent of the creditors' claims can the trustee pay?

b. How many cents on the dollar will creditors get on their claims?

c. How much will a creditor get who has a claim of $1,200?

SOLUTIONS a.

$12,000	proceeds from sale of assets
− 7,000	bankruptcy costs and preferred claims
$ 5,000	available for creditors

$\frac{\$5{,}000}{\$20{,}000} = 0.25 = 25\%$ percent of claims that can be paid

b. $0.25 \times 100¢ = 25¢$

c. $0.25 \times \$1{,}200 = \$300$

EXERCISE YOUR SKILLS

Find what percent of the creditors' claims the trustee can pay.

	Total Creditors' Claims	Cash Available for Creditors	Percent of Creditors' Claims Paid
1.	$ 34,000	$ 13,600	
2.	18,600	5,580	
3.	48,900	17,115	
4.	192,000	124,800	

Find how many cents on the dollar can be paid to the creditors.

	Total Creditors' Claims	Cash Available for Creditors	Cents on the Dollar of Creditors' Claims Paid
5.	$38,000	$24,320	
6.	55,200	19,320	
7.	98,000	41,160	
8.	18,100	2,896	

9. A luggage store is declared bankrupt. Creditors' claims total $92,000. After the assets are sold and bankruptcy costs are paid, $34,040 is left for creditors' claims. What percent of their claims will the creditors get?

10. A bankrupt company has debts totaling $12,425. The cash available for the creditors is $5,715.50. How many cents on the dollar will creditors get?

The trustee for a bankrupt mobile home builder paid off the builder's debts at the rate of 49¢ on the dollar. A creditor filed a claim with the trustee for $6,994.

11. What estimated amount will the creditor get? actual amount?

12. The creditors of a bankrupt firm are paid at the rate of 32.6¢ on the dollar. There are 40 creditors with total claims of $34,572. How much will a creditor get on a claim of $1,280?

PUT IT ALL TOGETHER

The New-Day Bakery was declared bankrupt, and its assets were sold for $69,600. The bakery also had $1,250 in cash in the bank. Legal costs of bankruptcy and other preferred claims totaled $19,050. The total of creditors' claims was $148,000. The Helping Hand Company had a claim for $300. The Eastern Supply Company had a claim against New-Day of $9,200.

13. How much money was available for all creditors?
14. What percent of each creditor's claim was paid?
15. How much did the Helping Hand Company get?
16. How much did Eastern Supply get?

MIXED REVIEW

1. Multiply 23.75 by 14.7.
2. Divide 8.6 by 0.04
3. Estimate the product of 4.79 × 30.17.
4. Estimate the quotient of $9,764 ÷ 19.8.
5. Round $0.7925 to the nearest tenth of a cent.
6. Round $14,582,589 to the nearest hundred thousand.
7. A water bed company went bankrupt, owing $150,800. The cash available for the creditors was $57,000. How many cents on the dollar did the creditors get on their claims?

The monthly payment on a $65,000, 30-year mortgage is $623.50. Over the 30 years of the mortgage,

8. What will be the total of the mortgage payments?
9. What total amount of interest will be paid?
10. The monthly charge for leasing a small car is $129. For all miles driven over 16,000 miles in 1 year, a 7¢ per mile charge is made. What is the yearly cost of leasing the car if it is driven 21,000 miles in a year?
11. A rental company bought 150 plastic chairs at $12.80, 240 wooden chairs at $16.40, and 320 cushioned chairs at $23.70. What was the average price paid per chair, to the nearest cent?
12. Rochelle Springer worked 21 hours at time-and-a-half pay last week. Her regular-time pay rate was $9.80 an hour. What was Rochelle's total overtime pay for the week?

SPREADSHEET TEMPLATE: Completing a Payroll Register

Companies should keep good records of their payrolls. They must be able to report on income taxes withheld for their employees. Load ET13 from your Spreadsheet Applications diskette to complete the payroll register for Martin-Allerton, Inc. Enter the data shaded in blue into your template as shown below. When you do, formulas inside the spreadsheet will calculate the total earnings, FICA taxes, total deductions, and net pay for each worker, and the totals for the company.

	A	B	C	D	E	F	G	H	I	J	K	L
1			PAYROLL REGISTER SPREADSHEET									
2	Emp.	Emp.	No. of	Reg.	O.T.	Hourly	Total	FICA	Fed.			Net
3	No.	Name	Allow.	Hours	Hours	Rate	Earn.	Tax	Inc. Tax	Other	Total	Pay
4	1	A. Orr	1	40	0	5.80	232	17.75	27	5.25	50.00	182.00
5	2	B. Pike	0	35	0	5.15	180.25	13.79	18	5.25	37.04	143.21
6	3	C. Quick	2	40	3	8.50	378.25	28.94	38	9.75	76.69	301.56
7	4	D. Ruiz	3	40	2	9.40	404.20	30.92	37	12.80	80.72	323.48
8	5	E. Saga	1	40	0	7.20	288	22.03	34	8.50	64.53	223.47
9		Totals					1482.70	113.43	154	41.55	308.98	1173.72
10		FICA Tax						.0765				

Answer these questions about your completed spreadsheet

1. Who received the largest amount of net pay for the week?
2. Who had the least federal income taxes withheld for the week?
3. What was the total net pay for the week?
4. What was the total gross pay for the week?
5. What was the total amount of "other" deductions?
6. If every employee works 5 hours overtime, what is the firm's total net pay for the week?
7. If all employees work a $37\frac{1}{2}$ hour week, with no overtime, what is the firm's total net pay for the week?
8. What is the spreadsheet doing arithmetically in cell H5?
9. Describe, then write a formula, showing what the spreadsheet is doing arithmetically in cell G8.
10. Write formulas that can be used to calculate Total Deduc. and Net Pay.

Design Your Own

Imagine that you work for a small company. Your boss has just purchased a computer which has both spreadsheet and database software loaded on it. He asks you to put the daily cash payments onto the computer. Each month, he wants you to report on how much was spent on cost of merchandise sold and operating expenses. Operating expenses are to be itemized into the following categories: Salaries, Office Supplies, Advertising, Rent, Postage, and Other.

Work in teams of two or more. Determine what kind of business you are. Then make up a month's worth of expenses. You should have a minimum of 10 expenses with at least one expense in each of the categories. Then create a spreadsheet template and enter your data into it. Next create a database with appropriate field names and enter the same expense records.

Answer these four questions about the data using first the spreadsheet. Then use your database's query feature to answer the same questions. List any differences you observe in working with the different programs..

11. What was the total monthly cost of merchandise sold?

12. What were the total operating expenses for the business?

13. Which operating expense was the highest for the month? the lowest?

14. Show the total of each operating expense in ascending order.

CHAPTER 13 SUMMARY AND REVIEW

VOCABULARY REVIEW

balance sheet *13-5*
bankruptcy *13-7*
capital stock *13-6*
cooperative *13-6*
cost of merchandise sold *13-1*
current assets *13-5*
declining balance method *13-3*
depreciation *13-3*
factory costs *13-1*
income statement *13-1*
inventory *13-1*
liabilities *13-1*
LIFO inventory method *13-5*
merchandise turnover rate *13-1*
modified accelerated cost recovery system (MACRS) *13-3*
net sales *13-1*
operating expense *13-1*
partnership *13-6*
profit *13-1*
retained earnings *13-6*
return on sales *13-1*
sum-of-the-years-digits method *13-3*

MATH SKILLS REVIEW

1. $300 is what percent (to the nearest tenth) of $4,500?
2. Find $\frac{2}{3}$ of $10,000 (Round to the nearest cent.)
3. $\frac{6}{8} \times \$15{,}987$
4. $4,553 ÷ $63,637 = ? percent, to the nearest cent.

APPLICATIONS

13-1
- Cost of Merchandise Sold = (Beginning Inventory + Purchases) − Ending Inventory.
- The sum of the costs of raw materials, direct labor, and factory expense is the total factory cost.
- Net Income or Loss = Gross Profit − Operating Expenses.
- Use net sales as the whole to compute a percentage analyses.
- Merchandise Turnover Rate = Cost of Merchandise Sold ÷ Average Inventory.

5. In March, Dillon's Frame Shop had a gross profit of $3,589. Operating expenses were $2,100 for wages and $1,823 for all other expenses. What was the net income or loss for the month?
6. The Weslyn Lamp Company has net sales last year of $300,000. The cost of merchandise sold for the year was $156,000, and operating expenses were $105,000. Find the percent of net sales represented by cost of merchandise sold, operating expenses, and net income.
7. Brinson Cable had these merchandise inventories: $125,000 on June 1; and $155,000 on June 30. The cost of merchandise sold for the month was $350,000. Find the equivalent yearly merchandise turnover rate.

13-2 ◆ The way a factory expense is distributed varies with the company and kind of expense. Rent for example may be distributed in proportion to the floor space used.

8. The 3 departments of Broeker's Electronics use this floor space: Sales \$30,000 ft^2.; Repair/Service, 20,000 $feet^2$, Administration, 10,000 ft^2. The annual maintenance cost of the building is \$48,000. It is distributed on the basis of the floor space of each department. How much should each department be charged annually?

13-3 ◆ The declining-balance method uses fixed rate depreciation for each year is applied to a declining balance.

◆ The sum-of-the-years-digits method of calculating depreciation provides the greatest depreciation in the first year.

◆ The modified accelerated cost recovery system (MACRS) must be used to calculate depreciation for most property placed in service after 1986. To use the MACRS method, use a table for the appropriate class life of the property.

9. A machine that cost \$40,000 decreases in value each year at the rate of 10%. Using the declining-balance method, what is the book value of the machine at the end of the fourth year?

13-4 ◆ Shipping costs depend on how you send the shipment, how far it is sent, the size of the shipment as well as insurance, special handling, C.O.D., and delivery.

◆ Advertising on television, radio, and newspapers lets people know about your product or service. The cost of an ad is usually based on a combination of frequency, size and audience reach.

10. Station KSBI has these rates for 60-second advertisements: daytime shows, \$800; prime-time shows \$1,300. You plan an advertising campaign that uses 4 60-second daytime ads and 2 60-second prime-time ads. How much will the campaign cost?

13-5 ◆ Find capital by subtracting liabilities from assets.

◆ Prepare a balance sheet by listing all assets on the left side and the liabilities and capital on the right side.

11. The Cycle Shop has \$22,000 in a checking account and \$72,000 in other assets. It owes \$17,000 on a loan, \$8,000 to an attorney, and \$58,400 in other liabilities. Find The Cycle Shop's capital.

13-6 ◆ To distribute net income (a) in proportion to partner's investment, find the part that each partner's investment represents, then multiply net income by the part. (b) by paying interest on investment, multiply the amount of each partner's investment by the rate. (c) in a fixed ratio, multiply the net income by the part or percent share each partner is to receive.

12. Crowe, Juliano, and Stevens are partners with investments of $26,000, $39,000, and $65,000, respectively. The partners have agreed that profits are to be shared in proportion to their investments. In 1 year, the firm had net income of $80,100. How much did each partner receive as a share of the net income?

- Find the total dividend paid by a corporation by multiplying the value of the capital stock by the percent dividend rate or by multiplying the number of shares by the dollar dividend rate.
- Calculate the amount kept in retained earnings by subtracting the total dividend paid from net income.

13. A corporation with capital stock of 120,000 shares at $75 par value had a net income of $360,000 for last year. The directors declared a dividend of $1.18 a share and kept the rest of the income in the retained earnings account. Find the amount paid as dividends and the amount kept in the retained earnings account.

13-7 ♦ Find the percent paid on creditors' claims by dividing the cash available for creditors by the total creditors' claims.

14. Wilkins Sporting Goods went bankrupt and had $14,400 in cash available to pay creditors. The company's debts were $80,000. How many cents on the dollar will be paid to creditors?

15. Under a bankruptcy agreement, the firm of Smith and Haynes must pay its creditors $0.57 on each dollar that is due them. How much would a creditor receive on a claim of $3,600?

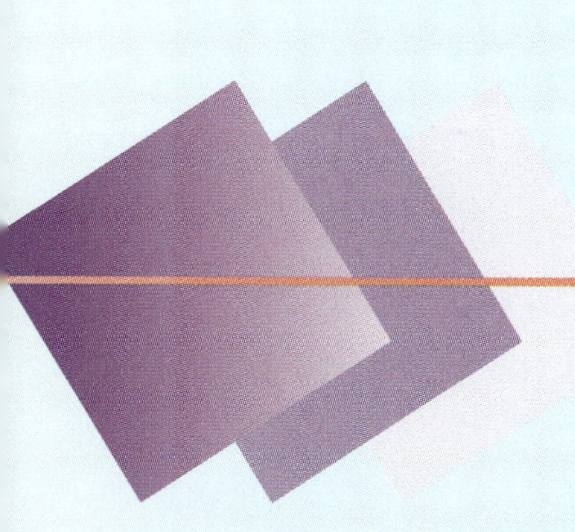

WORKPLACE KNOWHOW: **REASONING**

Reasoning skills include the ability to apply rules and principles to a new situation.

Assignment In this chapter, we have discussed profit and loss in business. Discuss how the principles of personal financial management and budgeting you learned earlier in the year relate to the principles of financial management and budgeting for a business. Discuss your ideas with your class.

Work Skills Portfolio You have applied percentages and ratios to business situations and learned about the costs of doing business and business analysis. Include an assessment of your skills in this area. If you have ever run or considered owning your business, you may want to include a statement about how you would handle the finances of your business. Also include an assessment of your reasoning skills.

If you completed the integrated project in your student notebook, you may want to include a copy of the worksheets in your portfolio.

CHAPTER 13 TEST

Math Skills

1. $132,000 − $97,000
2. $1,400 + $36,000 − $7,000
3. $1,500 − $60,000
4. $29,000 − $580,000
5. $463,180 − 2
6. 0.034 × $5,000,000
7. 12 × 0.863
8. $40,700 − $185,000
9. $\frac{1}{4}$ × $132,000
10. 15% × $850,000

Applications

11. A store owner has these assets: Cash, $4,760; Merchandise Inventory, $78,200; Equipment, $25,600; Supplies, $1,240. One bill of $15,070 and another bill of $7,213 are owed. How much is the owner's capital?
12. A store had net sales of $36,508.72 last month. If the cost of goods sold was $23,810.14 and the operating expenses were $7,834.28, what was the net income?
13. Aileen Uzin runs a retail store. Her net sales for the year were $162,000. The cost of the merchandise that she sold was $118,500. Her operating expenses were $28,920. Her net income was what percent of the net sales for the year?
14. Last year, a store's cost of merchandise sold was $366,600 and the store's average merchandise inventory was $15,600. What was the annual merchandise turnover rate?
15. The *Afton Star* charges $35 per column-inch for advertising. You want to place an ad that is 3 columns wide and 5 inches deep. Find the total cost for a 5-day advertising campaign.
16. Larkin and Mazern are partners. Their investments are $50,000 and $30,000, respectively. They share income in proportion to their investments. The net income last year was $53,600. What amount did Larkin receive?
17. Jan Ruell invested $32,000 in a business. She was to receive 5% on her original investment and share equally in any remaining net income. Her share of the remaining net income last year was $10,130. What were her total earnings for the year?
18. Farwell Corp. had capital stock consisting of 300 shares of preferred stock paying 6% dividends, and 2,000 shares of common stock. The par value of each type of stock is $100 per share. A dividend of $4,000 is to be paid to the shareholders. After the dividends have been paid to the preferred shareholders, how much will be paid as the dividend on each share of common stock?
19. A bankrupt company has $28,350 in cash available for creditors. The debts are $90,000. How many cents on the dollar will be paid to creditors?

CHAPTER 14

COMPUTER OPERATIONS

Companies that use computer systems to run their businesses must make sure their data is protected and their systems run efficiently. If they don't, they could lose millions of dollars or even go out of business because of lost information and work time.

A systems administrator is responsible for planning the procedures that keep the system running efficiently and safely. This includes job schedules, back-up procedures, and disaster recovery plans. Operators are responsible for executing an administrator's plans. Operators' work is task oriented and procedurally driven.

With computing environments changing rapidly, the distinction between operations and system administration often blurs. In some companies, the operator and the system administrator may even be the same person.

The primary responsibility of system administrators and operators is to manage and allocate the hardware and software resources of the system so that everyone's work gets done as efficiently as possible. The ability to think in integer mathematics is an important job skill because machines operate on integer (whole number) arithmetic.

Operators must also work well under pressure. When a computer system "crashes," an operator must think quickly and remain calm to identify the problem and get the system running again while following established procedures. An operator must also be able to recognize when a problem is outside his or her expertise and call in system programmers and administrators for support.

An operator's priority is to keep the system running, often 24 hours a day. When there is a problem or someone can't make it in to work his or her shift, operators must be willing to work past their regular work

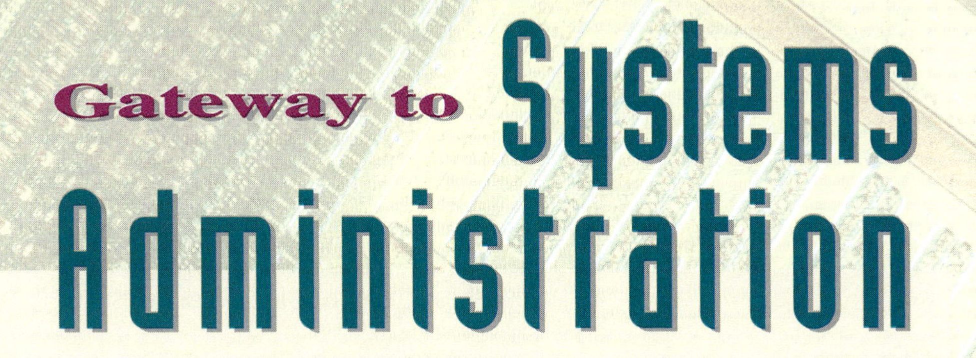

Gateway to Systems Administration

Elena, an operator at a large company, is making the transition from operator to system administrator. After high school, she learned to be an operator working for a company that installed computer systems. On-the-job experience and self-study helped her earn her present position.

For newcomers to today's job market, Elena recommends completing a two-year technical school program to learn the basics of software and computer structure. She believes the more an operator learns about software and hardware, the better his or her chances are for advancement. Elena's supervisor, Bill, agrees. He says that in addition to technical qualifications, he looks for people who have a "desire to learn new things."

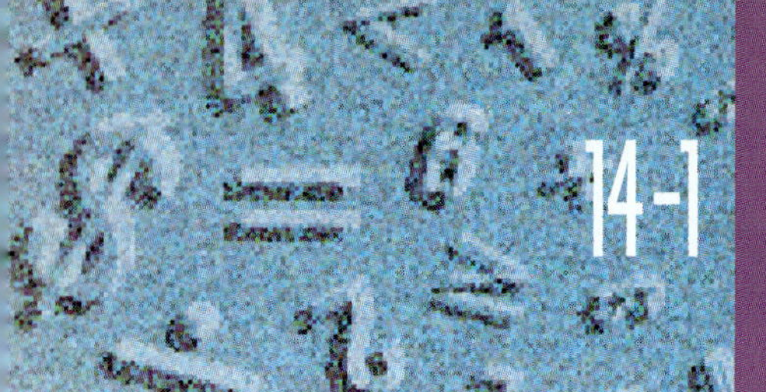

14-1 PURCHASING FOR A BUSINESS

OBJECTIVES

In this lesson, you will learn

- *the terminology of purchase invoices;*
- *to calculate invoice due dates and total amounts;*
- *to calculate cash and single trade discounts;*
- *to calculate shipping charges;*
- *to find the total due on an invoice; and*
- *calculate net purchases for a given time period.*

In business there are suppliers and consumers. While we think of ourselves as consumers, businesses are also consumers of raw materials and products.

◆ Name 3 things that you consider when you buy a product and rank them in order of importance. Next list 3 things that you think a business considers and rank them in order of importance. What are the similarities and differences?

WARM UP

a. Subtract 276.28 from 304.7

b. Divide 56.8 by 0.04.

c. Add 74.08, 531.7, 2.008, 52.76

d. $\$1.40 \times 560 = ?$

e. $24 \text{ m} = ? \text{ km}$

f. Estimate the quotient: $318 \div 37$

PURCHASING TERMINOLOGY

You learned in Chapter 13 that the cost of merchandise or goods sold is a major expense for most businesses. Most business-to-business transactions begin with a **purchase order** from the buyer which identifies what they want to order, the agreed upon price, and delivery terms. The purchase order may be written or verbal. After the order is filled, the seller issues a **purchase invoice.** Both documents include all the details of the purchase. When a purchase invoice is received, it is a good business practice to verify that the order is correct and that all charges are accurate. Check marks are often used to show that calculations have been checked for accuracy.

Illustration 14-1.1 shows a typical business invoice for Lagmeyer Hardware from RADEL Adhesives. In the first part of this lesson, you will learn purchasing terms from the Lagmeyer invoice. Then, you will see how the mathematics skills you have already learned can be used to calculate invoice due dates and amounts.

Remit payment to: Original invoice ***#OJL610***

RADEL ADHESIVES COMPANY
1631 Highway 55
Minneapolis, MN 55422-6083

Refer to:

Order Number	Date Entered	Date Shipped	Invoice Date	Terms of Sale
95231	6/7/--	6/8/--	6/10/--	2/10, 1/30, n/60

Shipping *FOB Minneapolis* Ship to: *Same address*
Sold to: *Lagmeyer's Hardware*
1575 Arrowhead Road
Fargo, ND 58103-2537

Quantity	Unit	Description	Unit Price	Amount
72	*Rolls*	*Transparent tape*	*$1.10*	*$79.20*
48	*Bottles*	*Rubber cement*	*1.25*	*60.00*
36	*Rolls*	*Masking tape*	*1.50*	*54.00*
24	*Cans*	*Wood glue*	*1.75*	*42.00*
		Total		**$235.20**
		Less 5% Trade Discount		**$ 11.76**
		Net Invoice		**233.44**
		Shipping		**6.00**
		Total Due		**$ 229.44**

PLEASE PAY ON INVOICE U.S. FUNDS ONLY
NO STATEMENT WILL BE SENT UNLESS REQUESTED
If paid by
6/20 Deduct $4.67
7/10 Deduct $2.33

Illustration 14-1.1. Purchase Invoice

While the placement of items on an invoice may vary, the information shown on the invoice is typical. Notice that the top portion of the invoice includes information about when the order was made.

The top portion of the bill also identifies the terms of the sale. **Terms of sale** specify how and when an invoice will be paid. Usually businesses sell to other businesses **on account.** This means the customer will be billed later for purchases. The time a purchaser has to pay a bill, usually 30 to 90 days, is called the *credit period.*

Businesses frequently offer **cash discounts** as a way to encourage their business customers who pay their invoices early. Cash discount terms are included in the terms of sale. For example, the terms of Jared's invoice are **2/10, 1/30, n/60.** That means Jared may deduct 2% from the invoice price if he pays it within 10 days. He can deduct 1% if he pays within 30 days of the date of invoice. The full amount is due within 60 days. Sometimes a term such as 3/10 EOM is used. This means that the buyer can claim a 3% discount if the bill is paid within 10 days after the **end of month** shown on the invoice.

The main portion of the invoice contains detailed information on the order. This includes the price of each item and the total amount due. Notice that both a unit price and an extended price are shown. The extended prices are totaled to find the invoice subtotal. Jared receives an $11.76 (5%) **trade discount.** Many businesses offer trade discounts. These are reductions in the list price to their business customers. The **invoice price**, also called the *net price* or *invoice cost,* is found by subtracting the trade discount from the total extended price on an invoice.

Critical Thinking Why do you think businesses offer other businesses trade discounts?

The next item on the invoice is shipping costs. Depending on the terms of the agreement, shipping charges may be paid by the buyer or the seller. This invoice tells us that the items are being shipped by freight, F.O.B. Minneapolis (the shipping point). **F.O.B.** is an acronym that stands for free on board. When the shipping point appears after F.O.B., the buyer (receiver of the goods) pays shipping. F.O.B. factory also means that the buyer pays all costs of shipping. When the destination point is shown after the acronym, the seller will pay the shipping charges. Jared pays the $6 shipping charges. Shipping charges are added to the invoice price to find the **net total** due.

Critical Thinking Based on what you see in the invoice and in this lesson, write formulas for calculating a trade discount and an invoice total.

CALCULATING INVOICE TERMS AND TOTALS

You learned how to find due dates for promissory notes and price extensions and discounts on sales slips. Now you will use these skills to check purchase invoice due dates and amounts.

Finding the Due Date of an Invoice The due date for receiving a discount depends upon the terms. This is found by counting ahead, from the date on the invoice, the number of days shown in the terms. For example, the invoice in Illustration 14-1.1 is dated June 10, and the terms are 2/10, 1/30, n/60.

- To earn a 2% cash discount, the invoice must be paid within 10 days of the invoice date.

  Invoice Date   June 10 + 10 days = June 20

- To earn a 1% discount, the invoice must be paid within 30 days of the invoice date.

30	days to due date to earn a 1% discount
− 20	days left in June
10	pay by July 10 to earn 1% discount

- To pay the bill on time, without being charged interest, you must pay within 60 days of invoice date.

60	**days to due date of invoice**
− 20	**days left in June**
40	
− 31	**days in July**
9	**days in August (August 9, due date)**

Finding Invoice Extensions Using Combinations of Fractional Equivalents To find the extended price of any item on an invoice, multiply the quantity times the unit price. The unit price of an item is sometimes a combination of the base and a fractional equivalent. When this is the case, you can use combinations of fractional equivalents to calculate invoice extensions quickly.

EXAMPLE Find the cost of 72 rolls of transparent tape at $1.10.

SOLUTION	EXPLANATION
$\$1.10 = \$1 + \frac{1}{10}$ of $1	• Find the base ($1) and fractional equivalent $\left(\frac{1}{10} \text{ of } \$1\right)$.
$72 \times \$1 = \$72$	• Multiply the number of items times the base.
$72 \times \frac{1}{10} = 7.20$, or $7.20	• Multiply the number of items times the fractional equivalent and add a dollar sign.
$\$72.00 + \$7.20 = \$79.20$	• Add the products.

You may want to memorize these commonly used equivalents.

$\$1.10 = \$1 + \frac{1}{10}$ of $1	$\$1.12\frac{1}{2} = \$1 + \frac{1}{8}$ of $1
$\$1.20 = \$1 + \frac{1}{5}$ of $1	$\$1.25 = \$1 + \frac{1}{4}$ of $1
$\$1.50 = \$1 + \frac{1}{2}$ of $1	$\$2.50 = \$2 + \frac{1}{2}$ of $1

Practice this technique by calculating the extended amounts for each item shown on Jared Lagmeyer's invoice. Check your answers against the extended prices shown on his invoice.

Calculating Trade Discounts Trade discounts are always based on the list price of the items being purchased. To find the trade discount, simply multiply the list price times the percent (rate of discount). To find the net or invoice price, subtract the discount from the total.

Rate of Discount × List Price = Trade Discount
List Price − Trade Discount = Invoice Price

EXAMPLE The list price of an outdoor grill in the catalog of the Cook Rite Company is $200. The discount given to the retailer is 40%. What invoice price will a retailer pay for this grill?

SOLUTION

List price	$200	
Trade Discount	− 80	(40% of $200)
Invoice Price	$120	

CHECK

	100% of list price	= List price
Subtract	40% of list price	= Trade discount
	60% of list price	= Invoice price

Shipping Charges Shipping charges are usually quoted on an invoice. They are sometimes referred to as "shipping and handling charges" or freight charges. Shipping charges are added to the invoice price. Any additional charges, such as insurance, are also added to the net invoice price to find the total amount due.

Finding Cash Price If a cash discount is offered and a customer pays an invoice within the discount period, the amount paid is called the **cash price.** The cash price is the invoice price plus shipping and other costs minus the cash discount. Cash discounts are not allowed on shipping charges. On Jared's bill RADEL specified the amount that could be deducted from the total bill when a cash discount was earned. It is the responsibility of the buyer to take the cash discount.

EXAMPLE An invoice with a net invoice amount of $2,400 is paid within the discount period. Credit terms are 2/30. The invoice is paid within the discount period. Find the amount to be paid.

SOLUTION

$0.02 \times \$2,400 = \$48$	Cash discount
$\$2,400 - \$48 = \$2,352$	Cash price, or amount paid

Finding the Rate of a Trade or Cash Discount

A rate is a comparison between two numbers. The rate of trade discount is always based on the list price. The rate is found by dividing the amount of the trade discount by the list price.

$$\textbf{Rate of Trade Discount} = \frac{\textbf{Trade Discount}}{\textbf{List Price}}$$

EXAMPLE: A power saw is listed in a wholesaler's catalog at $80. The saw is sold to retailers at an invoice price of $56. What is the rate of trade discount?

SOLUTION:

$\$80 - \$56 = \$24$	Amount of trade discount
$\$24 \div \$80 = 0.3$ or 30%	Rate of trade discount

The rate of cash discount is always based on the invoice price. The rate is found by dividing the amount of cash discount by the invoice price.

$$\textbf{Rate of Cash Discount} = \frac{\textbf{Cash Discount}}{\textbf{Invoice Price}}$$

EXAMPLE By taking advantage of a cash discount, a buyer paid $1,261 to settle a $1,300 invoice. Find the amount of the cash discount and the rate of cash discount.

SOLUTION

$1,300	Invoice price
− 1,261	Cash Price
$ 39	Discount amount

$39 ÷ $1,300 = 3% Rate of cash discount

EXERCISE YOUR SKILLS

1. What does terms of sale mean?
2. If an item mailed from The Horse Farm in Kentucky to the Big Sky Ranch in Colorado is sent F.O.B. CO, who pays the shipping charges?
3. Are cash discounts taken on shipping charges?
4. What is the difference between a trade discount and a cash discount?

Find the date on which each invoice must be paid.

	Invoice Date	Terms		Invoice Date	Terms
5.	August 16	10 days	6.	April 24	60 days
7.	March 5	30 days	8.	October 28	75 days
9.	November 14	90 days	10.	January 7	45 days

Mental Math Give each answer. Explain how you got each answer.

11. 13 @ $1.10
12. 28 @ $1.25
13. 22 @ $2.50
14. 41 @ $1.10
15. 24 @ $$1.12\frac{1}{2}$
16. 14 @ $1.50

The Leonard Supermarket bought the following items. The invoice was dated August 16; the terms were net, 45 days.

50 ice cube trays @ $1.67
48 food storage boxes @ $2.25
115 boxes trash bags @ $2.78
50 pair household gloves @ $2.45
18 measuring cups @ $1.93

17. Create an invoice and make the extensions.
18. Find the total amount of the invoice.
19. Find the due date.
20. How much would a retailer pay for 30 dozen paint brushes if the wholesale price is $54 a dozen, less 28%?

Mae Zurn, a store owner, gets prices on rolltop desks from two wholesale firms. The Rinz Company offers a desk for $650, less 30%. The Okemar Company offers the same desk for $800, less 40%.

21. Which wholesale firm has the lower price? How much is lower?

22. By paying $855 in cash, a retailer saves $45 on an invoice of oil filters. What is the rate of cash discount?

23. A catalog lists a tool chest at $96. The same tool chest is billed to the retailer at $64. Find the rate of discount.

24. A freezer listed at $450 in a wholesaler's catalog is sold to a retailer for $270. Find the rate of trade discount.

PURCHASES, RETURNS AND ALLOWANCES, NET PURCHASES

In addition to verifying the accuracy of their invoices, businesses must keep track of the total cost of all the goods they buy, called **purchases.** Purchases are tracked for a specified time period, such as by the week, month, or year.

A retailer may return some of the goods to the seller. Or, the seller may give the retailer an allowance because some of the goods were damaged. These returns and allowances decrease the cost of the purchases and should be deducted from the purchases. The amount left is the **net purchases.**

Purchases − Returns and Allowances = Net Purchases

EXERCISE YOUR SKILLS

For August, the records of The Wood Place showed the following purchases and returns. Copy and complete the table below.

	Items	Purchases	Returns	Net Purchases
25.	Shutters	$ 4,418	$135	
26.	Cabinets	12,876	562	
27.	Towel racks	1,317	35	
28.	Picture frames	2,089	172	
29.	Totals			

The table shows the purchases and returns of the Outdoor Shop. Copy and complete the table by adding a column for Net Purchases.

	Month	Purchases	Returns	Net Purchases
30.	January	$ 14,618	$ 316	
31.	February	17,830	387	
32.	March	38,512	571	
33.	April	46,984	1,309	
34.	May	53,011	1,470	
35.	June	38,737	698	
36.	Totals			

37. ♦ How can you check to be sure the net purchases amount is accurate?

MIXED REVIEW

1. Find 180% of $15.50
2. Multiply: 14.7 by 0.89
3. $1.00 is what part smaller than $2.50?
4. Find the cost of 48 items at $1.25.
5. Lexin, Inc. bought goods for $3,600, less 20% trade discount with terms of 2/10, n/30. The invoice was dated April 9 and was paid on April 19. What was the amount of cash paid?

A $200,000 term life insurance policy is sold at a rate of $1.07 per $1,000 of insurance.

6. What is the estimated annual premium?
7. What is the actual annual premium?

A survey found that six families had the following annual incomes: $40,000; $35,000, $58,000; $42,000, $40,000; $67,000. For this data, find the:

8. range
9. mean
10. median
11. mode

Jennifer Salems opens a sports memorabilia store. She kept track of the value of one particular card.

Year	1988	1989	1990	1991	1992	1993	1994
Value	$0.25	$1.50	$3.00	$9.00	$6.00	$7.50	$10.00

12. Make a line graph to show the changes in the card's value.

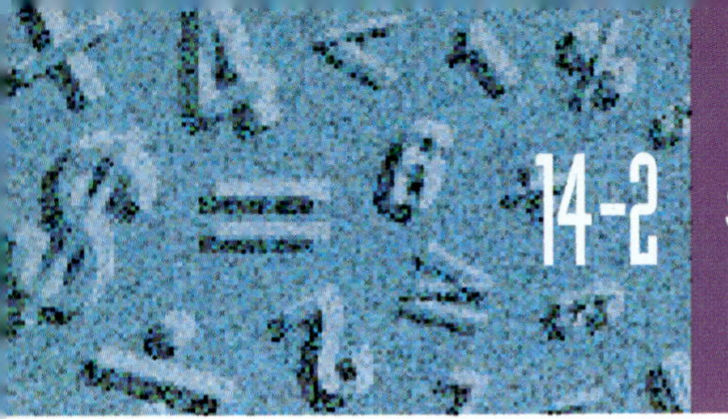

14-2 SERIES DISCOUNTS

SAVE AT BEST PRICE STORES
SAVAMART
Discount St
We will not be underpriced.

OBJECTIVES

In this lesson, you learn to:

- *calculate series discounts;*
- *calculate single discount equivalents; and*
- *find the rate of discount.*

◆ Have you ever seen advertised or been offered a product at a price less than the suggested retail price printed on the package? How can a retailer afford to sell for less than the suggested price? Make a list of reasons why retailers sell items at discount. Do you think they are losing money on those sales?

WARM *UP*

Solve for N

a. $N \times 0.6 = 4.8$

b. $186 \div N = 0.6$

c. $1\frac{1}{4} \times 180 = N$

d. $N - \$14.35 = \$100$

e. Multiply each by 0.01: \$16, \$96.80; \$821.

f. Divide each by 10: \$46.82; \$28; \$9,000

SERIES DISCOUNTS

In the previous lesson, you saw that businesses frequently offer business customers **trade discounts.** Trade discounts may be used to encourage retailers to buy in large quantities. For example, a discount of 3% may be given if a retailer buys 100 items; 5% if they buy 200, 8% if they buy 300; and so on. These are known as **quantity discounts.** Trade discounts are also common when items are purchased from a catalog. The price in the catalog is called the **list price.** You may recall from the last lesson that the list price is the price for consumers. The price the retailer pays is called the **invoice price, net price,** or **invoice cost.**

A retailer may be given a trade discount that has two or more discounts, called a **discount series** or a series of discounts.

To find the invoice price, the first discount is based on the list price. The second discount is based on the remainder after deducting the first discount. The third discount is based on the remainder after deducting the second discount, and so on.

EXAMPLE The quoted price of a furnace is $1,050, less 25%, 10%, and 5%.

SOLUTION

a.

$1,050.00	list price
− 262.50	first discount (25% or $\frac{1}{4}$ of $1,050)
$ 787.50	first remainder
− 78.75	second remainder (10% or $\frac{1}{10}$ of $787.50)
$ 708.75	second remainder
− 35.4375	third discount (5% or $\frac{1}{20}$ of $708.75)
$ 673.3125	third remainder (invoice price $673.31 rounded)

b.

$1,050.00	list price
− 673.31	invoice price
$ 376.69	trade discount

Note: Amounts are rounded to the nearest cent ONLY after all computations are done.

Critical Thinking Will the order in which you take the discounts change the final result?

EXERCISE YOUR SKILLS

Find the invoice price and the amount of trade discount. *Check your work by taking the discounts in a different order.*

	List Price	Trade Discounts	Invoice Price	Amount of Trade Discount
1.	$ 90	20%, 10%		
2.	125	20%, 10%, 10%		
3.	145	30%, 15%, 10%		

4. What is the invoice price of 6 cabinets that list at $130 each, with discounts of 30%, 20%, and 10%?
5. Eswell Manufacturing sells a ceiling fan for $185, with discounts of 20%, $12\frac{1}{2}$%, and 5%. Find the invoice price.
6. Morland Brothers offers to deliver an order of goods for $1,500 list price, less 25%, 20%, and 10%. Wick Wholesale offers the same goods at the same list price less 30%, 20%, and 5%. How much would be saved by taking the lower offer?

FINDING SINGLE DISCOUNT EQUIVALENT FOR A DISCOUNT SERIES

If you buy regularly from one vendor and always receive the same discount series, you can calculate the invoice price faster by using one discount that is equal to the series of discounts. That one discount is called the **single discount equivalent.** We will review three methods to find the single discount equivalent for a pair of running shoes that have a list price of \$80. They are offered to retailers with a series discount of 20%, 10%, and 10%.

METHOD 1 Use Percents to Find the Single Discount Equivalent for a Series Discount of 20%, 10%, and 10%.

Step 1

100%	list price
− 20%	first discount (20%, or $\frac{1}{5}$ of 100%)
80%	first remainder
− 8%	second discount (10%, or $\frac{1}{10}$ of 80%)
72%	second remainder
− 7.2%	third discount (10% or $\frac{1}{10}$ of 72%)
64.8%	third remainder, or invoice price

\$80 × 64.8% = \$51.84 invoice price.

Step 2

100%	list price
− 64.8%	invoice price
35.2%	single discount equivalent

METHOD 2 Use Tables to Find the Single Discount Equivalent for a series discount of 20%, 10%, and 10%.

The table below shows invoice price equivalents after discounts are taken.

Rate	5%	10%	15%	20%	25%	30%
5%	0.9025	0.855	0.8075	0.76	0.7125	0.665
5%, 5%	0.85738	0.81225	0.76713	0.722	0.67688	0.63175
10%	0.855	0.81	0.765	0.72	0.675	0.63
10%, 5%	0.81225	0.7695	0.72675	0.684	0.64125	0.5985
10%, 10%	0.7695	0.729	0.6885	0.648	0.6075	0.567

Illustration 14-2.1. Invoice Price Equivalents

Step 1: Find the 10%, 10% line under the 20% column. The invoice price equivalent is 0.648 or 64.8%.

Step 2: Multiply the list price, \$80, by 0.648. The product, \$51.84, is the invoice price of the running shoes.

Step 3: Check: 100% − 64.8% = 35.2% (single discount equivalent)

METHOD 3 Use Complements to Find the Single Discount Equivalent for a Series Discount of 20%, 10%, and 10%.

The **complement** of any discount rate is the difference between that rate and 100%. For example, if the discount rate is 30%, the complement of the discount rate is 70%.

Step 1: Multiply the complements of the discounts.
$0.80 \times 0.90 \times 0.90 = 0.648 = 64.8\%$

Step 2: *Subtract* the product from 100%.
$100\% - 64.8\% = 35.2\%$

EXPLANATION

35.2% is the single discount equivalent to 20%, 10%, 10%. If 35.2% is the rate of discount, then 64.8% must be the percent of invoice price.

CHECK: $64.8\% \times \$80 = \$51.84$ invoice price.

EXERCISE YOUR SKILLS

Use any method you wish to find the single discount equivalent for each. Use each method at least once. Show your work.

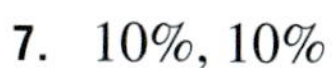

7. 10%, 10%
8. 25%, 20%
9. 20%, $12\frac{1}{2}\%$
10. 10%, $33\frac{1}{3}\%$
11. 25%, 20%, 5%
12. 30%, 20%, $12\frac{1}{2}\%$
13. A watch with a list price of $29.95 and series discounts of 15%, 10%. Find the invoice price.
14. A computer with a list price of $2,995.95 and series discounts of 20%, 5%, 5%. Find the invoice price.
15. A camera with a list price of $250 and series discounts of 30%, 20%, and 10%. Find the invoice price.

PUT IT ALL TOGETHER

♦ A business wants to purchase some new equipment. The owners identify three vendors and negotiate different terms of sale with each vendor. The terms are shown in the chart below. Assume the business can take the highest cash discount offered. Copy and complete the chart below. Then answer the following questions:

16. Which vendor offers the best package price?

17. If the purchase is made on December 14, by what date must the vendor pay the invoice?

	A	B	C
List Price	$800	$825	$965
Cash Discount	1/30; n60	2/30; n60	2/10; 1/30; n 60
Trade Discounts	5%	3%, 7%	10%
Shipping Terms	FOB Factory	FOB Factory	FOB Seller
Shipping & Handling	$125	$145	$125
Invoice Price			
Total Due			

MIXED REVIEW

1. Divide 76 by $2\frac{1}{2}$
2. $78 \times \$160 = ?$
3. $\$154 \div \$1.10 = ?$
4. Find interest on $280 for 4 months at 9%.
5. L&T Electronics will deliver a video camera to a retailer for $900 list price, less 25%, 10%, 10%. Inex will deliver a similar camera for $800, less 20%, 5%, 5%. Which company offers the retailer the best discount price? How much lower is that retailer's price?
6. Eve and Loretta are business partners. They agree to share net income and loss in a ratio of 5:4, with Eve getting the larger share. If their business earned a net income of $54,000 on sales of $612,000, what was Loretta's share of the net income?

14-3 PRICING FOR PROFIT

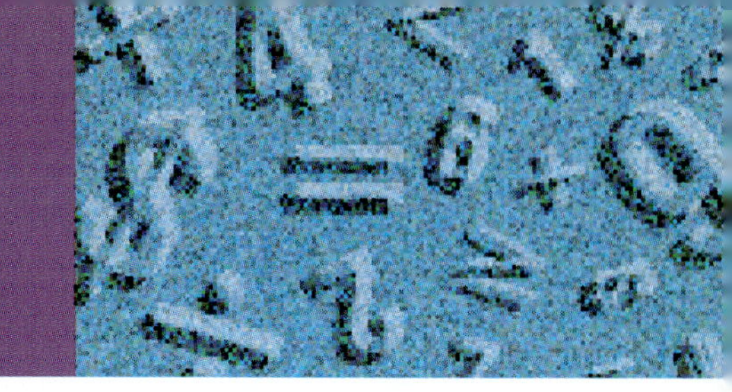

Name something you or your family purchase regularly. It may be a food item or a beverage, something you do for entertainment, or a cost involved in a favorite hobby. Write down the cost of the item. If the price of that item doubled, would you still buy it? Would you purchase it less frequently? If the price of that item was cut in half, would you buy more?

◆ Make a graph or chart that illustrates how price influences how much you buy of different kinds of products.

OBJECTIVES

In this lesson you will learn

- *the markup pricing techniques;*
- *to calculate dollar amounts of cost, markup or markdown, or selling price when one is unknown;*
- *to compute the rate of markup based on selling price and cost;*
- *to compute the percent of markdown based on selling price and marked price; and how to*
- *calculate gross and net profit.*

WARM UP

a. Gross Income − Cost of Goods Sold − Operating Costs =

b. Show 1.34 as a percent.

c. Multiply $28\frac{1}{2}$ by 14.

d. $162\frac{1}{2}\%$ of \$5.60 is ?

e. Add $9\frac{2}{3}$, $12\frac{3}{4}$, $15\frac{5}{6}$.

PRICING FOR PROFIT

One of the most important decisions merchandisers make is the selling price of products. An item priced too high may not sell enough. An item priced too low may not produce a net income. The selling price at which a product is sold involves many factors, including the price other companies are selling similar merchandise for and market demand. Generally speaking a business should sell a product at a price that covers the total cost and generates net income. By total cost, we mean all costs needed to purchase the item plus the operating expenses of the business. For example, suppose a business pays

$150 to purchase an item for resale. The operating expenses that can be allocated to that item are $20. To make a profit on that item, the retailer must sell that product for more than the total cost of $170.

One method used to set prices is **markup pricing.** With markup pricing, an amount is added to the cost of the goods to cover all other expenses plus a profit. This is known as the **markup** or **margin.** The **selling price** is the price at which the item is actually sold.

Markup Pricing as a Percent of Selling Price

The relationship of cost, markup, and selling price for a product can be shown mathematically by applying the basic pricing equation, or one of its variations:

Cost + Markup = Selling Price
Selling Price − Markup = Cost
Selling Price − Cost = Markup
Markup ÷ Selling Price = Rate(%) of Markup

EXAMPLE **Cost + Markup = Selling Price**
A dress costs $28 to manufacture. The factory adds $28 to the costs when the dress is sold to a wholesaler.

$28 + $28 = $56 selling price to wholesaler.

EXAMPLE **Selling Price − Markup = Cost**
A retail store sells a dress for $168. A $75 markup is added for dresses that sell over $150 and under $200. What was the cost of the dress to the retailer?

$168 − 75 = $93 cost

EXAMPLE **Selling Price − Cost = Markup**
The wholesaler who bought the dress from the manufacturer for $56 sold the dress to the retail store for $93. What was the wholesaler's markup?

$93 − 56 = $37

EXAMPLE **Rate of Markup = Markup ÷ Selling Price**
What was the rate of markup for the wholesaler who sold the dress to the store?

$37 ÷ $93 = 0.397 or 39.7%

Critical Thinking What is the rate of markup for the retail store?

Calculating Percentage Markups Based on Cost Markups can be stated as a percentage of selling price. When a markup is stated as a percentage of selling price, the selling price is the base. The selling price is always 100%. Sometimes you will want to find the rate of markup based on, or compared with, cost instead of selling price. In those situations, replace cost for selling price in the rate of markup formula. When a markup is stated as a percentage of cost, the cost is the base. Therefore, the cost is always 100% and the selling price is 100% plus the markup percentage. For example, if it cost $95 to produce dirt bikes that have a selling price of $185, first find the markup. Then divide the markup by the cost:

$185 − 95 = $90 markup
$90 ÷ 95 = 0.9473684 or 94.7% rate of markup

Since the cost is the base, the selling price is 194.7% of the cost, or $185. The percent is rounded to the nearest tenth.

Finding Missing Values in the Markup Equation As you saw in the dress example, when you know the value of two of the markup equation variables, you can find the third. When you are working with rate of markup, you can still find the missing variable by working with variations of the markup equation.

EXAMPLE **Find cost and markup when the selling price is given**

A supermarket is selling soup at $1.05 a can. The soup is marked up 15% *based on the selling price*. Find the cost and markup on the soup.

SOLUTION 1 Cost = (Selling Price % − Markup %) × Selling Price
Cost = (100% − 15%) × $1.05
Cost = 0.85 × $1.05 = $0.8925 or $0.89 rounded

Markup = Selling Price − Cost
Markup = $1.05 − $0.89 = $0.16

CHECK: Cost + Markup = Selling Price
$0.89 + $0.16 = $1.05

EXAMPLE **Find cost when price is known and markup must be based on selling price**

Jill Carr owns the Car Shack and sells a line of tires for $50 each. She knows that her markup must be at least 40% of the selling price to cover expenses and get the net income she wants. What should be the most she should pay for each tire?

SOLUTION Cost = (Selling Price % − Markup %) × Selling Price
Cost = (100% − 40%) × $50
Cost = 60% × $50 = $30

CHECK:
$50 × 40% = $20 — Find Markup
Cost + Markup = Selling Price
$30 + $20 = $50 — Substitute Values in Formula

EXPLANATION

To find out if retailers can sell an item in a price line for a profit, they must find the highest cost they can pay for the item and still receive their markup. The highest cost is found by subtracting the markup from the selling price.

EXAMPLE **Find selling price when the rate of markup and cost are known**

RB Computer Sales buys computer screens at \$90 each. The manager knows that the markup must be 40% of the selling price to cover expenses and give the business the net income that is wanted. What is the lowest price at which the computers can be sold and still make the specified rate of markup?

SOLUTION

Cost = (Selling Price % − Markup %) × Selling Price

\$90 = (100% − 40%) × Selling Price

$\frac{\$90}{0.60} = \frac{60\% \times \text{Selling Price}}{0.60}$ Divide each side by 60% (0.60)

\$150 = Selling Price

CHECK:

Cost = Selling Price − Markup

\$90 = \$150 − (\$150 × 40%)

\$90 = \$150 − \$60

\$90 = \$90

EXPLANATION

The selling price, 100%, is unknown. The markup is 40% of the selling price. The selling price is divided by 60%. Since the percent is based on the selling price, the selling price is the base, or whole, and is 100% of itself. The example shows that if 40% of the selling price is subtracted from the selling price, the difference, the cost, is \$90.

EXAMPLE **Find the selling price needed to generate the rate of markup desired when markup is based on cost**

A dealer buys 3-ring binders for \$5 per dozen. The dealer uses a *markup of 60% on the cost* to cover expenses and gain the net income wanted. What is the lowest price at which the binders can be sold and still earn the rate of markup wanted?

SOLUTION

Selling Price = % Cost + % Markup

Selling Price = 100% Cost + 60% Cost (Markup %)

Selling Price = 160% Cost

Selling Price = 160% × \$5 = \$8

CHECK:

Selling Price = Cost + Markup

\$5 × 60% = \$3 Find Markup

\$8 = \$5 + \$3 Substitute Values in Formula

Critical Thinking Find the rate of markup based on selling price for the last example. How does that compare with the rate of markup based on price? Why?

Illustration 14-3.1 shows markup formulas that have been derived from the basic markup equation. To use the chart to find missing values

1. Identify the known values and missing value.
2. Determine if the markup is based on cost or selling price.
3. Apply the appropriate formula.

When These Variables Are Known	and Markup is Based On	Find	Using This Formula
Markup Rate, Cost	Cost	Selling Price	Cost + (Markup % × Cost) or (100% + Markup %) × Cost
Markup Rate, Cost	Selling Price	Selling Price	$\frac{\text{Cost}}{1 - \left(\frac{\text{Markup \%}}{100}\right)}$
Selling Price, Cost	Cost	Markup Rate	$\frac{\text{Selling Price} - \text{Cost}}{\text{Cost}} \times 100\%$
Selling Price, Cost	Selling Price	Markup Rate	$\frac{\text{Selling Price} - \text{Cost}}{\text{Selling Price}} \times 100\%$
Markup Rate, Selling Price	Cost	Cost	$\frac{\text{Selling Price}}{1 + \left(\frac{\text{Markup Rate}}{100}\right)}$
Markup Rate, Selling Price	Selling Price	Cost	$\left(1 - \left(\frac{\text{Markup Rate}}{100}\right)\right) \times$ Selling Price

Illustration 14-3.1. Markup Formulas

EXERCISE YOUR SKILLS

Calculate the missing item, then find the markup, rate of markup on cost, and the rate of markup on selling price for each item.

Item	Cost	Selling Price	Markup	Rate of Markup On Cost	Rate of Markup On Selling Price
1.	$ 72.00		$ 18		
2.		225	75		
3.	720.00		144		
4.	110.25	147			

Use the markup equation or a variation of it to solve the following exercises.

5. Inazo Saga buys men's belts for a line with a selling price of $25. What is the most he can pay for this line if the markup must be at least 35% of the selling price?

6. At The Shoe Cellar, the markup on women's casual shoes that sell for $29.50 a pair is 38% of the selling price. Estimate then calculate the cost per pair to the store.

7. What is the most that a store owner should pay per dozen for scarves with a selling price of $27.98 each if the owner must make 34% gross profit on the retail price?

8. A buyer paid $157.50 for goods and $17.25 for delivery charges. What is the lowest price at which the goods may be sold if a 48% markup, based on cost, is wanted?

9. A produce store buys 60 crates of berries with 12 baskets per crate for \$15.50 per crate. The baskets without the berries weigh $\frac{1}{2}$ oz each. The store estimates that 24 baskets will spoil and must be thrown away. At what price per basket, to the nearest cent, must the rest be sold to make a markup of 40% on the cost of the 60 crates?

MARKDOWN PRICING

In some cases, retailers **mark down** or *discount* their **marked price,** or original selling price. The marked price is the price that is marked on the item. The selling price is always the price the item actually sold for. A price may be marked down on items at the end of a season or on items that are not selling well. Markdowns might also be done to attract customers to the store for a sale or to be more competitive. The markdown equation is similar to the markup equation. It will be useful to know if you ever need to mark down an item in a store, or want to calculate the rate of discount on an item you are interested in buying.

Markdown is stated as a percentage of marked price. The important markdown formulas correspond to the markup formulas. That is:

Marked Price − Markdown = Selling Price
Marked Price − Selling Price = Markdown
Rate of Markdown × Marked Price = Markdown
Rate of Markdown = Discount ÷ Marked Price

EXAMPLE A discount of 20% is given on baseball gloves with a marked price of \$45. Find the amount of the discount and the selling price.

SOLUTION	EXPLANATION
$0.20 \times \$45 = \$9$ discount	Multiply the marked price by the rate of discount. Remember, *discount* is another name for markdown.
$\$45 - \$9 = \$36$ selling price	Subtract the dollar discount from the marked price.

CHECK

$\$9 \div \$45 = 0.2$ or 20% rate of discount	The discount divided by the marked price equals the rate of discount.

EXERCISE YOUR SKILLS

	Item	Marked Price	Rate of Discount	Amount of Discount	Selling Price
10.	Telephone	\$ 259.90	10%		
11.	Roller skates	22.95	20%		
12.	Camera	170.00	$12\frac{1}{2}$%		

	Item	Marked Price	Rate of Discount	Amount of Discount	Selling Price
13.	Mower	289.00	25%		
14.	Guitar	370.00	$37\frac{1}{2}$%		
15.	Computer	1,599.95	40%		

16. In a shop window, a jogging outfit is marked "$99.45 —was $127.50." What is the rate of discount?

17. A store sells a line of bikes that are regularly priced at $89.95. At a seasonal sale, the bikes are marked down to $79.95. What is the rate of discount to the nearest tenth of a percent?

Critical Thinking Do you think retailers make money when they markdown prices for consumers?

CALCULATING GROSS PROFIT AND NET INCOME ON A SINGLE ITEM

As mentioned earlier, the price a business charges for a product is a strategic decision. Whether or not a product generated a profit is determined on actual sales records. Often businesses compare the actual revenue on individual items. They use that information to decide what products to produce or distribute and where to place their selling efforts.

EXAMPLE The Racquet Shop paid $2,420 for 100 pairs of tennis shoes and paid $165 to ship them to the shop. During the summer, the shop sold 80 pairs at $49 each. At the end of the summer, the remaining shoes were sold at $33 each. The Racquet Shop's income statement shows that total operating costs are 37% of net sales.

Find the store's actual gross profit, net income, rate of markup, and rate of gross profit and rate of net income.

SOLUTION **Step 1: Find the total revenue received.**

80 pairs @ $49 = $3,920
20 pairs @ $33 = + 660
Total Revenue = $4,580

Step 2: Find the cost of goods.

Invoice Price = $2,420
Shipping = + 165
Cost of goods = $2,585

Step 3: Find the gross profit or loss.

Net Sales	$4,580	*Note: If the cost is more than the*
Cost of Goods	− 2,585	*selling price, subtract the selling price*
Gross Profit	$1,995	*from the cost to find the gross loss.*

Step 4: Estimate operating expenses, then calculate net income.

Gross Profit	$1,995	
Operating Expenses	− 1,694.60	($4,580 × 0.37)
Net Income	$ 300.40	

Step 5: Calculate the rate of gross profit and the rate of net income.

Rate of Gross Profit $1,995 ÷ $4,580 = 0.4355 or 43.6%

Rate of Net Income $300.40 ÷ $4,580 = 0.0656 or 6.6%

EXERCISE YOUR SKILLS

18. The Byte Mart bought 25 computers for $14,750 and paid $275 for shipping charges. The store sold 15 of the computers at $1,299; 5 at $1,099, and the rest at $925. What was the store's gross profit on the lot?

19. Chang's Auto Store buys wheel covers at $29 and sells them at $49. The store figures its operating expenses at 26% of net sales. What is the net income per cover?

20. The Glass Factory bought windows that cost $950 each and found that the highest price they could sell them for was $1,250 each. Operating expenses for the purchase and sale of each window was estimated to be $394. What was the net income or net loss on each window?

21. Fred Baum, a retailer, buys can openers at $5 each plus $0.50 shipping charge. If he sells them at $12 each and his operating expenses are 25% of net sales, what is his net income on each can opener?

22. A wholesaler buys truck light bars at $53.90 a set and sells them at $70 a set. What is the rate of gross profit on the selling price?

MIXED REVIEW

1. $82 \times 2\frac{1}{2}$

2. $18 \div 2\frac{1}{4}$

3. The costs of manufacturing 500 VCRs were materials, $11,790; labor, $12,560; factory expense, $5,910. The VCRs were sold at $121.50 each. Selling and administrative expenses were estimated at 40% of net sales. What percent of the sales was the net income, to the nearest whole percent?

4. You bought a car for $7,500. After using it for 4 years, you bought a new car for $12,000 by trading in the old car and paying $8,900 in cash. What was the average annual depreciation of the old car?

5. On May 5, Sheila discounts at 12% banker's interest a 4-month noninterest-bearing note for $3,400. Find the proceeds of the note.

14-4 FINDING THE BREAK-EVEN POINT

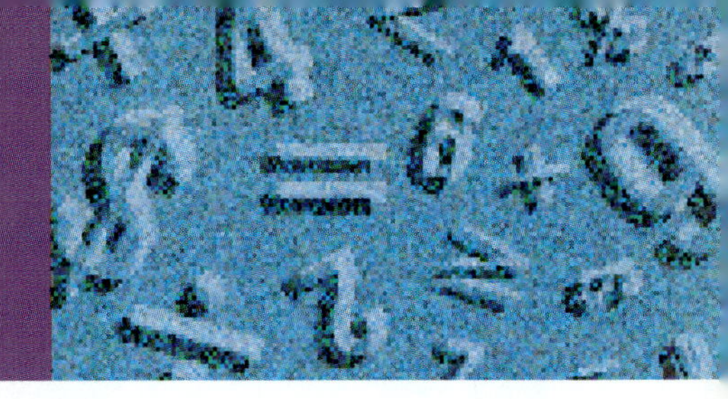

In the last lesson you learned some of the pricing questions a merchandiser must consider when pricing a product. Manufacturers must also set prices for products. Setting a price for a new product is especially difficult.

♦ Imagine that your company just invented a new product, never offered before. Make a list of questions you would ask to help determine the selling price of this new item.

OBJECTIVES

In this lesson you will learn to

- *use graphs and equations to calculate the break-even point for a product;*
- *calculate loss and absolute loss for a product; and*
- *calculate gross and net profit for a manufactured item.*

WARM UP

a. $\frac{2}{3} \times \frac{5}{8} =$

b. Divide 10 by $\frac{2}{5}$

c. Change 80% to a fraction in lowest terms.

d. Show 3.09 as a percent.

e. 350 cm + 12 m + 520 cm = ? m.

BREAK-EVEN POINT PRICING

The **break-even point** is the point at which income from sales equals the total cost of producing and selling goods. Any amount over the break-even point is profit. When a selling price covers the cost of goods, but not the operating costs, there is an **operating loss.** When the selling price is not enough to cover the cost of the goods as well, there is an **absolute loss.**

Break-even analysis is often used by manufacturers to determine how many units to produce, how much to spend to produce and sell the units, how many units they can expect to sell, and at what price they must sell the units to make the profit they want.

To find the break-even point, fixed costs and variable costs are considered. **Fixed costs** are overhead costs such as rent, salaries, heat, insurance, and advertising. Fixed costs are costs that remain the same no matter how much is produced and sold. **Variable costs** are costs such as raw materials, direct labor, and energy that vary or change with the amount of goods produced and sold. For example, one department of the Farber Company plans to produce a new picture frame that will be sold at \$10 per unit. The fixed costs for that department are \$12,000. Producing each frame will cost an estimated \$7 in variable costs.

Critical Thinking Why can you not simply divide the fixed costs by the number of items produced to find the break-even point?

You can write and graph these equations. If x is the number of frames sold, then

Income = $10x$
Total cost = \$12,000 + \$7x

The break-even point occurs when income equals total costs.

You can write

$10x = 12{,}000 + 7x$
$3x = 12{,}000$ (Subtract $7x$ from both sides.)
$x = 4{,}000$ (Divide by 3.)

The break-even point happens at 4,000 frames.

Compare the answer obtained algebraically with the one shown on the graph. On the graph, you can see that if less than 4,000 frames are sold, there is a loss. If more than 4,000 frames are sold, there is a profit.

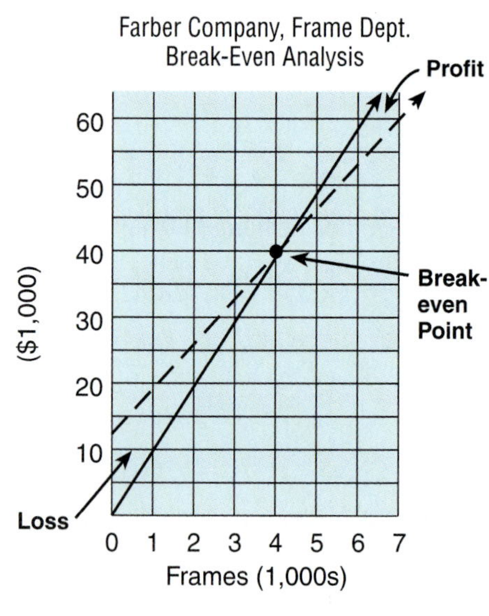

Illustration 14-4.2. Farber Company's Break-Even Point

EXERCISE YOUR SKILLS

1. Barrin, Inc., manufactured 5,000 football helmets. Their fixed costs of production were $105,000 and their variable costs were $195,000. What is the price per helmet at which Barrin's sales income would equal their total cost of production?
2. To make toy trucks, Babyco Corporation's variable costs were $800,000 and their fixed costs were $4 per truck. How many trucks must be sold at $20 each to break even?
3. Colbert, Inc., plans to make and sell 5,000 mufflers at an estimated total cost of $60,000. What is the minimum price at which they can make and sell each muffler and break even?

When planning the production of a new line of chairs, Oramac, Inc., estimated variable costs of $45 per chair and fixed costs of $150,000. The chairs will be sold for $75 each.

4. How many chairs must Oramac, Inc., sell to break even on these chairs?

Suppose Oramac, Inc., sells 10,000 of the chairs in the previous problem.

5. Find net sales.
6. Find the net profit.

A game company produced 3,000 video games and sold them at $10 each. Fixed costs were $9,000 and their variable costs were $5 per game.

7. What was the total production cost?
8. How many games did they have to sell to break even?
9. What was the net income from this operation?
10. Wordex Corporation plans to produce and sell 50,000 units of a computer printer. They estimate that their fixed costs will be $7,500,000 and their variable costs $9,800,000. What must be the minimum selling price per unit for Wordex to break even?
11. To produce and sell 20,000 books, Sci-Fi Publishers, Inc., will have fixed and variable costs totaling $250,000. The publisher wants to make a profit of $125,000. At what price must the books be sold to make the desired profit?

Hullins Inc., can produce and sell 30,000 fishing reels at a total cost of $288,000. Hullins wants to make a net profit of 25% on cost. If all units produced can be sold, what

12. is the break-even price at which each reel must be sold?
13. selling price per reel is needed to make the net profit wanted?

Critical Thinking How would you calculate a break-even point for a retail business?

PUT IT ALL TOGETHER

Profit and loss on an individual item for a manufacturer can be calculated in much the same way as for a merchant.

Sales Income − Total Factory Cost = Gross Profit
Gross Profit − Operating Costs = Net Income
Net Income ÷ Sales Income = Rate of Net Income

14. Goods costing $2,964 are sold for $3,900. The overhead expenses for the sale are $585. What is the rate of net income on the selling price?

15. An axle that a manufacturer sells for $88.75 has these manufacturing costs: materials, $19.50; labor, $25.90; factory expense, $8.15. What is the manufacturer's gross profit on each axle?

16. A manufacturer sells ceiling fans for $47.25. The manufacturing costs are: material, $12.19; labor, $17.71; factory overhead, $5.48. Selling and administrative expenses are calculated at 12% of the selling price. What is the manufacturer's net income on each fan?

MIXED REVIEW

1. $4.85 × 1,000 **2.** $2\frac{1}{4} \div 18$

3. Find the due date of a 35-day note dated April 2.

4. To make 16,000 personal computers, a factory has fixed costs of $3,600,000 and variable costs of $10,800,000. At what price must the computers be sold in order to break-even?

5. Tollen pays a $2,890 yearly equipment insurance premium. Each department is billed for their share of the premium based on the value of their department's equipment compared to the value of the entire company's equipment. The Painting Department's equipment is valued at $245,000. If all of Tollen's equipment is valued at $1,750,000, how much should the Painting Department be charged for insurance?

14-5 SALES RECORD

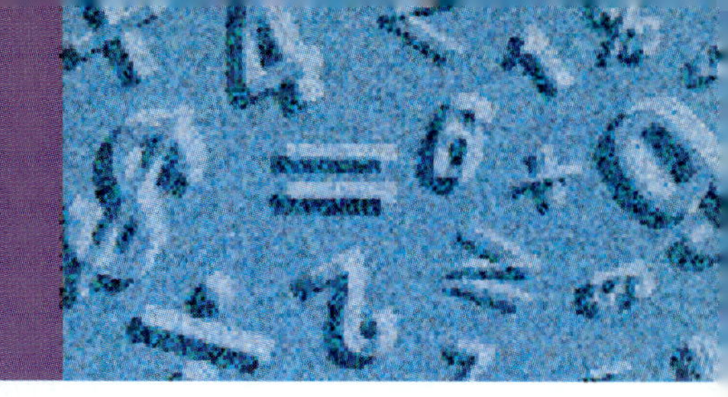

In Lesson 14-1, you learned about purchase orders and invoices. In addition to those important records, a business must develop a way of keeping track of its daily cash payments and receipts.

◆ Imagine that you have started your own business. Make a list of the financial records you would keep to record income and purchases.

OBJECTIVES

In this lesson, you will learn to

- *balance, or prove, the cash in the cash drawer of a cash register;*
- *use credit memos; and*
- *complete a charge account form for items purchased and returned on account.*

WARM UP

a. $166\frac{2}{3}\%$ of \$140 = ?

b. Find banker's interest on \$350 for 9 days at 16%

c. 0.98 times \$635 is?

d. $\frac{3}{8}$ less than \$26.88 = ?

e. \$450 less 10% and 5% = ?

CASH SALES RECORDS

Cash registers provide a place to keep cash and a means to record cash sales and payments. Employees who use cash registers are called cash register clerks or cashiers.

Modern electronic cash registers are really computer terminals. You have probably seen electronic cash registers in supermarkets. Each checkout counter has a register with a display screen and a scanner that may be connected to a computer. As items pass by, the scanner reads **bar codes** printed on the items. The bar codes tell the computer the department, brand, size, and price of each item bought. This information is shown on the display screen and printed on a cash register receipt so that customers can check their purchases.

The computer also finds the sales tax, totals the sale, and updates the store's inventory records. When the clerk keys in the amount of money received from the customer, the computer displays the correct change on the screen.

Proving Cash Cashiers put some money in the cash register drawer when they start work so that they can make change. This money is called a **change fund.** While they work, they take in and pay out cash. At the end of their work period they have to **prove cash.**

Proving cash means counting the money in the drawer and checking this amount against the cash register readings to see if the right amount is on hand. If you have less cash than you should, you are **cash short.** If you have more cash than you should, you are **cash over.**

For example, suppose Bess Turner worked as a cashier for Tri-County Markets. When she started work on March 16, she put $100 in her cash register drawer as a change fund. At the end of her work period, her register readings showed total cash received, $3,365.75, and total cash paid out, $72.20. When she counted the cash in her drawer, she found $3,395.55. So, she was $2 cash over. In proving cash, she used a **cash proof form** like the one shown in Illustration 14-5.1.

TCM TRI-COUNTY MARKETS
803 Blanchester Rd.
Augusta, ME 04330-7654

Cash Proof

Date *March 16, 19 --*

Change Fund	*100*	*00*
+ Register total of cash received	*3,365*	*75*
Total	*3,465*	*75*
− Register total of cash paid out	*72*	*20*
Cash that should be in drawer	*3,393*	*55*
Cash actually in drawer	*3,395*	*55*
Cash short		
Cash over	*2*	*00*

Cash register No. *8*

Cash Register Operator *Bess Turner*

Illustration 14-5.1. Cash Proof Form

Cash Sales Summary Records Cash sales and cash refunds for goods returned may be recorded using the cash register. Daily totals of cash sales and cash refunds may be summarized. The summary may show the totals for each department or for each salesclerk.

EXERCISE YOUR SKILLS

1. At the start of the day, July 3, Maude Olaf put $100 in change in her cash register. The cash register readings at the end of the day showed total cash received, $2,793.68, and total cash paid out, $29.50. The cash in the register at the end of the day was $2,863.38. Prepare a cash proof form.
2. When he opened his shop on March 9, Edwin Torinos put $75 in change in his cash register drawer. At the end of the day, the cash register readings showed that $3,812.34 had been taken in, that $213.88 had been paid out in cash refunds, and that Edwin had taken out $50 for his personal use. The cash on hand in the drawer at the end of the day was $3,624.75. Was the cash short or over? How much?

3. Ria had $100 in change in her cash register at the start of work on May 6. At the end of her work period, there was $2,988.68 in the cash register. The register totals showed that she received $3,307.43 and paid out $418.75 during the period. How much was the cash over or short?

The table below shows the cash sales and cash refunds for returns for the 4 departments of Klein's Drugstore for September 19.

4. Copy and complete the table. Prove total net cash sales by subtracting total cash refunds for sales returns from total cash sales.

Daily Summary—Cash Sales	September 19, 19—		
Department	**Cash Sales**	**Cash Refunds for Returns**	**Net Cash Sales**
Prescription Drugs	$1,798.22	$ 0.00	
Nonprescription Drugs	2,666.23	27.89	
Cosmetics	902.56	78.12	
Hygiene Items	613.55	43.49	
Totals			

The table shows the cash sales for 6 salespersons who worked at a store on August 17.

5. Copy and complete the table. Prove the total of the Net Cash Sales column by subtracting the totals of the cash refunds column from the cash sales column.

Daily Cash Sales Summary	August 17, 19—		
Salesperson	**Cash Sales**	**Cash Refunds**	**Net Cash Sales**
Allen	$ 308.25	$ 22.87	$
Brett	478.29	68.45	
Cruz	511.98	15.77	
Deng	278.41	12.39	
Eaton	645.90	34.85	
Farrell	198.34	5.89	
Totals			

ON ACCOUNT SALES RECORDS

When a seller sells goods to a buyer on credit, the seller gives the customer a sales slip, or sales invoice. A **sales invoice,** as you saw in Lesson 14-1 is a form that lists the goods sold and delivered to the buyer. The buyer calls this form a purchase invoice.

A sales invoice that Technipro Equipment Company, a wholesale computer equipment firm, sent to Computer Station, a retailer, is shown in Illustration 14-5.2.

TECHPRO EQUIPMENT COMPANY
7208 Central Avenue
Baltimore, MD 21207-3071

Date: *October 1, 19--*

To *Computer Station*
3078 Jefferson Street
Rockville, MD 20852-8671

Account No: *010-31-194-6*
Invoice No: *1078*

Shipped *VIA Truck*
Terms *n/30*

We have charged your account as follows:

50	*Printer Cables*	*12.78*	*639.00*
10	*Printers, Model No. 108*	*184.99*	*1,849.90*
3	*Printers, Model No. 288*	*908.40*	*2,725.20*
	Total		*5,214.10*

Illustration 14-5.2. Sales Invoice

TECHPRO EQUIPMENT COMPANY
7208 Central Avenue
Baltimore, MD 21207-3071

CREDIT MEMO

Date: *October 10, 19--*

TO *Computer Station*
3078 Jefferson Street
Rockville, MD 20852-8671

Account No: *010-31-194-6*
No: *143*

We have credited your account as follows:

Description	Unit Price	Total
Returned 3 damaged printer cables; Invoice No. 1078	*12.78*	*38.34*

Illustration 14-5.3. Credit Memo

Account:
Computer Station
3078 Jefferson Street
Rockville, MD 20852-8671

Acct. No: *010-31-194-6*
1-815-555-0155

Date	Description	Charges	Credits	Balance
Oct. 1, 19xx	*Balance Forward*			*$ 8,282.00*
Oct. 1, 19xx	*Invoice No. 1078*	*$5,214.10*		*13,496.10*
Oct. 10, 19xx	*Credit Memo No. 143*		*$ 38.34*	*13,457.76*
Oct. 14, 19xx	*Payment*		*8,282.00*	*5,175.76*
Oct. 31, 19xx	*Payment*		*5,175.76*	*0*

Illustration 14-5.4. Customer Account

The sales invoice may be completed by hand, with a typewriter, or on a computer system. The original invoice is sent to the buyer. Copies of the invoice are kept by the seller.

When merchandise bought on credit is returned, the seller does not return the buyer's money. Instead, the seller reduces the buyer's account balance by the amount of the return. The seller notifies the buyer about the reduction by sending the buyer a credit memorandum, or **credit memo.** For example, when Computer Station returned three defective printer cables from Invoice No. 1078, Technipro sent the credit memo shown in Illustration 14-5.3.

Customer Accounts Most people use credit cards, but businesses whose customers are other businesses often let them buy on credit, or on account. The store then keeps records for each customer to show how much each customer owes. A common form of **customer account** is shown in Illustration 14-5.4. This account shows the transactions between Technipro Equipment Company (the seller) and Computer Station.

EXERCISE YOUR SKILLS

6. Make an account form like the one in Illustration 14-5.4. Record these transactions between J. L. Sales, Inc., and a customer, Tenco Products, Inc. Tenco is located at 5302 Post Road, Warwick, RI 02886-9898. Record the final balance.

Aug.	1	The balance of Tenco's account was $2,353.90.
	8	Sold goods to Tenco, $1,685.49; Inv. No. 3369; terms, 30 days.
	16	Tenco returned $45.66 of the goods sold on Aug. 8; Credit Memo No. 388.
	20	Tenco paid the balance on Aug. 1, $2,353.90.
	26	Sold goods to Tenco, $2,275; Inv. No. 3672; terms, 30 days.
	31	Tenco paid the invoice of Aug. 8, less Credit Memo No. 388.
Sept.	12	Sold goods to Tenco, $1,216; Inv. No. 4180; terms 30 days.
	15	Tenco returned $120 of the goods sold on Sept. 12; Credit Memo No. 412
	25	Tenco paid Inv. No. 3672

7. Make an account form like the one in Illustration 14-5.4. Record these facts about Sterling Co.'s account with Gorbea Supply Company. Sterling's address is 708 Gower St., Greenville, SC 29611-3381.

June	1	The balance in Sterling's account was $4,116.53.
	3	Sterling paid the June 1 balance.
	5	Sold goods to Sterling, $3,642.39; Inv. No. 9708; terms, n/20.
	17	Sterling returned $488.12 of the goods on Inv. No. 9708; Credit Memo No. 398.
	18	Sold goods to Sterling, $1,785.84; Inv. #9934; terms, n/20.
	25	Sterling paid Inv. #9708, less Credit Memo #398.
	26	Sterling returned $21.72 of the goods on Inv. #9934; Credit Memo #416.

8. ♦ Work with a partner or small team. Set up a situation in which your members are businesses that buy and sell to one another. Develop your own cash receipts and payments record, account statements, and credit memos.

MIXED REVIEW

1. Add: $9\frac{2}{3}$, $12\frac{3}{4}$, $15\frac{5}{6}$
2. Divide: $23\frac{1}{3}$ by $1\frac{5}{9}$
3. Subtract: $153,935 - 59,056$
4. Multiply: $44\frac{1}{2}$ by $6\frac{1}{4}$
5. Trimer Co.'s account balance with MLR, Inc., was $5,893.11 on July 1. These facts were recorded in the account in July: July 2, Invoice #2977, $1,007.49; July 9, Credit Memo #125, $23.78; July 21, payment, $5,000. What was the balance of the account after these items were recorded?
6. A compact disk system listed at $1,280 is billed to the retailer at $768. What is the rate of trade discount?

14-6 WHAT IS INTERNATIONAL BUSINESS?

OBJECTIVES

In this lesson, you will learn to

- *determine trade surpluses or trade deficits for various countries; and*
- *calculate and compare the gross domestic product of various countries.*

In our discussions on business expenses, we have assumed that we are talking about **domestic business.** Domestic business is the manufacturing, purchasing, and selling of goods and services within a country. **International business** includes all the business activities necessary to manufacture, purchase, and sell goods and services across national borders. When people talk about **foreign trade,** they are talking about international business.

◆ Make a list of all the terms you have heard that relate to international business. For the next week, write any other words you hear or read that relate to international business.

WARM *UP*

a. Write "sixty thousand, forty five" as a number in standard form.

b. Write 5.2 million as a number in standard form.

c. Write 90,800,700,060 in words.

d. Divide $50,000,000 by 200.

e. Subtract: $205,092,623,523 − $95,099,253,863

THE INTERNATIONAL MARKETPLACE

Balance Of Trade

The two most important international business activities are exporting and importing. **Exporting** is the selling of goods (or services) produced within your country and sold to other countries. **Importing** is the buying of products produced outside your country. A measure of a country's international business activity is its *balance of trade.* **Balance of trade** is the difference between a country's exports and imports.

Critical Thinking What do you think are some of the advantages of exporting goods? of importing goods?

When foreign exports exceed imports, a country has a **trade surplus.** When foreign imports exceed exports, a country has a **trade deficit** and may have to borrow money or ask for credit from other countries, thus establishing **foreign debt.** As with individuals, governments must pay interest on money they owe another country. If the debt is too large, it can affect the country's economy by limiting the amount of money available for necessary work in that country. For example, a large debt could mean a country could not improve public buildings or roads, or provide services for its citizens, such as disability payments or medical care for the elderly.

Gross Domestic Product

The **gross domestic product (GDP)** of a nation measures the total market value of all final goods and services produced within the borders of a country. Exports are part of a country's GDP; imports are not.

GDP is usually reported on a yearly basis. Gross domestic product is an indication of the size of a country's economy. In this text, it will be calculated in United States dollars.

Per capita means per person. So **per capita GDP** shows the amount of goods and services produced per person by a country. Per capita GDP is one way to measure a nation's standard of living. For instance, in a recent year the per capita GDP for the United States was a little over $20,000 while the per capita GDP for India was $400. The average standard of living in the United States is significantly higher than the average standard of living in India.

EXAMPLE In a recent year, France had a GDP of $820 billion ($820,000,000,000) and a population of 57 million (57,000,000) people. Find the per capita GDP for France in that year.

SOLUTION Divide to find the per capita GDP.

$820,000,000,000 ÷ 57,000,000 = ?

Hint: Divide each number by 1,000,000 first, then use a calculator. Round to the nearest dollar.

$$\frac{\$820{,}000,\cancel{000{,}000}}{57,\cancel{000{,}000}} = \$14{,}385.96,\ \$14{,}386 \text{ rounded}$$

EXERCISE YOUR SKILLS

1. In a recent year Canada had imports of $124 billion and exports of $131 billion. Did Canada have a trade surplus or a trade deficit? How much?
2. In one year Mexico exported $27 billion in products and services and imported $23 billion. Did Mexico have a trade surplus or a trade deficit? How much?

3. In a recent year Sweden had a GDP of $15,700,000,000. If their population is 8 million people, what is the per capita GDP?

Copy and complete the table below.

	Country	Population (millions)	GDP (millions)	Per capita GDP
4.	Brazil	150	$ 375,000	$
5.	China	1,175	$ 350,000	$
6.	Japan	?	$2,844,000	$23,700.00
7.	Mexico	?	$ 194,850	$ 2,165.00
8.	U.S.A.	250	$	$22,000.00

9. Use the per capita GDP from Exercises 4–10 above and rank the countries from highest to lowest.

10. ◆ Find the current population and gross domestic product of several different countries, including the United States. Find the per capita gross national product for each and rank the countries from highest to lowest.

MIXED REVIEW

1. Find 15% of $25,050.
2. Find 10% of $593.
3. Write $6,004,040 in words.

To the nearest cent, an annual salary of $25,380 is equivalent to what

4. Monthly salary?
5. Weekly salary?
6. For 6 months, the Witiken Company kept track of the number of consecutive days their photocopier worked before it needed service. The days between service calls were: 25, 22, 19, 14, 40, 35, 15, 12. What was the mean number of days between service calls?

A vacuum cleaner with a marked price of $180 was sold for $135.

7. What was the amount of discount?
8. What was the rate of discount?
9. Pat Froelic is paid a salary of $280 a week and a commission of 2% on net sales. Last week her gross sales were $6,720 and customer returns were $1,120. How much did Pat earn last week?

14-7 INTERNATIONAL TIME, TEMPERATURE, AND MONEY

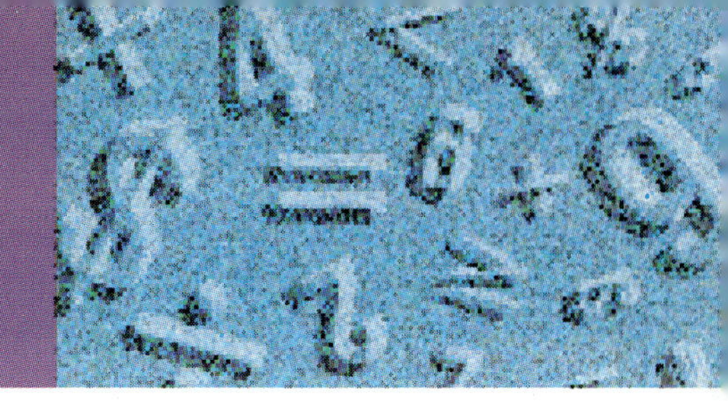

It is 3:45 P.M. in Wichita, Kansas. You have a message to call your boss, who has told you never to call after 9:00 P.M. Your boss is in Madrid, Spain. Is this a good time to call?

You have imported an exotic bird for display in your pet shop. The exporter tells you the temperature in the area the bird will be housed must always be 30°C or higher. At what Fahrenheit temperature should you set your thermostat?

You are offered a job in Canada. The offer is $800 Canadian dollars a week. Your current U.S. salary is $500 a week. How does the salary compare to what you are now making?

◆ Learning to translate time, temperature, and money in an international marketplace is reasonably easy, as you will learn in this lesson. Work in small groups to identify other kinds of differences you would have to deal with when doing business with people in other parts of the world. How do you think you could learn to deal with those differences?

OBJECTIVES

In this lesson you will learn to

- *calculate time in different time zones;*
- *convert Fahrenheit and Celsius temperatures; and*
- *convert foreign currencies.*

WARM UP

a. Add $17\frac{1}{8}$, $112\frac{3}{7}$, and $215\frac{4}{5}$.

b. Change 1 393 L to kiloliters.

c. Write 1.34 as a percent.

d. Multiply $5,680 by $\frac{1}{2}$%.

e. Find the median amount: $5, $6, $4, $4, $8, $12, $4, $5.

INTERNATIONAL TIME

Dealing with international travel and customers requires knowledge of international time zones and the international date line.

Time Zones The globe is divided into 24 standard time zones based on a 360° circle. The center of each time zone is designated by the standard meridians of longitude. Each time zone is spaced 15° apart. Greenwich, England is designated as the starting point, 0°, or the **prime meridian.** If you started in Greenwich and moved East, you would add one hour of clock time for each 15° of longitude or time zone. If you moved west from Greenwich, you would subtract one hour for every 15° of longitude, or time zone. You are probably familiar with the time zones of the United States. The mainland is divided into four time zones, eastern standard time (EST), central standard time (CST), mountain standard time (MST), and Pacific standard time (PST) zones. California is 3 time zones to the west of New York. If it is 1:00 in California, to find the time in New York, you would move three time zones to the east. The time in New York is 4:00. Illustration 14-7.1 shows the international time zones.

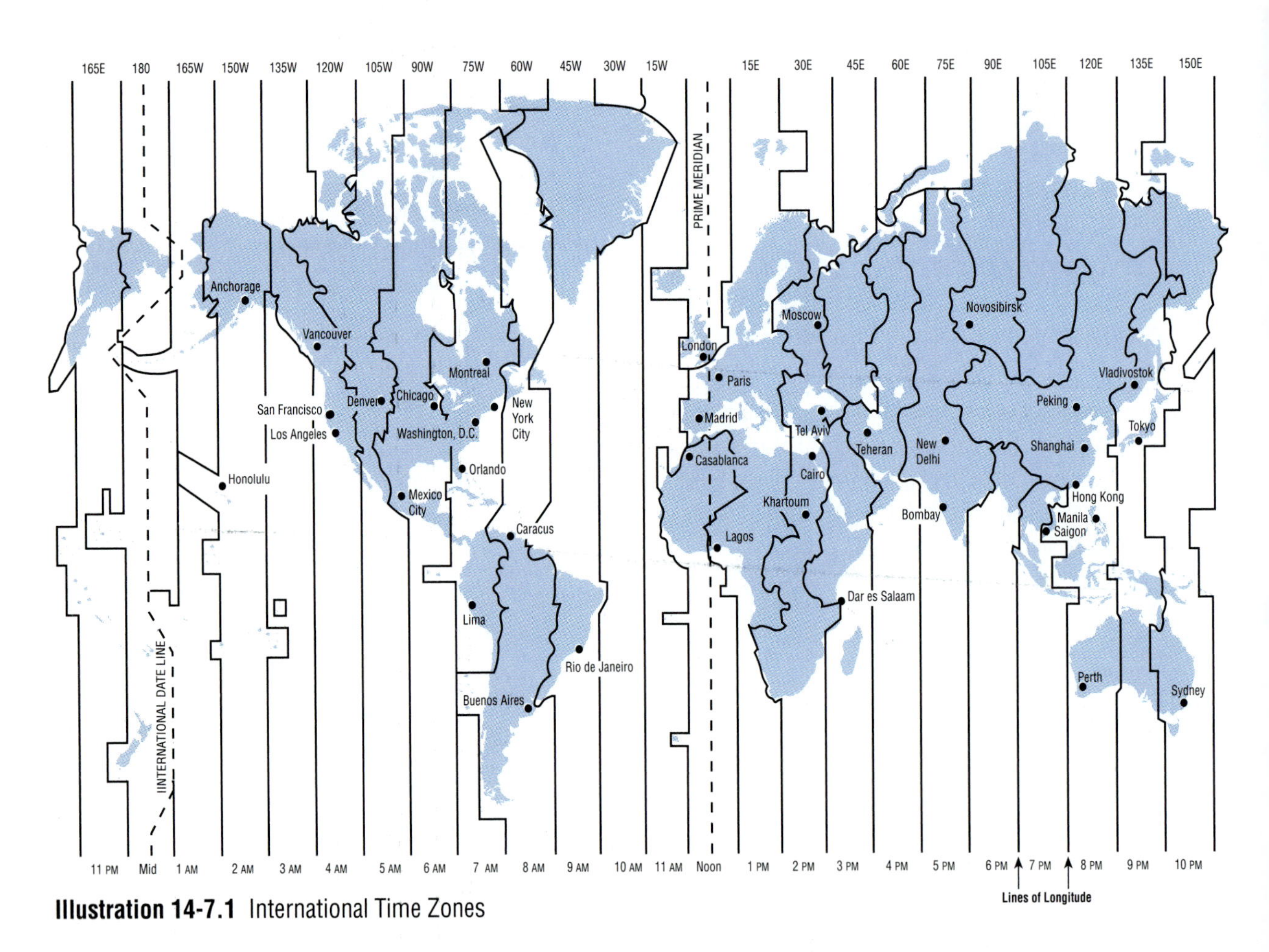

Illustration 14-7.1 International Time Zones

Critical Thinking If you are flying to New York from California, and your departure time is 7:30 P.M. eastern standard time and your flight time is 5 hours, what is your arrival time in California?

International Date Line The international date line is an imaginary line located in the middle of the Pacific Ocean. It is halfway around the world from Greenwich England. When you travel east and cross the date line, you lose a day. If you are travelling east around the world and it is May 15, when you cross the international dateline it will be May 14. If you were travelling west, when you cross the international date line, it would be May 16.

To determine the date and time in a particular country, calculate the time. Then check the map to see if you need to adjust the date.

EXAMPLE It is Monday at 4:00 P.M. in San Francisco. What time and day is it in Sydney Australia?

SOLUTION Look at the time zone map. Sydney is 6 time zones to the west of San Francisco. To find the time in Sydney, subtract 6 hours; it is 8 A.M. in Sydney.

Look at the map again. Moving west from San Francisco to Sydney, you would cross the International date line. You must add a day. Therefore when it is Monday at 4:00 P.M. in San Francisco, it is 8:00 A.M. Tuesday in Sydney.

EXERCISE YOUR SKILLS

1. It is 6:00 A.M. Wednesday in Chicago, Illinois. What is the date and time in Peking, China?
2. It is 4:00 A.M. Saturday in Orlando Florida. What day and time is it in Rio de Janeiro?
3. It is 12:00 noon Greenwich meridian time. What time is it in Cairo?
4. You are in Anchorage, Alaska. Your boss is in Bombay, India. She wants you to call at 5:00 P.M. Bombay time. What time do you have to make the call?
5. You are in Moscow, Russia. You want to call someone in Mexico City at 3:30 P.M. their time. What time is it in Moscow when you place the call?

TEMPERATURE

In Chapter 5, you were introduced to the metric system which is commonly used around the world. Another measurement system commonly used abroad, and also based on a base 10 system, is the Celsius temperature scale.

Illustration 14-7.2 shows Fahrenheit and Celsius thermometers.

As you can see, the Fahrenheit scale is based on a scale of 32°(freezing point) and 212°(boiling point). The Celsius system is based on a scale of 0°C (freezing point) to 100° (boiling point).

To convert between Celsius (C) and Fahrenheit (F) temperatures, you can use one of these two formulas.

$$F = 1.8C + 32$$
$$C = \tfrac{5}{9}(F - 32)$$

For example, we know that average body temperature in Fahrenheit degrees is 98.6. To find the Celsius equivalent:

$$\tfrac{5}{9}(F - 32) = C$$
$$\tfrac{5}{9}(98.6 - 32) =$$
$$\tfrac{5}{9} \times 66.6 = 37°C$$

To confirm your answer, convert back to degrees Fahrenheit

$$1.8C + 32 = F$$
$$1.8(37) + 32 =$$
$$66.6 + 32 = 98.6°F$$

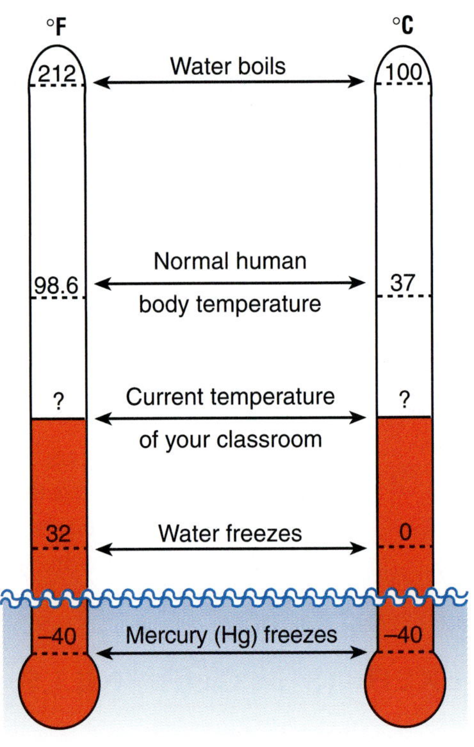

Illustration 14-7.2 Fahrenheit and Celsius thermometers.

EXERCISE YOUR SKILLS

6. The thermometers show that the freezing point of Mercury is −40° on both scales. Is this correct? Show your calculations.
7. You receive a shipment of items that includes the warning "Do Not Accept Shipment if Exposed to Temperatures Below 10°C." You know that the package was stalled for several days in a snow storm where temperatures never got above 20°F. Should you accept shipment?
8. You are sending cargo overseas. The cargo must be kept refrigerated at 45°F. Someone tells you that if you mark the package 90°C it will be okay. Is he correct? Explain your answer.

Many households and organizations conserve energy by keeping temperatures at 68°F in the winter and 72°F in the summer. What are the equivalent temperatures in degrees Celsius?

9. Winter
10. Summer

FOREIGN CURRENCIES

Most nations have their own kind of money or **currency.** Before you can buy something from another country, you have to buy their money. How much you have to pay for another country's currency is based on the **foreign exchange rate. Foreign exchange** is the process of changing or converting the currency of one country into the currency of another country. The **exchange rate** is the amount of currency one country can trade for one unit of the currency of another company. Exchange rates change daily. Illustration 14-7.3 shows the exchange rates one day in a recent year.

Currency	Country	Symbol	Value in U.S. Dollars	Units per U.S. Dollar
Deutsche mark	Germany	DM	$0.64	1.56 DM
dollar	Canada	$	$0.746	$1.34 Canadian
franc	France	Fr or F	$0.188	5.32 FR
pound	Great Britain	£	$1.59	£0.63
riyal	Saudi Arabia	SR	$0.267	3.745 SR
rupee	India	Re	$0.032	31.25 Re
yen	Japan	¥	$0.0104	96.15 ¥

Illustration 14-7.3. Sample Foreign Exchange Rates Chart

EXAMPLE 1 Suppose you bought a hamburger in Paris, France and it cost 20 francs. Use the foreign exchange rates shown in Illustration 14-7.3 to find the price in U.S. dollars.

SOLUTION Method 1: Using Value in U.S. Dollars

$20 \times 0.188 = \$3.76$

Method 2: Using Units per U.S. Dollar

$20 \div 5.32 = \$3.7594$ or $3.76

EXAMPLE 2 You bought a hamburger in Japan and paid $7 because beef is much more expensive in Japan. How many yen did you pay for the hamburger? Use Illustration 14-7.3.

SOLUTION Method 1: Using Value in U.S. Dollars

$7 \div 0.0104 = 673.077$ or 673 yen

Method 2: Using Units per U.S. Dollar

$7 \times 96.15 = 673.05$ or 673 yen

The value of currency, like most things, is affected by supply and demand. Currency exchange rates are affected by the country's economic condition and political stability.

In the early 1990s, Japan had a very strong economic condition and the value of the yen went up. In 1990, it took 135 yen to buy the same amount as one U.S. dollar would buy. In 1994, it took only 96 yen to buy the same amount as one U.S. dollar. Illustration 14-7.4 compares the values from 1991–1994.

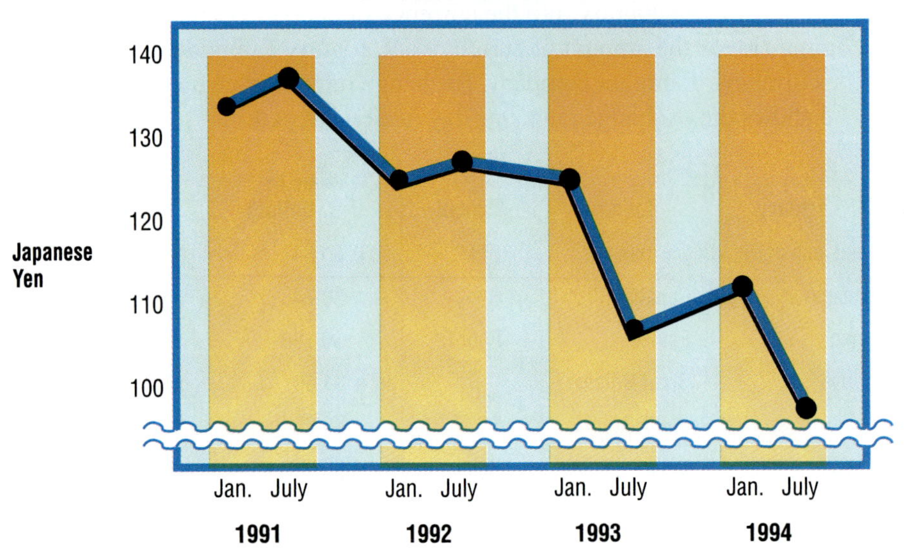

Illustration 14-7.4. Comparison of Value of Yen and Dollar 1991–1994

EXERCISE YOUR SKILLS

Use Illustration 14-7.3 for foreign exchange rates in these exercises. Round to the nearest dollar.

11. Your business bought a machine in Germany for 10,000 DM. How much did you pay for the machine in U.S. dollars?

12. Your company bought some computer equipment in Japan for $100,000. How many yen did the computer equipment cost?

An international company conducted business with several companies last month. The purchase prices of items are given in U.S. dollars. Find the amount to be paid in local currency for each company.

13. Tekni of Saudi Arabia, $50,000 for equipment

14. Soni Sarl of France, $250,000 for machinery

15. Alfreds, Ltd of Canada, $8,500 for office supplies

16. Sofo of India, $320,000 for equipment

17. A British company bought some equipment from an Indian company. One British pound was worth 50 rupees. The cost of the equipment was £35,000 and was paid for in rupees. How many rupees did the company pay?

You received an invoice from a company in Japan. For each item and the total, find the amount owed in U.S. dollars based on the currency chart shown. Round answers to nearest dollar.

18.	Photographic supplies	25,000 ¥
19.	Camera repair	4,000 ¥
20.	Photographic equipment	153,000 ¥
21.	Total	? ¥

22. ♦ Answer the questions at the beginning of the lesson. For the question on currency, use the current currency exchange rate found in the finance section of most newspapers.

PUT IT ALL TOGETHER

23. Based on what you have learned, do you think a strong dollar means that the United States would import or export more goods and services? What does that mean for the balance of trade? Include a visual or verbal explanation to support your conclusions.

MIXED REVIEW

1. $546{,}644 - 25{,}539$

2. $954 + 364 + 775 + 936$

3. 953×953

4. $950{,}526 \div 18$

5. Find 17% of 12,053. Round to the nearest whole number.

6. Jake Madison has storm doors installed at a cost of $520. He estimates that the storm doors will save 3% of his annual winter fuel bill. If his winter fuel bill averages $1,200 what is the payback period for the doors to the nearest whole year?

7. Linda Swan makes this deposit at her bank: (coins) 33 pennies, 25 nickels, 38 dimes, 25 quarters; (bills) 19 ones, 18 fives, 26 tens, 23 twenties; (checks) $153.47, $85, and $54.27. What is the total amount of Linda's deposit?

8. Pat bought these items at an office supply store: $3\frac{1}{2}$ dozen pencils @ $2.19 per dozen; 2 dozen boxes of correction tape @ $6.10 per box; $\frac{1}{2}$ carton of envelopes @ $12.73 per carton. What was Pat's total bill?

9. A real estate agent sold a 10-room house for Benjamin for $235,600. The listing price of the house was $250,000. The agent deducted a 6% commission. What net proceeds did Benjamin receive?

10. Your business bought a machine in Germany for 1,500 DM. How much did you pay for the machine in U.S. dollars?

14-8 LOOKING AT CAREERS

OBJECTIVES

In this lesson, you will learn about

- *career planning;*
- *compensation considerations in an international job market; and*
- *sources of further information about careers in general.*

As trade among nations expands, global career opportunities are increasing.

◆ Imagine your company asks you to relocate to another country. Make a list of the kinds of problems you might face.

WARM *UP*

a. 63,543 × 6,000 **b.** 592,641 ÷ 9

c. Find what percent (to the nearest tenth) 150 is of 6,260.

d. To the nearest whole dollar, find 0.3% of $15,952.

e. Find the mean and median of: 64, 75, 86, 67, 88, 97, 61, 64.

CAREERS IN A GLOBAL MARKET

Today's worker will likely change careers several, or even many, times in his or her work life. On-going development of work skills and job-search skills are critical. As you begin to plan your future, you can think about what type of work you like to do, what you are good at doing, and your personal goals. You also can think about what kinds of opportunities and challenges you will encounter in a global marketplace.

As the global nature of business increases, careers in some specialized fields will increase. For example, careers that combine science, engineering, and computer skills with knowledge of languages (especially Japanese, Chinese, and Spanish) might involve translating technical journals and preparing contracts for overseas jobs. Such careers are international in nature even if you never leave the United States.

Critical Thinking What kinds of skills do you need in an international market?

If you want to find a job or career outside the United States, certain documents are usually required. A **passport** is a government document that proves your citizenship. A **visa** is a stamp of endorsement issued by a country that allows a passport holder to enter that country. If you plan to work in a country, you may need a **work visa** or a **work permit.**

If you want to work outside the United States, be sure to become familiar with the customs and traditions in the country where you will be working. For example, in some Middle Eastern countries, pointing your index finger is considered impolite and vulgar. A nod of your head may mean "yes" in the United States, but it means "no" in Bulgaria and Greece.[1]

Compensation If you work abroad, you may receive an **expatriate bonus** in addition to your salary. This is a type of bonus paid to persuade an employee to work abroad. It helps make up for adjustment problems or hardships caused by living abroad.

You may also receive pay for moving expenses and other costs that you would not have if you did not work abroad. It may cost more to live abroad. For example, if you pay $400 per month in rent in the United States, the same type of housing may cost $596 in London, England. Therefore, your base pay in a foreign country is determined by what you would earn in the United States and then adjusted according to the country in which you work.

EXAMPLE Your base salary is $20,000 per year on your job. Your company asks you to relocate to Buenos Aires, Argentina. Based on the index of cost of living shown in Illustration 14-8.1, what will your base salary be while located abroad?

SOLUTION

$20,000	base salary in United States
× 1.35	cost-of-living adjustment
$27,000	base salary while in Argentina

Critical Thinking How would you determine the base salary in the U.S. if you were given the cost of living index and base salary in Argentina?

[1]Roger E. Axtell, editor, *Do's and Taboos Around the World,* A Benjamin Book, 1993.

INDEX OF COST OF LIVING IN SELECTED LOCATIONS AROUND THE WORLD

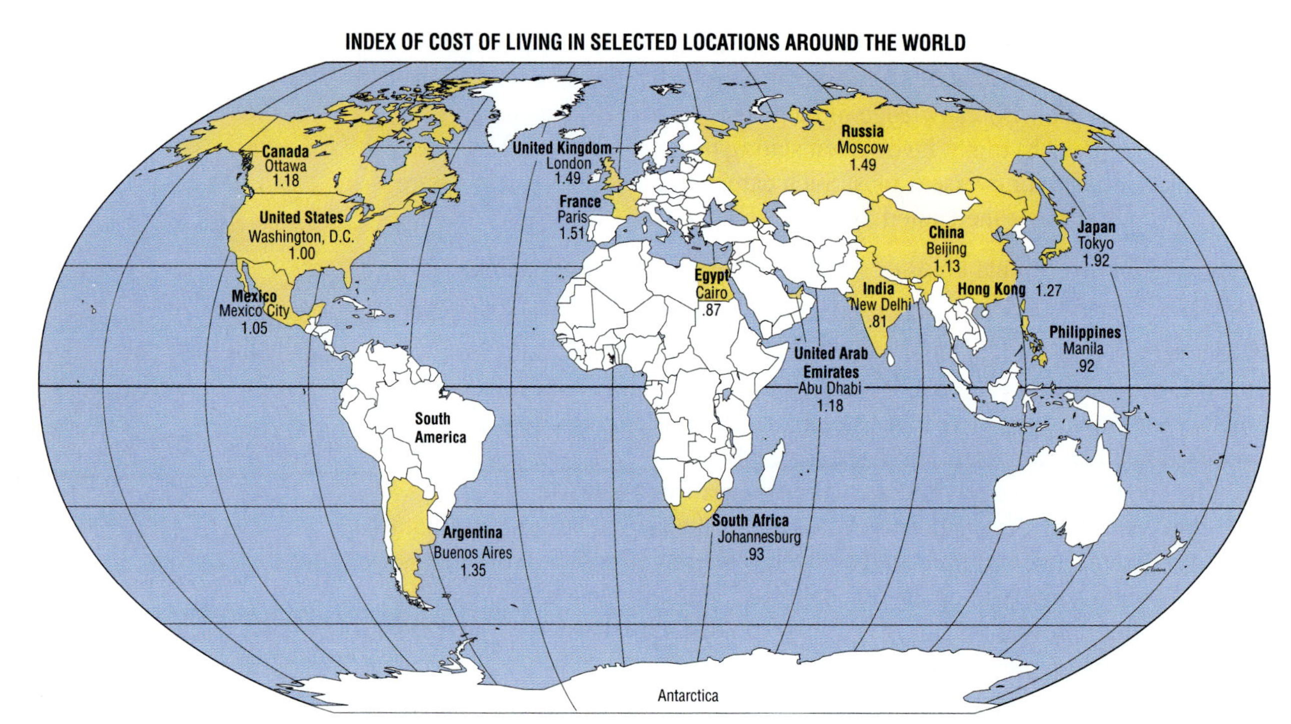

Illustration 14-8.1. Index of Cost of Living Around the World

Source: Adapted from *U.S. Department of State Indexes of Living Costs Abroad, Quarters Allowances, and Hardship Differentials—October 1992* (Washington, D.C.: Department of State Publication 9994, Bureau of Administration, Office of Allowances.)

PROBLEM SOLVING

Interviewing for Information

Frequently, to solve a problem, you will need more information than you have available. You can find information that will help solve job-related problems by asking for information and assistance. If the problem you are trying to solve is related to finding a job, tell friends and family, teachers, and professional people about your interests. Ask for their ideas about potential careers.

The technique is often called "interviewing for information" by career consultants. Look back at the Career Explores that begin each chapter of this book. What kinds of questions do you think were asked to gather the information? Did they provide you useful information? What questions do you have about those careers? Make a list of questions you have about a career that interests you.

Sources of Career Information

Your local public library or school library are two good places to start looking for information on careers. The *Occupational Outlook Handbook*, published by the Bureau of Labor Statistics, provides detailed information about many different career areas. Other helpful resources include the *Dictionary of Occupational Titles* and the *Occupational Outlook Quarterly*. If you are considering an international career, *International Careers*, published by Bob Adams, Inc., Publishers, 1990, is an excellent source. The United Nations, based in New York City, also will furnish information on its member countries.

Newspapers, magazines, and journals are another source of excellent information. If you have an idea of the area in which you want to work, investigate what professional journals are available.

You can also find out about careers and opportunities around the world by "surfing" the Internet. We'll show you how in the Explore Technology that ends this chapter.

EXERCISE YOUR SKILLS

Study the careers highlighted at the beginning of each chapter in this text.

1. Which career interests you most? Why?
2. Which career interests you least? Why?
3. What is important to you in a career?

Copy and complete the table. Use Illustration 14-8.1. Round answers to the nearest dollar.

	Location	Index	Base Salary in U.S.	Base Salary in Specified Location
4.	Paris, France		$25,000	
5.	Tokyo, Japan		$35,000	
6.	Johannesburg, South Africa			$20,000
7.	New Delhi, India		$45,000	

Look in the classified advertisements of a major Sunday paper.

8. Find a job that interests you. List the qualifications and experience needed for that job. Explain why you find the job interesting and how you could prepare for that type of work.
9. Find a job that involves international work of some type. What qualifications and experience are required for that job? Would you be interested in that job? Why or why not?
10. Write a description of your ideal career. Then list ways in which you can start preparing for this job now.

MIXED REVIEW

1. 9 030 m = ? km
2. 4.2 m = ? cm
3. $12\frac{1}{2}\%$ of $53,056 is?
4. Find the invoice price of a watch with a list price of $49.95 and series discounts of 15%, 10%.
5. Angie had $100 in change in her cash register at the start of work. At the end of her work period, there was $1,858.18 in the cash register. The register totals showed that she received $2,230.64 and paid out $273.48 during the period. How much was the cash over or short?

EXPLORING THE INTERNET

Finding and Accessing Information on the Web

The World Wide Web (WWW), or more simply the web, is part of the Internet. It is an ever changing collection of millions of documents stored on computers found throughout the world. These documents may contain text, images, sound, and movies.

Since the web contains a vast amount of information, some computer sites on the web provide you with *search tools* to help you find what you need. Search tools let you enter words that describe what you are looking for. The tools then search the web for documents that contain the information you seek. For example, if you wanted to find information about the trade deficit, you might connect to a web site called Yahoo!. When you are connected, you would see Illustration ET14-1, Yahoo!'s home page. A home page is the document displayed when you connect to a web site computer. Enter "trade deficit" in the white box, click on the search button, and the Yahoo! search tool will do your work for you.

The result of the search is shown in ET14-2—a list of 1,328 different documents on the web that contain information about the trade deficit. This search took seconds. Because the web is constantly changing, if you did the same search again, the list of documents might be different.

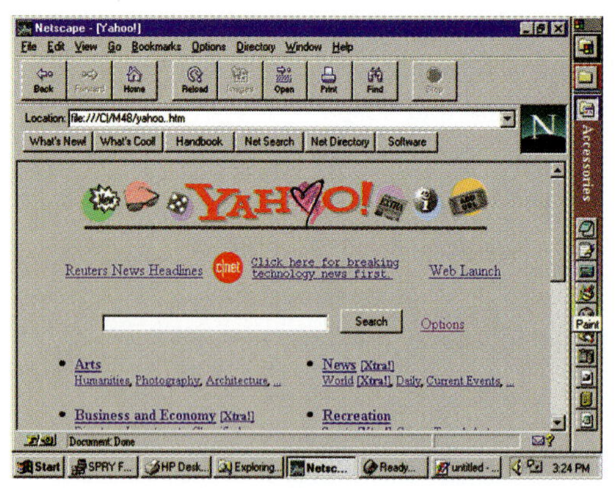

Illustration ET14.1.

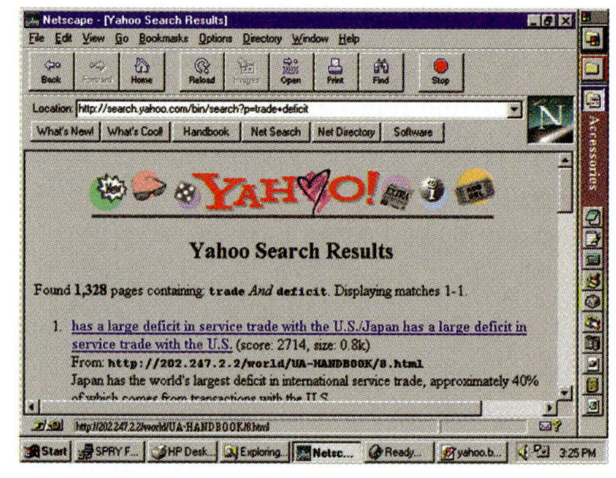

Illustration ET14.2.

Using, Printing and Saving Hypertext Documents

Web documents usually are hypertext documents. That means they contain links to other documents. These links may be text that is highlighted in some way, or they may be images. For example, all of the words shown underlined in blue in Illustrations ET14-2 are links to the documents that Yahoo found containing information about the trade deficit. To look at the documents, click on the links (underlined words in blue) on the list, one at a time. When you do, you will be transported, one at a time, to each document. You can use the features of your web browser to print or save a document as a file on your system.

Surfing to Home Pages on the Web

Many businesses, government agencies, schools, and individuals have home pages to which you can connect. Every web site home page has a web address so that you can find it. Even Applied Business Mathematics has a home page. Its address is:

http://www.thomson.com/swpco/internet/mb14nal.html

In computer talk, web addresses are called Uniform Resource Locators or URLs. In the URL for Applied Business Mathematics home page, http stands for hypertext transfer protocol. A protocol is a set of rules that are followed by two computers when they exchange information. The hypertext protocol is followed by computers using the World Wide Web to share documents.

The www stands for the World Wide Web. Thomson.com stands for Thompson Publishing Company. This part of the address tells you the computer system on which the home page is located. You can tell that it is a commercial web site because it contains .com. A site at a school or university would end in .edu. A government agency would end in .gov; a military site would end in .mil, and other organizations end in .org. Swpco and internet are subdirectories. Mb14nal.html is the actual home page document. Thus, the URL tells us that we must use the http protocol to find the document mb14nal.html at Thompson Publishing Company's World Wide Web computer system in the subdirectory /swpco/internet.

Searching For A Job on the Internet

Many have found that job searches on the Web have simplified the process. The job hunter can find lists of jobs, and employers can find resumes of prospective employees. Some companies have a permanent link to employment opportunities on their home page.

Suppose you are searching for a job as a marketing coordinator or facilities manager. Search the Web to find at least five relevant job openings in those fields in your geographic area. List the job title, company, location, and any relevant information about that position. You might start with the Online Career Center site, but don't limit yourself to just one site. There are numerous other on-line job services, including comprehensive sites to help prepare you for the job search before you start searching for positions.

DESIGN YOUR OWN

Find a topic in Chapter 14 that interests you and search for information about the topic on the web using a search tool. Then, prepare a brief report that lists at least 3 web addresses along with a brief description of what you found at each address.

CHAPTER 14 SUMMARY AND REVIEW

VOCABULARY REVIEW

balance of trade *14-6*
break-even point *14-4*
cash over *14-5*
cash short *14-5*
credit memo *14-5*
customer account *14-5*
expatriate bonus *14-8*
free-on-board (FOB) *14-1*
gross domestic product *14-6*
invoice price *14-1*
markdown *14-3*
markup *14-3*
net price *14-1*
net purchases *14-1*
purchase invoice *14-1*
sales invoice *14-1*
selling price *14-4*
trade discount *14-1*

MATH SKILLS REVIEW

14-1 ♦ Write each decimal as a mixed number in simplest form.

1. 1.10 **2.** 2.75 **3.** 5.5 **4.** 6.125

14-2 **5.** Write the complement of 32%

14-4 ♦ Solve for x:

6. $3x = 18$ **7.** $4x - 8 = 48$

APPLICATIONS

14-1 ♦ Rate of Discount × List Price = Trade Discount; List Price − Trade Discount = Invoice Price; Rate of Trade Discount = Trade Discount\List Price

♦ Rate of Cash Discount = Cash Discount\Invoice Price

♦ Purchases − Returns and Allowances = Net Purchases

8. You pay a $500 invoice dated September 19 on October 31. The terms are 2/15, 1/45, n/60. How much should you pay?

9. A television listed at $650 in a wholesale catalog is sold to a retailer for $552.50. Find the rate of trade discount.

10. Sweeney's Hardware purchases for July were $28,565, and the returns and allowances were $1,465. Find the net purchases.

14-2 ♦ You can calculate a single discount equivalent of a series discount using percents, a table, or the complement method.

11. What is the invoice price of 4 chairs that list at $60 each, with discounts of 30%, 20%, and 10%?

14-3 ♦ The basic pricing equation is Cost + Markup = Selling Price.

♦ When markup is based on cost, the selling price is = to cost plus markup %. When markup is based on selling price, cost % + markup % equals 100% (selling price).

♦ To find the missing variable in a markup equation, determine whether the markup is based on cost or selling price. Then use the appropriate formula from page 535 to find the missing variable.

- ♦ Markdown is stated as a percentage of selling price. The basic markdown formula is Marked Price − Markdown = Selling Price.
- ♦ Net income on a single item = Gross Profit − Estimated Operating Expenses
- ♦ Rate of Gross Profit = Gross Profit ÷ Net Sales
- ♦ Rate of Net income = Net Income ÷ Net Sales

12. What is the most a store can pay for a tie for a line of ties that sell for $20 if the markup must be 32% of the selling price?

13. A computer printer marked at $450 is marked down 20%. Specify the amount of discount and the selling price.

14-4 ♦ Sales income − Total Factory Cost = Gross Profit

♦ Gross Profit − Operating Costs = Net Income

14. A machine sells for $885 and has these manufacturing costs: materials, $195; labor, $259; factory expense, $85. What is the manufacturer's gross profit on each machine?

14-5

15. At the start of the day, Maureen put $200 in change in her cash register. The register readings at the end of the day showed total cash received, $3,929.23, and total cash paid out, $529.50. The cash in the register at the end of the day was $3,463.43. By how much was the cash short or over?

14-6

16. In a recent year, one country had imports of $294 billion and exports of $331 billion. What was that country's trade surplus or deficit?

17. In a recent year a country had a GDP of $575,000,000 and a per capita GDP of $2,500. Based on this information, what was the population of that country in that year?

14-7 ♦ The globe is divided into 24 time zones 15° apart. Add one hour for each 15° of longitude or time zone you move East. Subtract one hour for each 15° you move West.

♦ To convert between Celsius and Fahrenheit, use the appropriate formula: $F = 1.8C + 32$; $C = \frac{5}{9}(F - 32)$

♦ The exchange rate is the amount of currency one country can trade for one unit of the currency of another company.

18. It is 6 A.M. in New York City. What time is it in Lima, Peru?

19. Winter temperatures in the mid-west United States averaged −2°F one week. What was the equivalent temperature on the Celsius scale?

20. You bought equipment in Canada for $100,000 Canadian. How much did you pay in United States dollars? Use Illustration 14-7.3.

14-8 ♦ Base pay in a foreign country is your U.S. salary adjusted for the country in which you work using an international cost-of-living index.

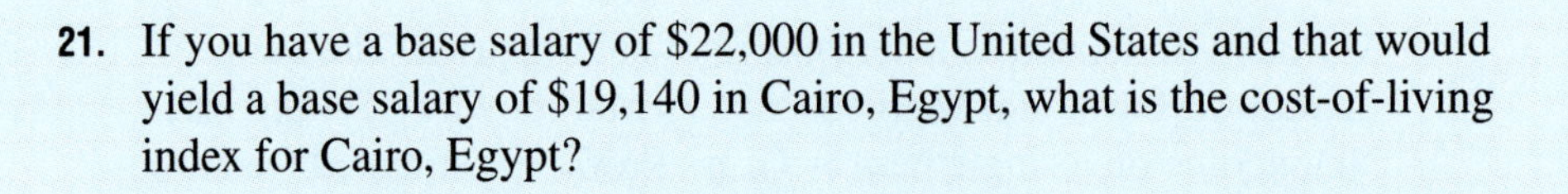

21. If you have a base salary of $22,000 in the United States and that would yield a base salary of $19,140 in Cairo, Egypt, what is the cost-of-living index for Cairo, Egypt?

WORKPLACE KNOWHOW: **WORKING WITH CULTURAL DIVERSITY**

The ability to work with cultural diversity includes working productively with people with different ethnic, social, and educational backgrounds. This ability requires an understanding of different cultures, a respect for the rights of others, a willingness to make judgements based on individual performance—not stereotypes, and a sensitivity to the concerns of other ethnic and gender groups.

Assignment Read the following information carefully. Work with a partner. One partner plays a Japanese businessperson, the other an American businessperson. The American is negotiating the selling price on an item that the Japanese business person wants to buy. Use what you've learned about pricing for profit and vendor discounts in the negotiation. Be prepared to present your situation to the class.

Bowing is the traditional Japanese greeting. However, the Japanese may greet you with a handshake. Their handshake may be weak, but this does not indicate a lack of assertiveness. If a Japanese person bows to you, bow to the same depth. The depth of the bow indicates the status relationship between you. As you bow, lower your eyes and keep your palms against your thighs.

Business cards are presented after the bow. A Japanese person will take time to study the card presented. The Japanese use both hands to present their cards. An American should read the card, and memorize the information. Ask for help in pronunciation if necessary.

Gift giving. Expect to receive a gift when you first meet a Japanese businessperson. Accept the gift with both hands. Do not open it unless you are asked to. The next time you meet the businessperson present a gift of equal value. Your gift should always be wrapped with pastel-colored paper with no bows or ribbons. The Japanese don't usually open gifts upon receiving them.

Gestures carry great meaning to the Japanese. Avoid dramatic gestures or expressions. Pointing is considered impolite. Instead, wave your hand, palm up toward an object.

When talking, Japanese stand farther apart than Americans do and rarely make direct eye contact. Silence is considered useful by the Japanese. They may not immediately answer your question. When talking with a Japanese businessperson use their last name plus the word *san,* which means Mr. or Ms. Don't suggest that the Japanese call you by your first name. Smile, even if you are upset or displeased.

After you have role played your situation, discuss how it felt to act and react differently.

Work Skills Portfolio Include an assessment of your purchasing and pricing skills, as well as your understanding of international trade issues. If you had access to the Internet, include the results of your searches. Also include in the portfolio your answers to questions dealing with your career interests and skills.

CHAPTER 14 TEST

Math Skills

1. \$186 less 5% + \$43 = ?
2. An invoice dated June 18 with net 15 terms is due?
3. \$196 less 15%, 7%, 4%.
4. The complement of 27% is ?
5. Cost = \$45; Markup = \$32; Selling Price = ?
6. Markup = \$32; Selling Price = \$77; Rate of Markup = ?
7. $4x - 14 + 32 = 5{,}846$. Find x.
8. $1.8 \times 42 - 32 = ?$

Applications

9. The Amos Supermarket bought the following items. The invoice was dated August 16; the terms were net 45 days. Find the due date and the total amount of the invoice.

   24 food storage boxes @ \$2.25
   100 boxes trash bags @ \$2.78
   50 storage bins @ \$12.45
   18 measuring cups @ \$1.93
   Total Due

10. A store window displays a chain saw with this sign: Was \$349.99—Now \$283.74. What is the rate of markdown, to the nearest tenth of a percent?
11. A toy store buys a brand name doll for \$18 each, total cost, and sells them for \$24. The store's operating expenses are estimated to be 21% of net sales. What rate of net income or loss based on selling price did the store make on each doll?
12. When he opened his shop on March 19, Judson Bruns put \$175 in change in his cash register drawer. At the end of the day, the cash register readings showed that \$812.34 had been taken in, that \$213.34 had been paid out in cash refunds, and that Judson had taken out \$20 for his personal use. The cash on hand in the drawer at the end of the day was \$755.75. By how much was he cash short or over?
13. One year a country has exports totaling \$45 million and imports totaling \$63 million. What was their trade deficit or surplus?
14. It is 6:00 P.M. in Cairo. What time is it in London, England?
15. It is 72°C. What is the equivalent temperature on the Fahrenheit scale?
16. You purchase an item in India for 500 Re. How much did you pay in American dollars?
17. You earn a base salary of \$53,000 in Buenos Aires, Argentina. If the cost-of-living index is 1.35, what is your base salary, to the nearest dollar, in the United States?

UNIT 4 TEST: CHAPTERS 12–14

◆ Choose the best definition for each term.

12-3 **1.** line graph
12-5 **2.** purchasing power of a dollar
13-2 **3.** return on sales
13-6 **4.** retained earnings
14-1 **5.** terms of sale
14-3 **6.** break-even point
14-8 **7.** passport

a. When income from sales equals the cost of producing and selling a product.
b. How and when an invoice is to be paid.
c. A document that allows you to work in another country.
d. A document that proves citizenship.
e. An account where net income is recorded on a regular basis.
f. How much a dollar can buy.
g. The middle number in a set of data.
h. Used to show changes over time.
i. Pure profit.

◆ Choose the best answer.

12-1 **8.** In the following set of numbers,

7, 3, 18, 9, 4, 22, 9, 16, 24

22 is the

a. mean **b.** mode **c.** range **d.** none of the above

12-4 **9.** Total monthly business expenses for Tim's Drywall Service is $1,000. Tim plans to display the allocation of expenses on a circle graph. To the nearest degree, how many degrees will be allocated to monthly transportation expenses of $200 a month?

a. 20° **b.** 2° **c.** 72° **d.** 7°

13-3 **10.** A drill press that cost $3,500 is expected to be used for 6 years and then traded in for $850. Using the sum-of-the digits-method, what will be the book value of the drill press at the end of the second year?

a. $1,120 **b.** $2,869.05
c. $2,649.15 **d.** $2,019.05

14-6 **11.** Which country has the lowest trade deficit? Country A: Imports: 14.6 billion; Exports: 12.3 billion; Country B: Imports 12.5 billion, Exports, 6.7 billion; Country C: Imports 22.5 billion; Exports, 24.3 billion. Country D: Imports: 17 billion, Exports 16.8 billion

a. Country A **b.** Country B
c. Country C **d.** Country D

◆ Solve each exercise.

12-2 **12.** The six faces of a perfectly shaped cube are marked 1, 2, 3, 4, 5, and 6. If you toss the cube, there is an equal chance that any one of the 6 numerals will show on top when it comes to rest. In 1 toss of the cube, what is the probability of tossing a numeral greater than 2?

13-1 **13.** For the last six months of a year, a shop made a gross profit of $74,600 on net sales of $145,000. Operating expenses for the period were $42,780. The net income was what percent of net sales?

13-5 **14.** The Keepsake Boutique had these assets and liabilities on July 1: Cash, $1,015; Merchandise inventory, $7,125; Land and buildings, $59,800; Furniture and equipment, $12,008; total debts, $10,040. Estimate the boutiques capital, then find the actual capital.

13-6 **15.** Ed and Lou, partners, invested $5,000 and $15,000 respectively in their business. The net income for 1 year was $12,016, which they share in proportion to their investments. For each partner find the net income actually received.

13-7 **16.** The A & J Sewing Center went bankrupt owing $74,600. The net cash available for the creditors was $11,190. What amount will be paid to a creditor with a claim of $5,250?

The Writer's Shop purchased supplies from Book Masters Services for $596.75. The invoice was dated November 12. The terms were 2/15, n/30.

14-1 **17.** When does the store have to pay their invoice in order to receive a cash discount? What cash discount can they claim?

14-2 **18.** What would The Home Shop pay for 15 dozen frying pans that list at $107 a dozen, with discounts of 20%, 15%, and 5%?

19. How much does a customer pay for a hand-held calculator marked $15.95 if a retail discount of 15% is given?

20. A basketball backboard, goal, and pole sell separately for $35, $19, and $59. If you buy them as a set, the price is $73.45. What rate of discount are you getting for the set?

14-5 **21.** In the morning, Lou put $100 in change in the cash drawer of his cash register. The readings at the end of the day showed total cash received $3,569.80 and total cash paid out $183.45. The cash on hand in the cash drawer at the end of the day was $3,486.25. Was the cash short or over? By how much?

14-7 **22.** If a country had a gross domestic product of $42 billion and a population of 21 million, what would be the per capital GDP for that country?

14-8 **23.** You have been asked by your company to move from the United States to Cairo, Egypt for one year. You are currently earning $43,480 a year. In exchange for agreeing to move, you negotiate an increase of 15% in your U.S. base salary. What will your base salary be while you are living in Cairo?

WORKSHOP 1: Number Theory

When you work with numbers as well as algebraic expressions, some properties always are true. For example, the product of 3×7 is the same as the product of 7×3 and zero times any number is zero. These properties are summarized in the table below. Remember, when two variables are written together, such as ab, it means $a \times b$.

Properties	Numerical Example	Algebraic Example
Associative Property of Addition and Multiplication	$(2 + 3) + 4 = 2 + (3 + 4)$ $(2 \times 3) \times 4 = 2 \times (3 \times 4)$	$(a + b) + c = a + (b + c)$ $(ab)c = a(bc)$
Commutative Property of Addition and Multiplication	$2 + 3 = 3 + 2$ $2 \times 3 = 3 \times 2$	$a + b = b + a$ $ab = ba$
Distributive Property of Multiplication over Addition	$2 \times (3 + 4) = (2 \times 3) + (2 \times 4)$	$a(b + c) = ab + ac$
Addition Property of Zero	$2 + 0 = 2$	$a + 0 = a$
Multiplication Property of One	$2 \times 1 = 1$	$a \times 1 = a$
Multiplication Property of Zero	$2 \times 0 = 0$	$a \times 0 = 0$

When you evaluate, or find the value of, numerical or algebra expressions, you must use the proper order of operations to get the correct answer.

Order of operations

- First, evaluate anything in parentheses.
- Next, multiply and divide in order from left to right.
- Then, add or subtract in order from left to right.

Use properties and order of operations to evaluate each expression.

1. $2 \times 5 + 3$

2. $2 \times (5 + 3)$

3. $(2 \times 5) + 3$

4. $9 \times 5 \times 0$

5. $9 + 5 + 0$

6. $8 + 9 + 12$

7. $12 - (6 \times 2)$

8. $(12 - 6) \times 2$

9. $12 - 6 \times 2$

10. $18 \times (14 - 9)$

11. $81 \div (9 \div 3)$

12. $81 \div 9 \div 3$

In the following, use $a = 5, b = 3$, and $c = 0$.

13. $ab - 8$

14. $ca + cb$

15. $6 \times (a + 4) - b$

WORKSHOP 2: Solving Equations with Addition and Subtraction

If you add 6 to some number, you can subtract 6 from that sum to get back to your original number. Addition and subtraction are inverse operations.

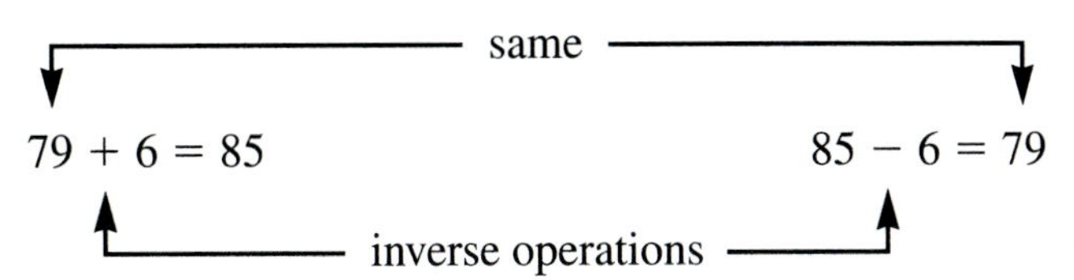

You can use inverse operations to solve equations.

$$x + 14 = 20$$
$$x + 14 - 14 = 20 - 14 \quad \leftarrow \text{Subtract 14 from each side.}$$
$$x = 6 \quad \leftarrow \text{Simplify.}$$

To check your solution, substitute it for x in the original equation.

$$x + 14 = 20$$
$$6 + 14 \ ? \ 20$$
$$20 = 20$$

Shelly earned \$18. That amount was \$7 less than David earned. How much did David earn?

Write an equation. (x is amount David earned.) Solve the equation. → $18 = x - 7$

$$18 + 7 = x - 7 + 7$$
$$25 = x$$

Check your solution. → $18 = 25 - 7$ ✓

Answer the question. → David earned \$25.

Solve and check each equation.

1. $x - 4 = 8$
2. $x + 4 = 8$
3. $x + 7 = 14$
4. $28 = x - 8$
5. $35 = x - 46$
6. $x + 40 = 76$
7. $127 + x = 147$
8. $657 = x + 168$
9. $x - 593 = 594$
10. $407 = x + 401$

Write an equation. Then solve the problem.

11. Maureen ran 3 blocks farther than Armando. Maureen ran 28 blocks. How far did Armando run?
12. Maria earned \$170. That amount is \$18 more than Ivan earned. How much did Ivan earn?

WORKSHOP 3: Solving Equations with Multiplication and Division

Multiplication and division are **inverse operations.**

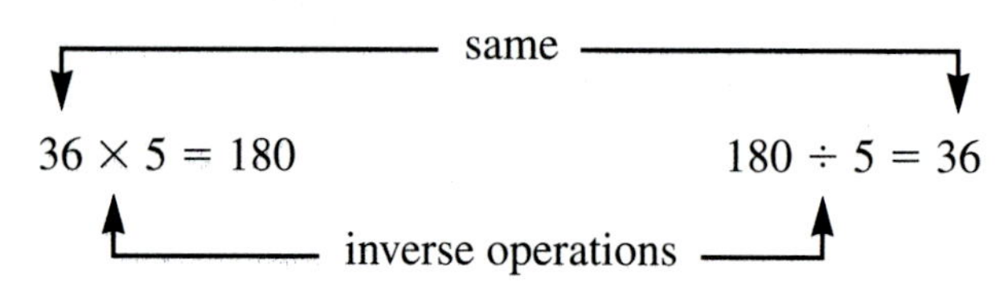

Another important relationship between multiplication and division involves reciprocals.

$180 \div 5 = 36$ $\qquad$ $180 \times \frac{1}{5} = 36$

reciprocals

To divide by a number, you can also multiply by the reciprocal of that number.

The reciprocal of any number, n, is $\frac{1}{n}$, $n \neq 0$. The product of a number and its reciprocal is 1.

$5 \times \frac{1}{5} = 1$ $\qquad$ $\frac{2}{3} \times \frac{3}{2} = 1$

You can use these properties to solve problems.

$3x = 285$

$\frac{3x}{3} = \frac{285}{3}$ Divide each side by 3.

$x = 95$ Simplify.

Here is another way to solve the same problem.

$3x = 285$

$\frac{1}{3}(3) = \frac{1}{3}(285)$

$x = 95$

Solve each equation.

1. $4x = 44$
2. $\frac{x}{8} = 24$
3. $\frac{x}{6} = 120$
4. $32x = 640$
5. $12 = \frac{2}{3}x$
6. $35 = \frac{7x}{5}$
7. $70 = \frac{140}{x}$
8. $21 = \frac{3}{7}x$
9. $x \div 1.5 = 9$
10. Mary Ellen watched television 6 hours one weekend. That was $\frac{2}{3}$ the amount of time her father watched television. How long did her father watch television that weekend?

WORKSHOP 4: Ratios and Proportions

When two quantities are compared, a **ratio** can be used. A ratio shows a comparison of two numbers by division. A ratio can be written in several ways:

2 to 3   2:3   2 out of 3 $\frac{2}{3}$

Two ratios are equivalent when they name the same number. $\frac{2}{3} = \frac{4}{6}$

When two ratios are written as an equation, a **proportion** is created. One way to determine whether two ratios are equivalent is to use cross products. In algebra the bullet symbol is often used in place of the multiplication sign (×). This is to avoid confusion between the × and the *x* which is commonly used as a variable in algebra.

$\frac{8}{3} = \frac{24}{9}$ $3 \bullet 24 = 72$
 $8 \bullet 9 = 72$

The cross products are equal. $\frac{8}{3} = \frac{24}{9}$ is a proportion.

You can use cross products to solve proportions.

EXAMPLE **Solve** $\frac{12}{15} = \frac{n}{120}$.

$\frac{12}{15} = \frac{n}{120}$ These are the cross products.

SOLUTION

$12 \bullet 120 = 15 \bullet n$ Cross products are equal.

$\frac{1440}{15} = \frac{15n}{15}$ Solve the equation.

$96 = n$

Solve each proportion.

1. $\frac{3}{4} = \frac{x}{24}$
2. $\frac{9}{5} = \frac{t}{25}$
3. $\frac{r}{6} = \frac{12}{2}$
4. $\frac{12}{18} = \frac{v}{24}$
5. $\frac{p}{100} = \frac{32}{25}$
6. $\frac{21}{x} = \frac{51}{68}$
7. $\frac{20}{80} = \frac{t}{100}$
8. $\frac{6}{n} = \frac{54}{180}$
9. $\frac{24}{48} = \frac{5}{r}$
10. $\frac{6}{2} = \frac{x}{11}$
11. $\frac{p}{8} = \frac{63}{14}$
12. $\frac{5}{8} = \frac{55}{x}$
13. $\frac{j}{18} = \frac{25}{90}$
14. $\frac{81}{17} = \frac{x}{51}$
15. $\frac{16}{m} = \frac{64}{20}$

WORKSHOP 5: Solving More Equations

Solving some equations requires more than one operation or step.

Miko scored 18 points in the game. That was six less than three times the number of points that Aaron scored. How many points did Aaron score?

$$3x - 6 = 18$$
$$3x - 6 + 6 = 18 + 6 \quad \leftarrow \text{Add 6 to each side.}$$
$$3x = 24 \quad \leftarrow \text{Simplify.}$$
$$\frac{3x}{3} = \frac{24}{3} \quad \leftarrow \text{Divide each side by 3.}$$
$$x = 8 \quad \leftarrow \text{Simplify.}$$

Some equations involve fractions.

$$5(19x + \tfrac{1}{2}) = 12$$
$$\frac{5(19x + \frac{1}{2})}{5} = \frac{12}{5} \quad \leftarrow \text{Divide each side by 5.}$$
$$19x + \frac{1}{2} = 2\frac{2}{5} \quad \text{Simplify.}$$
$$19x + \frac{1}{2} - \frac{1}{2} = 2\frac{2}{5} - \frac{1}{2} \quad \text{Subtract } \tfrac{1}{2} \text{ from each side.}$$
$$19x = \frac{19}{10} \quad \text{Simplify.}$$
$$(19x)\left(\frac{1}{19}\right) = \left(\frac{19}{10}\right)\left(\frac{1}{19}\right) \quad \text{Multiply each side by } \tfrac{1}{19}.$$
$$x = \frac{1}{10} \quad \text{Simplify.}$$

Solve each equation.

1. $2x + 5 = 13$

2. $3(x - 2) = 12$

3. $9 = \frac{1}{2}x + 3$

4. $6x - 11 = 37$

5. $\frac{x + 2}{6} = 1$

6. $\frac{x - 8}{7} = 16$

7. $5(2x + 6) = 100$

8. $\frac{3x - 8}{5} = 14$

9. $3\frac{1}{2} = \frac{2x + 3}{10}$

10. $\frac{1}{2}(3x - 24) = 60$

11. Mr. Williams received 1,392 votes. That was 14 less than twice the number Ms. Calloway received. How many votes did Ms. Calloway receive?

12. It costs \$22 per day plus \$0.12 per mile to rent a car. You rented a car for one day and the cost was \$50.80. How many miles did you drive the car that day?

WORKSHOP 6: Lines and Angles

In geometry, a **line** is a straight line that has no endpoints. A **line segment** is part of a line and has two endpoints. A **ray** is part of a line that has only one endpoint. Two lines intersect in a **point.**

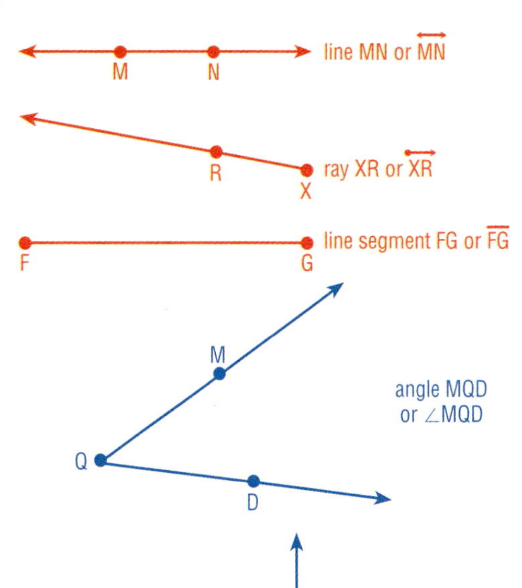

An **angle** is formed by two rays with a common endpoint. The rays are the **sides** of the angle. The symbol of angle is ∠. The common endpoint is the **vertex** of the angle. Angles are measured in degrees (°). A **protractor** is used to find the measure of an angle. Angles are also classified by their measure.

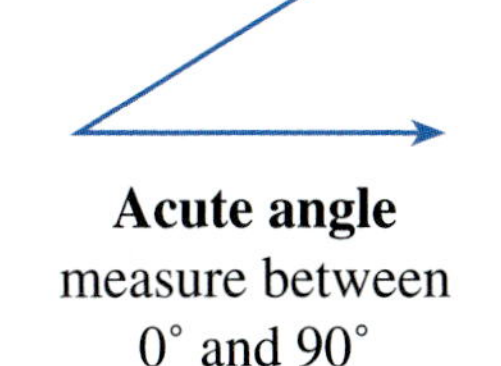

Acute angle
measure between
0° and 90°

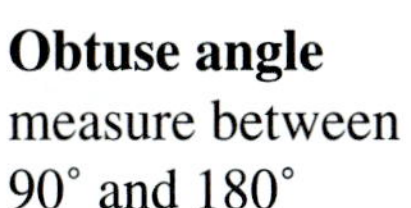

Obtuse angle
measure between
90° and 180°

Right angle
measure of 90°

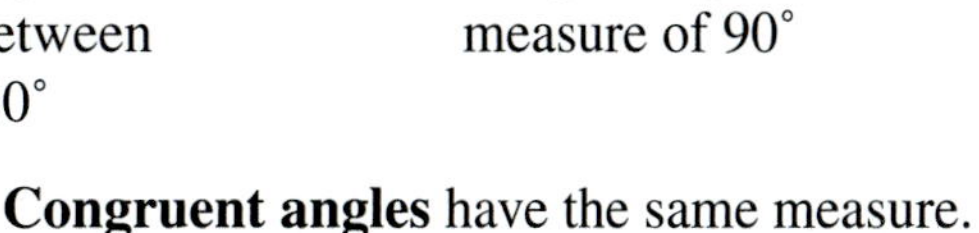

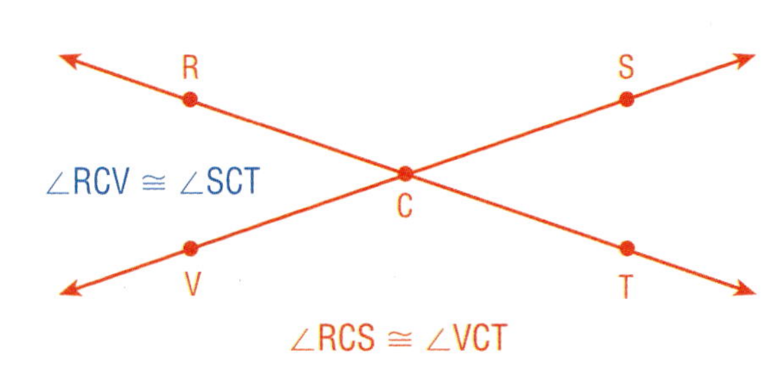

Congruent angles have the same measure. The symbol ≅ means "is congruent to."

When two lines intersect, two pair of **vertical** angles are formed. If the measure of the angles formed is 90°, the lines are **perpendicular** to each other. If two lines are in the same plane and never intersect, they are **parallel lines.** The symbol ∥ means "is parallel to."

Two angles that have a common side are called **adjacent angles.** Adjacent angles whose measures have a sum of 90° are **complementary angles.** Adjacent angles whose measures have a sum of 180° are **supplementary angles.**

Draw and label each of the following.

1. line AB
2. ray DC
3. line segment RW
4. ∠DFJ
5. ∠FDJ
6. two perpendicular lines
7. a right angle
8. an acute angle
9. two parallel lines
10. two complementary angles
11. two supplementary angles

SKILLS WORKSHOP 7: Polygons

Simple closed shapes are called **polygons.** The line segments that make the polygon are called the **sides** and the corners of the polygon are called the **vertices** (plural of vertex). A polygon is named by the number of sides it has.

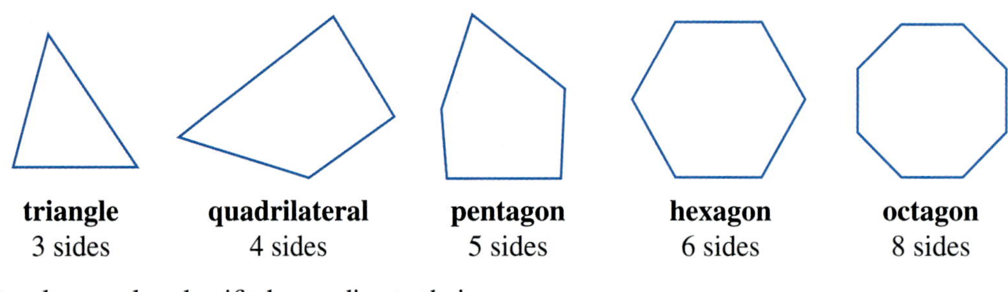

triangle 3 sides | **quadrilateral** 4 sides | **pentagon** 5 sides | **hexagon** 6 sides | **octagon** 8 sides

Triangles are also classified according to their sides or their angle measures. An **equilateral triangle** has three congruent sides. An **isosceles triangle** has at least two congruent sides. A **scalene triangle** has no sides congruent. An **acute triangle** has all acute angles. A **right triangle** has one 90° angle. An **obtuse triangle** has one obtuse angle. The sum of the angles of any triangle is 180°.

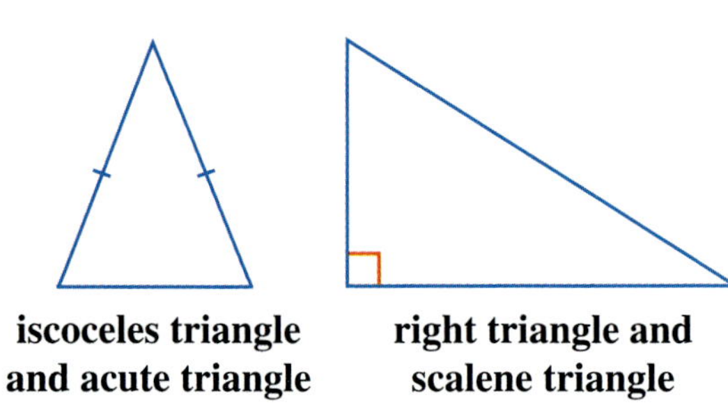

iscoceles triangle and acute triangle | **right triangle and scalene triangle**

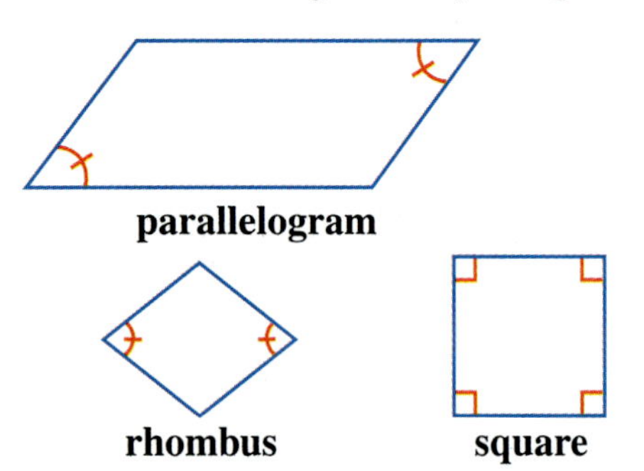

parallelogram

rhombus | **square**

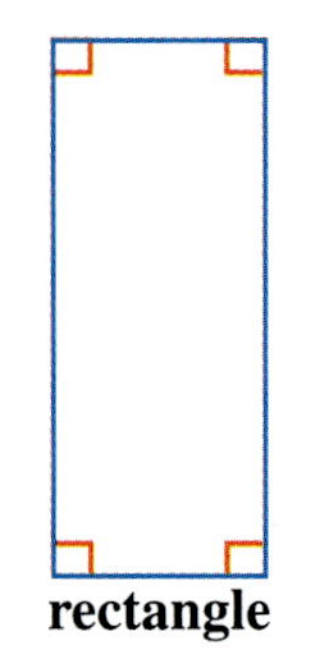

rectangle

A quadrilateral with opposite sides parallel is a **parallelogram,** with four right angles is a **rectangle,** or with four congruent sides is a **rhombus.** A quadrilateral with four right angles and four congruent sides is a **square.** The diagonals of a polygon connect non-adjacent vertices.

Give as specific name as possible for each figure. Congruent sides are shown in red. Congruent angles are marked with ∡.

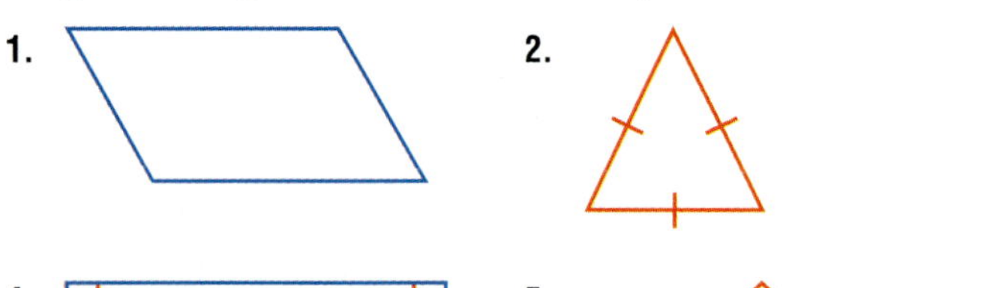

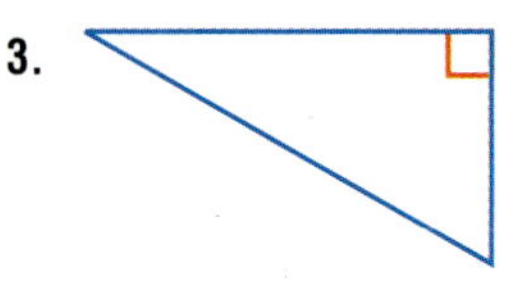

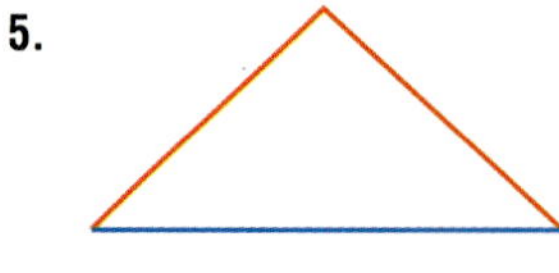

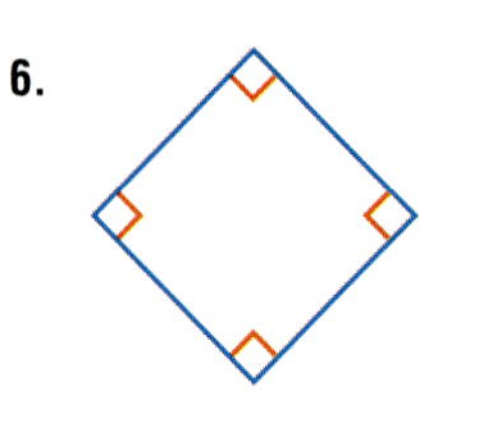

SKILLS WORKSHOP 8: Perimeter, Area, and Volume

The **perimeter** of a polygon is the sum of the measures of its sides. The perimeter of a circle is called its **circumference** and is the distance around the circle. The area of a polygon or circle is the amount of surface it covers and is measured in square units. The **volume** of a three-dimensional figure is the amount of space it occupies or the amount it can hold and is measured in cubic units.

When finding perimeter, area, or volume, you can use a formula.

Perimeter of a rectangle: $p = 2(l + w)$ if l = length and w = width.

Circumference of a circle: $C = 2\pi r$ if $\pi = 3.14$ and r = radius.

Area of a triangle: $A = \frac{1}{2}bh$ if b = base, h = height.

Area of a rectangle: $A = bh$ if b = base, h = height.

Area of a circle: $A = \pi r^2$ if $\pi = 3.14$ and r = radius.

Volume of a rectangular solid: $V = lwh$ if l = length, w = width, and h = height.

EXAMPLE Find the perimeter and area of a rectangle with a length of 25 inches and a width of 20 inches.

SOLUTION

$$p = 2(l + w) = 2(25 + 20) = 90 \text{ inches}$$

$$A = lw = 25 \times 20 = 500 \text{ square inches}$$

EXAMPLE Find the volume of a rectangular solid with a length of $3\frac{1}{2}$ inches, a height of 4 inches, and a width of $6\frac{3}{4}$ inches.

SOLUTION

$$V = lwh = 3\frac{1}{2} \times 4 \times 6\frac{3}{4} = 94\frac{1}{2} \text{ cubic inches}$$

1. Find the area of a triangle with a base of 10 centimeters and a height of 12 centimeters.
2. Find the perimeter and area of a square with sides of 12 inches.
3. Find the circumference and area of a circle with a radius of 10 centimeters.
4. Find the volume of a rectangular solid with a length of 3 feet, a height of 6 feet, and a width of 10 feet.
5. Find the perimeter and area of a rectangle with a length of 15.5 meters and a width of 22 meters.
6. Find the perimeter of a triangle with sides that are 5 yards, 6 yards, and 4 yards.

EXPLORE GEOMETRY

SKILLS WORKSHOP 9: Pythagorean Theorem

In a right triangle, the shorter sides are called **legs.** The side opposite the right angle is called the **hypotenuse.** The hypotenuse is the longest side of the triangle.

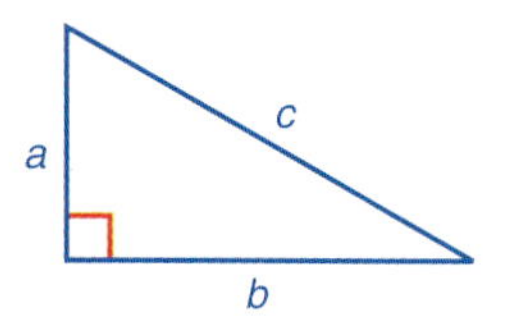

A special relationship exists between the lengths of the sides of any right triangle. The **Pythagorean Theorem** states this relationship.

> ### Pythagorean Theorem
> In a right triangle, the sum of the squares of the measures of the two legs is equal to the square of the measure of the hypotenuse.
>
> $$a^2 + b^2 = c^2$$

The converse of the Pythagorean Theorem is also true.

> ### Converse of the Pythagorean Theorem
> If the sum of the squares of the measures of two sides of a triangle is equal to the square of the measure of the third side, then the triangle is a right triangle.

Determine whether a triangle with sides of the given lengths is a right triangle.

Lengths of sides: 8, 10, 13

Check: Does $a^2 + b^2 = c^2$

$$8^2 + 10^2 \ ? \ 13^2$$
$$64 + 100 \neq 160$$

No, these sides do not make a right triangle.

Lengths of sides: 10, 24, 26

Check: Does $a^2 + b^2 = c^2$?

$$10^2 + 24^2 \ ? \ 25^2$$
$$100 + 576 = 676$$

Yes, these sides do make a right triangle.

Tell whether a triangle with sides of the given lengths is a right triangle.

1. 3, 4, 5
2. 6, 7, 8
3. 10, 12, 13
4. 20, 24, 25
5. 39, 42, 45
6. 14, 48, 50
7. 15, 17, 22
8. 15, 36, 39
9. 1.5, 1.5, 2.5
10. A right triangle has legs that are 15 centimeters and 20 centimeters long. Use guess-and-check to find the length of the hypotenuse of the triangle.

SKILLS WORKSHOP 10: Graphing Equations

A coordinate plane consists of two perpendicular number lines. The horizontal number line is the x-axis. The vertical line is the y-axis. The number lines intersect at the origin.

Points are located on coordinate plane by using ordered pairs. Point W is located at $(-3, 3)$. Point J is at $(2, -1)$. Notice that a positive number is either above or to the right of the origin. A negative number is either below or to the left of the origin. Notice that the order of the numbers makes a difference. Point A is at $(-1, 2)$ and that is a different location than Point J.

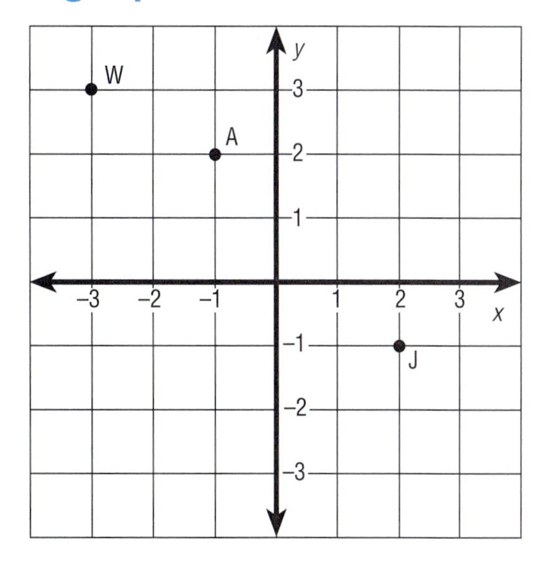

Suppose a boat is traveling at a constant rate of 7 miles per hour. The distance the boat travels, d, is related to the amount of time the boat travels, t, by the equation $d = 7t$. The time, t, is in hours. You can show this relationship with a graph.

Show solutions of the equation in a table.

t	d
1	7
2	14
3	21
4	28

The table provides ordered pairs to be graphed. (1, 7), (2, 14), (3, 21), and (4, 28).

There are no negative values since distance cannot be negative. The points are connected to form a line.

Let the horizontal axis represent the time and the vertical axis the distance.

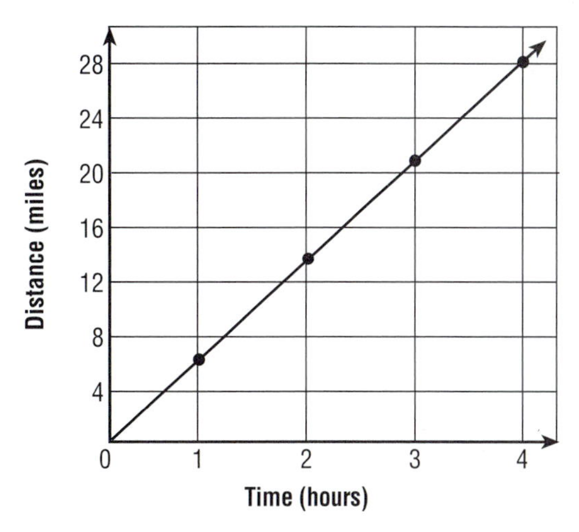

Graph each ordered pair.

1. $(2, 4)$
2. $(2, -4)$
3. $(-2, 4)$
4. $(-2, -4)$
5. $(0, -3)$
6. $(-1, 3)$
7. $(2, 0)$
8. $(-4, -3)$

Make a table of values for each equation. Then graph each equation. Use a separate coordinate plane for each equation.

9. $3x = y$
10. $x - y = 4$
11. $6x = y - 1$
12. $-2 = 3x - 2y$

TECHNOLOGY REFERENCE

USING ELECTRONIC SPREADSHEETS AND CALCULATORS

USING THE INTERNET

USING ELECTRONIC SPREADSHEETS AND CALCULATORS

Spreadsheets

Spreadsheet **software** is commonly used in business offices. Spreadsheet software lets you prepare tables, such as budgets and payroll registers, easily and quickly.

Spreadsheet software takes the place of a pad of paper and calculator to find the answers to common business problems. For example, spreadsheet software was used by Linda McClain to maintain the list of expenses shown on the computer screen in Illustration 1.

The spreadsheet has both columns and rows. The columns are identified by letter and the rows by number. Data are shown in **cells,** which are where columns and rows meet. For example, the word *Salaries* is found in Cell A5, or where Column A and Row 5 meet. The number 3000.00 is found in B5, or where Column B and Row 5 meet.

To develop the list of expenses, Linda created a spreadsheet that contained

1. A title: Expenses.
2. Column headings: Type and Amount.
3. Row labels: Salaries, Rent, Power, Postage, Telephone, and Other.
4. A formula for adding the expenses and getting the total.
5. The amount for each expense.

Linda entered a formula into Cell B12 and entered the label *Total* into Cell A12. The formula tells the spreadsheet program how to calculate the total of Column B. Formulas will be explained later in this appendix.

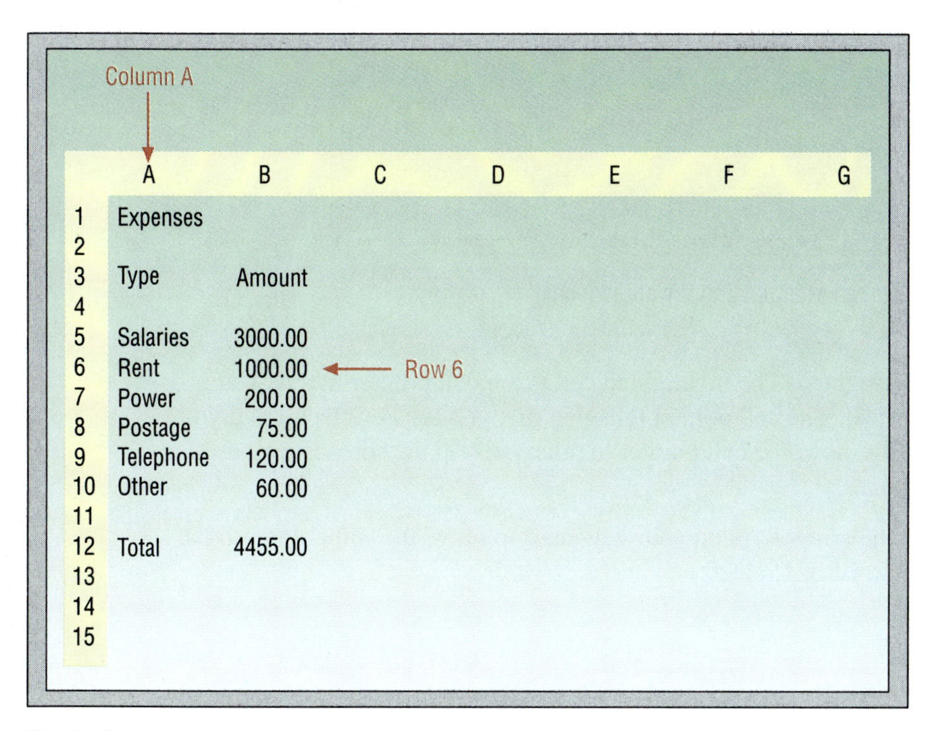

Illustration 1. List of expenses on a computer screen

Linda also entered the amounts for each expense in Column B, Rows 5–10. When she did this, the amount in the Total row was calculated automatically by the formula she had entered into Cell B12.

Spreadsheet Skill 1 Loading the Software Into Your Computer

Each type of computer and each spreadsheet program is somewhat different. Instructions to load spreadsheet programs into your computer will be supplied by your teacher. You should learn how to load your spreadsheet program and quit, or get out of, your spreadsheet program.

Spreadsheet Skill 2 Moving the Cell Pointer

After you load your spreadsheet program, your computer display screen should look similar to Illustration 2.

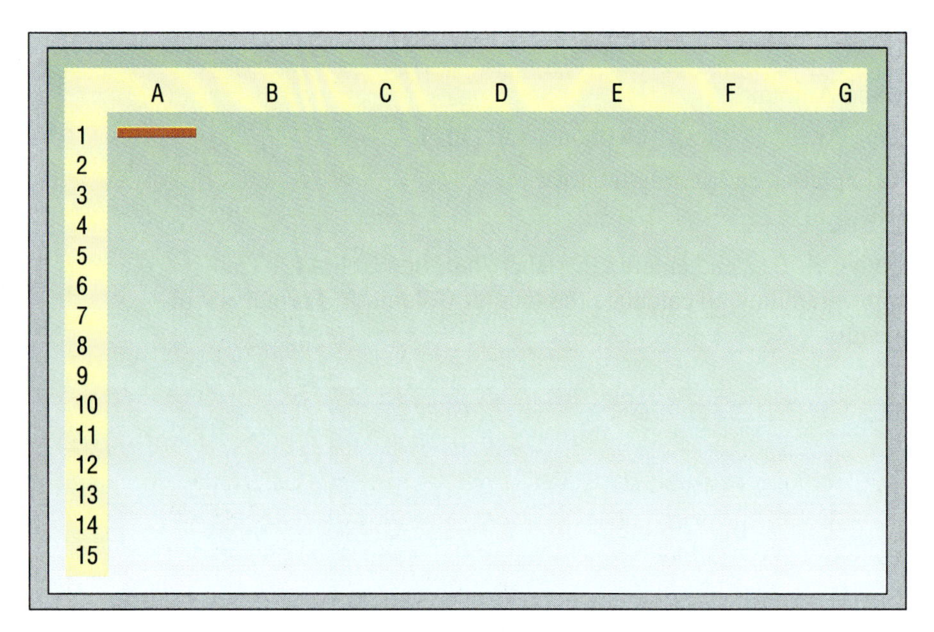

Illustration 2. The computer screen after loading the spreadsheet

Notice that Cell A1 is highlighted. The highlighted cell is the cell pointer. It tells you where you are in the spreadsheet. The cell pointer tells you that you are at Cell A1, and you may enter data into the cell. To move the cell pointer to other cells in the spreadsheet, use the arrow keys on your keyboard.

Application Problem Use the arrow keys on your keyboard to move the cell pointer to the following cells: A10, B10, B2, E5, C15, D20.

Spreadsheet Skill 3 Entering Alphabetic Data

You enter alphabetic characters into a cell by moving the cell pointer to the cell, keying the characters, and then pressing the Enter key.

For example, move the cell pointer to Cell A1. Key the title for the spreadsheet, Expenses. If you make mistakes while entering the characters, use the backspace key to erase the errors. Then, re-enter them correctly. When you are done, press the Enter key. This will insert the characters in Cell A1.

You may spot an error in the cell after pressing the Enter key. If you do, simply key the correct characters over the errors and press the Enter key. The correct characters will be inserted into the cell.

Application Problem Practice entering alphabetic characters by entering the column headings and row labels.

Move the cell pointer to Cell A3 using the arrow keys on your keyboard. Key the word *Type* and press the Enter key. Move the cell pointer to Cell B3, key the word *Amount,* and press the Enter key. You have now entered the headings for the two columns.

Move the cell pointer to Column A and enter Salaries in A5, Rent in A6, Power in A7, Postage in A8, Telephone in A9, and Other in A10. When the labels have been entered, enter the word *Total* in Cell A12 and press the Enter key. Do not forget to press the Enter key after each label has been keyed.

Spreadsheet Skill 4 Entering Numeric Data

You enter numbers into a cell in the same way as you entered alphabetic characters. You must move the cell pointer to each cell you want, enter a number, and press the Enter key. For example, move the cell pointer to Cell B5. Enter the amount for Salaries by keying 3000. Then press the Enter key.

Notice that you do not enter commas or dollar signs when entering numeric data. Some spreadsheet programs will show the salaries amount as 3000. Others will show it as 3000.00.

If the amount had been 3120.80, you would have had to enter the numbers 3120, then the decimal point, then 8, and then press the Enter key.

Application Problem Practice by entering the rest of the expense amounts. Enter 1000 in B6, 200 in B7, 75 in B8, 120 in B9, and 60 in B10. Press the Enter key after keying each number.

Spreadsheet Skill 5 Entering Formulas

A formula is a set of math steps that are performed by the spreadsheet program. For example, a formula might instruct the spreadsheet program to add the amount in one cell to the amount in another cell and place the answer in a third cell.

Suppose that you have entered $100 in Cell C1 and $50 in Cell C2. You could then move the cell pointer to Cell C3 and key this formula: +C1+C2.

When you press the Enter key, the total of cells C1 and C2, $150, would appear in Cell C3.

You are probably wondering why that first plus sign is needed. Why can't you just enter C1+C2? The answer is that if you keyed C1, the spreadsheet program would think that you were entering alphabetic characters and would not treat C1+C2 as a formula. By entering the + sign first, the program knows that you are entering a formula.

When you enter formulas, you tell spreadsheet programs to add, subtract, multiply, and divide by using these keys on your keyboard:

+ for addition.
− for subtraction.
* for multiplication.
/ for division.

Application Problem Practice by entering the formula to find the total of the Amount column in the Expenses spreadsheet. Move the cell pointer to Cell B12, where you want the total to be placed. Then enter the formula by pressing these keys in order: +B5+B6+B7+B8+B9+B10.

After entering the formula, press the Enter key. The total of Column B will appear in Cell B12 as soon as you do.

Spreadsheet Skill 6 Using Spreadsheet Functions in Formulas

Many spreadsheet programs let you enter formulas using special math functions such as SUM, COUNT, and AVERAGE. These functions tell the spreadsheet program to perform the function on a range of cells.

For example, to find the sum, or total, of cells B5 through B10, you might enter this formula: @SUM(B5..B10). The formula tells the spreadsheet program to add the amounts found in the range of cells B5 through B10.

Notice the @, or At symbol. Many spreadsheet programs look for the @ symbol to know when to perform a function. If you did not enter @ first, the spreadsheet would treat your formula as alphabetic characters since you would have started with the letters SUM. Notice also that you had to enter two periods between B5 and B10. Many spreadsheet programs use the periods to indicate a range.

Some spreadsheets look for the = sign instead of the @ sign. They may also use a colon instead of two periods to indicate a range. The formula to find the sum of cells B5 through B10 for a spreadsheet that uses this convention would be: =SUM(B5:B10)

Application Problem Practice by moving the cell pointer to Cell B12. Enter the formula to find the total of the Amount column: @SUM(B5..B10). Now press the Enter key and the new formula will replace the old one. Notice that the answer, 4455.00, remains the same.

Spreadsheet Skill 7 Saving the Spreadsheet

You should save it on your data disk so that you can use it again. When you are asked to name the spreadsheet, call it Expenses.

Application Problem Practice saving the Expenses spreadsheet several times so that you can do it easily.

USING A CALCULATOR AND COMPUTER KEYPAD

Calculators are commonly used in business. Many different models of calculators, both desktop and hand-held, are available. You must set several switches on your desktop calculator to get the results you want. You will need to look at the manual that came with the calculator for help in using it.

Decimal Placement and Rounding You should be familiar with the key and decimal features that allow you to set the desired decimal placement. The *decimal selector* sets the decimal places necessary for the numbers that will be entered. For example, if the decimal selector is set at 2, both the numbers entered and the answer will have two decimal places.

The *Floating (F) decimal setting* allows your answer to be displayed without any rounding to the maximum number of decimal places that your machine can display.

Some machines have an *automatic rounding* feature that rounds decimal answers to a predetermined level. This predetermined level is set by the *rounding selector* and may be shown as 5/4 and 5 (or a down arrow). At the 5/4 position, the calculator rounds the last desired digit up when the digit is 5 or greater. If the digit following the desired digit is less than 5, the last desired digit remains unchanged. If the rounding selector is set at the 5 (or down arrow) position, numbers past the selected position will be dropped without rounding.

Some calculators have an *add mode* setting. This setting is often used when working with dollar and cent amounts. The add mode setting automatically places 2 decimals in each entry, without pressing the decimal key. Check your machine and manual to identify its decimal place capacity, decimal, and rounding features.

The *GT* or *grand total switch* in the on position accumulates totals.

Kinds of Computer Keyboards

Even though several styles of keyboards for the IBM and compatible computers are found, there are two basic layouts. The standard layout and enhanced layout are shown in Illustration 3. On the standard keyboard the directional arrow keys are found on the numeric keypad, which is to the right of the keyboard. To use the numbers, press the key called *Num lock.* (This key is found above the "7" key.) When the Num Lock is turned on, numbers are entered when the keys on the keypad are pressed. When the Num Lock is off, directional keys (the up and down arrow, Home, Page Up, Page Down, End, Insert, and Delete keys) can be used.

On enhanced keyboards the directional keys are above and to the left of the numeric keypad. when using the keypad on an enhanced keyboard, Num Lock can remain on.

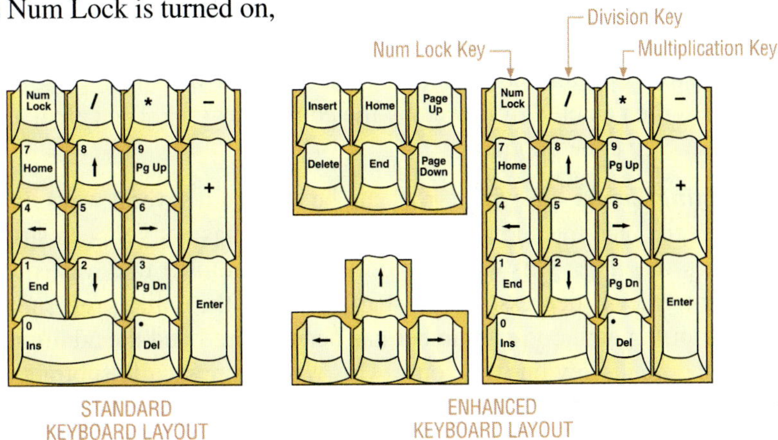

Illustration 3. Standard and enhanced keypads

The asterisk (*) performs a different function on the computer than the calculator. The asterisk on the calculator is used for the total while the computer uses it for multiplication.

Another difference is the division key. The computer key is the forward slash key (/). The calculator key uses the division key (÷).

PERFORMING MATHEMATICAL OPERATIONS WITH DESKTOP CALCULATORS

The basic operations of addition, subtraction, multiplication, and division are used frequently on a calculator.

Addition Each number to be added is called an *addend.* The answer to an addition problem is call the *sum.*

Addition is performed by entering an addend and striking the addition key (+). All numbers are entered on a calculator in the exact order they are given. To enter the number 4,455.65, strike the 4, 4, 5, 5, decimal, 6, and 5 keys in that order, and then strike the addition key. Commas are not entered. Continue in this manner until all addends have been entered. To obtain the sum, strike the total key on the calculator.

Subtraction The top number or first number of a subtraction problem is call the *minuend.* The number to be subtracted from the minuend is called the *subtrahend.* The answer to a subtraction problem is call the *difference.*

Subtraction is performed by first entering the minuend and striking the addition key (+). The subtrahend is then entered, followed by the minus key (−), followed by the total key.

Multiplication The number to be multiplied is called the *multiplicand.* The number of times the multiplicand will be multiplied is called the *multiplier.* The answer to a multiplication problem is called the *product.*

Multiplication is performed by entering the multiplicand and striking the multiplication key (×). The multiplier is then entered, followed by the equals key (=). The calculator will automatically multiply and give the product.

Division The number to be divided is called the *dividend.* The number the dividend will be divided by is called the *divisor.* The answer to a division problem is called the *quotient.*

Division is performed by entering the dividend and striking the division key (÷). The divisor is then entered, followed by the equals key (=). The calculator will automatically divide and give the quotient.

Correcting Errors Keying errors can be corrected several ways. If an incorrect number has been entered and the addition key or equals key has not yet been struck, strike the clear entry (CE) key one time. This will clear only the last number that was entered. However, if the clear entry key is depressed more than one time, the entire problem will be cleared on some calculators. If an incorrect number has been entered and the addition key has been struck, strike the minus key one time only. This will automatically subtract the last number added, thus removing it from the total.

PERFORMING MATHEMATICAL OPERATIONS ON COMPUTERS AND HAND-HELD CALCULATORS

On a computer keypad or a hand-held calculator, addition is performed in much the same way as on a desktop calculator. However, after the + key is depressed, the display usually shows the accumulated total. Therefore, the total key is not found. Some computer programs will not calculate the total until Enter is pressed.

Subtraction is performed differently on many computer keypads and hand-held calculators. The minuend is usually entered, followed by the minus (−) key. Then the subtrahend is entered. Pressing either the + key or the = key will display the difference. Some computer programs will not calculate the difference until Enter is pressed.

Multiplication and division are performed the same way on a computer keypad and hand-held calculator as on a desktop calculator. Keep in mind that computers use the * for multiplication and / for division.

CALCULATOR TIPS

Using the Memory Key

The memory key is a useful feature that can let you calculate answers more quickly.

Example: Using the Memory Key to Find Grand Totals

1. 14.56 + 12.35 _____
2. 20.13 + 87.43 _____

Grand Total _____

Solution:

a. Press these keys in order: [1] [4] [.] [5] [6]
b. Press the addition key: [+]
c. Press these keys in order: [1] [2] [.] [3] [5]
d. Press [=] and the sum, 26.91 appears in the display. Write down this number in the first Total column blank space.
e. Now press the memory *add* key: [M+] This adds the sum, 26.91, to the memory register. A little "M" will appear in the display. This "M" tells you that you have stored something in the memory register.
f. Clear your calculator by pressing the clear [CE/C] key twice. This will clear your calculator but *not* clear the memory register.
g. Now enter: [2] [0] [.] [1] [3]
h. Then press: [+]
i. Now enter: [8] [7] [.] [4] [3]
j. Press [=] and the sum, 107.56, will appear in the display. Write this number down in the second Total column blank space.

k. Now press: [M+] You have just added the second sum, 107.56, to the memory register.

l. Now press this key: [MR/C] and the sum, 134.47, will appear in the display. This is the sum of 20.13 + 87.43, the two amounts you added to the memory register. Write this number down in the grand total blank space.

MR/C stands for "memory recall/clear." This key lets you put what you have in the memory register in the calculator display. Once in the display, you can use it for other calculations.

m. Now press the memory recall/clear key again: [MR/C] This will clear the memory register. The little "M" in the display should go off.

Pressing the memory recall/clear key *once* puts the register contents in the display. Pressing the key *twice* clears the memory register. You can also clear the memory register *and* the display by pressing the AC, or all clear key.

You can use the memory register to solve many types of addition problems in business, such as finding the sum of the totals in a cash payment record with special columns.

Example: Using the Memory Key to Find Graduated Commission
In Lesson 4-7 you found Lori Well's graduated commission for $34,000 sales for one month. Her commission structure was 5% on the first $10,000 of monthly sales and 10% on all sales over $10,000.

Solution: To find Lori's commission using the memory key:

a. Find the first commission: press [1] [0] [0] [0] [0] [×] [5] [%]

b. Add the first commission to memory: press [M+]

c. Find the next commission: press [2] [4] [0] [0] [0] [×] [1] [0] [%]

d. Add the next commission to memory: press [M+]

e. Find total commissions: press [MR/C]

f. The answer, 2900, or $2,900 appears in the display.

Using the Percent Key

To find the number that is 5% greater or smaller than 40, use the percent key and either the plus or minus keys. For example:

Press [4] [0] [+] [5] [%] and 42 appears in the display.

To find what percent one number is greater or smaller than another, use the percent key again. For example, to find what percent 50 is greater than 40:

Press [5] [0] [−] [4] [0] [=] and 10 appears in the display.
Press [÷] [4] [0] [%] and the answer, 25 appears in the display.

Now all you have to do is add the percent sign.

To find a percent of a number fast, use the multiplication and percent keys. For example, to find 5% of 45

Press [5] [×] [4] [5] [%] and 2.25 appears in the display.

To multiply by a fraction of 1%, first change the fraction in the percent to a decimal. For example, to find $\frac{1}{2}$% of 45, change $\frac{1}{2}$% to 0.5%.

To find what percent one number is of another, use the division and percent keys. For example, to find what percent 1 is of 8

Press [1] [÷] [8] [%] and 12.5 appears in the display.

All you have to do is attach a percent sign.

USING THE INTERNET

How Do I Get Connected?

You may connect on the Internet in two ways. One way is to use a modem. The other is to connect directly to a larger computer that is connected to the Internet. The latter is the way usually available on college campuses and in large businesses. In either case, you will need an account with an Internet Provider to get connected. The Internet Provider may be:

1. An on-line service, such as America Online, CompuServe, Delphi, or Prodigy. These providers charge you to connect.
2. Universities, other educational institutions, statewide networks, or government agencies. These providers may not charge any connection fees to students or clients.
3. A commercial Internet provider, such as Pipeline. These providers also charge you to connect.

Using the Internet

Email Electronic mail, or email, is one of the most used features of the Internet. To send someone an email message, you must know or be able to find that person's email address. Email addresses have three parts: the person's name, the user's Internet Provider or computer system, and the type of organization the service represents. An example of an email message is:

jdoe@aol.com

jdoe is the email name for John Doe. The @ sign separates the email user's name from the rest of the address. Aol stands for America Online, the Internet Provider. Com stands for commercial. If your Internet Provider was an educational institution, com would be replaced with edu (jdoe@aol.edu). Here are some other replacements:

gov non-military US government agency
int International institutions
mil US military institutions
net Network resources
org Non-profit organizations

When the person you want to contact lives in another country, the email address will include the country in it. For example, au is added for Australia, jp for Japan, uk for the United Kingdom, and fr for France.

The World-Wide Web In addition to email, the Internet provides users with access to huge amounts of data found on computer systems throughout the world. The collection of these computer systems is called the World-Wide *Web* because the computer systems are networked to each other. To access and read the data on these computer systems, you use a web browser. A web browser is software that will let you connect to these computer systems and "browse," or read the information on them.

Usenet or News Groups A news group is a place where people interested in the same topic can send and share messages. Most news groups are managed by a person or organization; some are not. You join a news group by subscribing to the group, which is usually done through the software you are using for the news group. That software is often the same software you use for email. You send messages or *articles* to the group by *posting* email messages to the group. You read articles sent by others using your email software.

Netiquette (Internet rules of behavior and conduct) When you send email or post news group messages, you should follow the rules of effective communication. Make your messages courteous, clear, concise, concrete, correct, and complete. Before you post an email message to a news group, read (or *lurk*) many of the previous messages to get a feel for what is being discussed. Often the manager of a news group will create a FAQ (frequently asked questions) message. It is poor netiquette to ask a question of the news group that has already been answered in this FAQ message. Also, be careful about using humor or sarcasm in your messages. Without the tone of your voice or your facial expressions, your humor might be misinterpreted by other members of the news group.

Some other suggestions include:

1. Do not give your password to anyone through email or a news group message.
2. If you want to emphasize a word, enclose the word in asterisks. Write words in all capital letters *only* when you intend to *shout*.
3. Read the "Acceptable Use Policy" for your Internet Provider or news group.
4. Be careful about using graphics in your email. Graphics or pictures take a long time to transfer to the user. If the user is paying at an hourly rate, the user may be annoyed if you include unnecessary images in your messages.
5. Be careful about capitalizing letters before the @ sign in an email address. Most computer systems will consider JDOE@aol.com, or JDoe@aol.com a different person from jdoe@aol.com. Usually, it doesn't matter if you capitalize or don't capitalize letters after the @ sign.

You may have noticed that the Internet has a vocabulary all its own. Following is a short glossary of Internet terms. To learn more about the Internet, consult your school or local library. A short list of Internet books and periodicals follows the glossary.

Address The number sequence that identifies a unique user or computer on a network. Every computer and user on the Internet must have a different address for the system to know where to send E-mail and other computer data.

BBS Bulletin Board System A computerized meeting and announcement system that allows people to carry on discussions, upload and download files, and make announcements without the people being connected to the computer at the same time.

Bits Per Second A measurement of speed for network telecommunications systems. A bit is the smallest piece of information communicated by computers. The number of bits that pass through the modem each second.

Bookmark A means of storing favorite WWW site addresses for future use. *See* hotlists.

Client Any computer that is connected to a network, including the Internet. The client must have the software needed to connect to a network and use the network's resources.

Connections The software and hardware links between computers are called connections. The speed and compatibility of your computer's connection to your Internet host computer will determine the efficiency with which you can access Internet resources.

Cyberspace A term given to the electronic, computerized world of the Internet. Often called virtual reality. When you are on the Internet or the World Wide Web, you are in Cyberspace.

Distributed Network The Internet is a distributed network, meaning that there is not one central authority or group guiding its growth, use, and development. If one part of the Internet goes down, as in a natural disaster like an earthquake, Internet communications can be transferred to other lines of communication, and the downed portion of the network can be bypassed until repairs are made.

Domains A division or section of the Internet. For example, the military is one domain, education is another. Service providers like America Online and Prodigy have their own domains. Domains can be divided geographically by country, region, or state, or by other similarities such as business, commerce, government agencies, or private organizations.

Download To copy files, data, information, and software from a remote host computer to your computer or to another computer.

Finger An Internet software tool for locating people (via E-mail addresses) on other Internet sites.

Flames Rebukes sent by agitated Internet users to people who violate rules of Internet "netiquette," or who "spam" the Internet.

FTP File Transfer Protocol FTP software transfers files from one computer to another.

Gateways Tools that allow commercial E-mail software to communicate with each other.

Hits A hit is recorded anytime someone connects to a remote computer, site, or Home Page.

Home Page A Home Page is like an index that contains related information on a single topic in a hypertext WWW environment.

Host *See* servers.

Hotlists A list of WWW addresses. Some WWW browsers call their stored WWW addresses hotlists instead of bookmarks. *See* bookmarks.

Hypermedia Computer data that creates hypertext links between more than one kind of media. Types of media include video, pictures, graphics, animation, and text. *See* Multimedia.

Hypertext A system of information retrieval, where selected keywords are linked to text and other information in the same document or in another document.

Internet The name given to the current telecommunications system between networks of computers. The Internet is often called a "network of networks." The Internet will grow into the electronic superhighway of the future. The Internet is often called the Net because it is the largest computer network in the world.

Lurkers Newsgroup or E-mail users who read the articles posted by other group members but don't post their own articles.

Maillist (or Mailing List) A (usually automated) system that allows people to send E-mail to one address, whereupon their message is copied and sent to all of the other subscribers to the maillist.

Modem A simple communications tool that converts computer signals into signals that can travel over telephone lines.

Mosaic The popular and widespread WWW browser or client software.

Multimedia Systems that use more than one medium. A multimedia computer can utilize various types of media including video, sound, pictures, graphics, animation, and text.

Names A unique or different name is required for Internet users. In common speech, the words "name" and "address" are often used interchangeably. Technically speaking, a name involves the use of words (eugene@dixon.edu), and an address is a number like 158.95.6.2.

Network Two or more computers linked together to share information.

Newbies Term used to refer to new members of discussion groups or new users of the Internet.

Newsgroup Participants who follow threads of a particular topic with the help of a USENET newsreader.

Newsreader Software tool required to read and participate in USENET Newsgroups.

NIC Network Information Center Generally, any office that handles information for a network. The most famous of these on the Internet is the InterNIC, which is where new domain names are registered.

On-line When you are connected to and actively using a network.

Packets Bundles of data that are transmitted over the Internet.

PPP Point-to-Point Protocol One of the types of connections that allow Internet communications over a modem. PPP allows your computer to act like you have a direct connection to the Internet. *See* SLIP.

Resources Anything you can find on the Internet is a resource, including software, information, services, and people.

Saints People who provide help to new members of discussion groups.

Server Any computer providing network services and resources to other computers. Server computers are also called "hosts." Servers are the key computers in networks.

Search Engines Used to identify web sites that contain data pertinent to a query.

Site A computer connected to the Internet that contains information that can be accessed using a navigation tool such as a web browser or FTP. Also called Internet Providers.

SLIP Serial Line Internet Protocol One of the types of connections that allow Internet communications over a modem. SLIP allows your computer to act like you have a direct connection to the Internet. *See* PPP.

Spam Unwanted Internet garbage, particularly advertising on the public Internet.

Subject Line The title of a Newsgroup posting or article.

Subscribers Participants in a Newsgroup or E-mail discussion list.

Superhighway Another name for Cyberspace. Also, the name given to the Internet of the future.

Surfing Exploring the Internet. When you surf, you are looking for interesting information.

TCP Transmission Control Protocol TCP keeps track of every item in a packet or package that is transmitted over the Internet. If an item arrives broken or incomplete, TCP asks the host computer to send the packet over again.

Telnet Telnet provides the ability to log in to remote servers or host computers and to use their resources as if you were a computer terminal on those host computers.

Thread A sequence of messages on the same theme or topic.

Title The particular name of a Newsgroup within its category.

Transfer Rate The speed at which data is exchanged between computers.

Uniform Resource Locator An address of a WWW document. The URL, or address, is used by WWW browsers to locate the document, which may be stored on a computer system on the other side of the earth.

USENET A huge collection of computers that allow you to post, distribute, or publish Newsgroup articles. USENET is one of the most widely used Internet services.

WAIS Wide Area Information Servers A commercial software package that allows the indexing of huge quantities of information, and then makes those indexes searchable across networks such as the Internet. A prominent feature of WAIS is that the search results are ranked according to how relevant the "hits" are, and that subsequent searches can find similar topics and thus refine the search process. WAIS database searches take you inside the documents or files you're interested in.

WAN Wide Area Network Any internet or network that covers an area larger than a single building or campus.

Wizards Newsgroup experts.

WWW The World Wide Web, or W3 WWW is a system of computers that can share information by means of hypertext links.

Selected Internet Publications

Books

Ashton, Gary L., Karl Barksdale, Michael Rutter, and Earl Jay Stephens. Internet Activities: Adventures on the Superhighway. Cincinnati: South-Western Educational Publishing, 1995.

Dern, Daniel P. The New User's Guide to the Internet. New York: McGraw-Hill, 1993.

Turlington, Shannon R. Walking the World Wide Web: Your Personal Guide to the Best of the Web. North Carolina: Ventana Press, 1995.

Periodicals

The Internet Letter. NetWeek: Washington, D.C. 20045. helen@access.digex.com

Internet World. Meckler: Westport, Conn. 06880. meckler@jvnc.net.2

Measurement Equivalents and Conversion Table

U.S. Customary System

Length
1 inch
12 inches = 1 foot
36 inches = 3 feet = 1 yard

Volume or Capacity
1 pint = 1/2 quart = 1/8 gallon
2 pints = 1 quart = 1/4 gallon
4 pints = 2 quarts = 1/2 gallon
8 pints = 4 quarts = 1 gallon

Dry Measure
1 pint
2 pints = 1 quart
16 pints = 8 quarts = 1 peck
64 pints = 32 quarts = 4 pecks = 1 bushel

Weight
1 ounce (oz)
16 ounces = 1 pound (lb)
2000 pounds = 1 ton

Conversion Factors

Metric System	U.S. Customary System		
Meters	**Yards**	**Inches**	
1	1.094	39.37	
0.914	1	36	
Centimeters	**Inches**	**Feet**	
1	0.394	0.0328	
2.54	1	0.0833	
30.48	12	1	
Kilometers	**Miles**		
1	0.621		
1.609	1		
Grams	**Ounces**	**Pounds**	
1	0.035	0.0022	
28.35	1	0.0625	
453.59	16	1	
1000	35.274	2.205	
Kilograms	**Ounces**	**Pounds**	
1	35.274	2.205	
0.028	1	0.0625	
0.454	16	1	
Liters	**Pints**	**Quarts**	**Gallons**
1	2.114	1.057	0.264
0.473	1	0.5	0.125
0.946	2	1	0.25
3.785	8	4	1

Metric System

Length
1 meter (m) = 100 cm = 1000 mm

1 millimeter (mm)	= 0.001	m
1 centimeter (cm)	= 0.01	m
1 decimeter (dm)	= 0.1	m
1 dekameter (dkm)	= 10	m
1 hectometer (hm)	= 100	m
1 kilometer (km)	= 1000	m

Volume or Capacity
1 liter (L) = 100 cL = 1000 mL

1 milliliter (mL)	= 0.001	L
1 centiliter (cL)	= 0.01	L
1 deciliter (dL)	= 0.1	L
1 dekaliter (dkL)	= 10	L
1 hectoliter (hL)	= 100	L
1 kiloliter (kL)	= 1000	L

Weight
1 gram (g) = 100 cg = 1000 mg

1 milligram (mg)	= 0.001	g
1 centigram (cg)	= 0.01	g
1 decigram (dg)	= 0.1	g
1 dekagram (dkg)	= 10	g
1 hectogram (hg)	= 100	g
1 kilogram (kg)	= 1000	g

Common Conversion Factors
1 centimeter = 0.39 inches
1 meter = 39.4 inches
1 kilometer = 0.62 miles
1 gram = 0.035 ounces
1 kilogram = 2.20 pounds
1 liter = 1.06 quarts

PHOTO CREDITS

Chapter 1 Career Profile: Woman using laptop computer: ©Alan Brown/Photonics; **p. 5:** ©David Young-Wolff/Stock, Boston; **p. 19:** ©Bob Daemmrich/Stock, Boston; **p. 20:** ©Tony Freeman/PhotoEdit; **p. 30:** ©Herb Snitzer/Stock, Boston

Chapter 2 Career Profile: Service woman: ©Ken Rogers/West Light; Shipboard CIC station: ©Photri, Inc.; **p. 48:** ©Franz Edson/Tony Stone Worldwide; **p. 55:** ©Lawrence Migdale/Photo Researchers, Inc.; **p. 60:** ©Richard Hutching/PhotoEdit; **p.70:** ©Claus Meyer/Black Star; **p. 72:** ©Bob Daemmrich/Stock, Boston

Chapter 3 Career Profile: Piston on conveyer belt: ©Martin Rogers/Tony Stone Worldwide; Woman assembling hats: ©Bryce Flynn/Stock, Boston; Men in electronics factory: ©Lawrence Migdale/Photo Researchers, Inc.; **p. 83:** ©Jeff Greenberg/PhotoEdit; **p. 88:** ©Charles Gupton/Stock, Boston; **p. 99:** ©Mark Joseph/Tony Stone Images; **p. 107:** ©Jean Marc Barey/Photo Researchers, Inc.; **p. 114:** ©Bob Daemmrich/Stock, Boston

Chapter 4 Career Profile: House: ©Jeff Greenberg; **p. 133:** ©Photo courtesy of Siemens, Inc.; **p. 147:** ©Bachmann/Photo Researcher, Inc.; **p. 157:** ©Royuki Matusumoto/Black Star

Chapter 5 Career Profile: Scene from "Gone With The Wind": ©The Kobal Collection; Voice over actresses: ©Michael Grecco/Stock, Boston; Actor being made up: ©Mark Richards/PhotoEdit; **p. 175:** ©Story Litchfield

Chapter 6 Career Profile: Pastries: ©Richard Pasley/Stock, Boston; Chef: ©Tony Freeman/PhotoEdit; Waiter: ©Jeff Greenberg/PhotoEdit; **p. 200:** ©Gale Zucker/Stock, Boston; **p. 209:** ©Mark Richards/PhotoEdit; **p. 221:** ©Alan Brown/Photonics; **p. 226:** ©Jeff Greenburg (SW224-95S); **p. 231:** ©Tony Freeman/PhotoEdit; **p. 238:** ©Tom McCarthy/PhotoEdit

Chapter 7 Career Profile: Catalogs: ©Rhoda Sidney/Stock, Boston; Service desk employee: ©Alan Brown/Photonics; Customer service operator: ©Bruce Ayres/Tony Stone Images; **p. 225:** ©Henley and Savage/Tony Stone Worldwide; **p. 267:** ©Jeff Greenburg; **p. 270:** ©Bob Daemmrich/Stock, Boston; **p. 273:** ©Stephen Frisch/Stock, Boston

Chapter 8 Career Profile: Computerized sorter: ©Photo Researchers, Inc.; Man scanning boxes: ©Lee F. Snyder/Photo Researchers, Inc.; Man using forklift: ©Richard Pasley/Stock, Boston; **p. 288:** ©Bob Daemmrich; **p. 293:** ©Alan Brown/Photonics; **p. 298:** ©Phyllis Picardi/Stock, Boston; **p. 310:** ©William McCoy/Rainbow

Chapter 9 Career Profile: Doctor with young patient: ©Robert Llewellyn/Superstock; Radiology consultant: ©Frank Siteman/Stock, Boston; Stethoscope: ©David Wagner/PhotoTake NYC; **p. 323:** ©Diana Rasche/Tony Stone Worldwide; **p. 336:** ©Will McCoy/Rainbow

Chapter 10 Career Profile: Statue of justice: ©Reza Estakhrian/Tony Stone Images; Judge with lawyers: ©Billy E. Barnes/Tony Stone Images; Court reporter: ©Michael Newman/PhotoEdit; **p. 383:** ©David Young-Wolff/PhotoEdit

Chapter 11 Career Profile: Newsprint being fed: ©Richard Pasley/Stock, Boston; Printers viewing proofs: ©Michael Newman/PhotoEdit; **p. 403:** ©Kaluzny/Thatcher/Tony Stone Images; **p. 412:** ©TJ Florian/Rainbow; **p. 416:** ©Jon Riley/Tony Stone Worldwide; **p. 422:** ©Jeffrey Dunn/Stock, Boston; **p. 424:** ©Erik Von Fischer/Photonics

Chapter 12 Career Profile: Cds: ©Damien Lovegrove/Science Photo Library; Sound engineer: ©Mark Burnett/Stock, Boston; **p. 439:** ©Michael Newman/PhotoEdit; **p. 444:** ©Chip Henderson/Tony Stone Images; **p. 456:** PhotoEdit

Chapter 13 Career Profile: Men raising a house: ©Zigy Kaluzny; Man using a sander: ©Tim Davis/Photo Researchers, Inc.; Female carpenter: ©Bob Daemmrich/Stock, Boston; **p. 471:** ©Myrleen Ferguson/PhotoEdit; **p. 472:** ©Andy Sacks/Tony Stone Worldwide; **p. 474:** ©Jeff Issac Greenberg/Photo Researchers, Inc.; **p. 477:** ©Michael Ronsenfeld/Tony Stone Images; **p. 485:** ©Max Winter/Stock, Boston

Chapter 14 Career Profile: Computer light micrographic: ©Michael W. Davidson/Photo Researchers, Inc.; Computer operator: ©Bill Stanton/Rainbow

GLOSSARY

Absolute loss Occurs when the selling price does not cover the cost of goods.

Account A record of receipts and payments.

Addend A number to be added to another.

Addition A process of combining two numbers to give one number.

Adjusted gross income A tax term meaning gross income less adjustments.

Agent A person who legally acts for someone else.

Amortization Gradual repayment of a debt by regular payments of principal and interest.

Amount due at maturity The amount owed when note is due; the face plus interest on note; maturity value.

Amount financed The amount of credit given on an installment loan or purchase.

Annual depreciation A year's depreciation.

Annual membership fee A fee charged by credit card companies.

Annual percentage rate (APR) A percent that shows the ratio of finance charges to the amount financed.

Annual rate The rate for one year; the percent of the principal that is charged for one year's interest.

Appraised value An estimate of the value of property.

Approximate value A rounded amount; not exact.

Area An amount of surface. Example: for a rectangle, length times width.

Assessed value A value put on property as a base for figuring amount of tax.

Assessment roll A list of property, with assessed valuations.

Arithmetic average See mean.

Assets Things of value owned by a person or business.

ATM (Automatic Teller Machine) card An example of a type of debit card.

Automatic callback Telephone service that lets you return the last call received—whether you answered it or not.

Automatic teller machine (ATM) A computer system that lets you withdraw or deposit money in your bank account without a teller's help.

Average A single number used to represent a group of numbers.

Average annual depreciation The original cost less scrap value divided by years of use.

Balance The amount of money in an account; the difference between the two sides of an account.

Balance of trade A measure of a country's international business activity; difference between a country's exports and imports.

Balance sheet A statement showing the assets, liabilities, and capital of a person or business for a certain date.

Balloon payment mortgages Mortgage option where the entire monthly payment for the first 20–25 years goes to pay off the interest on a home; the payments for the last years pay off the principal.

Bank discount Interest collected in advance.

Bank statement A periodic report to a depositor showing deposits, payments, charges, and balance in a checking account.

Banker's interest Interest based on a 360-day year; ordinary interest.

Bankrupt Legally insolvent and unable to pay debts.

Bar code Patterns of printed bars that identify a stock item.

Base A number with which another is compared; a number that may be multiplied by a fraction or percent; in a numeration system, the number by which a digit is multiplied when moved one place to the left.

Base period A period in time with which comparisons are made.

Basic health coverage An insurance package that includes hospitalization benefits, surgical benefits, and medical insurance.

Beneficiary The person to whom life insurance benefits are paid when the insured dies.

Bodily injury insurance Auto insurance covering liability for injury to other persons.

Bond A form of long-term promissory note issued by a corporation or government.

Bondholder One who holds a corporation or government bond.

Book value Original cost less total depreciation to date.

Break-even point The point at which income from sales equals total costs of producing and selling goods.

Broker A dealer in stocks and bonds or real estate.

Broker's commission A fee paid to a broker; brokerage fee.

Budget A plan for using or spending income; an estimate of expenses.

Buying expense The cost of buying goods.

Byte A character stored in a computer system.

C Hundred.

Call forwarding A phone feature that lets you transfer a call to another number.

Call screening A phone feature that lets you block calls.

Call waiting A phone feature that lets you interrupt one call to talk to another caller.

Caller ID A phone feature that displays the number of the person who is calling.

Canceled check A check that has been marked "Paid" by the bank.

Capital The owner's claim to the assets of a business; proprietorship; owners' equity; investment; net investment; net worth.

Capital investment The original investment in property plus additions and improvements.

Capital stock Total value of the stock issued by a corporation.

Cash advance The cash received when money is borrowed on a credit card.

Cash discount Discount given for early payment of a bill.

Cash over When there is more cash on hand than there should be.

Cash payments record A written record of cash spent.

Cash price Invoice price less cash discount; amount paid when no credit is used.

Cash proof form A form used to check the amount of cash in a cash register.

Cash receipts record A written record of cash taken in.

Cash sales summary record A review of cash received and paid out, of income and expenses.

Cash register A device for storing cash and for recording cash received and paid out.

Cash short When there is less cash on hand than there should be.

Cash value The value of an insurance policy if it is canceled; cash surrender value.

Catalog price Same as list price.

Centimeter (cm) One-hundredth of a meter; about 0.3937 inches.

Central processing unit The brain of a computer system; the part that processes data.

Chance event The outcome of making a random choice.

Change fund Money placed in a cash drawer at the start of a period of time.

Charter A document, issued by a state, that creates a corporation and tells what the corporation can do.

Check A written order directing a bank to make a payment.

Check register A record of deposits and checks.

Circle graph A circle showing how parts relate to the whole and to each other.

Closing costs Expenses that must be paid when a home is purchased; settlement costs.

C.O.D. Cash on delivery; collect on delivery.

Coinsurance When the insured and the insurer share losses or costs.

Coinsurance clause A feature of insurance that requires coverage up to a specified percent of value to recover the full amount of a loss.

Collateral Personal property deposited or pledged as security for a loan.

Collateral loan A loan with a note that is backed up or secured by the deposit or pledge of personal property.

Collateral note A note that is backed up by collateral.

Collision insurance Auto insurance that covers damage to the insured's car from collision or upset.

Commission A payment to an agent or salesperson based on the value or quantity of goods bought or sold; broker's or brokerage fee.

Common denominator Any denominator that is shared by two or more fractions.

Common stock The ordinary stock of a corporation, paying no specified rate or amount of dividend.

Compatible numbers Numbers that make calculations easier.

Complement Applied to trade discounts, the difference between a given trade discount rate and 100 percent.

Compound amount The total in a savings account at the end of an interest period after compound interest is added.

Compound interest The difference between the original principal and the compound amount; interest figured on interest after it has been added to the principal.

Compounding interest Figuring and adding interest to make a new principal.

Comprehensive damage Insurance that covers damage to the insured's car from causes other than collision or upset.

Consumer loan Same as an installment loan.

Consumer price index A single number used to measure a change in consumer prices compared to a base year.

Cooperative A corporation whose customers are usually shareholders of the business.

Corporation A business owned by several people legally acting as one under a charter.

Cost of merchandise sold The amount paid by the seller for the goods sold; cost of goods sold.

Coverage The amount of insurance carried.

Credit card A card that identifies a customer who can buy on credit.

Credit memo A form that tells the buyer that the buyer's account has been reduced.

Credit period The length of time given to pay an invoice.

Creditor A person or firm to whom money is owed.

Cross product The result of multiplying a denominator by a numerator in a proportion.

Currency Something in circulation as a medium of exchange.

Current Assets Things of value that can be turned into cash or used in a year.

Customary system The system of measures used in the U.S.

Customer's account A record showing a customer's purchases, returns, payments, and balance owed.

cwt Hundredweight or hundred pounds.

Data Information used for base of reasoning or calculating. Plural for datum.

Database A collection of related information such as names and addresses that are stored electronically.

Date of maturity Same as due date.

Date of note The date on which the note was signed.

Death benefits Face amount of an insurance policy paid to the beneficiary.

Debit card A card that lets you pay for purchases using a terminal in a store.

Decimal (or decimal fraction) Represents a fraction with a denominator of 10, 100, 1,000, etc., and written with a decimal point; for example, 0.52 or 7.345.

Decimal point A point or period indicating the separation of a whole number and a decimal fraction.

Decimal rate A property tax rate shown as a decimal.

Decimal system A ten-digit system of numeration; a base-ten system.

Declare a dividend Distribute corporate profits to shareholders.

Declining-balance method A way of figuring depreciation at a fixed rate on a decreasing balance; fixed-rate method.

Deductible insurance policy A health or casualty insurance policy in which the insured pays the first part of any loss and the company pays the rest.

Deductions Allowable expenses that may be subtracted from income for tax purposes.

Demand loan A loan that must be repaid when the bank asks for payment.

Demand note The note that creates a demand loan.

Denomination The price on the face of a savings bond; the value on bills and coins.

Denominator The numeral below the line in a fraction, showing the number of equal parts into which a whole is divided.

Deposit To put money in a checking or savings account.

Deposit slip A form used to list all money deposited in a bank.

Depreciation The decrease in value caused by wear and aging.

Difference The missing addend that is found by subtracting a known addend from a sum or total.

Digit Any one of the symbols 0, 1, 2, 3, 4, 5, 6, 7, 8, 9.

Direct labor cost The cost of all production workers who work directly on products as they go through the factory.

Direct mail advertising Advertising sent by mail such as magazine subscriptions.

Disability income insurance Insurance that replaces part of the income lost if a worker is unable to work for an extended period of time because of illness or accident.

Discount A reduction in price (markdown) or in the amount of a bill; interest deducted in advance; the difference between market and par value of a bond.

Discount period The number of days given for a cash discount.

Discount rate Same as rate of discount; the percent of discount charged by a bank on a discounted loan; the percent of cash or trade discount given.

Discount series Same as series of discounts.

Discount sheet A form showing the discounts given from the catalog or list prices.

Discounting a note Signing a note and paying the interest (discount) immediately.

Dividend The number that is to be divided by another; earnings distributed to shareholders of a corporation or to holders of insurance policies.

Division The reverse of multiplication; a process of finding the unknown factor when one factor and the product are known.

Divisor The number that shows the size or number of groups into which a dividend is to be split.

Domestic business The manufacturing, purchasing, and selling of goods and services within a country.

Double-time pay Pay that is twice the regular pay.

Double-time rate A pay rate that is two times the regular rate.

Down payment The part of a price that is paid at the time of buying on the installment plan.

Due date The date on which a note must be paid; same as date of maturity.

Earned income Money received from working such as wages, salaries, and tips.

Effective rate of interest The annual rate of interest actually earned or paid.

Electronic funds transfer (EFT) When funds are withdrawn from one account and deposited into another using computers.

Employee A person who works for someone else.

Employee earnings card A form that shows the earnings, deductions, and net pay for one worker.

Employer A person or company that employs workers.

E.O.M. A term showing that a trade discount will be given on the invoice within a certain number of days after the end of the month.

Energy efficiency rating (EER) A measure of the efficiency of an electric appliance.

Equation Same as number sentence.

Equity The difference between what is owed on a home and its value. A person's stake in property.

Equivalent fractions Fractions that name the same number.

Estimate Find a rough answer.

Estimated tax Quarterly income tax paid by the self-employed.

Exact interest method Interest based on exact time and a 365-day year.

Exact value A quotient that is not rounded.

Exchange A place where stocks, bonds, or commodities are traded.

Exchange rate The amount of currency one country can trade for one unit of the currency of another country.

Excise tax A federal tax on telephone service, airplane tickets, and some other items. States may also levy excise taxes on certain goods.

Exemption An amount of income that is free from tax.

Expatriate bonus Money paid to persuade employees to work abroad.

Expense Money paid out; the cost of goods or services used up; a decrease in capital caused by the operation of a business.

Exporting Selling of goods or services produced within your country and sold to other countries.

Express A shipping service providing fast delivery of goods that are small and light in weight.

Extended coverage Insurance that adds wind, smoke, and hail coverage to a standard fire policy.

Extended-term life insurance Provides full coverage for a limited time to an insured person, bought with cash value.

Extension The total price of each quantity on a sales slip, found by multiplying the quantity by the unit price.

Face The amount stated on a business paper such as a note, bond, or insurance policy; face value; principal.

Face value Same as face.

Factor Each of the numbers in multiplication.

Factory expense The total cost of items such as rent, depreciation, heat, light, power, insurance, supplies, and indirect labor used in a factory.

Factory overhead Same as factory expense.

Federal income tax A progressive tax on the net income of individuals or businesses paid to the federal government.

Federal income tax return Form used to report and calculate income earned and taxes due to the federal government.

Federal unemployment tax act A payroll tax paid by employers to support the federal unemployment insurance program which is used to pay benefits to workers who are unemployed.

FICA tax A federal tax on employers, employees, and self-employed persons; social security tax.

Field A single piece of information in a database record.

Finance charge The sum of the interest and any other charges on an installment loan or purchase.

First in, first out (FIFO) A method used to find the value of ending inventory when the exact cost is not known. Merchandise purchased first is used first and the value of the ending inventory is based on the cost of the most recently purchased items.

Fiscal period The time for which net income or net loss is figured; financial period.

Fiscal year A fiscal period of twelve months starting on any date.

Fixed costs Overhead items such as rent, salaries, heat, and insurance that remain the same no matter how much is produced and sold.

Fixed-rate method Same as declining-balance method.

Fixed-rate mortgage A loan in which the interest rate stays the same over the life of the loan.

Floating decimal point A calculator feature that automatically puts the decimal point in the right place.

f.o.b. Free on board; a term used in price quotations to tell who will pay transportation costs.

Foreign exchange The process of changing or converting the currency of one country into the currency of another country.

Foreign exchange rate The amount paid for another country's currency.

Foreign trade Same as international business.

Fraction A symbol or name, such as $\frac{2}{3}$, $\frac{5}{4}$, $\frac{6}{2}$, standing for the quotient of two numbers or showing the division of the numerator (upper number) by the denominator (lower number); symbol or name for a fractional number; same as fractional numeral.

Fractional equivalent A fraction that shows what part one number is of another; for example, $\frac{1}{2}$ is the fractional equivalent of 50 cents on a base of $1.

Fractional number The number that results when one number is divided by another and the quotient is not a whole number.

Fractional numeral A symbol for a fractional number.

Freight A service for delivery of heavy, bulky goods.

Frequency The number of times an ad is scheduled to run.

Frequency distribution A table of numbers arranged in order and with the frequency, or how often they occur, tallied.

Fringe benefits Paid vacations, sick leave, retirement plans, insurance, and other benefits beyond the wage or salary.

Fuel adjustment rate A rate used to adjust your fuel costs to changes in the fuel company's costs for fuel.

Gas cost recovery rate A charge based on the cost per 100 cubic feet of gas purchased from suppliers based on current gas costs.

Girth A measurement around a package at its thickest part.

Graduated commission A pay system in which the rate of commission increases as the base increases.

Graduated payment mortgages (GPM) Mortgage that starts out with low initial monthly payments that rise gradually over the life of the loan.

Graduated rate scale Practice of charging a higher rate as more electricity and gas is used.

Graduated tax rate Tax rate increases as taxable income increases.

Gram (g) One one-thousandth of a kilogram; $\frac{1}{28}$ of an ounce.

Grand total The corner total on a columnar table. Used to check accuracy of vertical and horizontal addition.

Greatest common factor The largest number that will divide each term of a fraction without a remainder.

Gross The total without any deduction, such as gross profit, gross income; twelve dozen.

Gross domestic product (GDP) A system that measures the total market value of all final goods and services produced within the borders of a country.

Gross income Total income, before adjustments, deductions, or exemptions; total rent received from a tenant.

Gross loss The opposite of gross income; the amount by which costs exceed the selling price.

Gross pay The total of all pay earned.

Gross profit Net sales less cost of merchandise sold; margin.

Hardware Computer equipment such as terminals, disk drives, and printers.

Health insurance Protects policyholder from financial loss due to illness.

Health maintenance organizations (HMOs) A health insurance plan in which all medical care must be obtained from medical staff who participate in the HMO.

Hectare (ha) A square hectometer, about 2.5 acres.

Higher terms An equivalent fraction in larger terms.

Home depreciation A loss in the value of property caused by aging and use.

Home equity loan A loan using the owner's equity or stake in a home for collateral.

Homeowners insurance Fire and extended coverage on the home, plus theft and liability coverage.

Horizontal addition Adding across.

Horizontal bar graph A graph with bars running across.

Hourly rate An amount of pay for each hour worked.

Hundredweight (cwt) Hundred pounds.

Improper fraction A fraction whose numerator is equal to or greater than the denominator.

Income Revenue; money received.

Income statement A report showing sales, cost of merchandise sold, gross profit, operating expenses, and net income or loss; a profit and loss statement.

Income tax A federal, state, or local tax on income.

Income tax return A form showing how income taxes were figured.

Importing The buying of products produced outside your country.

Individual earnings record A payroll record that is kept for each employee.

Inflation A rise in the prices of goods and services.

Input device Equipment used to enter data into a computer system.

Insolvent A condition in which liabilities exceed assets, so debts cannot be paid.

Installment A partial payment on a loan or purchase.

Installment contract An agreement to pay the balance due on a purchased item in weekly or monthly payments.

Installment loan A loan repaid in partial payments; a consumer loan.

Installment plan A way of buying that usually requires a down payment, finance charges, and installment payments; a time-payment plan.

Insured The party whose life or property is insured.

Insurer The insurance company.

Interest The dollar cost of using a lender's money.

Interest-bearing note A note on which interest is to be paid.

International business All business activities necessary to manufacture, purchase, and sell goods and services across national borders; foreign trade.

International date line An imaginary line located in the middle of the Pacific Ocean.

Inventory A list of things on hand and their values.

Inverting the fraction A process of "switching" the numerator and denominator.

Investment Property bought for long-term use or for income or profit; the amount of money invested.

Invoice cost Same as invoice price.

Invoice price List price less trade discounts; invoice cost; net price.

Job expenses Money paid out of total job benefits for things such as travel, dues, tools.

Joint return An income tax return for a husband and wife together.

Kilobyte One thousand bytes, or characters.

Kilogram (kg) The basic metric unit of weight or mass; equal to about 2.2 pounds.

Kiloliter (kl) One thousand liters.

Kilometer (km) One thousand meters; about $\frac{3}{5}$ of a mile.

Kilowatt 1,000 watts of electric current.

Kilowatt-hours (KWH) The flow of 1,000 watts of electricity for one hour.

Labor force All persons who are working or looking for work.

Last in, last out (LIFO) A method to find the value of ending inventory when the exact cost is not known. Merchandise purchased last is used first so inventory is based on the costs of goods purchased first.

Lease A rental agreement; to rent.

Least common denominator The smallest denominator shared by two or more fractions.

Level payment plan A loan repayment plan in which all payments are the same amount.

Liabilities Creditors' claims to assets; debts that are owed.

Life insurance A source of protecting a surviving spouse and family from financial loss when the insured person dies.

Like fractions Fractions having the same denominator.

Limited payment life insurance Provides coverage for a lifetime, but premiums are paid for a fixed time only.

Line graph A graph on which lines connect dots to show values.

List price A price shown in a catalog or price list before trade discounts are deducted; a quoted price; catalog price.

Liter (L) The basic metric measure of capacity; slightly more than a quart.

Load fund A mutual fund with a commission charge.

Long lived assets Things of value, such as machinery, that will be used for more than one year.

Long lived liability Liability that will last for more than one year, such as a mortgage.

Lowest terms Occurs when no number other than 1 will divide both terms of a fraction.

M One thousand.

Modified accelerated cost recovery system (MACRS) A method of figuring depreciation based on the class life of property.

Magnetic ink character recognition (MICR) Special characters printed on checks to let computers process the checks.

Major medical insurance Supplements basic health coverage. Designed to help pay the hospital costs or other health care expenses due to a major illness or injury.

Maker One who writes a note and promises to pay.

Margin Same as gross profit; markup.

Markdown A reduction in marked or retail price; a discount.

Marked price The price on an item or attached tag; retail price.

Markup pricing An amount added to the cost of goods to cover expenses plus a profit.

Market price The price at which a stock or bond is sold; the price or value of an item in the open market.

Market value The price of a bond after it is issued; market price.

Markup The amount by which a selling price exceeds the cost; gross profit; margin.

Maturity date The date that marks the end of the term of a loan. When money must be repaid.

Maturity value Same as amount due at maturity.

Mean The sum of the numbers divided by the number of items; arithmetic average; a measure of central tendency.

Measures of central tendency The mean, median, and mode; averages.

Median The middle number in a group of numbers arranged in order.

Medicare A program that pays for hospital and medical bills for qualified retired workers.

Megabyte One million bytes, or characters.

Merchandise inventory A list that shows the cost or market value of goods on hand held for resale.

Merchandise turnover rate The number of times average inventory is sold in a period.

Meter (m) Basic metric unit of length or distance; equal to 39.37 inches.

Metric system A system of measures used in most of the world.

Metric ton (t) One thousand kilograms; about 1.1 Customary tons.

Mill One tenth of a cent.

Milligram (mg) One-thousandth gram.

Milliliter (mL) One-thousandth liter.

Millimeter (mm) One-thousandth meter; 0.03937 inches.

Mixed number A number having both a whole number and a fraction, such as $3\frac{1}{2}$.

Mode The number that occurs most frequently in a group of numbers.

Money market account A special savings account that earns a rate of interest based on the rate the federal government pays.

Monthly benefit The amount paid each month by social security to a retired or disabled worker and dependents.

Mortality table A table that shows the number of people who live to certain ages.

Mortgage A paper signed by a borrower that gives the lender the right to ownership of property if the borrower does not pay the principal or interest.

Multiplication A short way of adding two or more equal numbers.

Mutual fund An investment company that buys stocks and bonds of other companies.

Net The amount left after deductions have been subtracted.

Net asset value The value of a mutual fund share found by dividing net assets of the fund by outstanding shares.

Net assets The value of a mutual fund's assets less the money the fund owes to others.

Net income The amount left after subtracting operating expenses from gross profit; net profit.

Net job benefits The total value of the benefits received from a job less job expenses.

Net loss When operating expenses are greater than gross profit, net loss equals operating expenses less gross profit.

Net pay The remaining pay after deductions have been subtracted from total or gross wages; take-home pay.

Net proceeds For an agent, gross proceeds or selling price less commission and expenses; in stock sales, market price less commission, taxes, and fees.

Net price Same as invoice price.

Net profit Same as net income.

Net purchases Purchases less purchases returns and allowances.

Net sales Sales less sales returns and allowances.

No-load funds Mutual funds that sell shares without charging commission.

No-par stock Stock issued without a stated value on the certificate.

Non-interest-bearing note A note without interest.

Note Same as promissory note.

Number sentence A statement that says two values are equal.

Numeral A symbol or name for a number.

Numerator The numeral above the line in a fraction, indicating how many of the equal parts of the fraction are shown.

Offer price What the buyer pays for a mutual fund share; the net asset value of the share plus commission, if any.

Office equipment Office furniture and machinery expected to last a year or more.

Office supplies Items, such as paper and paper clips, that are used up quickly in an office.

On account A term indicating that payment is to be made at a later date, or describing a partial payment on an amount owed.

Open sentence A number sentence that has a missing numeral.

Operating expenses The cost of items such as rent, salaries, and advertising that decrease profits and so must be subtracted from gross profit; overhead.

Operating loss When a selling price covers the cost of goods, but not the operating costs.

Output device Equipment used to get information out of a computer system.

Outstanding check A check issued but not yet received and paid by the bank.

Overhead Same as operating expenses.

Overtime Time worked over or beyond regular time.

Paid-up life insurance A smaller amount of insurance for life, with no more premiums, bought with cash value.

Parcel post The common name for fourth-class mail.

Partnership A business owned by two or more persons, without a charter, who agree to share profits and losses.

Partnership agreement A written contract between partners that tells who will do what, and how profits and losses will be shared.

Par value The face value of a bond or stock.

Passbook A savings account record.

Passport A government document that proves citizenship.

Payback period The number of years it takes to save enough money to pay back the cost of an item.

Payee The one to whom a note or check is payable.

Payroll register A record showing earnings, deductions, and net pay of all workers in a business.

Per capita Per person.

Per capita GDP Used to indicate a nation's standard of living.

Percent Per hundred; by the hundred, or out of a hundred; shows the comparison or ratio of any number to one hundred.

Piece An item produced.

Piece rate A wage system in which workers are paid by the number of pieces produced.

Place value The value of a digit determined by its position.

Policy An insurance contract.

Policy loan A loan of part or all of the cash value of a policy made by an insurance company to an insured.

Preferred Provider Plans (PPOs) Insurance plan that combines features of traditional plans and HMOs. Members may use preferred providers or, for a higher fee, doctors outside the network.

Preferred Providers A network of health care providers who have agreed to take a specified amount of money for their services.

Preferred stock Stock that gets first choice in distributed profits and a stated rate of dividend.

Premium The amount paid for insurance; a bond selling above par value is selling "at a premium."

Price list A sheet or booklet that lists the names and prices of items for sale.

Prime meridian for time zones. Time zone 0°, starting point at Greenwich, England.

Principal The one for whom an agent acts; the face of a note; the amount on which interest is paid.

Probability The likelihood that an event will occur; a branch of mathematics concerned with predicting chance events.

Proceeds The amount the borrower gets on a discounted note; the amount received from a sale.

Product The result of multiplication.

Profit and loss statement Same as income statement.

Promissory note A written promise to repay money at a certain time; a note.

Proper fraction A fraction in which the numerator is smaller than the denominator.

Property damage insurance Auto insurance covering damage to property of others.

Property tax A tax on real estate.

Proportion A statement showing that two ratios are equal.

Pro rata In proportion; proportionately.

Prove cash Count cash on hand and check accuracy against the record of cash received and paid out.

Purchase invoice An itemized statement showing goods bought, with quantities, descriptions, prices, and charges; called a sales invoice by the seller.

Purchase order A document that identifies what was ordered, the price, and delivery terms.

Purchases The total merchandise bought during a period.

Purchasing power of dollar A measure of how much a dollar will buy.

Quantity discount A trade discount based on the amount purchased.

Quarterly Once every three months or quarter of a year.

Quota A fixed amount of sales above which commission is paid.

Quotation The price and terms on which a seller offers to sell; the published market price of a stock, bond, or commodity.

Quotient The figure that shows how many times the divisor is included in the dividend.

Random sample A few items selected by chance from the whole group.

Range The difference between the highest and lowest numbers in a group.

Rate The percent that is found by dividing the amount of interest, discount, depreciation, dividend, or income by the base.

Rate of depreciation Depreciation shown as a percent.

Rate of discount The percent of interest charged on a note and deducted when the loan is made; the percent of price reduction.

Rate of dividend The amount of dividend divided by the par value.

Rate of interest Interest shown as a percent.

Ratio A way of comparing numbers, such as 1, 2 divided by 3, 2:3.

Raw materials The cost of materials used in manufacturing that are part of the final product.

Reach Indicates how often people hear or see an ad.

Real estate Land and things attached to it, such as buildings and trees.

Receipt An official record of a transaction.

Reciprocal The reciprocal of a given number is the number whose product with the given number is 1. For example, the reciprocal of $\frac{3}{2}$ is $\frac{2}{3}$; of 5 is $\frac{1}{5}$; of $\frac{1}{5}$ is 5.

Reconciliation statement A form showing how the checkbook and bank statement balances are made to agree.

Reconciling The process of bringing the checkbook balance and bank balance into agreement.

Record A unit of information in a database. A record is made up of fields.

Rectangle graph A graph that uses vertical or horizontal rectangles to show how parts relate to the whole and to each other.

Redemption value The amount you receive when a savings bond is turned in for cash.

Refinance Use a new loan to pay off an old loan.

Regular pay The amount of pay expected in a regular work day or week.

Regular time The number of hours of work expected in a regular work day or week; straight time.

Remainder A leftover in division; the part of the dividend that is left over when the divisor is not contained a whole number of times in the dividend.

Renters policy Similar to homeowners insurance, but not covering loss of building or apartment.

Replacement cost policy Insurance that pays the cost of replacing a property at current prices.

Resale value The amount you receive when you sell an asset, such as a car.

Retail price Same as marked price.

Retained earnings Net income that is kept by a corporation rather than distributed to shareholders.

Revenue An increase in capital resulting from operations of the business.

Reverse addition A way of checking addition by adding a column of numbers in the opposite direction of the first addition.

Rounding The process by which unwanted digits on the right of a number are dropped.

Round down Process of estimating in which a product will be less than the actual product.

Round up Process of estimating in which a product will be greater than the actual product.

Rule of five rounding method Find the place to which you wish to round and look at the digit to the right. If it is 5 or more, round up; if it is less than 5, round down.

Salaried Employee An employee who is paid a fixed amount of money for each time period worked, such as a day, week, or month.

Salary A fixed amount of pay for a week, month, or year.

Sales The total amount of merchandise sold over a period of time, expressed in selling prices.

Sales invoice An itemized statement showing goods sold; called a purchase invoice by the buyer.

Sales price Same as selling price.

Sales returns and allowances The dollar amount of goods sold that were later returned for refunds or for which credit was given because of damage; a decrease in sales.

Sales slip A form showing the details of a sale.

Sales tax A tax charged by a city, county, or state on the sale of items or services and collected by sellers from buyers.

Savings account Money put into an account at a bank, credit union, or savings and loan to keep safe and earn interest.

Savings bond A series EE bond sold by the federal government.

SEC fee A fee charged by the Securities and Exchange Commission on stock sold.

Security Something given to guarantee that a debt will be paid or a promise kept.

Self-employment tax A social security tax paid by self-employed persons.

Selling and administrative expenses A manufacturer's costs for salespeople, advertising, office salaries, and other operating costs.

Selling price The price actually paid by a buyer; the sales price.

Series of discounts Two or more trade discounts to be applied in succession; a discount series.

Service charge A bank charge or deduction for handling a checking account; a charge in addition to interest on an installment purchase.

Settlement costs Same as closing costs.

Shareholder One who owns shares of corporation stock; a stockholder.

Short-term policy A property insurance policy issued for less than a year.

Signature card A bank form that identifies a bank account owner.

Simplest form See lowest terms.

Single discount equivalent The one discount equal to a series of discounts.

Social security tax Same as FICA tax.

Software Computer programs, a set of instructions that tells the computer what to do.

Sole proprietorship A business owned by one person; proprietorship.

Speed dialing A phone feature that lets you dial a number using only one or two buttons.

Square meter (m^2) The basic unit of area in the metric system; equal to 1.196 square yards.

Standard deduction A fixed amount that can be deducted from taxable income. Used in place of itemized deductions.

Standard fire policy Covers loss by fire and lightning only.

Stock Goods or supplies on hand; shares of ownership in a corporation.

Stock certificate A paper issued to a shareholder that shows on its face the number of shares it represents.

Stockholder Same as shareholder.

Stock transfer tax A tax on stock sold.

Storage device A computer system component, such as a disk or tape drive, on which data and programs can be kept.

Straight commission A pay system in which commission is the only pay; there is no other wage or salary.

Straight life insurance Premiums are paid and the insurance is in force for the whole life of the insured.

Straight-line method A way of figuring depreciation that spreads the total depreciation evenly over the life of the item.

Straight time Same as regular time.

Subtotal On a sales slip, the sum of the extensions before taxes are added.

Subtraction The reverse of addition.

Sum The result of addition; the total; an amount of money.

Sum-of-the-years-digits method A variable-rate way of depreciating that provides decreasing amounts of depreciation as an item ages.

T (Customary ton) Two thousand pounds; equal to 0.91 metric tons.

t (metric ton) One thousand kilograms; equal to 1.102 Customary tons.

Take-home pay Same as net pay.

Taxable income The amount used to figure income tax with a tax-rate schedule.

Tax return A form showing the facts used to figure income tax due; an income tax report.

Tax table income The amount used to find income tax in a tax table.

Term life insurance Premiums are paid and insurance is in force for a fixed time only, with no cash values.

Term of discount The time during which a bank holds a discounted note; date of discount to date of maturity.

Terms The numerator and denominator of a fraction.

Terms of sale Tell when an invoice must be paid and what discounts are given.

Therm A measurement of the heat value of gas

Three-way calling A phone feature that enables you to talk to more than one person at a time.

Time For a note, the length of time for which the money is borrowed.

Time-and-a-half-pay Pay that is one and a half times the regular pay rate.

Time-and-a-half-rate A rate that is one and a half times the regular rate.

Time card A record showing when a worker arrived and left for regular time and for overtime.

Time clock A device used to stamp a time card.

Time-deposit account Money in a savings or certificate of deposit account.

Time payment plan Same as installment plan.

Tips Money received by employees from customers.

Total Same as sum.

Total cost The cost of producing or purchasing goods plus the operating cost of a business. Also the invoice price and buying expense.

Total factory cost The sum of the costs of raw materials, direct labor, and factory expense.

Total job benefits Gross pay plus fringe benefits.

Trade deficit When foreign imports exceed exports.

Trade discount A reduction or discount given from a catalog or list price; a discount given "within the trade."

Trade-in value The amount you get for your old car or other asset when buying a new car or other asset.

Transit number A bank identification number shown on a check as the top part of a fraction.

Unemployment rate The percent of the labor force not working.

Uninsured motorists insurance Coverage that protects against damage to your car or injury to persons in the car caused by a driver who carries no insurance.

Unit price The price of a single measure, such as $1 per quart or $2 per kilogram.

Unlike fractions Fractions that have different denominators, such as $\frac{1}{2}$ and $\frac{1}{5}$

U.S. individual income tax A federal tax on income.

Variable costs Costs such as raw materials, direct labor, and energy that vary or change with the amount of goods produced and sold.

Variable-rate mortgage A loan in which the interest rate may change over the life of the loan.

Vertical addition The usual way of adding up and down.

Vertical bar graph A graph with bars running up and down.

Visa Stamp of endorsement issued by a country that allows a passport holder to work in a country.

Wages The total pay for a day or week of a worker paid on an hourly rate basis.

Weighted Average Method A method used to find the value of ending inventory when the exact cost is not known. Inventory is priced at the average price per unit of the beginning inventory plus the cost of all purchases during the fiscal year.

Whole life insurance Life insurance that gives coverage for a lifetime, and in which premiums are paid for a lifetime.

Whole numbers Whole units, such as 1 and 6.

Withdraw To take money from a bank account, such as a savings or certificate of deposit account.

Withholding allowance An allowance for a person used to reduce the amount of tax withheld from pay.

Withholding tax A deduction from pay for income tax.

Yield The rate of income on an investment, such as 9 percent yield on bonds. $40,000, 13%, 30-Year Mortgage, with $442.48 Monthly Payment

Unit 1: Managing Your Money

CHAPTER 1

Lesson 1-1

1. $4,385.69 **3.** $1,029.19 **5.** 40
7. 39 **9.** 43 **11.** 50
13. 7 **15.** 24 **17.** 58
19. $154 **21.** $62.70 **23.** $250.26
25. $220.20 **27.** $1,066.76 **29.** $770.40

Lesson 1-2

1. $348.82 **3.** $11,209.53 **5.** $956.90
7. F **9.** F **11.** T
13. T **15.** F **17.** T
19. F **21.** F **23.** 7
25. 5 **27.** 2 **29.** 5
31. 12 **33.** 14 **35.** 4
37. 6
39. thirty-four thousand nine hundred sixty-seven
41. 1 **43.** 5 **45.** 500
47. 1
49. 900 + 80 + 7; nine hundred eighty-seven
51. 7,000 + 600 + 5; seven thousand six hundred five
53. 50,000 + 6,000 + 700 + 80 + 3; fifty-six thousand seven hundred eighty-three
55. 234
57. 61,890; sixty-one thousand eight hundred ninety

Lesson 1-3

1. grand total 46 **3.** grand total 395
5. grand total 1,469 **7.** grand total, $245.86
9. $8,873.61 **11.** $885.12
13. $975 **15.** $949.29
17. $11,685.49

Lesson 1-4

1. $363.24
3. Bills, $3,220; Coins, $108.82; 4 cks; Net dep, $3,592.66
5. 22 **7.** 30 **9.** 118.89
11. $786.88 **13.** $2,539 **15.** $999.99
17. $466.66 **19.** $666.47 **21.** $581.80
23. $31.34 **25.** $529.90 **27.** 12.4 lbs
29. $1.27 **31.** $276,293.98 **33.** $67.86
35. $1,070.30 **37.** $315.15 **39.** $191.83

Lesson 1-5

1. $488.89 **3.** $855.05

Lesson 1-6

1. $872.79 bank balance **3.** $801.55 bank balance
5. $227.24 bank balance **7.** $700.69 bank balance
9. $598.74 bank balance

Lesson 1-7

1. $386.30 **3.** $671.49 **5.** $272.96
7. $598.44 **9.** $885.35

CHAPTER 2

Lesson 2-1

3. 26.52 **5.** 58.433
7. 5.6544 **9.** 870
11. 0.06 **13.** 3¢
15. $7.29 **17.** $61.00
19. $40.01 **21.** $68.00
23. $1,000.00 **25.** 8.3; 8.27; 8.274
27. 62.9; 62.87; 62.868 **29.** 10.2; 10.20; 10.200
31. 900,000; 930,000; 935,000; 935,000
33. 4,000 customers **35.** 7, $9
37. 60, 12 **39.** 60¢, 900
41. $50 \times 70 = 3,500$ **43.** $50 \times 40 = 2,000$
45. $0.8 \times 0.5 = 0.4$ **47.** 2 × $.80 = $1.60
49. $40 \times 20 = 800$ **51.** $30 \times 50 = 1,500$
53. $0.5 \times .3 = 0.15$ **55.** 70×90; 6,300
57. .80 × .60; 0.48¢ **59.** .30 × $7; $2.10
61. 8,000 × 50¢; 400,000¢ or $4,000
63. 14.72 **65.** $7,267.80
67. $12,364.44 **69.** Round down
71. 290; 2,900; 29,000 **73.** 49; 490; 4,900
75. $38.80; $388; $3,880
77. $131.20; $1,312; $13,120
79. 40¢; 400¢; 4,000¢ **81.** 435¢; 4,350¢; 43,500¢
83. 7.4 **85.** 38

Lesson 2-1 ***continued***

87. 1500¢ **89.** $60
91. $68,000 **93.** 18,400
95. $1,480 **97.** $644
99. $15.60 **101.** $21.80
103. 730¢ **105.** 1,300¢
107. $24 **109.** 2,730¢
111. $7.14 **113.** 750¢
115. 3,900¢ **117.** $51.10

Lesson 2-2

1. $80.00 **3.** $306 **5.** $280
7. $256 **9.** $360 **11.** $1,440 total
13. $364 **15.** $778 **17.** $19,500
19. $21,840 **21.** $57,600
23. 290 total pieces; $464.00 gross pay
25. 303 total pieces; $469.65
27. $87.04 **29.** $61.75 **31.** $76.50
33. $107.10 **35.** $119.10

Lesson 2-3

1. 27 **3.** 8 **5.** 1
7. FFP **9.** PFF **11.** PFF
13. 7 **15.** 28 **17.** 50
19. 7 **21.** 25 **23.** 0.735
25. 0.905 **27.** 1.219 **29.** 0.515
31. 200 **33.** 1 **35.** 8.599
37. $0.03 **39.** $0.38
41. 79.6 points per game
43–51. Answers will vary.
53. $76 **55.** 0.35 **57.** 550
59. 540 **61.** 750 **63.** 0.68
65. $0.80 **67.** 0.07 **69.** $0.03
71. 0.0704 **73.** 0.09 **75.** 0.1047
77. 0.3667

Lesson 2-4

1. $6 **3.** $1,288 **5.** $18,700
7. $672 **9.** $7 **11.** $1,107
13. $369 **15.** $910 **17.** $21,696
19. $83 **21.** $106 **25.** $544.75

CHAPTER 3

Lesson 3-1

1. Fractional Number **3.** Fractional Number
5. Fractional Number **7.** $\frac{5}{8}$
9. $\frac{24}{4}$ **11.** 36 ÷ 72 **13.** 9 ÷ 3
15. $\frac{1}{6}$ **17.** $\frac{1}{20}$ **19.** $\frac{5}{16}$
21. $\frac{8}{35}$ **23.** $\frac{8}{15}$ **25.** $\frac{6}{8}, \frac{9}{12}, \frac{12}{16}, \frac{15}{20}$
27. $\frac{4}{4}, \frac{6}{6}, \frac{8}{8}, \frac{10}{10}$ **29.** $\frac{1}{2}$ **31.** $\frac{5}{6}$
33. $1\frac{1}{2}$ **35.** $2\frac{1}{4}$ **37.** 1
39. $\frac{1}{2}$ **41.** $\frac{5}{9}$ **43.** 1
45. $\frac{3}{10}$ **47.** $\frac{5}{4}$ **49.** $\frac{1}{14}$
51. $\frac{1}{5}$ **53.** 2 **55.** 1
57. 9 **59.** 64 **61.** $\frac{9}{20}$
63. $18 **65.** $28.80 **67.** $7.60
69. $\frac{1}{4}$ **71.** $2.20 **73.** $400.60

Lesson 3-2

1. $\frac{4}{8}, \frac{6}{8}$ **3.** $\frac{3}{12}, \frac{4}{12}, \frac{2}{12}$ **5.** $\frac{3}{5}$
7. $\frac{7}{8}$ **9.** $\frac{3}{4}$ **11.** $\frac{1}{7}$
13. $\frac{2}{3}$ **15.** $\frac{1}{4}$ **17.** $\frac{2}{3}$
19. $1\frac{2}{5}$ **21.** $1\frac{5}{7}$ **23.** $1\frac{3}{4}$ hours
25. $\frac{1}{3}$ yard **27.** 8 **29.** 15
31. 8 **33.** 30 **35.** $1\frac{1}{6}$
37. $\frac{3}{4}$ **39.** $\frac{5}{8}$ **41.** $\frac{5}{8}$
43. $\frac{11}{16}$ **45.** $\frac{1}{6}$ **47.** $1\frac{1}{12}$
49. $2\frac{1}{6}$ **51.** $\frac{3}{8}$ yard

Lesson 3-3

1. 16 to 8 **3.** 16 to 24 **5.** 3 to 1
7. 4 to 3 **9.** 200 to 1 **11.** 9 to 2
13. 1 to 6 **15.** $5,250
17. Wes, $5,700; Jared, $3,800
19. $22,500; $1,500 **21.** 15
23. 2 **25.** $546 **27.** $237\frac{1}{2}$ pounds
29. 329 miles **31.** estimates may vary, $36
33. 7 to 10 **35.** 3 to 7

Lesson 3-4

1. 4 **3.** 4 **5.** $2\frac{1}{2}$
7. $3\frac{3}{4}$ **9.** $2\frac{1}{4}$ **11.** 37
13. $35\frac{1}{4}$ **15.** No **17.** $338.30

19. 7
21. $73.50
23. 37; 2
25. 35; 4
27. $285.60
29. $99
31. $111.20
33. 7
35. $104.48
37. $520
39. 38
41. $45
43. $665
45. 2
47. $27
49. 38 × $10.30 = $391.40 regular-time pay
51. $468.65

Lesson 3-5

1. $\frac{11}{5}$
3. $\frac{17}{3}$
5. 56
7. 297
9. $20\frac{5}{8}$
11. 441
13. 1,284
15. $16\frac{1}{2}$
17. 408
19. 63
21. $3\frac{1}{2} \times 26 = 91$ gallons
23. $\frac{4}{5}$
25. $6\frac{2}{5}$
27. 24
29. $1\frac{3}{5}$
31. 25
33. $4\frac{43}{50}$
35. $14\frac{11}{12}$
37. $12\frac{7}{8}$
39. $20\frac{19}{24}$
41. $5\frac{1}{6}$
43. $5\frac{17}{24}$
45. $\frac{9}{10}$
47. $\$8 \times 27\frac{1}{2} = \$220$ earned in three days
49. $311

Lesson 3-6

1. 20
3. 21
5. 35
7. 49
9. 15
11. $300.35
13. $21,424
15. 24
17. $18
19. $15
21. $40
23. $36
25. $31,590
27. $387,750
29. $1,650,000
31. $\frac{5}{12}$
33. $\frac{5}{8}$
35. $\frac{5}{4}$ or $1\frac{1}{4}$
37. 1
39. $\frac{1}{2}$
41. $\frac{6}{7}$
43. $\frac{1}{4}$
45. $\frac{11}{12}$
47. $\frac{4}{3}$ or $1\frac{1}{3}$
49. $\frac{\$1,500}{\$25,000} = \frac{1}{16}$
51. $\frac{1}{3}$
53. $\frac{1}{11}$
55. $\frac{4}{5}$
57. $\frac{7}{12}$
59. $\frac{1}{3}$
61. $\frac{2}{5}$
63. $\frac{1}{7}$
65. $\frac{1}{9}$
67. $\frac{1}{13}$
69. $\frac{5}{6}$

CHAPTER 4

Lesson 4-1

1. 0.21
3. 0.8
5. 0.74
7. 0.011
9. 6.39
11. 0.556
13. 0.727
15. 0.833
17. 0.267
19. 0.147
21. 0.906
23. 0.046
25. $5.74
27. $\frac{9}{20}$
29. $\frac{1}{40}$
31. $\frac{5}{8}$
33. $\frac{17}{20}$
35. 8.5
37. 20.8
39. 3,300
41. 2,760
43. $18
45. $5,310
47. $2,878

Lesson 4-2

1. $\frac{100}{100} = 1.00 = 100\%$
3. $\frac{500}{100} = 5.00 = 500\%$
5. 58%
7. 3%
9. 28.12%
11. 400%
13. 50%
15. 48%
17. 18.75%
19. 93.75%
21. 7.1%
23. 83.3%
25. 1.78
27. 0.061
29. 0.008
31. 0.1575
33. $\frac{3}{25}$
35. $\frac{1}{20}$
37. $\frac{3}{5}$
39. $1\frac{1}{4}$
41. $\frac{3}{1,000}$
43. $\frac{3}{8}$

Lesson 4-3

1. $34
3. $44
5. $0
7. $7
9. $36.34
11. $19.08
13. $57.32
15. $22.15
17. $153
19. $33.28
21. $25.52
23. $33.11
25. $25.35
27. $33.35
29. $56.40
31. $40
33. $282.43
35. $96.61; $239.13
37. $91.71; $241.36

Lesson 4-4

1. $12
3. $4.80
5. $160
7. $68
9. $21.70
11. $630
13. 50%
15. 40%
17. 200%
19. 120%
21. $4\frac{1}{2}\%$
23. $33\frac{1}{3}\%$
25. $14
27. $27
29. $4
31. $26
33. $50
35. $400
37. $870
39. $92
41. $150
43. $180
45. 0.4
47. $4.75
49. $150
51. $4; $1; $9; $7; $2.05
53. $800
55. $8
57. $1
59. 0.005
61. 0.00625
63. $\frac{1}{2}\%$
65. $\frac{1}{3}\%$
67. $\frac{7}{8}\%$

Lesson 4-5

1. $8,379
3. $500
5. $64.65
7. $360
9. $6,710.91
11. $16,649.28
13. $16,116
15. $21,840
17. $29,484
19. $27,533
21. $27,257.40
23. Ellis Inc.; $836.60

Lesson 4-6

1. $550
3. $90
5. $630
7. $400
9. $60
11. $60
13. $9.60
15. 25%
17. 25%
19. 25%
21. 10%
23. 25%
25. 200%
27. 250%
29. 12% increase
31. 14% increase
33. approx 33%

Lesson 4-7

1. $156
3. $58.80
5. $746.10
7. $592.25
9. $1,100
11. $207
13. $614
15. $2,505
17. $2,940
19. 7%
21. 15%
23. $5\frac{1}{2}$%
25. 15%
27. 6%
29. Varied; $6,000
31. $2,700; $2,887; $42,113
33. $6,013; $6,013; $79,887
35. $62,190

CHAPTER 5

Lesson 5-1

3. 10
5. 1 000
7. 10
11. 10
13. 1 000
15. 0.01
17. 0.001
19. 0.1
21. 0.6
23. 300
25. 70
27. 0.7
29. 0.041
31. 6 200
33. 3 000 cm
35. 1 040 mm
37. 2 400 m
39. 8.9 cm
41. 770 cm
43. 3.6 m
45. 206 mm
47. 1.42 m
49. 420 km
51. 128.4 m
53. 1 000 mm
55. 24 cm
57. 300 km
59. 510 cm
61. 13 rolls
63. 588 km
65. 196 km
67. 1,378 ft
69. no

Lesson 5-2

7. 10 000
9. 100
11. 5 000
13. 16 000
15. 12 m^2
17. 24 ha
19. 97 000 cm^2
21. 4 cm^2 or 400 mm^2
23. 0.4 ha
25. 140 m^2
27. 10 000 m^2
29. 1.5 km^2
31. 0.25m^2
33. 1 524 ha
35. 571 m^2
37. 875 acres
39. about 900
41. est. may vary; 900 cm^2
43. 37 210 cm^2
45. 3.721 m^2
47. 7

Lesson 5-3

5. 10 times
7. 1 000
9. 0.001
11. 1
13. 0.32
15. 113.83L
17. 14.344L
19. 9.98 kL
21. 24 L
23. 341 mL
25. 20 kL
27. 1.5L
29. 5 cans
31. 625 cases
33. 56.5L
35. $48.10
37. 1 000
39. 1 000
41. kg or metric ton; g or mg
43. 1 000
45. 1 000
47. 1 000
49. 0.9
51. 750
53. 0.4
55. 1.3
57. 1 500
59. 4.2
61. 17.4 kg
63. 0.13 kg
65. 20 loads
67. 1 600 kg
69. 104 kg
71. 208.45
73. 2.8 oz
75. $0.34

UNIT 2: SPENDING WISELY

CHAPTER 6

Lesson 6-1

1. $\frac{24}{25}$
3. 34%
5. 13% rounded
7. 7%; $4,865
9. 12%; $8,340
11. 14%, $9,730
13. 6.1%
15. $426,250

Lesson 6-2

1. $25; $24.45 **3.** $2.10; $2.24
5. $39.95 **9.** $17.99
11. Subtotal: $48.01 **13.** $2
15. $4 — may vary **17.** $53.34
19. $23.85 **21.** $0.44
23. $3.85

Lesson 6-3

1. $3.21 **3.** $\frac{3}{4}$
5. $8.20 **7.** $38
9. $13.99 **11.** $105.12 actual
13. $4,694.32 actual **15.** $1230.01 actual
17. $8 **19.** $15
21. $10 **23.** $15
25. $45 **27.** $18
29. $6 **31.** $11
33. $64 **35.** $14
37. $28

Lesson 6-4

1. $1.40 **3.** $2.63 **5.** $\frac{1}{2}$ ft
7. $\frac{3}{8}$ lb **9.** $\frac{1}{2}$ qt **11.** $\frac{3}{4}$ gal
13. $\frac{2}{3}$ dz **15.** $3.65 **17.** $2.27
19. $1.92 **21.** $7.56 kg **23.** $0.20/L
25. $0.83 **27.** $0.34 **29.** $0.97

Lesson 6-5

1. $7.80 **3.** $10.04 **5.** $12.30
7. $281.19 **9.** $1.01 **11.** Brand B
13. $0.67 **15.** $12 **17.** 7 days
19. 4 years, 8 months **21.** $34

Lesson 6-6

1. 647 **3.** $2.81 **5.** $74.80
7. $147.34 **9.** $7.63 **11.** 551
13. $2.40 **15.** $62.11 **17.** $195.44
19. $10.11 **21.** $117.01 **23.** $57.98
25. 120 **27.** $69.27 **29.** $72.73
31. 137 **33.** $2.73 **35.** 111
37. $49.64 **39.** $52.12 **41.** 126
43. $2.54 **45.** 73 **47.** $37.87
49. $39.76

Lesson 6-7

1. $348.30 **3.** 15% **5.** 315 KWH
7. $1,002.60 **9.** $1,146 **11.** 27 years
13. 1.4 years **15.** $0.55
17. $0.22; $80.30 **19.** $0.53 **21.** $0.11

Lesson 6-8

1. $26.17 **3.** $159.51 **5.** $37.32
7. $156.91 **9.** $209.83 **11.** $5.43
13. 40 gallons **15.** 7,300 gallons **17.** 600
19. $0.55

Lesson 6-9

1. $44.39 **3.** $70.43 **5.** $1.49
7. $0.82 **9.** $3.52 **11.** $3.68
13. $2.22 **15.** 40%

CHAPTER 7

Lesson 7-1

1. $18,000 **3.** $2,700
5. $1,400 **7.** $10,325
9. a. $539.84; b. $69,561.60
11. a. $714.25; $182,130 **13.** $1,298.88
15. $199.32 **17.** $19,046.52
19. $1,296.08

Lesson 7-2

1. $1,506 **3.** $1,872.50
5. $4,800 **7.** $16,078
9. $9,800 **11.** buying; $1,346
13. $920 less **15.** Yes.

Lesson 7-3

1. $16,400 **3.** $36,300 **5.** $7,950
7. $6,600 **9.** $2,100 **11.** $3,645
13. $3,214 **15.** $2,731.52 **17.** $1,450
19. 16% **21.** 20% **23.** $7,956

Lesson 7-4

1. $5,227 3. $4,910 5. $4,724
7. $10,500 9. $2,800 11. $14,016
13. $982.80 15. $13,326

CHAPTER 8

Lesson 8-1

1. $4,160 3. $510 5. $555.30
7. $989.38 9. $2,800 11. $1,159.50
13. $486 15. $1,689.60 17. $6,763.35
19. 0.8053 21. 0.09078 23. 0.02334
25. 0.10479 27. 0.0123
29. $314,300; 0.014
31. $1,375,000 actual
33. $1,189 35. $6,100,000 37. $3,040
39. Assessed Value: $40,000; Total Due: $2,280

Lesson 8-2

1. $0.04 3. $0.01
5. $0.06 7. $1.25
9. $0.64 11. $137.70
13. $7.29; $255.99 15. $0.42; $209.88
17. $2.23; $81.77

Lesson 8-3

1. $15.76 3. $39.49
5. $29.13 7. $2,964.07
9. $88,800 11. $4,200 may vary
13. $1,287.60 15. $421.60
17. $0 19. $98.60
21. $811 23. $1,666
25. $1,799 27. $1,166
29. $1,546 31. $252

Lesson 8-4

1. $17,268 3. $52,967.35 5. $20,807
7. $28,101 9. $2,089 11. $3,865
13. $1,330 15. $2,074 17. $30,781
19. $28,762 21. $3,341 23. $644
25. $1,950 27. $4,000 29. $159
31. $170 33. $384 35. $392

Lesson 8-5

1. $589.90 3. $3,054.80 5. $44,492
7. $6,900 9. $263 11. $169
13. $814 15. $1,831.68

CHAPTER 9

Lesson 9-1

1. $83.70 3. $466.68 5. $339
7. $2,040 9. 59% 11. $491,000
13. $1,902.25 15. $23,628 17. $33,000
19. $1,556.45

Lesson 9-2

1. $2,915 3. $1,998 5. $5,365
7. $290 9. $932.40 11. $4,320
13. $120 15. $405 17. $257.60
19. $4,311

Lesson 9-3

1. $194 3. $171, may vary
5. $392 7. $381, rounded
9. $294 11. $114; $1,600; $3,200
13. $94, $1,200; $2,400 15. $3,450
17. 25% 19. $720
23. $31 25. $18,000
27. $31,000
29. Amera, $20,000; Genie, $0.
31. $6,000 33. $11,000
35. $17,500 37. $2,586

Lesson 9-4

1. $815 2. $667 3. $1,793
5. $394 7. $359.80 9. $368

UNIT 3: SAVING AND BORROWING MONEY

CHAPTER 10

Lesson 10-1

1. $506.82 3. $3.75
5. $2.45 7. $878.37
9. $71.32, actual 11. $512

13. $718.40 **15.** $1,328.81; $628.81
17. $1,195.62; $195.62 **19.** $19
21. $50.95; 5.10% **23.** $143.50; 10.25%

Lesson 10-2

1. $46.50 **3.** $117.68 **5.** $35
7. $360 **9.** $3,883.33 **11.** $8.25

Lesson 10-3

1. $108; $408 **3.** $137.70; $857.70
5. $2,076.25 **7.** $26.63; $27
9. $26.63; $27 **11.** $27.81; $28.19
13. $295.89 **15.** $109.50
17. 11% **19.** $10\frac{1}{2}\%$
21. $1,970 **23.** $8,700
25. $496.67 **27.** $970.13

Lesson 10-4

1. $2.19 **3.** $20.03 **5.** $3.51
7. $22.19 **9.** $41.10 **11.** $13.81
13. $4.72 **15.** March 12 **17.** July 31
19. April 4 **21.** March 16 **23.** March 18
25. 66 days **27.** 108 days **29.** 71 days
31. 90 days **33.** $5,172.60

Lesson 10-5

1. $1,770; $270 **3.** $263.50; $13.50
5. 20 weeks **7.** 6% **9.** $515
11. $866 **13.** $42.75 **15.** $9.75
17. $11.79 **19.** between 15% and $15\frac{1}{4}\%$
21. $92 **23.** between $12\frac{3}{4}\%$ and 13%
25. $4,445 **27.** $11.13

Lesson 10-6

1. $1,312,000 **3.** $625,710 **5.** $210,000
7. $6.90 **9.** $299.54 **11.** $3.77
13. $11.52

CHAPTER 11

Lesson 11-1

1. $980 discount **3.** $1,060 premium
5. $9720.50 discount **7.** $1,041.25 premium
9. $2,760 **11.** $9,630
13. $4,750 **15.** $2,240
17. $4,725 **19.** $4,675
21. $9,512.50 **23.** $7,070
25. $5,976 **27.** $7\frac{1}{4}\%$
29. $925 **31.** $997.50
33. $2,000 **35.** $112.56
37. $360 **39.** $420
41. $280 **43.** $1,160
45. $325 **47.** $22.50
49. $4,400; $400 **51.** $31,725; $3,150
53. 10.7% **55.** 8.3%
57. 11.2% **59.** 11.1%
61. $17,120 **63.** $32,000
65. 10.7%

Lesson 11-2

1. $2,867.60 **3.** $1,575.33 **5.** $2,256.36
7. $1,646.80 **9.** $1,398.72 **11.** $1,290
13. Selway Foods **15.** Amliss, $2 **17.** 130,000
19. Borco **21.** Selway Foods **23.** 7%
25. $720 **27.** $130 **29.** 6.2%
31. 2.5% **33.** 5.7% **35.** 6.2%
37. 0.9% **39.** $22,728 **41.** $7,480
43. $564.24 **45.** $2,197.32 **47.** $19,149.89
49. $70.85 profit **51.** $779.14 loss
53. $19 loss **55.** $22.40

Lesson 11-3

1. $370.80 **3.** $876.84
5. $3,603.30 **7.** 1,188.707
9. 288.351 shares **11.** 20.202
13. 2.6% **15.** $0.13; 2.0%
17. $1.65; 8.5% **19.** 7.8%
21. $600.40 profit **23.** $4,004.94

Lesson 11-4

1. $940 **3.** $330 **5.** $1,110
7. 8.1% **9.** 5.9% **11.** 10.7%
13. 7.6% **15.** 13% **17.** $525
19. $1,210 **21.** $2,340 **23.** $195
25. $458 **27.** $40,000 **29.** 8.2%

Unit 4: Business Mathematics

CHAPTER 12

Lesson 12-1

1. 16.4; 17; 12
3. 9,645; 8,579; $17,054
5. $8.18
7. $8
9. $415
11. $410
13. 23; 23; 22; 6
15. 13.5; 14; 14; 9
17. $0.016; $0.016; $0.015

Lesson 12-2

1. $\frac{1}{2}$
3. $\frac{2}{3}$
5. $\frac{4}{9}$
7. 0
9. $\frac{1}{2}$
11. $\frac{1}{5}$
13. 0
15. $\frac{4}{5}$
17. $\frac{1}{3}$
19. $\frac{1}{4}$
21. 36
23. 24 times
25. 4%
27. 85%
29. 40%

Lesson 12-3

1. Saturday
3. Friday & Saturday
5. $38,000
7. Electrical & Plumbing
13. Sept: $36,000; Oct: $28,000; Nov. $30,000; Dec. $18,000
15. Feb. & Dec.
17. Mar. to Apr.; Nov. to Dec.

Lesson 12-4

1. Hardware, 180°; Software, 108°; Training, 54°; Repair, 18°
3. Newspapers, 30%; Radio, 20%; Yearbook ads, 5%; Circulars, 25%; Mailings, 15%; Misc., 5%
5. Raw Materials, 10%, 36°; Direct Labor, 25%, 90°; Overhead, 30%, 108°; Delivery & Test, 15%, 54°; Training, 20%, 72°
7. Musicland sales for November broken down by type (records, cassettes, and CDs).
9. 70%

Lesson 12-5

1. 5.4%
3. 1990; 5.4%
5. 37.9%
7. 0.766
9. 0.12
11. 0.308
13. 1991
15. 144.5
17. Energy
19. 32.8%
21. $14,450
23. Black 16-19 year-olds; 38.9%
25. 106

CHAPTER 13

Lesson 13-1

1. Sales − Returns & allowances = Net Sales; $145,846
3. Cost of goods sold = Beginning finished goods inventory + cost of goods produced during quarter − ending finished goods inventory. $84,615; Gross Profit = Net sales − cost of goods sold. $51,105
5. $1,089,121.90
7. $1,240,876.22
9. 48.4%
11. 10.3%
13. 73.60
15. 40%
17. 4.5
19. $48,400

Lesson 13-2

1. $2,290
3. $21,000
5. $7.20
7. $7,770 actual
9. $11,000
11. $621
13. Answers will vary.
15. $71.27
17. Printers: $6,750; Copiers: $4,500; Assembly: $4,050; Finishing: $2,700
19. $97.50

Lesson 13-3

1. $13,122
3. $1,395; $1,206.54
5. $3,640; $3,014.28
9. $33,000; $21,333.34
11. $273,000
13. $0
15. $12,960
17. $80,000

Lesson 13-4

1. Express mail
3. Yes
5. No, girth is greater than maximum
7. $11,100
9. $1,645
11. $6,394.50
13. $255 savings for all three ads.
15. $3,718.74; $309.90
17. Yes he makes a profit. Gross Revenue = $6,000 Total Expenses = $4,318.74; Net Profit = $1,681.26

Lesson 13-5

1. $62,530 3. $57,505 5. $37,900
7. Assets = $63,210; Liab. = $9,200; Capital = $54,010
9. Assets: $126,320; Liabilities, $9,640; Capital, $116,680; Total Liability and Capital, $126,320
11. 3:1

Lesson 13-6

1. $2,100; $6,300
3. Dujovny, $4,075; Ching, $4,890
5. Daniels, $33,000; Multon, $49,500
7. A: $12,300; B: $10,700
9. A: $21,520; B: $17,920; C: $19,570
11. $4,424; $4,746; $5,061 13. $2.40
15. $0.59 share, actual 17. $7
19. $59,650 21. $74,720
23. $222

Lesson 13-7

1. 40% 3. 35%
5. $0.64 7. $0.42
9. 37% 11. $3,427.06, actual
13. $51,800 15. $105

CHAPTER 14

Lesson 14-1

1. Agreed on price and discounts for a purchase.
3. No 5. August 26 7. April 4
9. February 12 11. $14.30 13. $55
15. $27
17. Subtotals: ice cube trays, $83.50; storage boxes, $108; trash bags, $319.70; household gloves, $122.50; measuring cups, $37.74.
19. September 30
21. The Rinz Company price is $25 less.
23. 33.33% 25. $4,283 27. $1,282
29. $19,796 31. $17,443 33. $45,675
35. $38,039

Lesson 14-2

1. $64.80; $25.20 3. $77.65; $67.35
5. $123.03 7. 19%
9. 30% 11. 43%
13. $22.91 15. $126
17. December 24

Lesson 14-3

1. $90; 25%; 20% 3. $864; 20%; $16\frac{2}{3}$%
5. $16.25 7. $221.60
9. $1.87 11. $4.59; $18.36
13. $72.25; $216.75 15. $639.98; $959.97
17. 11.1% 19. $7.26
21. $3.50

Lesson 14-4

1. $60 3. $12 per muffler
5. $750,000 7. $24,000
9. $6,000 11. $18.75
13. $12 15. $35.20

Lesson 14-5

1. $0.80 short 3. $0, total was correct.
5. Grand total: $2,260.95 7. Final balance: $1,764.12

Lesson 14-6

1. surplus, $7 billion 3. $1,962.50
5. $297.87 7. 900 million
9. Japan, U.S., Brazil, Mexico, China

Lesson 14-7

1. Thursday, 8:00 P.M. 3. 2 P.M.
5. 11:30 P.M.
7. No, package should have remained above 50°F.
9. 20°C 11. $6,400
13. 187,266 SR
15. $11,394.10 Canadian dollars
17. 1,750,000 Re 19. $42
21. $1,893

Lesson 14-8

1. Answers will vary. 3. Answers will vary.
5. 1.92; $67,200 7. 0.81; $36,450
9. Answers will vary.

INDEX

INDEX

INDEX

INDEX

INDEX

INDEX